High-Voltage Test and Measuring Techniques

Wolfgang Hauschild · Eberhard Lemke

High-Voltage Test and Measuring Techniques

Second Edition

 Springer

Wolfgang Hauschild
Dresden, Germany

Eberhard Lemke
Dresden, Germany

ISBN 978-3-030-07359-6   ISBN 978-3-319-97460-6   (eBook)
https://doi.org/10.1007/978-3-319-97460-6

Foreword to the First Edition

Most textbooks on high-voltage (HV) engineering published in the recent years are focused on general aspects of this field but not on the specifics of HV test and measuring techniques provided in this book. This topic is mainly experimentally based and essential for the wide range of present and future challenges, due to the increasing use of renewable power, the wider application of cable systems as well as the erection of long-distance ultrahigh voltage (UHV) lines using not only alternating but also direct transmission voltages.

Therefore, researchers and engineers engaged in HV test and measuring techniques are developing new equipment, instruments, and procedures. For a general basis, international organizations as CIGRE, IEC, and IEEE summarize the results of research work and provide commonly accepted rules, guides, and standards. Many researchers, designers, and technicians engaged in the field of HV engineering are not well familiar with the approaches prepared and introduced by the abovementioned organizations. In this situation, this book will close a gap and contribute to a better understanding of the advanced technique recently developed and adopted for quality assurance testing and diagnostics of HV insulation. Moreover, the book is a help for students to get well-understandable information on today's tools for insulation testing and diagnostics. Another main application will be the training, further education and individual learning of engineers.

In this context, it should be noted that great progress has been made in developing HV test systems including the associated measuring equipment which are the main topics of the book written by Hauschild and Lemke. In summary: The book gives a complete introduction and an overview of the state-of-the-art HV test and measuring techniques in close connection to practical aspects. For me, great work has been done by the authors, which I know since the beginning of the 1970s when I visited the HV Institute in Dresden for the first time. Thereafter we became good partners and close friends. I met the authors periodically, mainly when participating in various working groups of CIGRE and IEC where Wolfgang Hauschild was especially engaged in the field of HV test technique and Eberhard Lemke in the field of HV measuring technique. Their outstanding work and fruitful cooperation with the HV Institute of the Graz University of Technology has been recognized by awarding both with the degree of a "Doctor honoris causa" in 2007 and 2009, respectively.

Michael Muhr

Graz Technical University

Graz, Austria

Paris, France   Chairman of Cigre AG HV Test Techniques
November 2013

Preface to the First Edition

More than a century after its beginning, high-voltage (HV) engineering still remains an empirical field. Experimental investigations are the backbone for the dimensioning of electrical insulations and indispensable for quality assurance by type, routine and commissioning tests as well as for insulation condition assessment by monitoring and diagnostic tests. There is no change in sight for such empiric procedures. The application of higher transmission voltages, improved insulation materials, and new design principles require the further development of HV test and measuring techniques. The relevant bodies of experts in *CIGRE, IEC* and *IEEE* provide commonly accepted standards and guides of HV testing adapted to both, the needs and the level of knowledge.

Coming from the Dresden School of HV engineering of *Fritz Obenaus* and *Wolfgang Mosch,* the authors have been lucky to follow and to contribute to the development of HV test techniques for half of a century. This book is based on that experience and shall reflect the actual state of the art of HV test and measuring techniques. According to our intention, the book shall close a gap in the international literature of HV engineering and lead to a better understanding of the relevant IEC and IEEE standards. It is hoped our text will fill the needs of designers, test field and utility engineers as well as those of senior undergraduate and graduate students and researchers. Today, many engineers who are confronted with or even engaged in HV testing did not have an in-depth education in HV engineering. Therefore, the book is intended to support the individual learning as it is useful for further training courses, too.

After an introduction related to the history and the position of HV test techniques within electric power engineering, the general basis of test systems and test procedures, the approval of measuring systems and the statistical treatment of test results are explained. In separate chapters for alternating, direct, impulse, and combined test voltages, respectively, their generation, their requirements, and their measurements are described in detail. Because partial discharge and dielectric measurements are mainly related to alternating voltage tests, separate chapters on these important tools are arranged after that of alternating test voltages. The book closes with chapters on HV test laboratories and on-site testing.

The cooperation with many experts from all over the world has been a precondition for writing this book. We are grateful to all of them, but we can mention only a few: We got our stamping at the HV Laboratory of Dresden Technical University and acknowledge the cooperation of its staff, represented by *Eberhard Engelmann* and *Joachim Speck*. We consider our membership in the expert bodies of *CIGRE 33 (later D1), IEC TC 42* and *IEEE-TRC and ICC* as a school during our professional life. We have got numerous suggestions from this work on HV testing as well as from discussions with the members. We are grateful to *Dieter Kind, Gianguido Carrara, Kurt Feser, Arnold Rodewald, Ryszard Malewski, Ernst Gockenbach, Klaus Schon, Michael Muhr* and all others who are not mentioned here. Of course, the daily work in our companies has been connected with many technical challenges of HV test techniques. As they have always been mastered in our reliable teams, we would like to express our sincere thanks to both, the management and the staff of *Highvolt Prüftechnik Dresden GmbH* and *Doble-Lemke GmbH*. Thanks to *Harald Schwarz and Josef Kindersberger*, who appointed Wolfgang Hauschild to a lectureship on HV test techniques at Cottbus Technical University respectively on Munich Technical University. This required a suitable structure for the subject which is also used in this book. For the careful proof-reading of the manuscript and the helpful advices, we thank our friends *Jürgen Pilling* and *Wieland Bürger*. We would be grateful for further suggestions and critics of the readers of this book.

Dresden, Germany   Wolfgang Hauschild
October 2013   Eberhard Lemke

Preface to the Second Edition

The recent years after the first edition of this book has been published are characterized by many developments in electric power generation, transmission, and distribution, e.g., the increasing application of renewable energy, the extensions of the AC transmission voltages to the UHV level >800 kV, the wider application of HVDC power transmission, also by using cable systems, and improved methods of diagnostics and condition assessment. All these advances are of consequence for the high-voltage test and measuring technique. The second edition of this book shall reflect the trend in HV testing and should be understood as a contribution to the present impetus of high-voltage engineering in general.

Also for this second Edition, we have been supported by many colleagues and mention *Dr. Ralf Pietsch, Günter Siebert* and *Uwe Flechtner*. Especially, we acknowledge the cooperation with *Dr. Christoph Baumann, Petra Jantzen* and *Sudhany Karthick* of Springer Nature.

Dresden, Germany
September 2018

Wolfgang Hauschild
Eberhard Lemke

Acknowledgement

Due to the generous aid by HIGHVOLT Prüftechnik Dresden GmbH, the book has got its colored appearance. Furthermore, all photographic figures and three-dimensional drawings without reference are supplied by the HIGHVOLT archives. Our sincere thanks are related to the management, especially to *Bernd Kübler, Thomas Steiner and Ralf Bergmann*, for their permanent support of our project.

Contents

Abbreviations

AC	Alternating current (in composite terms, e.g., AC voltage)
ACIT	HV units for feeding induced voltage tests
ACL	Accredited Calibration Laboratory
ACRF	HVAC series resonant circuit of variable frequency
ACRL	HVAC series resonant test circuit of variable inductance
ACT	HVAC test circuit based on transformer
ACTF	HVAC test circuit of variable frequency based on transformers
ADC	Analog–digital converter
AE	Acoustic emission
AMS	Approved measuring system
C	Capacitance
CD	Committee Draft (IEC)
CH	Channel
CRO	Cathode ray oscilloscope
DAC	Damped alternating current (in composite terms, e.g., DAC voltage)
DC	Direct current (in composite terms, e.g., DC voltage)
DCS	Directional coupler sensor
DNL	Differential nonlinearity
DSP	Digital signal processing
EMC	Electromagnetic compatibility
GIL	Gas-insulated (transmission) line
GIS	(1) Gas-insulated substation
	(2) Gas-insulated switchgear
GST	Grounded specimen test
GUM	ISO/IEC Guide 98-3:2008
HF	High frequency
HFCT	High-frequency current transformer
HV	High voltage (in composite terms, e.g., HV tests)
HVAC	High alternating voltage
HVDC	High direct voltage

IEC	International Electrotechnical Commission
IEEE	Institute of Electrical and Electronic Engineers (USA)
IGBT	Insulated gate bipolar transistor
INL	Integral nonlinearity
IVPD	Partial discharge measurement at induced AC voltage
IVW	Induced voltage withstand test
L	Inductance
LI	Lightning impulse (in composite terms, e.g., LI test voltage)
LIC	Chopped lightning impulse
LIP	Liquid-impregnated paper (insulation)
LSB	Least significant bit
LTC	Life time characteristic (or test)
LV	Low voltage
M/G	Motor–generator (set)
ML	Maximum likelihood
MLM	Multiple level method
MS	Measuring system
MV	Medium voltage (do not mix-up with the dimension "Megavolt"!)
NMI	National Metrology Institute
OLI	Oscillating lightning impulse
OSI	Oscillating switching impulse
PD	Partial discharge (in composite terms, e.g., PD measurement)
PSM	Progressive stress method
R	Resistor
R&D	Research and development
RF	Radio frequency
RIV	Radio interference voltage
RMS	Reference measuring system
rms	Root of mean square
RoP	Record of performance
RVM	Return voltage measurement
SFC	Static frequency converter
SI	Switching impulse (in composite terms, e.g., SI test voltage)
TC	Technical Committee (of IEC)
TDG	Test data generator
TDR	Time domain reflectometry
THD	Total harmonic distortion
TRMS	Transfer reference measuring system
UDM	Up-and-down method
UHF	Ultrahigh frequency
UHV	Ultrahigh voltage (in composite terms, e.g., UHV laboratory)
V	Voltage
VHF	Very high frequency
X	Reactance
XLPE	Cross-linked polyethylene
Z	Impedance

Symbols

A	Area
a	Distance
α	Phase angel
ß	Overshoot magnitude
C	Capacitance
C_i	Impulse capacitance
C_l	Load capacitance
c	Velocity of light
D	Dielectric flux density
d	Diameter
dV	Voltage drop (DC)
Δf	Bandwidth
ΔT	Error of time measurement
ΔV	Voltage reduction (DC)
δ	(1) Air density
	(2) Weibull exponent
	(3) Ripple factor
	(4) Loss angle (tan δ)
δV	Ripple voltage (DC)
E	Electric field strength
e	(1) Elementary charge ($e = 1.602 \times 10^{-19}$ As)
	(2) Basis of natural logarithm ($e = 2.71828\ldots$)
ε	Permittivity ($\varepsilon_0 = 8{,}854 \times 10^{-12}$ As/Vm)
ε_r	Relative permittivity
η	(1) 63% quantile (Weibull and Gumbel distributions)
	(2) Utilization or efficiency factor
F	(1) Scale factor
	(2) Coulomb force
F_p	Polarization factor
F(f)	Transfer function

$F(x)$	Distribution function
f	Frequency
f_m	Rated frequency
f_t	Test frequency
f_0	(1) Natural frequency
	(2) Centre frequency (narrowband PD measurement)
f_1	Lower frequency limit
f_2	Upper frequency limit
Φ	Magnetic flux
φ	Phase angle
G	Current density
g	Parameter for atmospheric corrections
$g(t)$	Unit step response
H	(1) Magnetic field strength
	(2) Altitude
h	Humidity
I	Current
I_m	Rated current
I_{sc}	Short-circuit current
i_L	Discharge current
K	Coverage factor for expanded uncertainty
K_t	Atmospheric correction factor
k	(1) Parameter for atmospheric corrections
	(2) Fixed factor
k_d	Constant in life time characteristic
k_e	Field enhancement factor
$k(f)$	(1) Test voltage factor
	(2) Test voltage function for LI evaluation
k_1	Air density correction factor
k_2	Humidity correction factor
κ	Conductivity
L	(1) Inductance
	(2) Likelihood function
M	Pulse magnitude (PD measurement)
m	Estimated mean value
μ	Theoretical mean value
μ	Permeability ($\mu_0 = 0.4\,\pi \times 10^{-6}$ Vs/Am $= 1{,}257 \times 10^{-6}$ Vs/Am)
μ_r	Relative permeability
n	(1) Life time exponent
	(2) Number (e.g., of electrons)
ω	Angular frequency
P	Active test power
P_F	Feeding power
P_m	Dipole moment
P_N	Natural power of a transmission line

P_R	Loss power of a resonant circuit
p	(1) Probability
	(2) Pressure
p_0	Reference pressure
Q	(1) Charge
	(2) Quality factor (resonance circuit)
q	(1) Charge of a PD pulse
	(2) Charge of a leakage current pulse
R	(1) Resistance
	(2) Ratio between two results
R_d	Damping resistance
R_f	Front resistor
R_t	Tail resistor
r	(1) Ratio (e.g., divider or transformer)
	(2) Radius
S	(1) Reactive test power
	(2) Steepness (LI/SI test voltage)
S_f	Scale factor
S_{50}	50 Hz equivalent test power
s_g	Mean square deviation (estimation of standard deviation)
σ	Standard deviation
T	Duration (AC period)
T_C	Time to chopping
T_N	Experimental response time
T_R	Residual response time
T_T	Duration of overshoot
T_1	Front time of LI voltage
T_2	Time to half-value of impulse voltages
t	(1) Temperature
	(2) Time
t_s	Settling time
t_t	Test time
t_0	Reference temperature
τ	Time constant
U	Expanded uncertainty
U_{cal}	Expanded uncertainty of calibration
U_M	Expanded uncertainty of measurement
u	Standard uncertainty
u_A	Type A standard uncertainty
u_B	Type B standard uncertainty
V	Voltage
V_B	Maximum of base curve (LI voltage)
V_E	Extreme value of recorded curve (LI voltage)
V_e	PD extinction voltage
V_F	Feeding voltage

V_i (1) PD inception voltage
        (2) Impulse voltage

V_k Short-circuit voltage (test transformer)

V_m (1) Highest voltage of equipment, rated voltage
        (2) Arithmetic mean (DC)

V_{max} Maximum of DC voltage

V_{min} Minimum of DC voltage

V_n Nominal voltage

V_{peak} Peak voltage

V_r Return or recovery voltage

V_{rms} Root mean square value of voltage

V_T Test voltage value

$V(v)$ Performance function

V_Σ Cumulative charging voltage

V_0 (1) Line-to-ground voltage
        (2) Initial voltage for a test
        (3) Charging DC voltage

V_1 Primary voltage of a test transformer

V_2 Secondary voltage of a test transformer

V_{50} 50% breakdown voltage

v Variance

$v(t)$ Time-depending voltage

v_k Short-circuit impedance of a test transformer

w Number of turns of a winding

W Energy

W_i Impulse energy (of impulse voltage generator)

X Reactance

X_{res} Short-circuit reactance of a transformer

Z Impedance

Z_L Surge impedance of a transmission line

Chapter 1
Introduction

Abstract High-voltage (HV) test and measuring techniques are considered in most general HV text books (e.g. Kuechler 2009; Kuffel et al. 2007; Beyer et al. 1986; Mosch et al. 1988; Schufft et al. 2007; Arora and Mosch 2011). There are teaching books on HV test techniques for students (Marx 1952; Kind and Feser 1999) as well as few text books on special fields, e.g. on HV measuring technique (Schwab 1981; Schon 2010, 2016). It is the aim of this book to supply a comprehensive survey on the state of the art of both, HV test and measuring techniques, for engineers in practice, graduates and students of master courses. A certain guideline for this is the relevant worldwide series of standards of the Technical Committee 42 (TC42: "High-Voltage and High-Current Test and Measuring Techniques") of the *International Electrotechnical Commission (IEC)*, largely identical with the corresponding standards of the Institute of *Electrical and Electronic Engineers (IEEE)*. This introduction contains also the relation between HV test and measuring techniques and the requirements of power systems with respect to the increasing transmission voltages and the principles of insulation coordination. Furthermore, HV testing for quality assurance and condition assessment in the life cycle of power equipment is investigated.

1.1 Development of Power Systems and Required High-Voltage Test Systems

Within the last 125 years, the development of transmission voltages of power systems from 10 to 1200 kV has required a tremendous development of high-voltage (HV) engineering. This includes, e.g. the introduction of many new insulating materials and technologies, the precise calculation of electric fields, the knowledge about the phenomena in dielectrics under the influence of the electric field and the understanding of electric discharge processes. Nevertheless, as an empirical technical science, HV engineering remains closely related to experiments and verifications of calculations, dimensioning and manufacturing by HV tests. The reasons for that are, e.g. unavoidable defects of the structure of technical insulating

© Springer Nature Switzerland AG 2019
W. Hauschild and E. Lemke, *High-Voltage Test and Measuring Techniques*,
https://doi.org/10.1007/978-3-319-97460-6_1

materials, imperfections of technical electrodes, but also failures of production and assembling. Therefore, in parallel to the development of HV engineering, national and international standards for HV testing have been developed, as well as equipment for the generation of test voltages and for measurements of (and at) these voltages.

Since the early beginning of the wider application of electrical energy, its transmission from the place of generation (power station) to that of consumption (e.g. industry, households, public users) influences the energy cost remarkably. The transportable power of a high-alternating voltage (HVAC) overhead line is limited by its surge impedance Z_L to a power transfer capability of approximately

$$P_L = V^2/Z_L. \tag{1.1}$$

Whereas the surge impedance ($Z_L \approx 250\ \Omega$) can only be influenced within certain limits by the geometry of the overhead line, the power transfer capability is mainly determined by the height of the transmission voltage, e.g. the power transfer capability of a 400 kV system is only a quarter of that of an 800 kV system. Consequently, increasing energy demand requires higher HVAC transmission voltages. Remarkable increases in the ratings *of HVAC transmissions are* to 123 kV in 1912 in Germany, to the 245 kV level in 1926 (USA), to 420 kV in 1952 (Sweden), to 800 kV in 1966 (Canada and Russia) and to UHV (1000–1200 kV in 2010, China) (Fig. 1.1).

Because at direct voltage, no surge impedance becomes effective; the limitation of the power transfer capability is mainly caused by the current losses. For identical

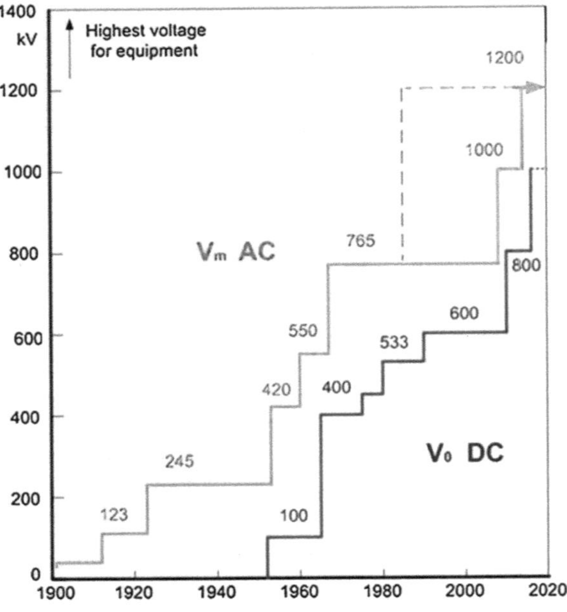

Fig. 1.1 History of HVAC and HVDC transmission systems

rated voltages, the HVDC power transfer capability is about three times higher than that at HVAC. This means that one 800 kV HVDC line with an efficiency of 94% replaces three 800 kV HVAC overhead lines with an efficiency of only 88% (Swedish Power Cycle 2009). But *HVDC transmission* requires expensive converter stations. Therefore, the application of HVDC transmission has been limited to very long transmission lines, where the cost reduction for the line compensates the higher station cost. The present cost reduction of power electronic elements, efficient HVDC cable production and other technical advantages of HVDC transmission has triggered worldwide activities in that field (Long and Nilsson 2007; Gockenbach et al. 2007; Yu et al. 2007). The historical development (Fig. 1.1) shows that the 1000 kV level is reached in China now, but the next levels above 1000 kV or more are under preparation (IEC TC115 2010).

The HV test and measuring techniques have to be able to test components which belong to both HVAC and HVDC power systems. But additionally, also the kind of insulation to be tested determines the kind of the test equipment. The "classical" insulating materials (air, ceramics, glass, oil, paper) are completed by insulating gases, e.g. SF_6 for gas-insulated substations and transmission lines (GIS, GIL) (Koch 2012), and synthetic solid materials, e.g. epoxy resin for instrument transformers and polyethylene for cables (Fig. 1.2) (Ghorbani et al. 2014). In the future, environmental viewpoints require a wider application of cables and GIL for power transmission of both, AC and DC voltages.

It is assumed that the present technology of *UHV DC transmission* would even allow the erection of a global-spanning supergrid to interconnect regional networks (Fig. 1.3) (Gellings 2015). This enables the world-wide exchange of electric power,

Fig. 1.2 History of the application of insulating materials

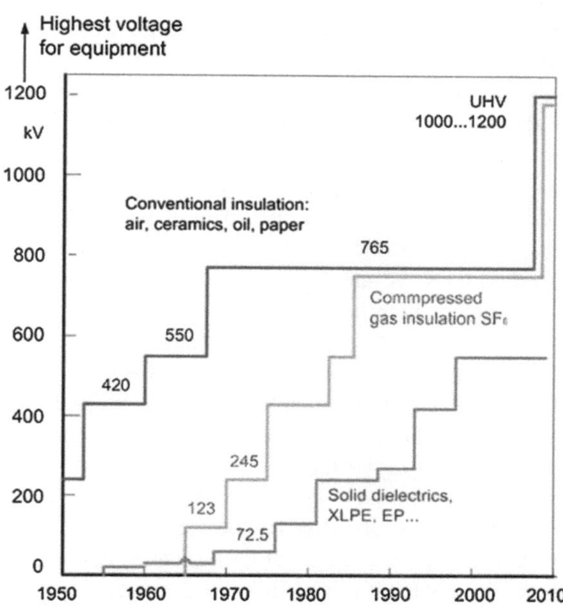

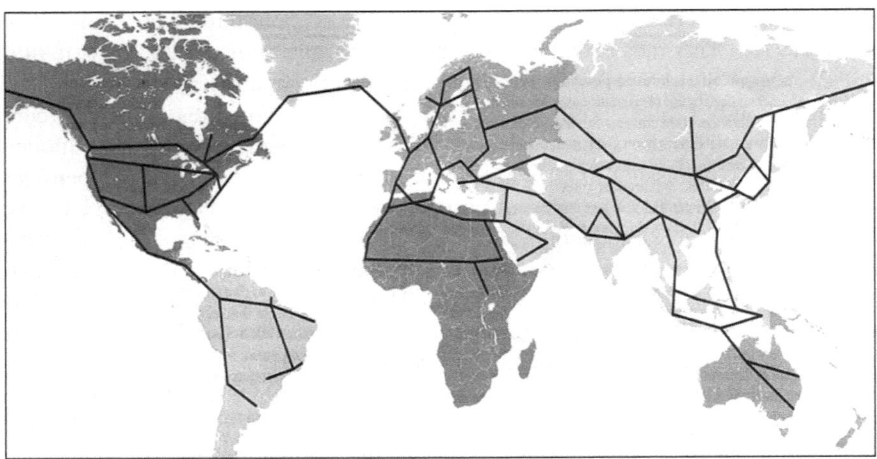

Fig. 1.3 The idea of a global-spanning super-grid

maintains the balance between supply and demand as well as secures grids against electronic and physical attacks. Planned and even constructed EHV/UHV regional networks, e.g. in China, Europe or around the Mediterranean can be understood as first steps to a supergrid which would also be a challenge to UHV testing!

The basic principle of HV testing expresses that test voltages stresses shall represent the characteristic stresses in service (IEC 60071-1). When electric power transmission started, not all these stresses were known. Furthermore, the kind and height of stresses depend on the system configuration, the used apparatus, the environmental conditions and other influences. The historical development of HV testing is closely related to the development of and the knowledge on power systems. It can be characterized by the following steps:

HV testing started in the first decade of the twentieth century with *alternating test voltages* of power frequency (50 or 60 Hz) (Spiegelberg 2003). The test voltages have been generated by test transformers, later also by transformer cascades (Fig. 1.4a, see also Sect. 3.1). It was assumed that suited *HVAC tests* would represent all possible HV stresses in service. Of course, HVDC equipment has been tested with DC voltages generated by DC generators (Fig. 1.4b, see also Sect. 6.1).

But independently on the performed HVAC tests, equipment has been destroyed in power systems, e.g. as a consequence of lightning strokes, which caused *external over-voltages*. These overvoltage impulses are characterized by front times of few microseconds and tail times of several ten microseconds. Based on that knowledge, tests with *lightning impulse (LI) test voltages* (front time $\approx 1...2$ µs, time to half-value $\approx 40...60$ µs) have been introduced in the 1930s. For LI voltage testing, suited generators have been developed (Fig. 1.4c, see also Sect. 7.1).

Another 30 years later, it has been found that *internal over-voltages* lead to lower breakdown voltages of long air gaps than LI or AC voltage stresses. They are caused by switching operations in the power system. Their durations lay between

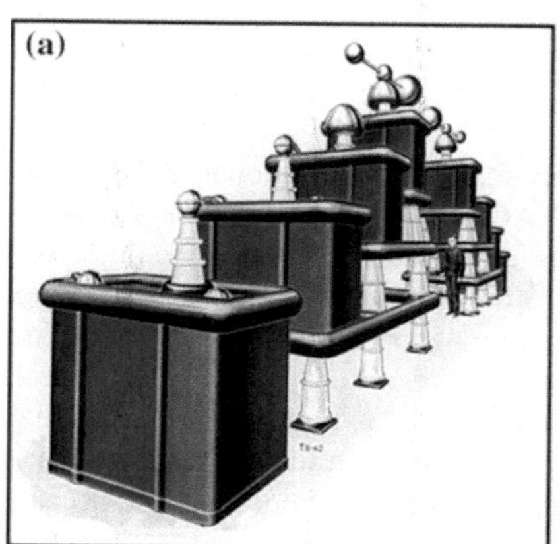

Fig. 1.4 Historical test voltage generators **a** world's first 1000 kV cascade transformer (Koch and Sterzel Dresden 1923) **b** 1000 kV DC test voltage generator (Koch and Sterzel Dresden 1936) **c** 2000 kV LI test voltage generator (Koch and Sterzel Dresden 1929) **d** 7000 kV LI/SI test voltage generator (TuR Dresden 1979)

some hundreds of microseconds and few milliseconds. As a consequence, *switching impulse (SI) test voltages* have been introduced in the 1960s. *SI test voltages* can be generated by the same type of generators as *LI test voltages* (but with larger HV electrodes for better control of the electric field, Fig. 1.4d, see also Sect. 7.1) or by test transformers.

Again 30 years later, it has been found that disconnector switching of gas-insulated substations (GIS) causes oscillating *over-voltages of very fast front* (VFF, several ten nanoseconds) which may harm the GIS insulation itself, but also attached equipment. Whereas a test with *very fast front (VFF) test voltage* has been introduced for GIS, it is under discussion for other components of power systems (Sect. 7.1.5).

The mentioned overvoltages are superimposed on the operational voltages. The traditional HV testing of components of HVAC power systems must not consider the operational voltage, only for special cases, e.g. disconnectors or three phase GIS busbars, the superposition plays a role. Therefore *"mixed voltages"* of two voltage components have been introduced. Depending on the position of the insulation in a test, one distinguishes between *"combined test voltages"* for three-pole test objects (e.g. disconnectors) and *"composite test voltages"* for two-pole test objects (e.g. polluted insulators), for details see Chap. 8. In case of HV testing of components of HVDC power systems, composite test voltages, play a very important role because of the space charge generation at DC voltages.

1.2 The International Electrotechnical Commission and Its Standards

The International Electrotechnical Commission (IEC) is the worldwide organization for international standards on electrical engineering, electronics and information technology. It has been founded in 1906, and its first president was the famous physicist Lord Kelvin. Today, about 60 national committees are IEC members. During its first years, IEC tried to harmonize the different national standards. But now, more and more national committees contribute to maintaining existing or establishing new IEC standards which are later overtaken as national and regional standards (e.g. CENELEC Standards of the European Union). This book refers mainly to IEC Standards and mentions also relevant standards of the US organization *"Institute of Electrical and Electronic Engineers"* (*IEEE*) which play an important role in some parts of the world. IEEE publishes also *"IEEE Guides"* which may overtake the role of missing text books. The IEEE Guides supply only recommendations and no requirements as standards are doing. The actual trend shows a closer cooperation between IEC and IEEE for the harmonization of IEC and IEEE Standards.

The structure of IEC (in 2014) is given in Fig. 1.5: The national committees send delegates to the IEC Council which is the IEC parliament and controls the IEC activities performed by the IEC Executive Committee. The Executive Committee is supported by three management boards, one of them related to IEC Standards. For the different fields of the IEC activities, the Management Board is supported by special groups. The active standardization work is done by Technical Committees (TC) and Subcommittees (SC). Each TC or SC is responsible for a certain number of standards of a special field. Existing IEC Standards are observed by Maintenance

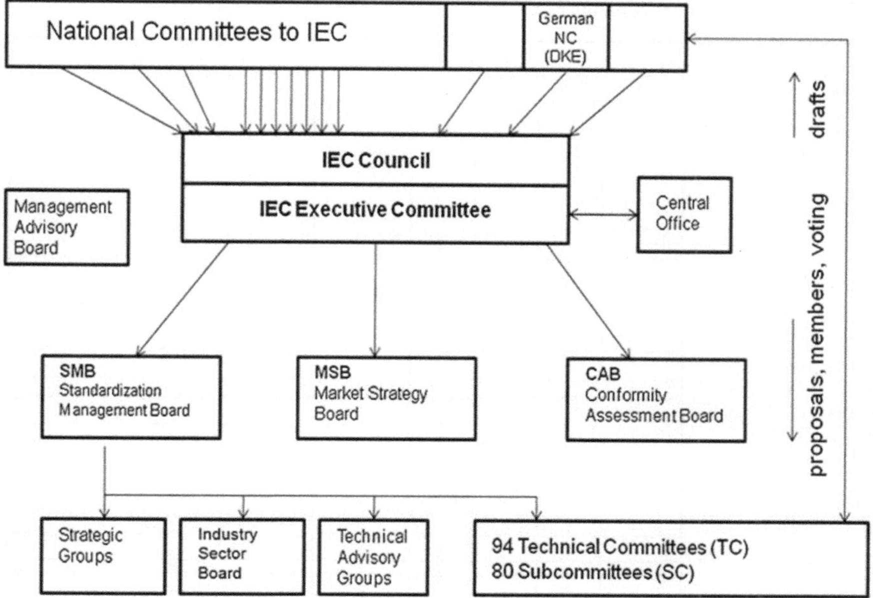

Fig. 1.5 Structure of the International Electrotechnical Commission (IEC)

Groups (MG); new IEC Standards are established by Working Groups (WG) based on proposals of National Committees. Each National Committee can be a TC/SC member or observer (or not attend the activities of certain TCs and SCs) and sends members to the active WGs.

There are TCs which have to maintain IEC Standards important for power systems and all types of apparatus. These are so-called *horizontal standards* (e.g. on insulation coordination or HV test techniques); the related TCs are shown in the first column of Table 1.1. IEC Standards related to apparatus or equipment (e.g. tests on transformers, GIS ore cables) are called *"vertical" (or apparatus) standards*. When a vertical standard is developed, all relevant horizontal standards shall be considered. Vice versa, during the development of a horizontal standard, the requirements of different apparatus should be known. The co-operation between horizontal TC's and vertical (apparatus) TC's requires improvement. Vertical TC's should better contribute to the activities of the horizontal committees and then apply horizontal standards consequently.

This book is closely related to the tasks of the TC 42 "High-voltage and high-current test techniques". It explains the scientific and technical background of the TC 42 standards, but cannot replace any of them. Rather, it should be understood as an application guide to the relevant IEC Standards and stimulate their application.

Table 1.1 The technical committees for horizontal and vertical standards

Horizontal Technical Committees for systems and basic tasks	Vertical Technical Committees TC / SC for apparatus and equipment								
	Rotating Machines TC 2	Power Transformers TC 14	Switchgear TC 17	Cables TC 20	Power Electronics for T & D SC 22F	Capacitors TC 33	Insulators TC3 6	Arresters TC 37	Instrument Transformers TC38
TC 1 Terminology									
TC 8 System Aspects									
TC 28 Insulation Coordination									
TC 42 HV Test Techniques									
TC 77 Electrom. Compatibility									
TC 104 Environment. Conditions									
TC 115 HVDC Transmission									
TC 122 UHV AC Transmission									

1.3 Insulation Coordination and Its Verification by HV Testing

In service, an electrical insulation is stressed with the *operational voltage* (including its temporary increase, e.g. in case of a load drop) and with the over-voltages mentioned above. The reliability of a power system has to be guaranteed under all possible stresses of its insulations. This is realized by the *insulation coordination* and described in the relevant group of IEC Standards (IEC 60071).

Insulation coordination is the correlation of the *withstand voltages* of different apparatus in a power system among each other and with the characteristics of *protective devices*. Today, protective devices (IEC 60099-4 2009) are *mainly metal oxide arresters (MOA)*, and partly conventional silicon carbide arresters with internal gaps and protection air gaps are still in use. An ideal protective device conducts electric current for voltages above the *protection level* and is an insulator below that voltage (A MOA is near to that characteristic). In the design of a power system, protective devices are installed at sensitive points, guaranteeing the protection level and protecting the insulation from excessive over-voltages.

The *insulation level* of the apparatus is selected in such a way that it is—under consideration of economic viewpoints—by a safety margin above the protection level. The insulation levels are defined by values of the relevant test voltages. The insulation under test must withstand the test voltage in a certain procedure. Usually, the AC or DC test voltage procedure is a 1-min stress (see Sects. 3.6 and 6.5); an impulse voltage test consists of a number of impulses defined according to the kind of insulation (see Sect. 7.3.2). Tables 1.2 and 1.3 deliver the test voltages for apparatus of AC three-phase power systems, depending on its *highest voltage for equipment* V_m (rms value of the phase-to-phase voltage).

For equipment of $V_\mathrm{m} = 3.6\text{–}245$ kV, the AC voltage test covers also withstand against internal (switching) over-voltages and no SI impulse voltage withstand test is specified. For equipment of $V_\mathrm{m} = 300\text{–}1200$ kV, the switching impulse test covers for air insulations also the AC voltage test; for internal insulations, the AC test voltages are specified in the relevant apparatus standards.

Table 1.2 Standard insulation levels for HVAC equipment $V_m = 3.6\text{–}245$ kV (IEC 60071-2: 2006)

Highest voltage for equipment V_m kV (rms, phase-to-phase)	Short-duration AC withstand voltage V_t kV (peak/$\sqrt{2}$, phase-to-earth)	LI withstand voltage V_t kV (peak value)
3.6	10	20
	10	40
7.2	20	40
	20	60
12	28	60
	28	75
	28	95
24	50	95
	50	125
	50	145
36	70	145
	70	170
72.5	140	325
123	(185)	(450)
	230	550
145	(185)	(450)
	230	550
	275	650
170	(230)	(550)
	275	650
	325	750
245	(275)	(650)
	(325)	(750)
	360	850
	395	950
	460	1050

Explanation

Usually, the phase-to-earth withstand voltages are also applied to phase-to-phase insulation. If the values in brackets are considered too low, additional phase-to-phase withstand voltage tests are needed

Table 1.3 Standard insulation levels for HVAC equipment $V_m = 300$–1200 kV (IEC 60071-2: 2006), (IEC 60071-2-Amendment 2010)

Highest voltage for equipment V_m kV (rms, phase-to-phase)	SI withstand voltage kV (peak value)			LI withstand voltage[c] V_t kV (peak value)
	Longitudinal insulation[a]	Phase-to-earth insulation	Phase-to-phase insulation[b]	
300	750	750	1125	850
	750	750	1125	950
	750	850	1275	950
	750	850	1175	1050
362	850	850	1275	950
	850	850	1275	1050
	850	950	1425	1050
	850	950	1425	1175
420	850	850	1360	1050
	850	850	1360	1175
	950	950	1425	1175
	950	950	1425	1300
	950	1050	1575	1300
	950	1050	1575	1425
550	950	950	1615	1175
	950	950	1615	1300
	950	1050	1680	1300
	950	1050	1680	1425
	950	1175	1763	1425
	1050	1175	1763	1550
800	1175	1300	2210	1675
	1175	1300	2210	1800
	1175	1425	2423	1800
	1175	1425	2423	1950
	1175	1550	2480	1950
	1300	1550	2480	2100
1200	1425	1550	2635	2100
	1425	1550	2635	2250
	1550	1675	2764	2250
	1550	1675	2764	2400
	1675	1800	2880	2400
	1675	1800	2880	2550

Explanations

[a]Longitudinal insulation means the insulation between different parts of the grid, realized e.g. by disconnectors and tested with combined voltages (see Sect. 8.1) The column gives only the value of the SI voltage component of the relevant combined voltage test. The peak value of the AC component of opposite polarity is $(V_m \cdot \sqrt{2}/\sqrt{3})$

[b]This is the peak value of the combined voltage in the relevant combined SI/AC voltage test
[c]These values apply to both, phase-to-earth and phase-to-phase insulation. For longitudinal insulation, they apply as the standard rated LI component of the relevant combined voltage test, while the peak of the AC component of opposite polarity is $0.7 \cdot (V_m \cdot \sqrt{2}/\sqrt{3})$
Note Each apparatus has a nominal voltage (e.g. $V_n = 380$ or 400 kV), but the insulation is designed for insulation coordination according to the highest voltage of a group of nominal voltages. This voltage is also called rated voltage (IEC 60038-2009). For the mentioned nominal voltages, the insulation is designed and must be tested according to the rated voltage $V_m = 420$ kV

Table 1.4 Simplified example for the selection of withstand test voltages for three protection levels

Highest voltage for equipment V_m	AC test	SI test	LI test	LIC test (transformers only)	Application to insulation
420 kV		850 kV	1050 kV	1175 kV	External insulation (atmospheric air)
	(630kV)		1175 kV	1300 kV	Internal insulation (SF$_6$, oil, solids)
		950 kV	1175 kV	1300 kV	External insulation
	(680 kV)		1300 kV	1425 kV	Internal insulation
		1050 kV	1300 kV	1425 kV	External insulation
	(680 kV)		1425 kV	1570 kV	Internal insulation

Note Consider that most test voltages are applied between phase and ground. The reference value for that is the *line-to-ground voltage* $V_0 = V_m/\sqrt{3}$, respectively, its peak voltage $V_p = \sqrt{2}\, V_0$

For one and the same rated voltage, different protection levels can be applied depending on the required reliability, safety and/or economy. The example of Table 1.4 shows this for the rated voltage $V_m = 420$ kV: The three main lines represent three different protection levels, and each line is divided into two lines applicable for external (air) and internal insulation. The AC test voltages are related to internal insulation only and given in the relevant apparatus standard.

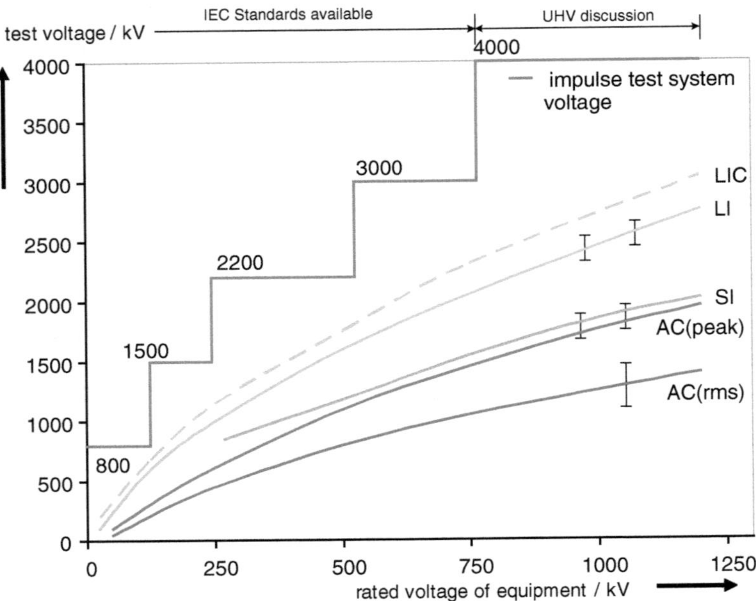

Fig. 1.6 Highest withstand test voltages for HVAC equipment and selection of impulse voltage test systems

A diagram of the test voltage values versus the rated voltages (Fig. 1.6) shows that the SI test voltage (peak value) is identical with the AC test peak voltage (Tables 1.2, 1.3, and 1.4 show AC rms values!). The chopped LI test voltages (LIC) are 10% higher than the full LI test voltages. The diagram may help in the selection of the HV test systems required for a HV test field (See Sect. 9.1).

Example: For a transformer test field, the selection of the rated voltage of an impulse voltage test system (equal to the cumulative charging voltage of the generator, see 7.1.1) shall be shown. Outgoing from the highest test voltage (LIC in Fig. 1.6), one has to consider that the utilization factor for large test objects may go down to $\eta = 0.85$. Furthermore, for internal development tests, a test voltage 20% higher than the LIC withstand voltage might be necessary. This means the rated voltage of the impulse voltage test system should be by a factor k = 1.2/0.85 $\approx$ 1.4 higher than the highest LIC test voltage. This means that for the rated voltage V_m = 800 kV, a 3000 kV impulse test system is sufficient. If a later extension of the test capability to 1200 kV equipment is planned, a 4000 kV test system should be considered. The selection of impulse test systems according to Fig. 1.5 is recommended.

For HVDC connections, no rated voltages exist, because the nominal voltages and currents of the present point-to-point HVDC connections are optimized according to the available power electronic components (One can assume that rated voltages will be introduced, when HVDC grids (CENELEC 2010) are realized). The relevant standard on insulation coordination (IEC 60071-5: 2002) does not deliver test voltages but only formulas which allow the calculation of test voltage

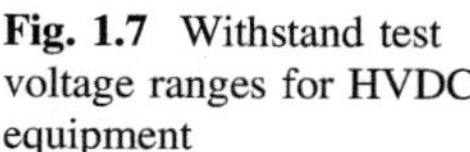
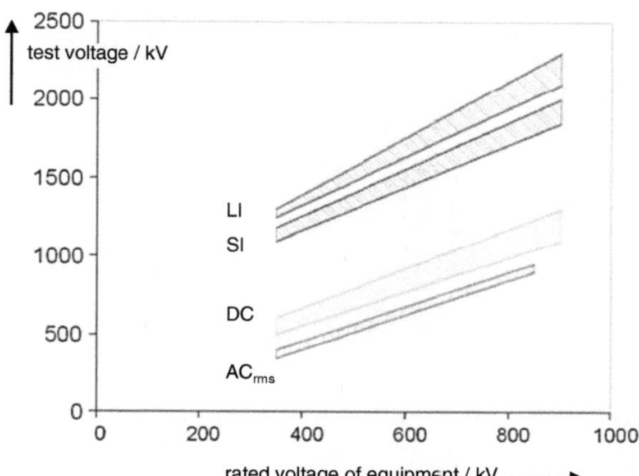

Fig. 1.7 Withstand test voltage ranges for HVDC equipment

ranges from the nominal voltages of the HVDC connection (Fig. 1.7). For HVDC equipment, the difference between LI and SI test voltages is lower than for HVAC equipment. Considering the lower efficiency of the SI voltage generation ($\eta \leq 0.75$), the selection of impulse voltage test systems should take the required SI test voltages into consideration.

1.4   Tests and Measurements in the Life Cycle of Power Equipment

The principles of insulation coordination are only applied to new equipment and verified in factory tests. These include *type tests* and *routine tests*. Both tests are *quality tests* of the insulation; a successful type test demonstrates the correct design according to the test voltages (Tables 1.2 and 1.3), and a successful routine test verifies the correct production according to the confirmed design. The two tests are not the only tests in the *life cycle* of the insulations of power equipment (Fig. 1.8).

Development tests at model insulations are often performed before the insulation is finally designed. When type and routine tests are successfully performed, the power equipment is transported to the site and assembled there. It may happen that defects of the insulation are caused by transportation and assembling. Also some huge apparatus (e.g. power transformers) cannot be transported as complete units. The final assembling takes place not in the factory, but on site. Therefore, additional *quality acceptance tests* (*commissioning tests*) or even the routine test must be performed on site with mobile HV test systems. This test should always be performed in relation to factory tests (Fig. 1.8).

After the correct quality is confirmed in a successful on-site test, the equipment is overtaken to the user who has the full responsibility for all further (diagnostic)

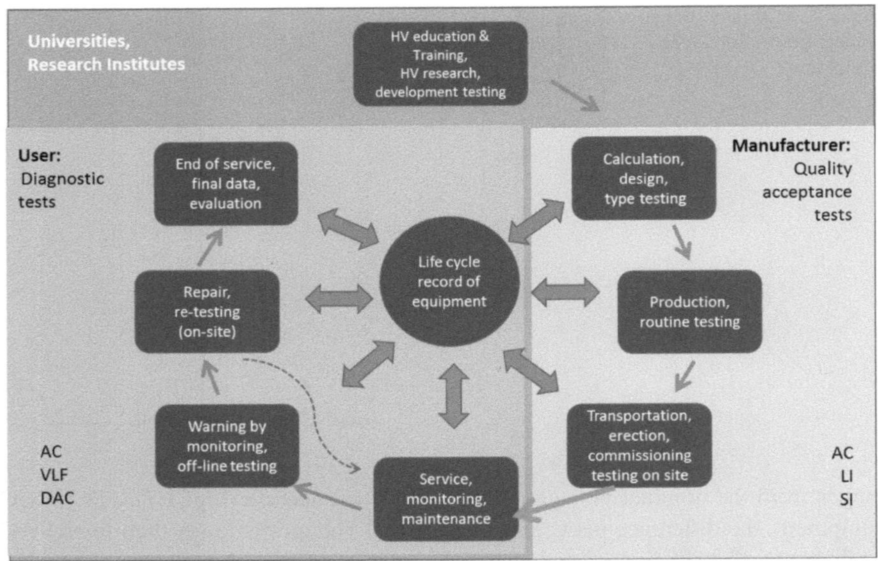

Fig. 1.8 Tests and measurements in the life cycle of HV insulation

tests and measurements now. It will be commissioned and has to operate under electrical, thermal, mechanical and/or environmental influences. These influences cause an ageing process of the insulation until—after tens of years—the end of the life cycle is reached. In the past, the duration of the life cycle has been estimated and the end of use has been defined by the user independently on the real condition of the equipment. To reduce the life cycle cost, tests and measurements have been introduced for *condition assessment* of the insulation and estimation of the remaining life time (Zhang et al. 2007; Olearczyk et al. 2010; Balzer et al. 2004). In opposite to the quality tests, these tests for condition assessment shall be called *diagnostic tests*. There are no standards for diagnostic tests, only recommendations by organizations (like CIGRE or IEEE) who provide technical guides, by service providers or by equipment suppliers.

Both quality and diagnostic tests require a test voltage application and measurements (in minimum a voltage measurement, but very often partial discharge or dielectric measurements). A withstand test is a *direct test*, which is directly related to the insulation capability of the test object. A healthy insulation which passes the test has a high withstand voltage. A defective insulation has a low withstand voltage and fails the test (This is much better than it fails in service!). It must be considered that a voltage stress may cause life-time consumption. An insulation shall be designed in such a way that the life-time consumption of the healthy insulation is negligible during a withstand test, whereas a defective insulation breaks down. When a withstand test is completed by a parallel partial discharge (PD) measurement (see Sect. 2.5), it can be excluded that such a successful *"PD-monitored" withstand test* has caused or enlarged insulation defects.

When a diagnostic test is decided according to measurements of a single parameter or a set of parameters (preferably of partial discharges), the test result must be compared with pre-given limits. These limits shall be related by experience or any physical model to the remaining life time. The sharpness of such an *indirect test* is lower than that of a monitored withstand test. Sometimes, it is published that withstand tests are "destructive" and diagnostic measurements are "non-destructive". Such qualifying terms are not suited to describe the quality and sharpness of a diagnostic HV test.

Diagnostic measurements performed at operational voltage in service are called *on-line monitoring* (CIGRE TF D1.02.08 2005) (Monitoring is not only related to dielectric measurement; there is also monitoring of voltage, current, thermal or mechanic parameters). Automatic monitoring delivers a warning when the measured parameter exceeds a preset limit.

This brings one back to Fig. 1.7: In service, online monitored data deliver the data trend, describe the situation and may supply data for maintenance. In case of a warning by the monitoring system, the reason of the defect must be clarified. Often, the monitored data are not sufficient for a clarification. In that case, a more detailed investigation is necessary, e.g. by an (off-line) on-site test including appropriate measurements. This test with a separate test voltage source enables a withstand voltage test (with a test voltage value well adapted to the age of the insulation) and the measurement of the parameters depending on the applied voltage (instead of only one fixed voltage at monitoring). After the condition of the insulation has been clarified, it might be decided to repair the equipment in the factory or on site. Then, one or several loops in the scheme appear, and one has to go back to on-site testing and service again. At the end of the life cycle, the equipment is dismantled and also this may deliver some data for future development.

The data of all tests and measurements during the life cycle must be recorded in the—preferably electronic—life cycle record of the equipment. The life cycle record is the most important document of the equipment, which delivers the trend of parameters and enables qualified decisions. As all stages of the life cycle are connected together, it must be stressed that quality and diagnostic testing have a common physical background. This is often forgotten, when quality testing and diagnostic testing/monitoring are considered separately. The only reference for all HV quality on-site tests of new equipment is the factory testing based on the test voltages of the insulation coordination. This may include that also the test voltages for diagnostic testing of service-aged equipment should be in close relation to stresses in service.

Chapter 2
Basics of High-Voltage Test Techniques

Abstract High-voltage (HV) testing utilizes the phenomena in electrical insulations under the influence of the electric field for the definition of test procedures and acceptance criteria. The phenomena—e.g., breakdown, partial discharges, conductivity, polarization and dielectric losses—depend on the insulating material, on the electric field generated by the test voltages and shaped by the electrodes as well as on environmental influences. Considering the phenomena, this chapter describes the common basics of HV test techniques, independent on the kind of the stressing test voltage. All details related to the different test voltages are considered in the relevant Chaps. 3–8.

2.1 External and Internal Insulations in the Electric Field

In this section definitions of phenomena in electrical insulations are introduced. The insulations are classified for the purpose of high-voltage (HV) testing. Furthermore environmental influences to external insulation and their treatment for HV testing are explained.

2.1.1 Principles and Definitions

When an electrical insulation is stressed in the electric field, ionization causes electrical discharges which may grow from one electrode of high potential to the one of low potential or vice versa. This may cause a high current rise, i.e., the dielectric looses its insulation property and thus its function to separate different potentials in an electric apparatus or equipment. For the purpose of this book, this phenomenon shall be called "*breakdown*" related to the stressing voltage:

Definition: The breakdown is the failure of insulation under electric stress, in which the discharge completely bridges the insulation under test and reduces the voltage between electrodes to practically zero (collapse of voltage).

© Springer Nature Switzerland AG 2019

W. Hauschild and E. Lemke, *High-Voltage Test and Measuring Techniques*,

https://doi.org/10.1007/978-3-319-97460-6_2

Note In IEC 60060-1 (2010) this phenomenon is referred to as *"disruptive discharge"*. There are also other terms, like *"flashover"* when the breakdown is related to a discharge over the surface of a dielectric in a gaseous or liquid dielectric, *"puncture"* when it occurs through a solid dielectric and *"sparkover"* when it occurs in gaseous or liquid dielectrics.

In homogenous and *slightly non-homogenous fields* a breakdown occurs when a critical strength of the stressing field is reached. *In strongly non-homogenous fields, a local stress concentration causes a localized electrical partial discharge (PD)* without bridging the whole insulation and without breakdown of the stressing voltage.

Definition: A partial discharge is a localized electrical discharge that only partly bridges the insulation between electrodes, for details see Chap. 4.

Figure 1.2 shows the application of some important insulating materials. Till today atmospheric air is applied as the most important dielectric of the external insulation of transmission lines and the equipment of outdoor substations.

Definition: External insulation means air insulation including the outer surfaces of solid insulation of equipment exposed to the electric field, to atmospheric conditions (air pressure, temperature, humidity) and to other environmental influences (rain, snow, ice, pollution, fire, radiation, vermin).

External insulation recovers its insulation behaviour in most cases after a breakdown and is then called a *self-restoring insulation.* In opposite to that, the *internal insulation* of apparatus and equipment—such as transformers, gas-insulated switchgear (GIS), rotating machines or cables—is more affected by discharges, often even destroyed when a breakdown is caused by a HV stress.

Definition: Internal insulation of solid, liquid or gaseous components is protected from direct influences of external conditions such as pollution, humidity and vermin.

Solid and liquid- or gas-impregnated laminated insulation elements are *non-self-restoring insulations*. Some insulation is *partly self-restoring*, particularly when it consists e.g., of gaseous and solid elements. An example is the insulation of a GIS which uses SF_6 gas and solid spacers. In case of a breakdown in an oil- or SF_6 gas-filled tank, the insulation behaviour is not completely lost and recovers partly. After a larger number of breakdowns, partly self-restoring elements have a remarkably reduced breakdown voltage and are not longer reliable.

The insulation characteristic has consequences for HV testing: Whereas for HV testing of external insulation, the atmospheric and environmental influences have to be taken into consideration, internal insulation does not require related special test conditions. In case of self-restoring insulation, breakdowns may occur during HV tests. For partly self-restoring insulation, a breakdown would only be acceptable in the self-restoring part of the insulation. In case of non-self restoring insulation no breakdown can be accepted during a HV test. For the details see Sect. 2.4 and the relevant subsections in Chaps. 3 and 6–8.

The test procedures should guarantee the *accuracy* and the *reproducibility* of the test results under the actual conditions of the HV test. The different test procedures necessary for external and internal insulations should deliver comparable test results. This requires regard to various factors such as

- random nature of the breakdown process and the test results,
- polarity dependence of the tested or measured characteristics,
- acclimatisation of test object to the test conditions,
- simulation of service conditions during the test,
- correction of differences between standard, test and service conditions, and
- possible deterioration of the test object by repetitive voltage applications.

2.1.2   *HV Dry Tests on External Insulation Including Atmospheric Correction Factors*

HV dry tests have to be applied for all external insulations. The arrangement of the test object may affect the breakdown behaviour and consequently the test result. The electric field at the test object is influenced by *proximity effects* such as distances to ground, walls or ceiling of the test room as well as to other earthed or energized structures nearby. As a rule of thumb, the *clearance* to all external structures should be not less than 1.5 times the length of the possible discharge path along the test object. For maximum AC and SI test voltages above 750 kV (peak), recommendations for the minimum clearances to external earthed or energized structures are given in Fig. 2.1 (IEC 60060-1:2010). When the necessary clearances are considered, the test object will not be affected by the surrounding structures.

Atmospheric conditions may vary in wide ranges on the earth. Nevertheless, HV transmission lines and equipment with external insulations have to work nearly everywhere. This means on one hand that the atmospheric service conditions for HV equipment must be specified (and for these conditions it must be tested), and on

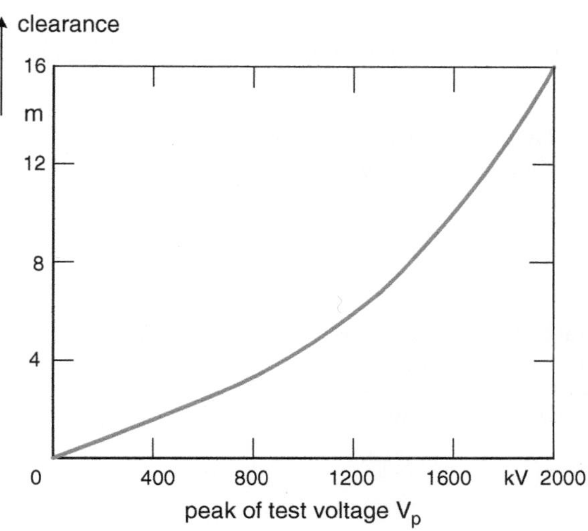

Fig. 2.1 Recommended clearances between test object and extraneous energized or earthed structures

the other hand the test voltage values for insulation coordination (IEC 60071:2010) must be related to a *standard reference atmosphere*:

- temperature $t_0 = 20\ °C$ (293 K)
- absolute air pressure $p_0 = 1013$ hPa (1013 mbar)
- absolute humidity $h_0 = 11\ g/m^3$

Note. It shall be mentioned, that for the correction of atmospheric conditions to reference conditions different procedures are proposed in different IEC Standards of HV equipment. It is a future task harmonizing the different procedures. This book follows the IEC 60060-1: 2010.

The temperature shall be measured with an expanded uncertainty $t \leq 1\ °C$, the ambient pressure with $p \leq 2$ hPa. The absolute humidity $h/g/m^3$ can be directly measured with so-called ventilated dry-and wet-bulb thermometers or determined from the relative humidity R and the temperature $t/°C$ by the formula (IEC 60060-1:2010):

$$h = \frac{6.11 \cdot R \cdot e^{\frac{17.6 \cdot t}{273 + t}}}{0.4615 \cdot (273 + t)} \tag{2.1}$$

If HV equipment for a certain altitude shall be designed according to the pressure-corrected test voltages, the relationship between altitude H/m and pressure p/hPa is given by

$$p = 1013 \cdot e^{\frac{-H}{8150}} \tag{2.2}$$

A test voltage correction for air pressure based on this formula can be recommended for altitudes up to 2500 m. For more details see Pigini et al. (1985), Ramirez et al. (1987) and Sun et al. (2009). The temperature t and the pressure p determine the *air density* δ, which influences the breakdown process directly:

$$\delta = \frac{p}{p_0} \cdot \frac{273 + t_0}{273 + t}. \tag{2.3}$$

The air density delivers together with the air density correction exponent m (Table 2.1) the air density correction factor

$$k_1 = \delta^m. \tag{2.4}$$

Table 2.1 Air density and humidity correction exponents m and w according to IEC 60060-1:2010	g	m	w
	<0.2	0	0
	0.2–1.0	$g(g - 0.2)/0.8$	$g(g - 0.2)/0.8$
	1.0–1.2	1.0	1.0
	1.2–2.0	1.0	$(2.2 - g)(2.0 - g)/0.8$
	>2.0	1.0	0

The humidity affects the breakdown process especially when it is determined by partial discharges. These are influenced by the kind of test voltage. Therefore, for different test voltages different humidity correction factors k_2 have to be applied, which are calculated with the parameter k and the humidity correction exponent w

$$k_2 = k^w,$$ (2.5)

with

$$
\begin{aligned}
DC: \quad & k = 1 + 0.014(h/\delta - 11) - 0.00022(h/\delta - 11)^2 && \text{for } 1\ g/m^3 < h/\delta,\ < 15\ g/m^3, \\
AC: \quad & k = 1 + 0.012(h/\delta - 11) && \text{for } 1\ g/m^3 < h/\delta < 15\ g/m^3, \\
LI/SI: \quad & k = 1 + 0.010(h/\delta - 11) && \text{for } 1\ g/m^3 < h/\delta < 20\ g/m^3.
\end{aligned}
$$

The correction exponents m and w describe the characteristic of possible partial discharges and are calculated utilizing a parameter

$$g = \frac{V_{50}}{500 \cdot L \cdot \delta \cdot k},$$ (2.6)

with

V_{50}	*Measured or estimated 50% breakdown voltage at the actual atmospheric conditions, in kV (peak),*
L	*Minimum discharge path, in m,*
δ	*Relative air density and*
k	*Dimension-less parameter defined with formula (2.5).*

Note 1: For withstand tests it can be assumed $V_{50} \approx 1.1 \cdot V_t$ (test voltage). Then Table 2.1 or Fig. 2.2 delivers the exponents m and w depending on the parameter g (Eq. 2.6).

Note 2: Consider the limitations of the applicability of the Eqs. (2.1)–(2.6), especially for the altitude (Eq. (2.2)) and for the humidity (Eq. (2.5)).

According to IEC 60060-1:2010 the atmospheric correction factor

$$K_t = k_1 \cdot k_2,$$ (2.7)

shall be used to correct a measured breakdown voltage V to a value under standard reference atmosphere

$$V_0 = V/K_t.$$ (2.8)

Vice versa when a test voltage V_0 is specified for standard reference atmosphere, the actual test voltage value can be calculated by the converse procedure:

$$V = K_t \cdot V_0.$$ (2.9)

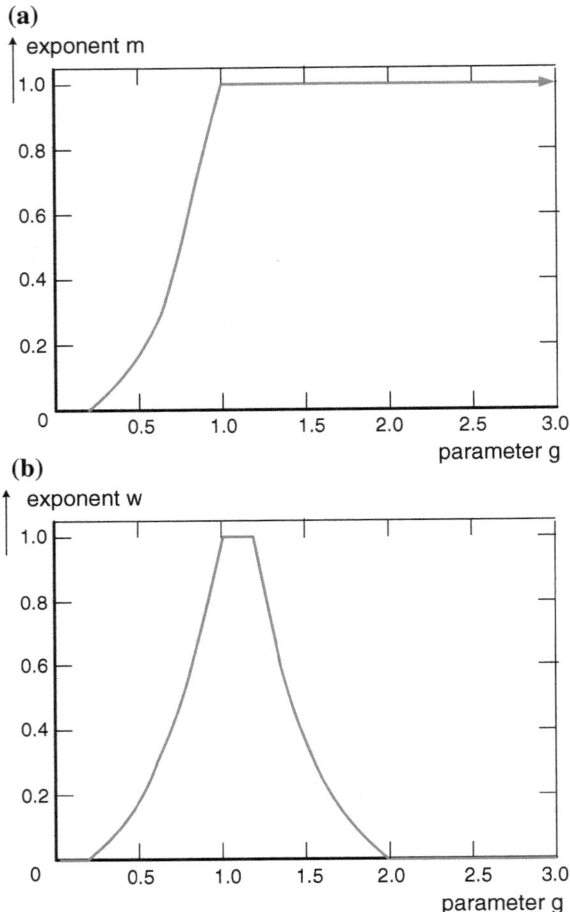

Fig. 2.2 Correction exponents according to IEC 60060-1:2010. **a** m for air density. **b** w for air humidity

Because the converse procedure uses the breakdown voltage V_{50} (Eq. 2.6), the applicability of Eq. (2.9) is limited to values of K_t close to unity, for $K_t < 0.95$ it is recommended to apply an iterative procedure which is described in detail in Annex E of IEC 60060-1:2010.

It is necessary to mention that the present procedures for atmospheric corrections are quite far from being perfect (CIGRE WG D1.36/Draft 2017; Wu et al. 2009): The humidity correction is limited only to air gaps and not applicable to flashovers directly along insulating surfaces in air. The reason is the different absorption of water by different surface materials. Furthermore the attention is drawn to the limitations of the application of humidity correction to $h/\delta \leq 15$ g/m^3 (for AC and DC test voltages), respectively $h/\delta \leq 20$ g/m^3 (for LI and SI test voltages). This means the procedure for tropic countries is incomplete. The clarification of the humidity correction for surfaces as well as the extension of their ranges requires further research work (Mikropolulos et al. 2008; Lazarides and Mikropoulos 2010, 2011).

In general, the atmospheric correction for altitudes above 2500 m is not yet described in the standard (Ortega et al. 2007; Jiang et al. 2008; Jiang et al. 2008). Nevertheless, also the available correction to and from reference atmospheric conditions is important in HV testing of external insulation as it should be shown by two simplified examples:

Example 1 In a development test, the 50% LI breakdown voltage of an air insulated disconnector [breakdown (flashover) path L = 1 m, not at the insulator surface] was determined to V_{50} = 580 kV at a temperature of t = 30 °C, an air pressure of p = 980 hPa and a humidity of h = 12 g/m³. The value under reference atmospheric conditions shall be calculated:

Air density	$\delta = (995/1013) \cdot (293/303) = 0.95$
Parameter	$k = 1 + 0.010 \cdot ((12/0.95) - 11) = 1.02$
Parameter	$g = 580/(500 \cdot 1 \cdot 0.95 \cdot 1.02) = 1.20$
Table 2.1: delivers the air density correction exponent	$m = 1.0$
And the humidity correction exponent	$w = (2.2 - 1.2)(2.0 - 1.2)/0.8 = 1.0$
With the density correction factor	$k_1 = 0.95$
And the humidity correction factor	$k_2 = 1.02$
One gets the atmospheric correction factor	$K_t = 0.95 \cdot 1.02 = 0.97$
Under reference conditions the 50% breakdown voltage is	$\mathbf{V_{0-50} = 580/0.97 = 598}$ kV

Example 2 The same disconnector shall be type tested with a LI voltage of V_0 = 550 kV in a HV laboratory at higher altitude under the conditions t = 15 °C, p = 950 hPa and h = 10 g/m³. Which test voltage must be applied?

Air density	$\delta = (950/1013)(293/288) = 0.954$
Parameter	$k = 1 + 0.010((10/0.95) - 11) = 0.995$
Parameter	$g = 598/(500 \cdot 1 \cdot 0.95 \cdot 0.995) = 1.265$
Table 2.1: delivers the air density correction exponent	$m = 1.0$
And the humidity correction exponent	$w = (2.2 - 1.265)(2.0 - 1.265)/0.8 = 0.86$
With the density correction factor	$k_1 = 0.954$
And the humidity correction factor	$k_2 = 0.995^{0.86} = 0.996$
One gets the atmospheric correction factor	$K_t = 0.954 \cdot 0.996 = 0.95$
Under the actual laboratory conditions the test voltage is	$\mathbf{V = 550 \cdot 0.95 = 523}$ kV

The two examples show, that the differences between the starting and resulting values are significant. The application of atmospheric corrections is essential for HV testing of external insulation.

2.1.3 HV Artificial Rain Tests on External Insulation

External HV insulations (especially outdoor insulators) are exposed to natural rain. The effect of rain to the flashover characteristic is simulated in artificial rain (or *wet*) *tests* (Fig. 2.3). The artificial rain procedure described in the following is applicable for tests with AC, DC and SI voltages, whereas the arrangement of the test object is described in the relevant apparatus standards. The influence of rain on the LI voltage breakdown can be neglected.

The test object is sprayed with droplets of water of given resistivity and temperature (Table 2.2). The rain shall fall on the test object under an angel of about $45°$, this means that the horizontal and vertical components of the *precipitation rate* shall be identical. The precipitation rate is measured with a special collecting vessel with a horizontal and a vertical opening of identical areas between 100 and 700 cm^2. The rain is generated by an artificial rain equipment consisting of nozzles fixed on frames. Any type of nozzles which generates the appropriate rain conditions (Table 2.2) can be applied.

Fig. 2.3 Artificial rain test on an 800 kV support insulator. *Courtesy* HSP Cologne

Table 2.2 Conditions for artificial rain precipitation

Precipitation condition	Unit	IEC 60060-1:2010 range for equipment of $V_m \leq 800$ kV	Proposed range for UHV equipment $V_m > 800$ kV
Average precipitation rate of all measurements:			
• Vertical component	(mm/min)	1.0–2.0	1.0–3.0
• Horizontal component	(mm/min)	1.0–2.0	1.0–3.0
Limits for any individual measurement and for each component	(mm/min)	±0.5 from average	1.0–3.0
Temperature of water	(°C)	Ambient temperature ± 15 K	Ambient temperature ± 15 K
Conductivity of water	(µS/cm)	100 ± 15	100 ± 15

Note 1 Examples of applicable nozzles are given in the old version of IEC 60-1:1989-11 (Fig. 2, pp. 113–115) as well as in IEEE Std. 4–1995.

The precipitation rate is controlled by the water pressure and must be adjusted in such a way, that only droplets are generated and the generation of water jets or fog is avoided. This becomes more and more difficult with increasing size of the test objects which requires larger distances between test object and artificial rain equipment. Therefore, the requirements of IEC 60060-1:2010 are only related to equipment up to rated voltages of $V_m = 800$ kV, Table 2.2 contains an actual proposal for the UHV range.

The reproducibility of wet test results (*wet flashover voltages*) is less than that for dry HV breakdown or withstand tests. The following precautions enable acceptable wet test results:

- The water temperature and resistivity shall be measured on a sample collected immediately before the water reaches the test object.
- The test object shall be pre-wetted initially for at least 15 min under the conditions specified in Table 2.2 and these conditions shall remain within the specified tolerances throughout the test, which should be performed without interrupting the wetting.

Note 2 The pre-wetting time shall not include the time needed for adjusting the spray. It is also possible to perform an initial pre-wetting by unconditioned tap water for 15 min, followed without interruption of the spray by a second pre-wetting with the well conditioned test water for at least 2 min before the test begins.

- The test object shall be divided in several zones, where the precipitation rate is measured by a collecting vessel placed close to the test object and moved slowly over a sufficient area to average the measured precipitation rate.
- Individual measurements shall be made at all measuring zones considering also one at the top and one near the bottom of the test object. A measuring zone shall have a width equal to that of the test object (respectively its wetted parts) and a

maximum height of 1–2 m. The number of measuring zones shall cover the full height of the test object.

- The spread of results may be reduced if the test object is cleaned with a surface-active detergent, which has to be removed before the beginning of wetting.
- The spread of results may also be affected by local anomalous (high or low) precipitation rates. It is recommended to detect these by localized measurements and to improve the uniformity of the spray, if necessary.

The test voltage cycle for an artificial rain test shall be identical to that for a dry test. For special applications different cycles are specified by the relevant apparatus committees. A density correction factor according to Sect. 2.1.2, but no humidity correction shall be applied.

Note 3 IEC 60060-1:2010 permits one flashover in AC and DC wet tests provided that in a repeated test no further flashover occurs.

Note 4 For the UHV test voltage range, it may be necessary to control the electric field (e.g., by toroid electrodes) to the artificial rain equipment and/or to surrounding grounded or energized objects including walls and ceiling to avoid a breakdown to them. Also artificial rain equipment on a potential different from ground might be taken into consideration.

Under extreme, especially tropical conditions, the precipitation rate might be higher than given in Table 2.2. Even water jets cannot be excluded. For such conditions, special tests might be necessary, as e.g. described by Yuan et al. (2015). Furthermore the flashover voltage reduces with increasing precipitation rate and with increasing water conductivity (CIGRE WG D1.36, 2017). Also HVDC insulation in artificial rain tests require special considerations (Zhang et al. 2016).

2.1.4 HV Artificial Pollution Tests on External Insulation

Outdoor insulators are not only exposed to rain, but also to pollution caused by salt fog near the sea shore, by industry and traffic or simply by natural dust. Depending on the position of a transmission line or substation, the surrounding is classified in several different *pollution classes* between low (*surface conductivity* $\kappa_s \leq 10\ \mu S$) and extreme ($\kappa_s \geq 50\ \mu S$) (Mosch et al. 1988). The severity of the pollution class can also be characterized by the equivalent salinity (SES in kg/m^3), which is the salinity [content of salt (in kg) in tap water (in m^3)] applied in a salt-fog test according to IEC 60507 (1991) that would give comparable values of the leakage current on an insulator as produced at the same voltage by natural pollution on site (Pigini 2010).

Depending on the pollution class, the artificial pollution test is performed with different intensities of pollution, because the test conditions shall be representative of wet pollution in service. This does not necessarily mean that any real service condition has to be simulated. In the following the performance of typical pollution tests is described without considering the representation of the pollution zones. The pollution flashover is connected with quite high pre-arc currents supplied via the wet and polluted surface from the necessary powerful HV generator (HVG). In a pioneering work, Obenaus (1958) considered a flashover model of a series connection of the pre-arc discharge with a resistance for the polluted surface. Till today the *Obenaus model* is the basis for the selection of pollution test procedures and the understanding of the requirements on test generators (Slama et al. 2010; Zhang et al. 2010). These requirements to HV test circuits are considered in the relevant Chaps. 3 and 6–8. Pollution tests of insulators for high altitudes have to take into consideration not only the pollution class, but also the atmospheric conditions (Jiang et al. 2009).

The test object (ceramic insulator) must be cleaned by washing with tap water and then the salt-fog pollution process may start. Typically the pollution test is performed with subsequent applications of the test voltage which is held constant for a specified test time of at least several minutes. Within that time very heavy partial discharges, so-called pre-arcs, appear (Fig. 2.4). It may happen that the wet and polluted surface dries (This means electrical withstand of the tested insulator and passing the test) or that the pre-arcs are extended to a full flashover (This means failing the test.). Because of the random process of the pollution flashover, remarkable dispersion of the test results can be expected. Consequently the test must be repeated several times to get average values of sufficient confidence or to estimate distribution functions (see Sect. 2.4). Two pollution procedures shall be described.

The *salt-fog method* uses a fog from a salt (NaCl) solution in tap water with defined concentrations between 2.5 and 20 kg/m^3 depending on the pollution zone. A spraying equipment generates a number of fog jets each generated by a pair of nozzles. One nozzle supplies about 0.5 l/min of the salt solution, the other one the compressed air with a pressure of about 700 kPa which directs the fog jet to the test object. The spraying equipment contains usually two rows of the described double nozzles. The test object is wetted before the test. The test starts with the application of the fog and the test voltage value which should be reached—but not overtaken— as fast as possible. The whole test may last up to 1 h.

The *pre-deposit method* is based on coating the test object with a conductive suspension of Kieselgur or Kaolin or Tonoko in water ($\approx$ 40 g/l). The conductivity of the suspension is adjusted by salt (NaCl). The coating of the test object is made by dipping, spraying or flow-coating. Then it is dried and should become in thermal equilibrium with the ambient conditions in the pollution chamber. Finally the test object is wetted by a steam-fog equipment (steam temperature $\leq$ 40 °C). The surface condition is described by the *surface conductivity* (µS) measured from the current at two probes on the surface (IEC 60-1:1989, Annex B.3) or by the equivalent amount of salt per square centimetre of the insulating surface [so-called salt deposit density (S.D.D.) in mg/cm^2]. The test can start with voltage application before the test object is wetted, or after wetting, when the conductivity has reached its maximum. Details depend on the aim of the test, see IEC 60507:1991.

Fig. 2.4 Phases of a
pollution flashover of an
insulator. Courtesy of FH
Zittau, Germany

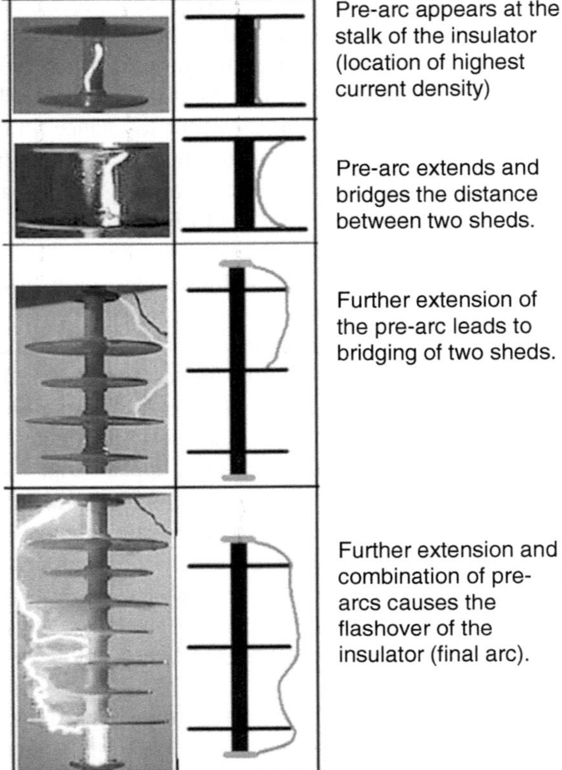

Both described procedures can be performed with different aim of the pollution test:

- determination of the withstand voltage for a certain insulator of specified degree of pollution and a specified test time,
- determination of the maximum degree of pollution for a certain insulator at a specified test voltage and specified test time.

Composite insulators have—beside of many mechanical advantages—the advantage of the hydrophobicity of their surface (Baer et al. 2016). Consequently they have proven their superiority to ceramic insulators for HVAC and HVDC applications (Yang et al. 2012; Abbasi et al. 2014), but require additional tests (IEC 62217, 2012).

Pollution tests require separate *pollution chambers*, usually with bushings for the connection of the test voltage generator. Because of the salt fog and humidity, the HV test system itself is outside the chamber under clean conditions. Inside a salt-fog chamber, the clearances around the test object should be ≥ 0.5 m/100 kV but not less than 2 m. When no pollution chamber is available, also tents from plastic foil may be applied, to separate the pollution area from the other areas of a HV laboratory.

2.1.5 Hints to Further Environmental Tests and HV Tests of Apparatus

There are also other environmental HV tests, e.g., under *ice* or *snow*. They are made with natural conditions in suitable open-air HV laboratories or in special climatic chambers (Sklenicka et al. 1999; Farzaneh and Chisholm 2014; Hu et al. 2016; Yin et al. 2016; Taheri et al. 2014). Other environmental influences which are simulated in HV tests are UV light (Kindersberger 1997), sandstorms (Fan and Li 2008) and *fire* under transmission lines (Peng et al. 2016). The HV test procedures for apparatus and equipment are described in the relevant "vertical" standards, examples are given in the chapters of the different test voltages.

2.1.6 HV Tests on Internal Insulation

In a HV test field, the HV components of test systems are designed with an external indoor insulation usually. When internal insulation shall be tested, the test voltage must be connected to the internal part of the apparatus to be tested. This is usually done by bushings which have an external insulation. For reasons of the insulation co-ordination or of the atmospheric corrections, cases will arise that the HV withstand test level of the internal insulation exceeds that of the external insulation (bushing). Then the withstand level of the bushing must be enhanced to permit application of the required test voltages for the internal insulation. Usually special "test bushings" of higher withstand level, which replace the "service bushings" during the test, are applied. A further possibility is the immersion of the external insulation in liquids or compressed gases (e.g., SF_6) during the test.

In rare cases, when the test voltage level of the external insulation exceeds that of the internal insulation a test at the complete apparatus can only be performed when the internal insulation is designed according to the withstand levels of the external insulation. If this cannot be done, then the apparatus should be tested at the internal test voltage level, and the external insulation should be tested separately using a dummy.

Internal insulation is influenced by the ambient temperature of the test field, but usually not by pressure or humidity of the ambient air. Therefore, the only requirement is the temperature equilibrium of the test object with its surrounding when the HV test starts.

2.2 HV Test Systems and Their Components

This section supplies a general description of HV test systems and their components, which consist of the *HV generator*, the *power supply unit*, the *HV voltage measuring system*, the *control system* and possibly additional measuring equipment, e.g., for PD or dielectric measurement. In all cases the test object cannot be neglected, because it is a part of the *HV test circuit.*

A *HV test system* means the complete set of apparatus and devices necessary for performing a HV test. It consists of the following devices (Fig. 2.5).

The *HV generator* (HVG) converts the supplied low or medium voltage into the high test voltage. The type of the generator determines the kind of the test voltage. For the generation of high alternating test voltages (HVAC), the HVG is a test transformer (Fig. 2.6a). It might be also a resonance reactor which requires a capacitive test object (TO) to establish an oscillating circuit for the HVAC generation (see Sect. 3.1). For the generation of high direct test voltages (HVDC) the HVG is a special circuit of rectifiers and capacitors (e.g., a Greinacher or Cockroft-Walton generator, Fig. 2.6b, see Sect. 6.1), and for the generation of high lightning or switching impulse (LI, SI) voltages, it is a special circuit of capacitors, resistors and switches (sphere gaps) (e.g., a Marx generator, Fig. 2.6c, see Sect. 7.1).

The test object does not only play a role for HVAC generation by resonant circuits, there is an interaction between the generator and the test object in all HV test circuits. The voltage at the test object may be different from that at the generator because of a voltage drop at the HV lead between generator and test object or even a voltage increase because of resonance effects. This means the voltage must be measured directly at the test object and not at the generator (Fig. 2.5). For this voltage measurement a sub-system—usually called *HV measuring system*—is connected to the test object (Fig. 2.7a, see Sect. 2.3). Further sub-systems, e.g., for dielectric measurement, can be added. Up to few 10 kV such systems can be designed as compact units including voltage source (Fig. 2.7b). Very often PD

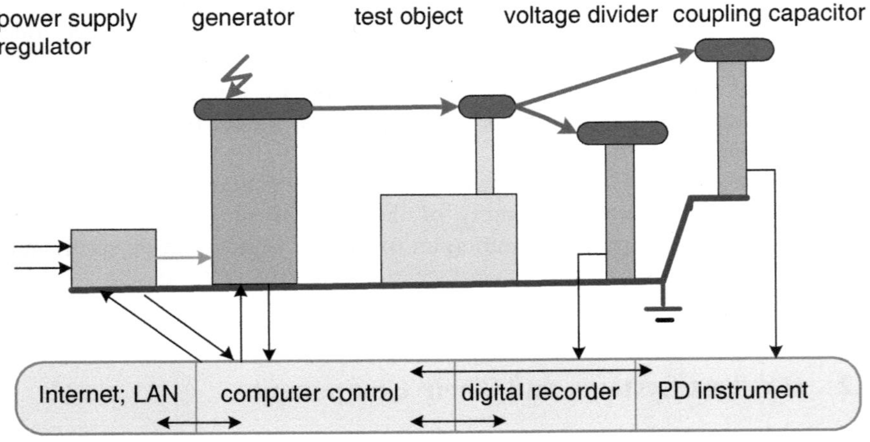

Fig. 2.5 Principle circuit of a HV test system

Fig. 2.6   HV generators. **a** For AC test voltage 1000 kV at Cottbus Technical University. **b** For DC test voltage 1500 kV at HSP Cologne. **c** For LI/SI test voltages 2400 kV at Dresden Technical University

measurements are performed during a HVAC test. For that a *PD measuring system* is connected to the AC test system (Fig. 2.7c). All these systems consist of a HV component (e.g., voltage divider, coupling or standard capacitor), measuring cable for data transfer and a low-voltage instrument (e.g., digital recorder, peak voltmeter, PD measuring instrument, tan delta bridge).

All the components of a HV test system and the test object described above form the HV circuit. This circuit should be of lowest possible impedance. This means, it should be as compact as possible. All connections, the *HV leads* and the *ground connections* should be straight, short and of low inductance, e.g., by copper foil (width 10–25 cm, thickness depending of current). In HV circuits used also for PD measurement, the HV lead should be realized by PD-free tubes of a diameter appropriate to the maximum test voltage. Any loop in the ground connection has to be avoided.

The necessary power for the HV tests is supplied from the power grid—in case of on-site testing also from a Diesel-generator set—via the *power supply unit* (Fig. 2.5). This unit consists of one or several *switching cubicles* and a *regulation unit* (regulator transformer or motor-generator set or thyristor controller or frequency converter). It controls the power according to the signals from the control system in such a way, that the test voltage at the test object is adjusted as required for the HV test. For safety reasons the switching cubicle shall have two circuit breakers in series; the first switches the connections between the grid and the power supply unit (power switch), the second one that between the power supply unit and the generator (operation switch). For the reduction of the required power from the grid in case of HVAC tests on capacitive test objects, the power supply unit is often completed by a fixed or even adjustable *compensation reactor*.

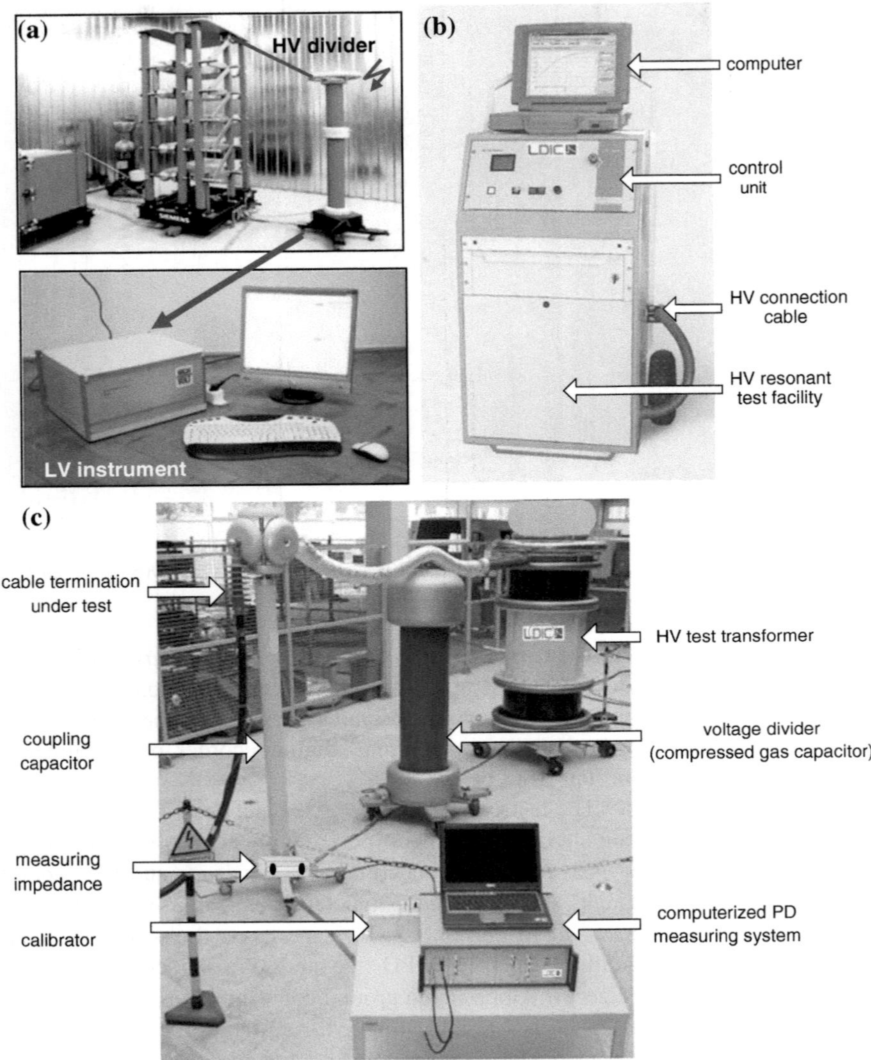

Fig. 2.7 Measuring systems. **a** Voltage measuring system with digital recorder including impulse generator. **b** Compact capacitance/loss factor measuring system with integrated AC voltage source (Courtesy of Doble-Lemke). **c** Partial discharge measuring system including AC voltage test circuit. Courtesy of Doble-Lemke

When the generator is the heart of HV test system then the control and measuring sub-system—usually called *control and measuring system* (Fig. 2.8; Baronick 2003)—is its brain. Older controls were separated from the measuring systems and the adjustment of the test voltage was manually made by the operator (The brain was that of the operator). As a next step, programmable logic controllers

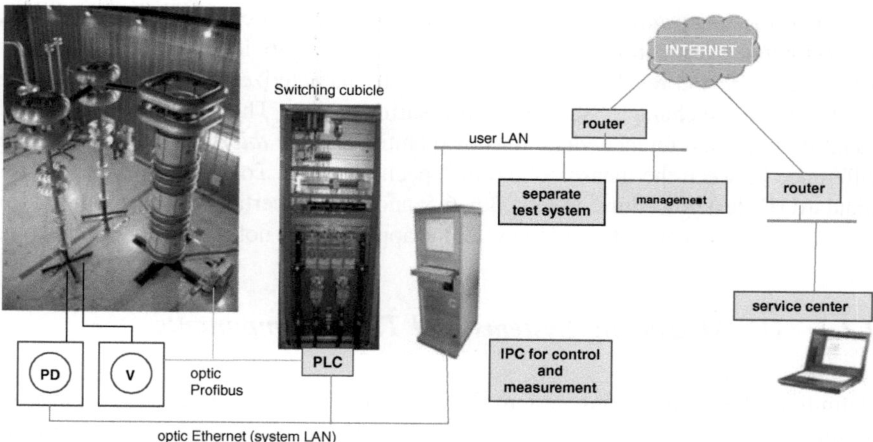

Fig. 2.8 Computerized control and measuring system

have been introduced. Now a state-of-the-art control system is a *computer control* which enables the pre-selection of the test procedure with all test voltage values, gives the commands to the power supply unit, overtakes the data from the measuring systems, performs the test data evaluation and prints a test record. In that way one operator can supervise very complex test processes. The test data can be transferred to a local computer network (LAN) e.g., for combining with test data from other laboratories or even to the Internet. The latter can also be used in case of technical problems for *remote service.*

A HV test system is only complete when it is connected to a *safety system* which protects the operators and the participants of a HV test. Among others, the safety system includes a fence around the test area which is combined with the electrical *safety loop.* The test can only be operated when the loop is closed, for details see Sect. 9.2.

2.3 HV Measurement and Estimation of the Measuring Uncertainty

This section is related to voltage measurement and describes *HV measuring systems*, their *calibration* and the estimation of their *uncertainties of measurements*. Precise measurement of high test voltages is considered to be a difficult task for many years (Jouaire et al. 1978; "Les Renardieres Group" 1974). This situation is also reflected by the older editions of the relevant standard IEC 60060-2. For good practice in HV test fields, this Sect. 2.3 on HV measurement and uncertainty estimation is closely related to the newest edition of the standard IEC 60060-2:2010. A detailed description of the basics and the state of HV impulse measuring technique is given by Schon (2010, 2013).

The terms *"uncertainty"*, *"error"* and *"tolerance"* are often mixed up. Therefore, the following clarification seems needed: The *uncertainty* is a parameter which is associated with the result of a measurement. It characterizes the dispersion of the results due to the characteristics of the measuring system. The *error* is the measured quantity minus a reference value for this quantity and the *tolerance* is the permitted difference between the measured and the specified value. Tolerances play a role for standard HV test procedures (Sects. 3.6, 6.5, and 7.6). Uncertainties are important for the decision, whether a measuring system is applicable or not for acceptance testing.

2.3.1   HV Measuring Systems and Their Components

Definition: A HV measuring system (MS) is a "complete set of devices suitable for performing a HV measurement". Software for the calculation of the result of the measurement is a part of the measuring system (IEC 60060-2:2010).

A HV measuring system (Fig. 2.9) which should be connected directly to the test object consists usually of the following components

- A *converting device* including its HV and earth connection to the test object which converts the quantity to be measured (*measurand*: test voltage with its voltage and/or time parameters) into a quantity compatible with the measuring instrument (low-voltage or current signal). It is very often a voltage divider of a

Fig. 2.9 HV measuring system consisting of voltage divider, coaxial cable and PC-based digital recorder

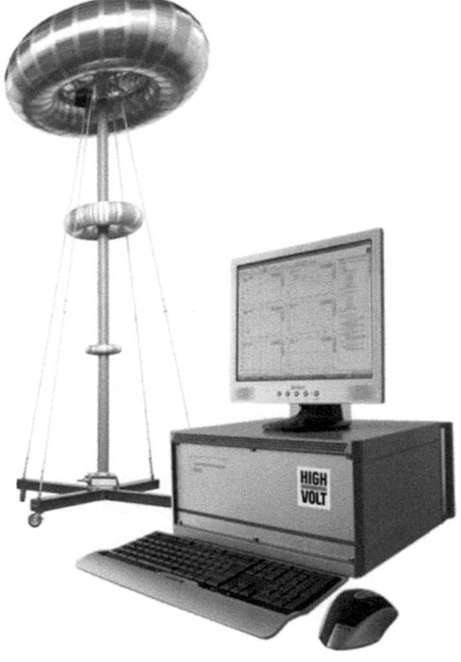

type depending on the voltage to be measured (Fig. 2.10). For special application also a voltage transformer, a voltage converting impedance (carrying a measurable current) or an electric field probe (converting amplitude and time parameters of an electric field) may be used. The clearances between the converting device and nearby earthed or energized structures may influence the result of the measurement. Such *proximity effects* shall be considered by the uncertainty estimation (see Sect. 2.3.4). To keep the contribution of the proximity effect to the uncertainty of measurement small, the clearances of movable converting devices should be as those recommended for the test object (see Sect. 2.1.2 and Fig. 2.1). If the converting device is always in a fixed position and the measuring system is calibrated on site, the proximity effect can be neglected.

- A *transmission system* which connects the output terminals of the converting device with the input terminals of the measuring instrument. It is very often a coaxial cable with its terminating impedance, but may also be an optical link which includes a transmitter, an optical cable and a receiver with an amplifier. For special application also cable connections with amplifiers and/or attenuators are in use.

- A *measuring instrument* suitable to measure the required test voltage parameters from the output signal of the transmission system. Measuring instruments for HV application are usually special devices which fulfil the requirements of the IEC Standards 61083 (part 1 and 2 for LI/SI test voltages has been published, part 3 and 4 for AC/DC voltages is under preparation). The conventional analogue peak voltmeters are replaced by digital peak voltmeters and more and more by digital recorders (Fig. 2.11). Digital recorders measure both test voltage and time parameters. This is mandatory for LI/SI test voltages, but more and more also for AC/DC test voltages with respect to changes in time by voltage drop (see Sects. 3.2.1 and 6.2.3.2), harmonics (AC, see Sect. 3.2.1) or ripple (DC, see Sect. 6.2.1).

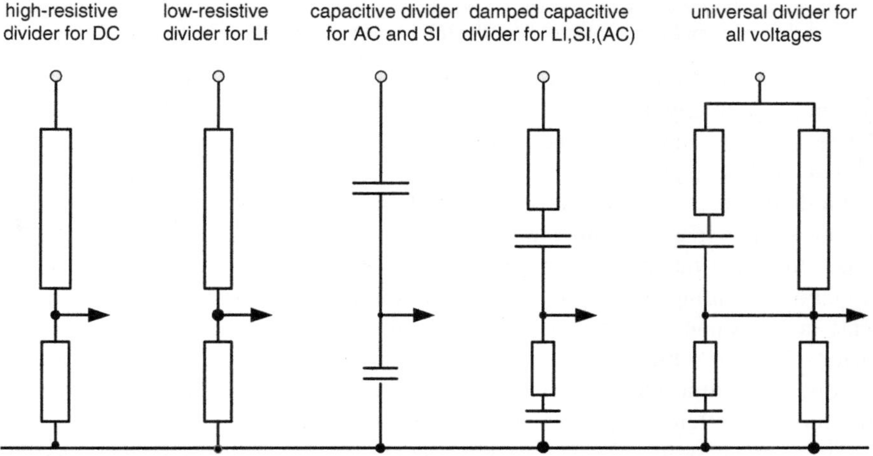

Fig. 2.10 Kinds and applications of voltage dividers

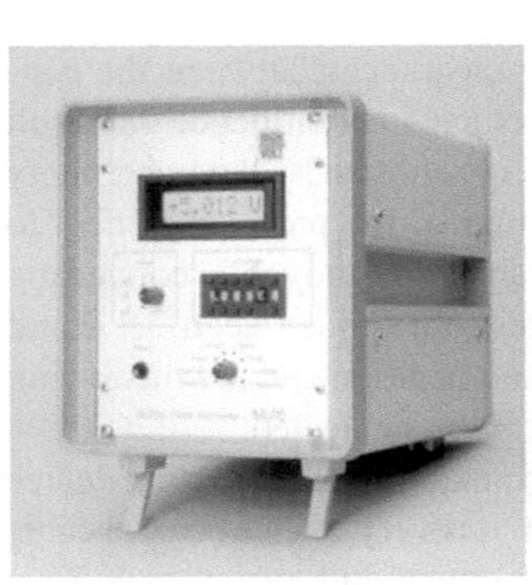

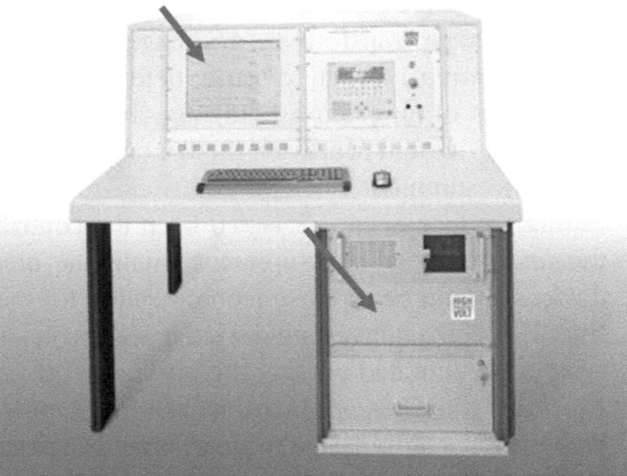

(a) stand-alone device   (b) marked components built into a control desk

Fig. 2.11 Instruments for HV measurement. **a** Digital AC/DC peak voltmeter. **b** LI/SI digital recorder

Each HV measuring system is characterized by its *operating conditions*, as they are the rated operating voltage, the measurement ranges, the operating time (or kind and number of LI/SI voltage applications) and the environmental conditions. The dynamic behaviour of a measuring system can be described as an output signal depending on frequency (*frequency response* for AC and DC voltage measuring systems, Fig. 2.12a) or on a voltage step (*step response* for LI/SI voltage measuring systems, Fig. 2.12b) or by a sufficiently low uncertainty of LI/SI parameter measurement within the nominal epoch of the measuring system.

Note The nominal epoch of an impulse voltage, which will be explained more in detail in Sects. 7.2 and 7.3, is the range between the minimum and the maximum of the relevant LI/SI time parameter for which the measuring system is approved. The nominal epoch is derived from the upper and lower tolerances of the front time parameter of the impulse voltage.

All these rated values have to be supplied by the manufacturer of the measuring system (respectively its components) after type and routine tests. They should fit to the requirements and the conditions of the HV test field where the measuring system shall be applied.

Furthermore each voltage measuring system is characterized by its *scale factor*, this means the value by which the reading of the instrument must be multiplied to obtain the input quantity of the HV measuring system (voltage and time parameters). For measuring systems that display the value of the input quantity directly, the scale factor is unity. In that case—and transmission by a coaxial cable—the scale factor of the instrument is the inverse of the scale factor of the converting device. A correctly terminated coaxial cable has the scale factor unity, other types of transmission systems may have one different from unity.

The scale factor must be calibrated to guarantee a voltage measurement traceable to the National Standard of measurement. The *calibration* consists of two main parts.

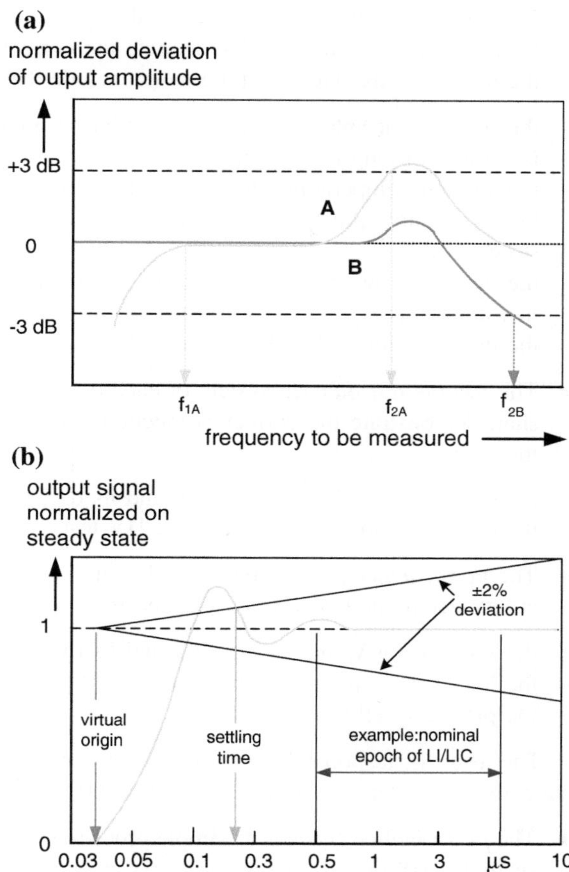

Fig. 2.12 Response of measuring systems (IEC 60060-2:2010). **a** Frequency response (*curve A* with lower and upper limit frequency, *curve B* with upper frequency limit, related to AC/DC measurement). **b** Unit step response after a step-voltage input (related to LI/SI voltage measurement, see Sect. 7.3)

On one hand the value of the scale factor shall be determined including the necessary dynamic behaviour. On the other hand, the uncertainty of the HV measurement shall be estimated. When the uncertainty and the dynamic behaviour are within the limits given by IEC 60060-2:2010, the *approved measuring system (AMS)* is applicable for measurement in an accredited HV test field (see the relevant Chaps. 3, 6 and 7).

2.3.2 Approval of a HV Measuring System for an Accredited HV Test Field

A HV measuring system is qualified for the use in an accredited HV test field by several successful tests and checks described in IEC 60060-2:2010. It becomes an "*AMS*" when it has passed the following tests and checks:

- The *type test* by the manufacturer on the system or its components on sample(s) from the production shall demonstrate the correct design and conformity with the requirements. These requirements include the determination of

  – the scale factor value, its linearity and its dynamic behaviour,
  – its short and long term stability,
  – the ambient temperature effect, i.e., the influence of the ambient temperature,
  – the proximity effect, i.e., the influence of nearby grounded or energized structures,
  – the software effect i.e., the influence of software on the dispersion of measurements,
  – the demonstration of withstand in a HV test.

- The *routine test* on each system or each of its components by the manufacturer shall demonstrate the correct production and the conformity with the requirements by

  – the scale factor value, its linearity and its dynamic behaviour and also
  – the demonstration of withstand in a HV test.

- The *performance test* on the "complete measuring system" shall characterize it at its place in the HV test field "under operation conditions" by determination of

  – the scale factor value, its linearity and its dynamic behavior,
  – the long term stability (from repetitions of performance tests) and
  – the proximity effects.

The user is responsible for the performance tests and should repeat them annually, but at least once in 5 years (IEC 60060-2:2010).

- The *performance check* is a "simple procedure"—usually the comparison with a second AMS or with a standard air gap (see Sect. 2.3.5)—"to ensure that the most recent performance test is still valid". The user is responsible for the performance checks and should repeat them according to the stability of the AMS, but at least annually (IEC 60060-2:2010).

The mentioned single tests are described together with the uncertainty estimation in Sect. 2.3.4. For the reliable operation of the measuring system a *HV withstand test* of the converting device is necessary as a type test and, if the clearances in the laboratory of use are limited, also in a first performance test. The usually required withstand test voltage level is 110% of the rated operating voltage of the converting device. The test procedure shall follow those typical for the relevant test voltages (Sects. 3.6, 6.5 and 7.6). In case of a converting device for outdoor application, the type test should include an artificial rain test.

This first performance test shall also include an *interference test* of the transmission system (coaxial cable) and the instrument of LI/SI measuring systems disconnected from the generator, but in their position for operation. This test generates an interference condition at the short-circuited input of the transmission system by firing the related impulse voltage generator at a test voltage representative for the highest operating voltage of the measuring system. The interference test is successful, when the measured amplitude of the interference is less than 1% of the test voltage to be measured.

The results of all tests and checks shall be reported in the *"record of performance"* of the measuring system, which shall be established and maintained by the user of the AMS (IEC 60060-2:2010). This record shall also contain a detailed technical description of the AMS. It is the right of an inspector of an acceptance test of any apparatus to see the record of performance for the used HV measuring system.

As required in performance tests, the scale factor, the linearity and the dynamic behaviour of a complete measuring system can be determined by different methods. The most important and preferred method is the comparison with a *reference measuring system (RMS)*, in the following called *"comparison method"* (IEC 60060-2:2010) and described in the following subsection.

> **Note** An alternative is the *"component method"* which means the determination of the scale factor of the measuring system from the scale factors of its components (IEC 60060-2:2010). The scale factor of the components can be determined by the comparison with a reference component of lower uncertainty or by simultaneous measurements of input and output quantities or by calculation based on measured impedances. For each component, the uncertainty contributions must be estimated similar to those for the whole system qualified by the comparison method. Then these uncertainties of components must be combined to the uncertainty of measurement.

2.3.3 Calibration by Comparison with a Reference Measuring System

The assigned scale factor of a measuring system shall be determined by calibration. Using the comparison method, the reading of the measuring system (AMS, index X) is compared for approval with the reading of the *reference measuring system* (RMS, index N) (Fig. 2.13). Both measuring systems indicate the same voltage V, which is the reading multiplied with the relevant scale factor F:

$$V = F_N \cdot V_N = F_X \cdot V_X \tag{2.10}$$

This simple equation is the *model equation* for the later uncertainty estimation (see Sect. 2.3.4) which delivers the scale factor of the measuring system under calibration:

$$F_X = (F_N \cdot V_N)/V_X \tag{2.11}$$

> **Note** Because the usual symbol of the uncertainty is the letter "u" or "U" (ISO/IEC Guide 98-3:2008), for the voltage the symbol "V" is used.

For practical cases it is recommended to arrange the two dividers in the same distance from a support which is directly connected to the test voltage generator.

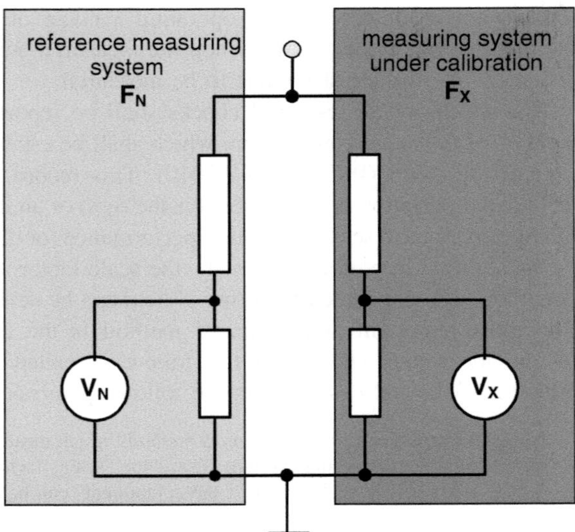

Fig. 2.13 Principle arrangement of the measuring systems for calibration using the comparison method

This symmetric arrangement works well when the two voltage dividers have about the same size (Fig. 2.14). All HV and earth connections shall be without loops and as short and straight as possible.

When the rated voltage of the RMS is higher or equal to that of the system under calibration, one can assume ideal conditions, because the calibration can be performed in minimum at $g = 5$ voltage levels including the lowest and highest of the assigned operating range (Fig. 2.15). In this case the calibration includes also the linearity test.

However, as RMS are not available up to the highest test voltages, IEC 60060-2:2010 allows that the comparison may be made at voltages as low as 20% of the assigned measurement range. An additional linearity test shows that the calibrated scale factor is applicable up to the upper limit of the measurement range which is often the rated operating voltage (see Sect. 2.3.4). In that case the symmetric arrangement of the two measuring systems as in Fig. 2.14 is impossible, but one should use sufficient clearances (Fig. 2.1) that the RMS is not influenced by the often much larger system under calibration.

The used RMS shall have a calibration traceable to national and/or international standards of measurement maintained by a *National Metrology Institute* (NMI) (Hughes et al. 1994; Bergman et al. 2001). This means that RMS calibrated by a NMI or by an *accredited calibration laboratory* (ACL) with NMI accreditation are traceable to national and/or international standards. The requirements to RMS are given in Table 2.3. The calibration of RMS can be made with *transfer RMS* (TRMS) of lower uncertainty ($U_M \leq 0.5\%$ for voltage and $U_M \leq 3\%$ for impulse time parameter measurement). The *traceability* is maintained by inter-comparisons of RMS's of different calibration laboratories (Maucksch et al. 1996).

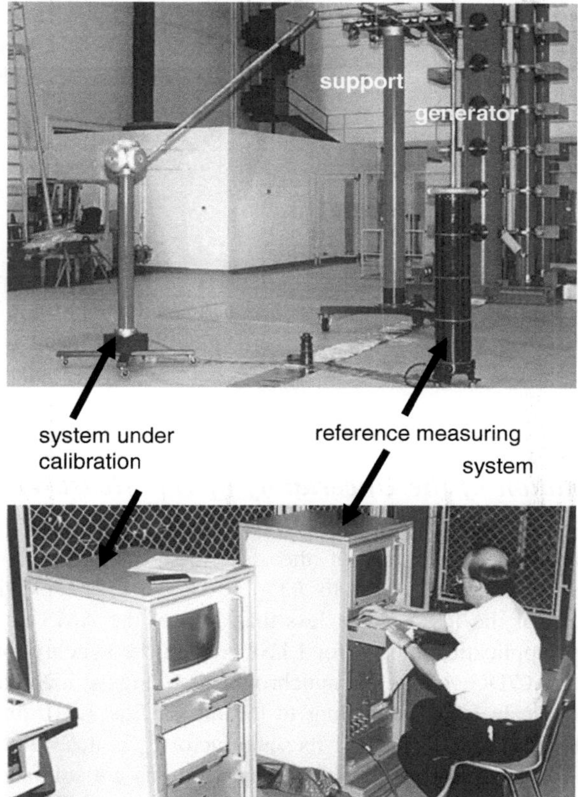

Fig. 2.14 Practical arrangement for LI voltage calibration using the comparison method. Courtesy of TU Dresden

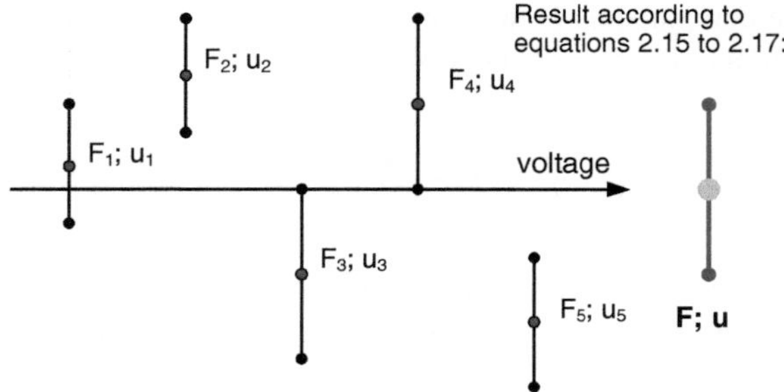

Fig. 2.15 Calibration over the full voltage range (IEC 60060-2:2010)

Table 2.3 Requirements to reference measuring systems (RMS)

Test voltage	DC (%)	AC (%)	LI, SI (%)	Front-chopped LIC (%)
Expanded uncertainty of **voltage** measurement U_M	1	1	1	3
Expanded uncertainty of **time parameter** measurement U_{MT}	–	–	5	5

Calibrations can be performed by accredited HV test laboratories provided that a correctly maintained RMS and skilled personnel are available and traceability is guaranteed. This may be possible in larger test fields, the usual way is to order calibration by an ACL.

2.3.4 Estimation of the Uncertainty of HV Measurements

The calibration process consists of the described comparisons with $n \geq 10$ applications on each of the $g = 1$ to $h \geq 5$ voltage levels provided the rated operating voltage of the RMS is not less than that of the AMS under calibration (Fig. 2.15). One application means for LI/SI voltage the synchronous readings of one impulse, for AC/DC voltage the synchronous readings at identical times. From each reading the scale factor according to the model Eqs. (2.10 and 2.11) is calculated and for each voltage level V_g, its scale factor F_g is determined as the *mean value* of the n applications (usually $n = 10$ applications are sufficient):

$$F_g = \frac{1}{n} \sum_{i=1}^{n} F_{i,g}. \tag{2.12}$$

Under the assumption of a *Gauss normal distribution* the dispersion of the outcomes of the comparisons is described by the *relative standard deviation* (also called "variation coefficient") of the scale factors F_i:

$$s_g = \frac{1}{F_g} \sqrt{\frac{1}{n-1} \sum_{i=1}^{n} \left(F_{i,g} - F_g\right)^2}. \tag{2.13}$$

The standard deviation of the mean value F_g is called the "*Type A standard uncertainty u_g*" and calculated for a Gauss normal distribution by

$$u_g = \frac{s_g}{\sqrt{n}}. \tag{2.14}$$

After the comparison at all $h \geq 5$ voltage levels V_g, the calibrated *AMS scale factor F* is calculated as the mean value of the F_g:

$$F = \frac{1}{h}\sum_{g=1}^{h} F_g,\tag{2.15}$$

with a Type A standard uncertainty as the largest of those of the different levels

$$u_A = \max_{g=1}^{h} u_g.\tag{2.16}$$

Additionally one has to consider the *non-linearity* of the scale factor by a Type B contribution

Note *Type A uncertainty* contributions to the standard uncertainty are related to the comparison itself and based on the assumption that the deviations from the mean are distributed according to a Gauss normal distribution with parameters according to (2.12) and (2.13) (Fig. 2.16a), whereas the *Type B uncertainty* contributions are based on the assumption of a *rectangular distribution* of a width $2a$ with the mean value $x_m = (a_+ + a_-)/2$ and the standard uncertainty $u = a/\sqrt{3}$ (Fig. 2.16b), details are described in IEC 60060-2:2010 and below in this subsection.

$$u_{B0} = \frac{1}{\sqrt{3}} \cdot \max_{g=1}^{h} \left| \frac{F_g}{F} - 1 \right|.\tag{2.17}$$

When the RMS rated operating voltage is lower than that of the AMS under calibration, IEC 60060-2:2010 allows the comparison over a limited voltage range ($V_{RMS} \geq 0.2\ V_{AMS}$) using only $a \geq 2$ levels. The comparison shall be completed by a *linearity test* with $b \geq (6 - a)$ levels (Fig. 2.17). Then the scale factor F is estimated by

$$F = \frac{1}{a}\sum_{g=1}^{a} F_g,\tag{2.18}$$

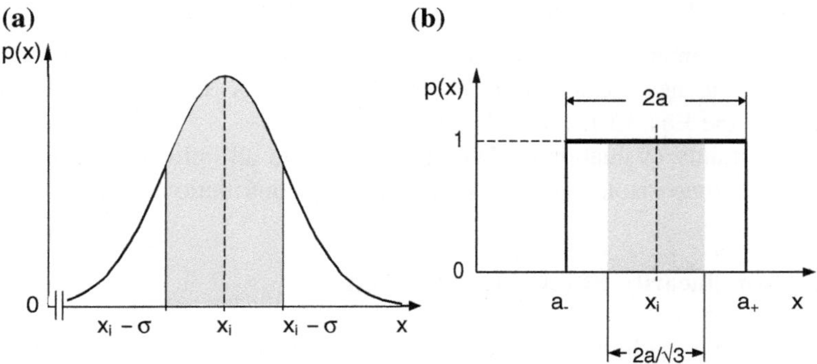

Fig. 2.16 Assumed density distribution functions for uncertainty estimation **a** Gauss normal density distribution for Type A uncertainty. **b** Rectangular density distribution for Type B uncertainty

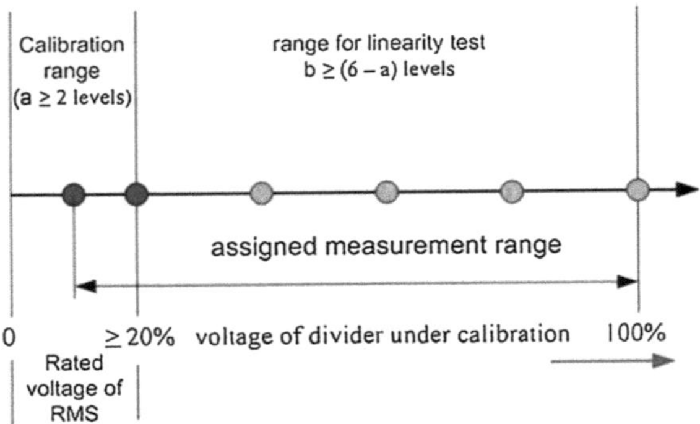

Fig. 2.17 Calibration over a limited voltage range and additional linearity test (IEC60060-2:2010)

the standard uncertainty by

$$u_A = \max_{g=1}^{a} u_g,\tag{2.19}$$

and the non-linearity contribution of the calibration by

$$u_{B0} = \frac{1}{\sqrt{3}} \cdot \max_{g=1}^{a} \left| \frac{F_g}{F} - 1 \right|.\tag{2.20}$$

An additional non-linearity contribution comes from the range of the linearity test and shall be calculated as described below under "*non-linearity effect*".

Example 1 A 1000 kV LI voltage measuring system (AMS) with a scale factor of $F_{X0} = 1000$ (calibrated 3 years ago) has shown peak voltage deviations of more than 3% by comparison with a second AMS during a performance check. Therefore, it has to be calibrated by comparison with a RMS. A 1200 kV LI reference measuring system (RMS) is available for that calibration. It has been decided to perform the comparison at $g = 5$ voltage levels with $n = 10$ applications each. The RMS is characterized by a scale factor $F_N = 1025$ and an expanded uncertainty of measurement of $U_N = 0.80\%$. Table 2.4 shows the comparison at the first level. As a result, one gets the scale factor F_1 for that first voltage level ($g = 1$) and the related standard deviation s_1 and standard uncertainty u_1.

Table 2.5 summarizes the results of all five comparison levels and delivers—as a final result—the new scale factor F_X and the Type A standard uncertainty u_A according to the Eqs. (2.15) and (2.16).

The uncertainty evaluation of Type B is related to all influences different from the statistical comparison. It includes the following contributions to the uncertainty.

2.3.4.1 Non-linearity Effect (Linearity Test)

When the AMS is calibrated over a limited range, the linearity test is used to show the validity of the scale factor up to the rated operating voltage. It is made by comparison with an AMS of sufficient rated voltage or with the input (DC) voltage of a LI/SI test voltage generator (when the AMS is related to these voltages) or with

Table 2.4 Comparison at the first level $V_1 \approx 0.2\ V_r$

No. of application	RMS measured voltage (V_N/kV)	AMS measured voltage (V_X/kV)	Scale factor F_i (Eq. 2.11)
$i = 1$	201.6	200.8	1.0291
2	200.7	200.9	1.0240
3	201.4	200.9	1.0276
4	199.9	199	1.0296
5	201.2	199.9	1.0317
6	201.3	200.3	1.0301
7	200.9	200.4	1.0276
8	201.3	200.4	1.0296
9	201.2	199.9	1.0317
$n = 10$	200.6	200.7	1.0245
Result by Eqs. (2.12)–(2.16)			**$F_1 = 1.0286$** **$s_1 = 0.25\%$** **$u_1 = 0.08\%$**

Table 2.5 Scale factor and Type A uncertainty estimation (results of comparison at the five voltage levels)

No. voltage level g	V_X/V_{Xr} (%)	Scale factor F_g	Standard deviation s_g (%)	Standard uncertainty u_g (%)
$g = 1$ (Example!)	20	1.0286	0.25	0.08
2	39	1.0296	1.94	0.61
3	63	1.0279	1.36	0.43
4	83	1.0304	2.15	0.68
$h = 5$	98	1.028	1.4	0.44
Result		**New scale factor** **$F_X = 1.0289$**		**Type A uncertainty** **$u_A = 0.68\%$**

a standard measuring gap according to IEC 60052:2002 or with a field probe (see Sect. 2.3.6). It does not matter when the linearity test shows a ratio R different from the scale factor, it is only important that it is stable over the range of the linearity test (Fig. 2.18). If this is guaranteed also other methods to investigate the linearity could be applied. The maximum deviation of the investigated $g = b$ ratios $R_g = V_x/V_{CD}$ (V_{CD} is the output of comparison device) from their mean value R_m delivers the Type B estimation of the standard uncertainty (Fig. 2.18) related to non-linearity effects:

$$u_{B1} = \frac{1}{\sqrt{3}} \cdot \max_{g=1}^{b} \left| \frac{R_g}{R_m} - 1 \right|. \tag{2.21}$$

Fig. 2.18 Linearity test with a linear device in the extended voltage range (IEC 60060-2:2010)

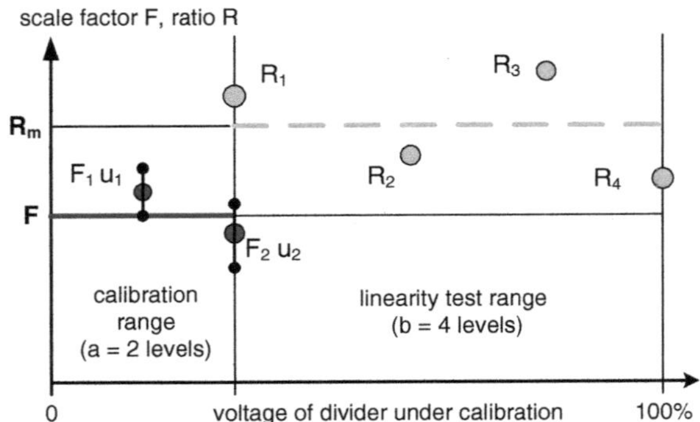

2.3.4.2 Dynamic Behaviour Effect

For the investigation of the dynamic behaviour it is recommended to determine the scale factor of the AMS at $i = k$ different values within a frequency range or within a range of impulse shapes both representative for its use (e.g., for the rated frequency range or the nominal epoch). Then the related standard uncertainty contribution is evaluated from the maximum deviation of an individual scale factor F_i from the nominal scale factor F:

$$u_{B2} = \frac{1}{\sqrt{3}} \cdot \max_{i=1}^{k} \left| \frac{F_i}{F} - 1 \right|. \tag{2.22}$$

The dynamic behaviour can also be investigated by the unit step response method. For details see Sect. 7.4.1.

2.3.4.3 Short-Term Stability Effect

The short-term stability is often determined by the *self-heating* of the AMS, especially its converting device. The test shall be performed at rated operating voltage, it starts with the determination of the scale factor F_1 when the test voltage is reached and is terminated with a new determination of the scale factor F_2 when the pre-defined test time, usually the anticipated *time of use* or the assigned operating time, is over:

$$u_{B3} = \frac{1}{\sqrt{3}} \cdot \left| \frac{F_2}{F_1} - 1 \right|. \tag{2.23}$$

The short time contribution to the measuring uncertainty should be given in the manufacturer's data of components.

2.3.4.4 Long-Term Stability Effect

A starting value for the contribution of the long-term stability may also be given by the manufacturer. Then it can also be determined for the time of use T_{use} from the change of the scale factor within the time of two *performance checks* (from F_1 to F_2 made at times T_1 respectively T_2, often the projected time of use is $T_{\text{use}} = T_2 - T_1$):

$$u_{B4} = \frac{1}{\sqrt{3}} \cdot \left| \frac{F_2}{F_1} - 1 \right| \cdot \frac{T_{\text{use}}}{T_2 - T_1}. \tag{2.24}$$

2.3.4.5 Ambient Temperature Effect

Usually the measuring system is specified for a certain temperature range. The scale factor is determined for the minimum and maximum temperature of that range. The larger deviation F_T from the nominal scale factor F is used to estimate the standard uncertainty contribution:

$$u_{B5} = \frac{1}{\sqrt{3}} \cdot \left| \frac{F_T}{F} - 1 \right|. \tag{2.25}$$

Often the uncertainty contribution related to the temperature effect within the specified temperature range may be taken from manufacturer's data.

2.3.4.6 Proximity Effect

The uncertainty contribution due to nearby earthed structures my be determined from the scale factors F_{min} and F_{max} at minimum and maximum distances from those structures:

$$u_{B6} = \frac{1}{\sqrt{3}} \cdot \left| \frac{F_{\text{max}}}{F_{\text{min}}} - 1 \right|. \tag{2.26}$$

The proximity effect for smaller HV measuring systems is often investigated by the manufacturer of the converting device and can be taken from the manual.

2.3.4.7 Software Effect

When digital measuring instruments, especially digital recorders, are used, a correct measurement is assumed when *artificial test data* (which are given in IEC 61083-2:2011) are within certain tolerance ranges, also given in IEC 61083-2:2011. It should not be neglected that there may be remarkable standard uncertainty contributions caused by that method. The assumed uncertainty contribution by the software is only related to the maximum width of these tolerance ranges T_{oi} given in IEC 61083-2:

$$u_{B7} = \frac{1}{\sqrt{3}} \cdot \max_{i=1}^{n}(T_{oi}). \tag{2.27}$$

Note Only those tolerance ranges T_{oi} of artificial test data similar to the recorded impulse voltage must be taken into consideration.

Example 2 The AMS characterized in the first example is investigated with respect to the Type B standard uncertainty contributions. For the uncertainty estimation of the calibration also the standard uncertainties of the RMS which are not included in its measuring uncertainty must be considered. Table 2.6 summarizes both and mentions the source of the contribution.

2.3.4.8 Determination of *Expanded Uncertainties*

IEC 60060-2:2010 recommends a simplified procedure for the determination of the expanded uncertainty of the scale factor calibration and of the HV measurement. It is based on the following assumptions which meet the situation in HV testing:

- *Independence*: The single measured value is not influenced by the preceding measurements.
- *Rectangular distribution*: Type B contributions follow an rectangular distribution.
- *Comparability*: The largest three uncertainty contributions are of approximately equal magnitude.

Table 2.6 Type B uncertainty contributions

Uncertainty contribution	Symbol of contribution	Uncertainty contribution for RMS	Uncertainty contribution for AMS
Non-linearity effect Eq. (2.22)	u_{B1}	Included in calibration: $u_N = U_N/2 = 0.4\%$	Included in calibration u_A
Dynamic behaviour effect Eq. (2.23)	u_{B2}	Included in calibration	0.43% from deviation within nominal epoch
Short-term stability effect Eq. (2.24)	u_{B3}	Included in calibration	0.24% from deviation before and after a 3 h test
Long-term stability effect Eq. (2.25)	u_{B4}	Included in calibration	0.34% from consecutive performance tests
Ambient temperature effect Eq. (2.26)	u_{B5}	0.06% because outside of specified temperature range	0.15% from manufacturers data
Proximity effect Eq. (2.27)	u_{B6}	Included in calibration	Can be neglected because of very large clearances
Software effect Eq. (2.28)	u_{B7}	Included in calibration	Can be neglected because no digital recorder applied

Note IEC 60060-2:2010 does not require the application of this simplified method, all procedures in line with the ISO/IEC Guide 98-3:2008 (GUM) are also applicable. In the Annexes A and B of IEC 60060-2:2010 a further method directly related to the GUM is described.

The relation between the standard uncertainty and the calibrated new scale factor can be expressed by the term $(F \pm u)$ which characterizes a range of possible scale factors (Not to forget, F is a mean value and u is the standard deviation of this mean value!). Under the assumption of a Gauss normal density distribution (Fig. 2.17a) this range covers 68% of all possible scale factors. For a higher confidence, the calculated standard uncertainty can be multiplied by a "covering factor" $k > 1$. The range $(F_X \pm k \cdot u)$ means the scale factor plus/minus its "expanded uncertainty" $U = k \cdot u$. Usually a coverage factor $k = 2$ is applied which covers a confidence range of 95%.

To determine first the *expanded uncertainty of the calibration* U_{cal}, the standard uncertainty u_N of measurement of the RMS from its calibration, the Type A standard uncertainty from the comparison and the Type B standard uncertainties related to the reference measuring system are combined according to the geometric superposition:

$$U_{cal} = k \cdot u_{cal} = 2 \cdot \sqrt{u_N^2 + u_A^2 + \sum_{i=0}^{N} u_{BiRMS}^2} \qquad (2.28)$$

The *expanded uncertainty of calibration* appears on the calibration certificate together with the new scale factor. But in case of a HV acceptance test, the *expanded uncertainty of a HV measurement* is required. When the AMS is calibrated and all possible ambient conditions are considered (ambient temperature range, range of clearances, etc.), then the expanded uncertainty of HV measurement can be pre-calculated by the standard uncertainty of the calibration u_{cal} and the Type B contributions of the AMS u_{BiAMS}

$$U_M = k \cdot u_M = 2 \cdot \sqrt{u_{\mathrm{cal}}^2 + \sum_{i=0}^{N} u_{\mathrm{BiAMS}}^2}. \qquad (2.29)$$

The pre-calculated expanded uncertainty of measurement should also be mentioned on the calibration certificate together with the pre-defined conditions of use. The user of the HV measuring system has only to estimate additional uncertainty contributions when the HV measuring system has to operate outside the conditions mentioned in the calibration certificate.

Example 3 For the calibrated AMS the expanded uncertainties of calibration and HV measurement shall be calculated under the assumption of certain ambient conditions mentioned in the calibration certificate. The calculation uses the results of the two examples above:

Calibration results:		
Reference measuring system (RMS):		
RMS: measuring uncertainty	$U_N = 0.80\%$	$u_N = 0.4\%$
RMS: temperature effect		$u_{B5} = 0.06\%$
Calibration by comparison		$u_A = 0.68\%$
Expanded calibr. uncertainty (95% confidence, $k = 2$)		$U_{cal} = 1.58\%$
Standard uncertainty of calibration		$u_{cal} = 0.79\%$
HV measurement:		
AMS: non-statistical influences		$u_{B2} = 0.43\%$
		$u_{B3} = 0.24\%$
		$u_{B4} = 0.\,4\%$
		$u_{B5} = 0.15\%$
Expanded uncertainty of measurement (95% confidence)		$U_M = 2.10\%$
Precise measurement result:		$V = V_x\,(1 \pm 0.021)$

The HV measuring system shall be adjusted according to its new scale factor of $F = 1.0289$ (Table 2.5), possibly with a change of the instrument scale factor to maintain the direct reading of the measured HV value on the monitor. IEC 60060-2:2010 requires an uncertainty of HV measurement of $U_M \leq 3\%$. Because of $U_M = 2$, $10\% < 3\%$ (Example 3), the system can be used for further HV measurement. But it is recommended to investigate the reasons for the relatively high expanded uncertainty U_M for improvement of the measuring system.

2.3.4.9 Uncertainty of Time Parameter Calibration

IEC 60060-2:2010 (Sect. 5.11.2) describes a comparison method for the estimation of the expanded uncertainty of time parameter measurement. Furthermore in its Annex B.3, it delivers an additional example for the evaluation according to the ISO/IEC Guide 98-3:2008. Instead of the consideration of the dimensionless scale factor for voltage measurement, the method applies to the time parameter (e.g. the LI front time T_{1X}) itself, considers the deviation from the nominal time parameter T_{1N} measured by the reference measuring system as negligible and gets from the comparison directly the mean error ΔT_1,

$$\Delta T_1 = \frac{1}{n}\sum_{i=1}^{n}\left(T_{1X,i} - T_{1N,i}\right), \tag{2.30}$$

the standard deviation

$$s(\Delta T_1) = \sqrt{\frac{1}{n-1}\sum_{i=1}^{n}\left(\Delta T_{1,i} - \Delta T_1\right)^2},$$
(2.31)

and the Type A standard uncertainty

$$u_A = \frac{s(\Delta T_1)}{\sqrt{n}}.$$
(2.32)

The Type B contributions to the measuring uncertainty of time parameters are determined as maximum differences between the errors of individual measurements and the mean error of the time parameter T_1 for different LI front times, e.g., the two limit values of the nominal epoch of the measuring system.

For external influences, the procedure of the Type B uncertainty estimation follows the principles described above for voltage measurement Eqs. (2.22–2.27). For the expanded uncertainty of time calibration and time parameter measurement an analogous application of Eqs. (2.28) and (2.29) is recommended.

A performance test includes the calibration of the scale factor, for impulse voltages also of the time parameters, and the described full set of tests of the influences on the uncertainty of measurement. The data records of all tests shall be included to the record of performance. The comparison itself and its evaluation can be aided by computer programs (Hauschild et al. 1993).

2.3.5   HV Measurement by Standard Air Gaps According to IEC 60052:2002

The breakdown voltages of uniform and slightly non-uniform electric fields, as e.g., those between sphere electrodes in atmospheric air, show high stability and low dispersion. Schumann (1923) proposed an empirical criterion to estimate the critical field strength at which self-sustaining electron avalanches are ignited. If modified, this criterion can also be used to calculate the breakdown voltage V_b of uniform fields versus the gap spacing S. For a uniform electric field in air at standard conditions the breakdown voltage can be approximated by the empirical equation

$$V_b/\text{kV} = 24.4\left[S + \left(\frac{S}{13.1\ \text{cm}}\right)^{0.5}\right].$$
(2.33)

This equation is applicable for sphere gaps if the spacing is less than one-third of the sphere diameter Fig. (2.19).

Based on such experimental and theoretical results, *sphere-to-sphere gaps* are used for peak voltage measurement since the early decades of the twentieth century (Peek 1913; Edwards and Smee 1938; Weicker and Hörcher 1938; Hagenguth et al.

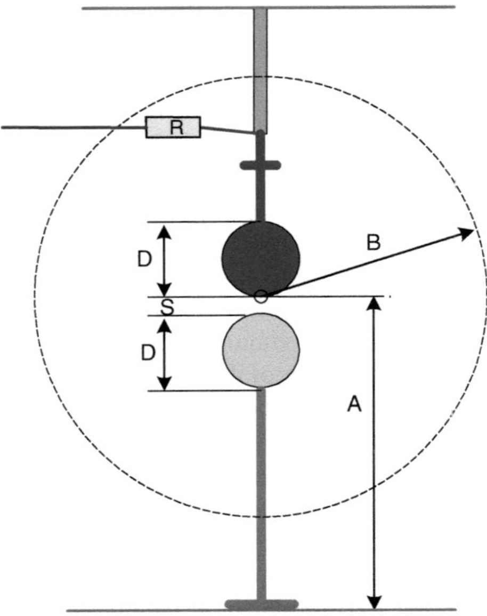

Fig. 2.19 Vertical measuring sphere gap (explanations in the text)

1952) and led to the first standard of HV testing, the present IEC 60052:2002. Meanwhile it is fully understood that this applicability is based on the so-called streamer breakdown mechanism, e.g., Meek (1940), Pedersen (1967), and breakdown voltage-gap distance characteristics of sphere gaps can also be calculated with sufficient accuracy (Petcharales 1986).

For a long time *measuring sphere gaps* with gap diameters up to 3 m formed the impression of HV laboratories. But the voltage measurement by sphere gaps is connected with the breakdown of the test voltage therefore their application is not simple. Furthermore they need a lot of clearances (see below), well maintained clean surfaces of the spheres and atmospheric corrections (see Sect. 2.1.2) for measurement according to the standard.

Today they are not used for daily HV measurement and do not play the same important role in HV laboratories as in the past. Their main application is for *performance checks* of AMSs (see Sect. 2.3.2) or linearity checks (see Sect. 2.3.4). For acceptance tests on HV apparatus the inspector may require a check of the applied AMS by a sphere gap to show that it is not manipulated. For these applications mobile measuring gaps with sphere diameters $D \leq 50$ cm are sufficient.

The IEC Standard on voltage measurement by means of sphere gaps has been the oldest IEC standard related to HV testing. Its latest edition IEC 60052 Ed.3:2002 describes the measurement of AC, DC, LI and SI test voltage with horizontal and vertical sphere-to-sphere gaps with sphere diameters $D = (2 \ldots 200$ cm) and one of the spheres earthed (Fig. 2.19 and 2.20). The spacing S for voltage measurement is required $S \leq 0.5\ D$, for rough estimations it can be extended up to $S = 0.75\ D$. The surfaces shall be smooth with maximum roughness below 10 μm and free of irregularities in the region of the sparking point.

The curvature has to be as uniform as possible, characterized by the difference of the diameter of no more than 2%. Minor damages on that part of the hemispherical surface, which is not involved in the breakdown process, do not deteriorate the performance of the measuring gap. To avoid erosion of the surface of the sphere after AC and DC breakdowns, pre-resistors may be applied of 0.1–1 M'Q.

Surrounding objects may influence the results of sphere gap measurements. Consequently the dimensions and *clearances* for standard air gaps are prescribed in IEC 60052 and shown in Figs. 2.19 and 2.20. The required range of the height A above ground depends on the sphere diameter, and is for small spheres $A = (7 \ldots 9) \cdot D$ and for large spheres $A = (3 \ldots 4) \cdot D$. The clearance to earthed external structures depends on the gap distance S, and shall be between $B = 14\ S$ for small and $B = 6\ S$ for large spheres.

The dispersion of the breakdown voltage of a measuring gap depends strongly from the availability of a free starting electron, especially for gaps with $D \leq 12.5$ cm and/or measurement of peak voltages $U_p \leq 50$ kV. Starting electrons can be generated by photo ionization (Gänger 1953, Kuffel 1959; Kachler 1975). The necessary high energy radiation may come from the far ultra-violet (UVC) content of nearby corona discharges at AC voltage, or from the breakdown spark of the open switching gaps of the used impulse generator, or a special mercury-vapour UVC lamp with a quartz tube.

Note In the past, even a radioactive source inside the measuring sphere has been applied. For safety reasons this is forbidden now.

Table 2.7 gives the relationship of the measured breakdown voltage U_b depending on the distance S between electrodes for some selected sphere diameters $D \leq 1$ m which are mainly used for the mentioned checks, for other sphere diameters see IEC 60052:2002. A voltage measurement with a sphere gap means to establish a relation between an instrument at the power supply input of the HVG (e.g., a primary voltage measurement at the input of a test transformer) and the

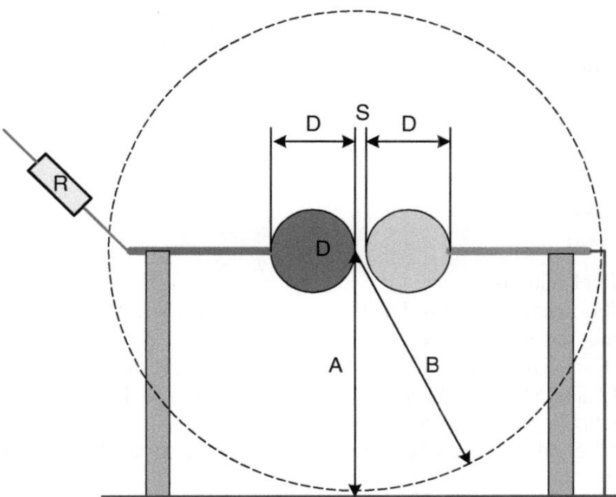

Fig. 2.20 Horizontal measuring sphere gap

known breakdown voltage of the standard measuring gap in the HV circuit depending on its gap distance D (Table 2.7). This is similar to the calibration by comparison (Sect. 2.3.3).

For *AC voltage measurement* a progressive stress test (see Sect. 2.4) delivers 10 successive breakdown voltage readings by the instrument. Their mean value (Eq. 2.12) and the relative standard deviation (Eq. 2.13) are determined. The voltage shall be raised sufficiently slowly to allow accurate readings. The mean value characterizes the breakdown voltage according to the gap parameters (D, S). When the standard deviation is $\leq 1\%$, one can assume that the measuring gap was correctly maintained and the relative expanded uncertainty of measurement is $\leq 3\%$.

Note With $n = 10$ measurements and a standard deviation of 1% one gets a standard uncertainty of $u = 0.32\%$ (Eq. 2.14). This means there are about 1.2% for the other contributions to the standard uncertainty when the expanded uncertainty ($k = 2$) shall be $\leq 3\%$ (Eq. 2.29).

Table 2.7 Peak value of breakdown voltages of selected standard sphere gaps

Gap distance	50% breakdown voltage V_{b50}/kV at sphere diameter D/mm[b]							
S/mm	100		250		500		1000	
	AC, DC[a], −LI, −SI	+LI, +SI	AC, DC[a], −LI, −SI	+LI, +SI	AC, DC[a], −LI, −SI	+LI, +SI	AC, DC[a], −LI, −SI	+LI, +SI
5	16.8	16.8						
10	31.7	31.7	31.7	31.7				
15	45.5	45.5	45.5	45.5				
20	59	59.0	59.0	59.0	59.0	59.0		
30	84	85.5	86.0	86.0	86.0	86.0	86.0	86.0
50	123	130	137	138	138	138	138	138
75	(155)[c]	(170)	195	199	202	202	203	203
100			244	254	263	263	266	266
150			(314)	(337)	373	380	390	390
200			(366)	(395)	460	480	510	510
300					(585)	(620)	710	725
400					(670)	(715)	875	900
500							1010	1040
600							(1110)	(1150)
750							(1230)	(1280)

Explanations

[a]For measurement of DC test voltages >130 kV standard sphere gaps are not recommended, apply rod–rod gaps and see Eq. (2.34)

[b]For correctly maintained standard sphere gaps, the expanded uncertainty of measurement of AC, LI and SI test voltages is assumed to be $U_M \approx 3\%$ for a confidence level of 95%. There is no reliable value for DC test voltages

[c]The values in brackets are for information, no level of confidence is assigned to them

For *LI/SI voltage measurement,* the pre-selected breakdown voltages (*D, S* in Table 2.7) are compared e.g., with charging voltage of the impulse voltage generator. The 50% breakdown voltages U_{50} are determined in a multi-level test of $m = 5$ voltage levels with $n = 10$ impulse voltages each (see Sect. 2.4), and the corresponding reading is taken as the pre-selected reading. When the evaluated standard deviation is within 1% for LI and 1.5% for SI voltages it is assumed that the measuring gap works correctly.

For *DC voltage measurement,* sphere gaps are not recommended because external influences as dust or small fibres are charged in a DC field and cause a high dispersion. Therefore, a *rod–rod measuring gap* shall be applied if the humidity is not higher than 13 g/m^3 (Feser and Hughes 1988; IEC 60052:2002). The rod electrodes of steel or brass should have a square cross section of 10–25 mm for each side and sharp edges. When the gap distance S is between 25 and 250 cm the breakdown is caused by the development of a streamer discharge of a required average voltage gradient $e = 5.34$ kV/cm. Then the breakdown voltage can be calculated by

$$V_b/\text{kV} = 2 + 5.34 \cdot S/\text{cm}. \tag{2.34}$$

The length of the rods in a vertical arrangement shall be 200 cm, in a horizontal gap 100 cm. The rod–rod arrangement should be free of PD at the connection of the rods to the HV lead, respectively to earth. This is realized by toroid electrodes for field control. For a horizontal gap the height above ground should be ≥ 400 cm. The test procedure is as that for AC voltages described above.

2.3.6 Field Probes for Measurement of High Voltages and Electric Field Gradients

The ageing of the insulation and thus the reliability of HV apparatus is mainly governed by the maximum electrical field strength. Even if the field distribution in dielectric materials can well be calculated based on the Maxwell equations using advanced computer software, the validity of the theoretical results should be validated experimentally. For this purpose *capacitive sensors,* commonly referred to as field probes, can be used. The field distribution, however, may substantially be affected by the presence of such field probes which should thus be designed as small as possible to minimize the field distortion und thus the inevitable measuring uncertainty (Les Renardieres Group 1974; Malewski et al. 1982). In specific cases, however, field probes can be designed such that the field is not disturbed, as in the case of coaxial electrode configurations representative for bushings, power cables and SF$_6$ switchgears. Moreover, field probes can be integrated in the earth electrode of a plane-to-plane electrode arrangement of Rogowski profile, as sketched in Fig. 2.21. Under this condition the voltage applied to the HV electrode can simply be deduced from the field strength appearing at the sensing electrode.

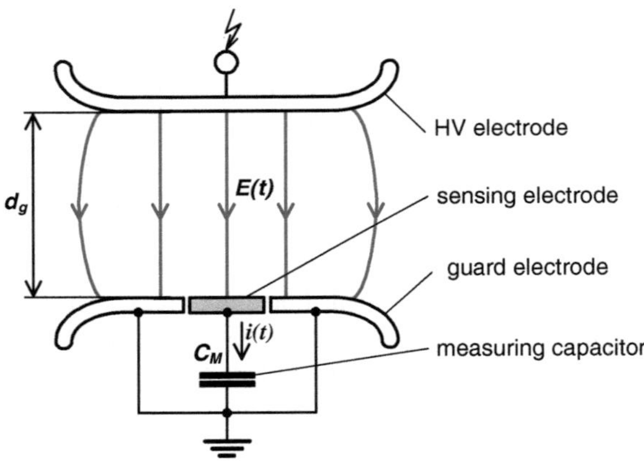

Fig. 2.21 Principle of a field probe for measuring high alternating voltages

The fundamental measuring principle is based on the first Maxwell equation which reads:

$$\mathrm{rot}\,\vec{H} = \partial\vec{D}/\partial t + \vec{G} \tag{2.35}$$

with $\vec{H}$—vector of the magnetic field strength, $\vec{D}$—vector of the electric displacement flux density, $\vec{G}$—vector of the current density at the sensor electrode.

For gaseous dielectrics the conductivity is extremely low so that the density of the conductive current given by the second term in Eq. (2.35) can be neglected:

$$\mathrm{rot}\,\vec{H} = \partial\vec{D}/\partial t, \tag{2.36a}$$

In contrast to this the conductivity of the sensing electrode is extremely high so that for this case the first term in Eq. (2.35) can be neglected:

$$\mathrm{rot}\,\vec{H} = \vec{G}, \tag{2.36b}$$

Combining the Eqs. (2.36a and 2.36b) and substituting the displacement flux density by the electrical field strength, i.e., $\vec{D} = \varepsilon \cdot \vec{E}$, one gets

$$\vec{G} = \partial\vec{D}/\partial t = \varepsilon \cdot \partial\vec{E}/\partial t, \tag{2.37}$$

with ε—permittivity of the dielectric between both electrodes.

For the here considered homogenous field configuration the field gradient is directed perpendicular to the surface of the sensing electrode. Thus, instead of the vector presentation the simple scalar presentation valid for the one dimensional configuration is applicable. Consequently the current $I(t)$ captured by the sensor can simply be expressed by the current density G multiplied with the area A of the sensing electrode:

$$I(t) = A \cdot G = A \cdot \varepsilon \cdot dE(t)/dt. \tag{2.38}$$

To convert the current induced at the sensor surface into an equivalent voltage signal $V_m(t)$, it is a common practice to connect the sensor via a measuring capacitance C_M to earth potential, see Fig. 2.21. As this provides a capacitive voltage divider, the time-dependent voltage $V_h(t)$ applied to the HV electrode can simply be deduced from the voltage $V_m(t)$ measured across C_M using the following equation:

$$V_h(t) = \frac{d_g \cdot C_m}{A \cdot \varepsilon} \cdot V_m(t) = S_f \cdot V_m(t), \qquad (2.39)$$

with S_f—scale factor, d_g—gap distance.

In principle the capacitance C_M shown in Fig. 2.21 could also be replaced by a resistor denoted in the following as R_m. Under this condition Eq. (2.38) can be expressed as

$$V_m(t) = R_m \cdot A \cdot \varepsilon \cdot dE(t)/dt. \qquad (2.40)$$

From this follows for the interesting peak value of the high voltage V_{hp} applied to the top electrode

$$V_{hp} = \frac{d_g}{R_m \cdot A \cdot \varepsilon} \cdot \int_0^t V_m(t)dt = \frac{d_g}{R_m \cdot A \cdot \varepsilon \cdot 2\pi f} \cdot V_{mp} = S_f \cdot V_{mp} \qquad (2.41)$$

That means, the scale factor S_f is inversely proportional to the test frequency f, so that not only R_m but also the test frequency f must exactly be known to deduce the peak value of the applied high voltage from the measured low voltage. In this context it has to be taken care that superimposed harmonics may cause severe measuring errors.

Example Consider an arrangement according to Fig. 2.21 in ambient air, i.e. $\varepsilon_0 = 8.86$ pF/m. Assuming a gap distance $d_g = 10$ cm and an area of the sensing electrode of $A_s = 10$ cm^2 as well as a capacitance of $C_m = 2$ nF, one gets the following scale factor:

$$S_f = \frac{d_g \cdot C_m}{A \cdot \varepsilon_0} = \frac{(10 \text{ cm}) \cdot (2 \text{ nF})}{(10 \text{ cm}^2) \cdot (8.86 \text{ pF/m})} = 22.6 \times 10^3.$$

If, for instance, a low voltage of $V_m = 5$ V appears across C_m, this is caused by an applied high voltage of $V_{hp} = 113$ kV.

Substituting the capacitor C_m by a measuring resistor of $Rm = 500$ kΩ and assuming a test frequency $f = 50$ Hz, one gets the following scale factor:

$$S_f = \frac{d_g}{R_m \cdot A_s \cdot \varepsilon_0 \cdot 2\pi f} = 18 \times 10^3.$$

The curves plotted in Fig. 2.22, which are based on the above calculations, enable a simple determination of the applied high voltage V_{hp} from the measured low voltage V_m. In this context it has to be taken into account that the scale factor is also dependent on the test frequency, if a measuring resistor is used.

The main drawback of the arrangement shown in Fig. 2.21 is that this is only capable of measuring the field gradients adjacent to earth potential. To measure also arbitrarily oriented field vectors in the space between HV and LV electrodes, *spherical sensors* are commonly employed to prevent the ignition of partial discharges (Feser and Pfaff 1984). As illustrated in Fig. 2.23 (left), the surface of such a sphere electrode is subdivided into six partial sensors to receive the three cartesian components of the electromagnetic field. To minimize the inevitable field disturbance caused by the metallic parts providing the field probe, this is battery powered and the whole components required for signal processing are integrated in the hollow sphere electrode. Moreover, the captured and processed signals are transmitted to earth potential via a fiber optic link. An essential benefit of the spherically shaped probe is that the admissible radius depending on the field strength to be measured can be calculated without great expenditure.

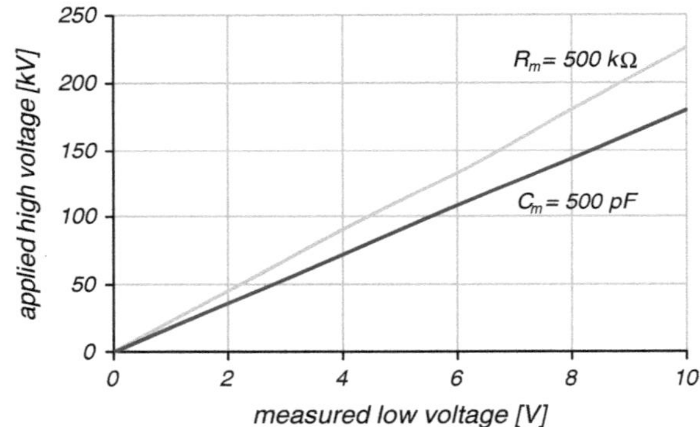

Fig. 2.22 Applied high voltage versus low voltage measurable across a capacitive respectively a resistive measuring impedance using the parameters given in the text

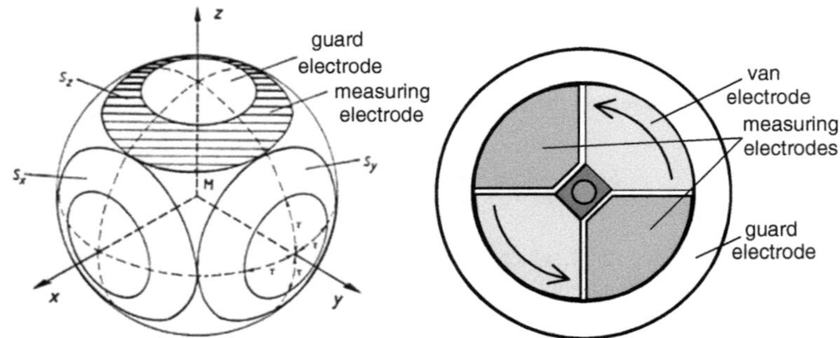

Fig. 2.23 Field probes using fixed and rotating electrodes

Practically realized spherical field probes are capable of measuring field strengths up to about 1 kV/cm at measuring frequencies ranging between approx. 20 Hz and 100 MHz. In this context it should be noted that the application of field probes is not only restricted to field strength measurements. If calibrated with a reference measuring system this tool can also be employed for high voltage measurements, provided the field is free of space charges, i.e. corona discharges must be prevented in the surroundings of the field probe.

As the field probe provides in principle a capacitive sensor, it has to be taken into account that the induced charge and thus the measurable current $i(t)$ is the consequence of a time-dependent displacement flux and thus correlated with the time-dependent field strength $E(t)$. Thus, only time-dependent voltages, such as LI, SI, AC and other transients are capable of inducing a measurable displacement current. To measure also DC voltages, the desired alternating displacement current could be generated by shielding the sensing electrode periodically by means of a rotating electrode at earth potential. This approach, as schematically shown in Fig. 2.23 (right), is applied by the so-called *field mill* (Herb et al. 1937; Kleinwächter 1970). Here the sensing electrode is established by two half-sectioned discs providing the measuring electrodes, which must well be insulated from each other, while the vane electrode is connected to the guard electrode at earth potential. Due to the rotation of the vane electrode in front of the measuring electrodes these are exposed to an alternating displacement flux. Thus an alternating current correlated to the electrostatic field strength will be induced, which is amplified and indicated accordingly.

Another option of field probe measurements is the use a mechanical method, which uses the so called Coulomb force discovered in 1884. For this purpose the measuring impedance C_M shown in Fig. 2.21 is replaced by a sensitive force measurement system, as originally applied by Kelvin in 1884 for absolute DC voltage measurement. Considering a homogeneous field, the force attracting the sensing electrode of area A, if subjected to a potential gradient E, can be expressed by:

$$F_e = \frac{1}{2} \cdot \varepsilon \cdot A \cdot E^2. \tag{2.42}$$

Example If, for instance, a high voltage $V_h = 100$ kV is applied to the top electrode and the gap distance amounts $d_g = 10$ cm, then the field strength at the sensing electrode achieves 10 kV/cm. Inserting these values in Eq. (2.42), the force attracting the sensing electrode becomes $F_e \approx 3.5 \times 10^{-2}$ N ≈ 3.6 p.

The torsion on account of the attracted sensing electrode is amplified and indicated by a spot light and mirror system. As the Coulomb force is proportional to the quadratic value of the field gradient and thus also proportional to the quadratic value of the applied test voltage, its indication becomes independent on the polarity. Thus, *electrostatic voltmeters* are not only applicable for DC voltage measurements but also for measuring the rms value of HVAC test voltages, see Sect. 3.4.

2.4   Breakdown and Withstand Voltage Tests and Their Statistical Treatment

Electrical discharges and breakdown of insulations are stochastic processes which must be described by statistical methods. This subsection gives an introduction to the planning, performing and evaluation of HV tests on a statistical basis. It describes tests with voltages increasing up to the breakdown ("progressive stress method", PSM) and with multiple application of pre-given voltages and estimation of breakdown probabilities ("multiple level method", MLM, "up-and-down method", UDM). Life-time tests (LTT) of insulation are performed with pre-given voltages, but progressive test durations and can be evaluated accordingly. Also standardized HV withstand tests are described and valuated from the viewpoint of statistics. The subsection submits first tools for the application of statistical methods and supplies hints to the special literature, e.g. (Hauschild and Mosch 1992).

2.4.1   *Random Variables and the Consequences*

The phenomena of *electrical discharges*—as most others in nature, society and technology—are based on stochastic processes and characterized by their randomness (Van Brunt 1981; Hauschild et al. 1982). This is often ignored and only an average trend is considered in order to interpret a relationship being investigated. Quite often, however, it is not the mean value, but an extreme value that determines the performance of a system. In technology this is often taken into account by applying a "safety factor". A rather better approach is the statistical description of stochastic phenomena. Therefore, *HV tests* shall be selected, performed and evaluated on a statistical basis: They are *random experiments (trials)* and described by *random variables* (sometimes also called "random variates").

When a pre-given *constant voltage stress*—e.g., a certain LI test voltage—is applied to an insulation—e.g., an air gap—one can observe the random event "breakdown" (A) or the complimentary event "withstand" ($A*$). The relative breakdown frequency $h_n(A)$ is the relation between the number of breakdowns k and the number of applications n

$$h_n(A) = k/n. \tag{2.43}$$

The relative *withstand frequency* follows to

$$h_n(A*) = (n - k)/n = 1 - h_n(A). \tag{2.44}$$

The relative frequency depends on the number of performed tests (often called *sample size*) and the respective test series as shown for the breakdown frequency in Fig. 2.24. The relative frequencies vary around a fixed value and reach it as the limiting value "*breakdown probability*" p:

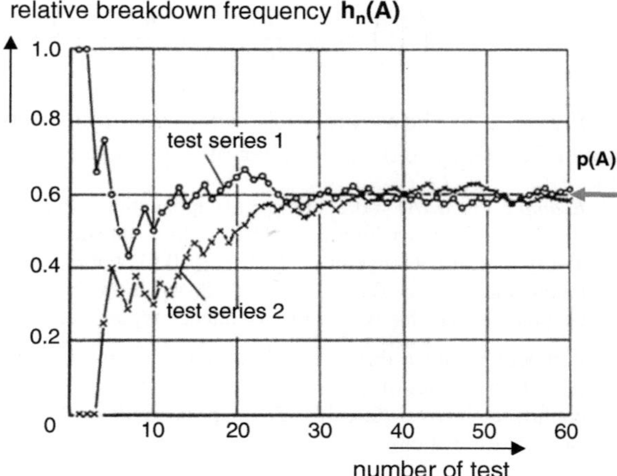

Fig. 2.24 Relative breakdown frequencies of two test series depending on the number of tests performed

$$\lim_{n \to \infty} h_n(A) = p. \tag{2.45}$$

The *probability of a withstand q* follows accordingly

$$\lim_{n \to \infty} h_n(A*) = q. \tag{2.46}$$

Because a characteristic value of withstand cannot be measured ("Nothing happens!"), the withstand probability is determined from the complementary breakdown probability $q = 1 - p$. Consequently the statistical definition of a withstand voltage is a voltage which causes a breakdown with low probability, usually $p \leq 0.10$. The relative breakdown frequency is a *point estimation* of the breakdown probability.

The larger the number of applications for the estimation of the frequency, the better is the adaptation of the estimate to the true, but unknown probability (Fig. 2.24). A confidence estimation delivers a feeling for the accuracy of the estimation by the width of the calculated confidence region. This region covers the true but unknown value of p with a certain *confidence level*, e.g., $\varepsilon = 95\%$. From the sample the upper and the lower limit of the confidence region are determined on the basis of the assumption of a theoretical distribution function, in this case based on the *binomial distribution function* and the Fisher (F) distribution as test distribution.

Explanation: The binomial distribution is based on two complementary events A and A^* (as breakdown and withstand) occurring with the known probabilities p and q (Bernoulli trial). The binomial distribution indicates the probability P $(X = k)$ with which the event A will occur k-times in n independent trials.

$$P(X = k) = \binom{n}{k} p^k (1 - p)^{n-k} \tag{2.47}$$

where $k = 0, 1, 2, \ldots, n$ and

$$\binom{n}{k} = \frac{n!}{k! \cdot (n-k)!} = \frac{n \cdot (n-1) \cdot \ldots (n-k+1)}{1 \cdot 2 \cdot 3 \cdot \ldots \cdot k}$$

Figure 2.25 shows the 95% *confidence limits* depending on the relative breakdown frequency and the number of applications (sample size).

Example For $h_n(A) = 0.7$ and $n = 10$ applications, Fig. 2.25 delivers the lower limit $p_l = 0.37$ and the upper limit $p_u = 0.91$. With a statistical confidence $\varepsilon = 95\%$, the real probability is within a range of $0.37 \leq p \leq 0.91$. For $n = 100$ applications one would get the much smaller range $0.60 \leq p \leq 0.78$. Again, with increasing sample size the estimation becomes better, this means the confidence region becomes smaller.

It should be noted that a confidence interval can also be calculated when no breakdown ($k = 0$) in a series of n stresses occurs. For the case ($n = 20$; $k = 0$) Fig. 2.25 delivers the confidence interval (0; 0.16). The upper confidence limit can be used for further comparisons or estimations.

When a constant voltage test has been performed for the estimation of the breakdown probability, it must be checked whether the test is "independent" or not. *Independence* means that the previous voltage applications have no influence on the result of the application under consideration. This can be shown by the investigation of the trend (Table 2.8). A sample of $n = 100$ is subdivided into five samples of $n^* = 20$ each. If the relative frequencies of the sub-samples scatter around that of the whole sample, it could be considered as independent (case a). If there is a clear trend (case b) it would be dependent and any statistical evaluation is forbidden. Independence can only be ensured due to an improvement of the test procedure,

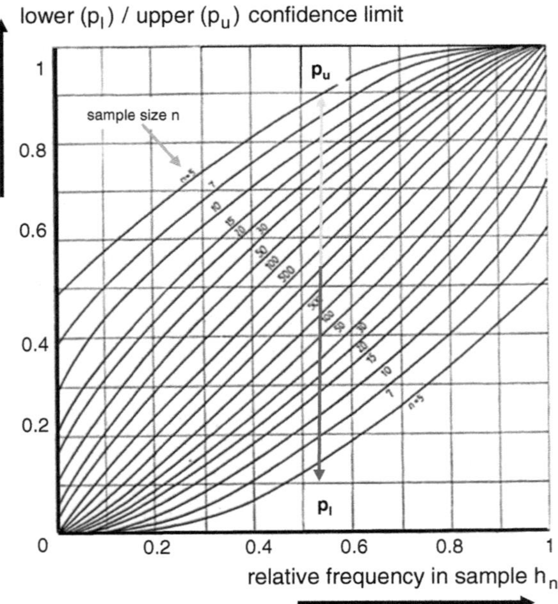

Fig. 2.25 Confidence limits of the breakdown probability for a confidence level $\varepsilon = 95\%$. Depending on sample size

Table 2.8 Check of independence of two tests with sample size $n = 100$

(a) Independent sample: Sphere-plane gap in atmospheric air (withstand: -; breakdown: x)	breakdown frequency	
	h_{20}	h_{100}
- x x x - - - x - x x x x x x x - x - x	0.65	
x - x- - x x x x x - x x- x x x x x x	0.75	
- x x x x - x x x x x x - x - x x x –	0.70	
x x x - x x - x - x - - - x - x x x - x	0.6	0.68
x x - x x x x - - x - x - x x x x x x -	0.7	
(b) Dependent sample: As above, but enclosed air in a tank (withstand: -; breakdown: x)	h_{20}	h_{100}
x x x x x x x x x x x x x x x x x x x - x	0.95	
x - x- x x - x x - - x x x - x - - x x -	0.55	
- - - x x x - x - - - x - x x - x- x x	0.5	
- x x - - - - - - - x - - - x - x - - x x	0.35	0.48
- - - - - - x - - - - - - - - - - - - -	0.05	

e.g., breaks of sufficient duration between single stresses. Further details of constant voltage tests are described in (Hauschild and Mosch 1992).

A sequence of constant voltage tests at different voltage levels (n $\geq$ 5) means the application of the multiple level method (MLM) which is described in detail in Sect. 2.4.3.

There is a second group of HV breakdown tests with increasing stress, e.g., continuously raising AC or DC test voltages or LI or SI voltages raising in steps until breakdown (Fig. 2.26). Now the breakdown is sure, but the height of the breakdown voltage is random. This group of tests is called "progressive stress tests", PSM. The random variable is the breakdown voltage V_b. But in life-time tests at constant voltage, the time-to-breakdown T_b becomes the random variable. In both cases we have a continuous variable.

Note In case of stepwise increasing voltages, the starting value may be varied to get continuous outputs (realizations) of the random variable.

The evaluation of progressive stress tests follows the typical treatment of *random variables* in mathematical statistics as described in many text books as e.g., by Mann et al. (1974), Müller et al. (1975), Storm (1976) or Vardeman (1994). Descriptions especially related to HV tests are given by Lalot (1983), Hauschild and Mosch (1984) (in German) and (1992) (in English), Carrara and Hauschild (1990) and Yakov (1991) as well as in Appendix A of (IEC 60060-1:2010). A brief introduction to PSM evaluation is given in Sect. 2.4.2.

The distribution of a random variable X with realizations x_i found in a progressive stress test is described by a *distribution function*. It is defined by

Fig. 2.26 Procedures of progressive stress tests. **a** Continuous increasing AC or DC voltage. **b** Stepwise increasing AC or DC voltage. **c** Stepwise increasing LI or SI voltage

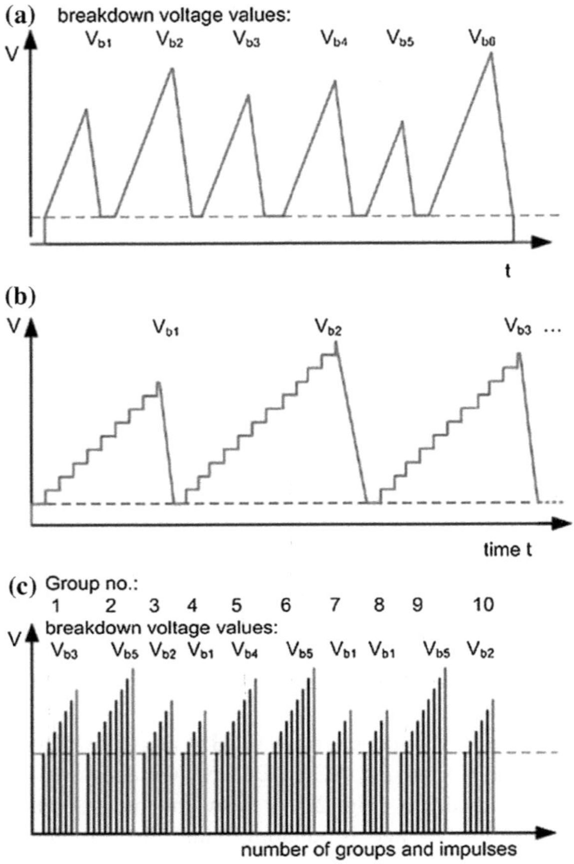

$$F(x_i) = P(X < x_i), \tag{2.48}$$

and it indicates the probability P with which the random variable X will assume a value below the considered value x_i. A distribution function (Fig. 2.27) is any mathematical function with the following properties:

- $0 \leq F(x_i) \leq 1$ (realizations between impossible and sure events),
- $F(x_i) \leq F(x_{i+1})$ (monotonously increasing),
- $\lim_{x \to -\infty} F(x) = 0$ and $\lim_{x \to +\infty} F(x) = 1$ (boundary conditions).

Note Instead of the distribution function also the density function delivers a complete mathematical description, but it is not meaningful for HV test evaluation.

A distribution function is characterized by parameters describing the mean value and the dispersion, sometimes in addition to the position of the function. The evaluation of a progressive stress test means the selection of a well adapted

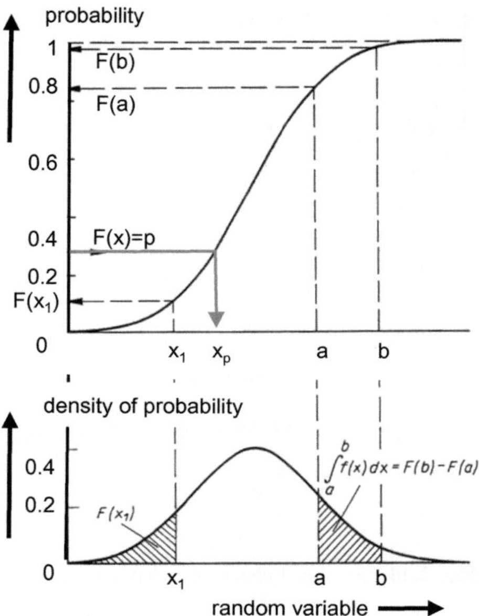

Fig. 2.27 Distribution and density function with the definition of quantile x_1 and probability interval $(F(b) - F(a))$

type of distribution function and the estimation of its parameters. The parameter estimation can be made as a point or a confidence estimation based on formulas of functional parameters (e.g., the mean value), quantiles (realization of the random variable related to a pre-given probability) or intervals (difference between two quantiles) (Fig. 2.27).

2.4.2 HV Tests Using the Progressive Stress Method

The *progressive stress method* (PSM) with continuous increasing voltage shall be considered for an electrode arrangement in SF_6 gas (Fig. 2.26a). The initial voltage v_0 must be low enough to avoid any influence on the result, the rate of rise of the voltage shall be so that a reliable voltage measurement can be performed, and the interval between two individual tests shall guarantee the *independence* of the realizations v_b. The independence may be checked by a graphic plot of the measurements. Other independence tests are described in the above mentioned literature.

Example Fig. 2.28 shows the sequence of four test series at four different pressures of the SF_6 gas. A series is considered to be independent if the realizations fluctuate in a random manner around a mean value. A dependence must be assumed, if there is a falling, raising or periodically fluctuating tendency. According to this simple rule, the series at gas pressures of 0.40, 0.25, and 0.15 MPa can be considered as independent and are well suited for a statistical evaluation. That at 0.10 MPa is dependent and shall not be statistically evaluated. The reason for the dependence should be clarified and the series repeated under improved conditions.

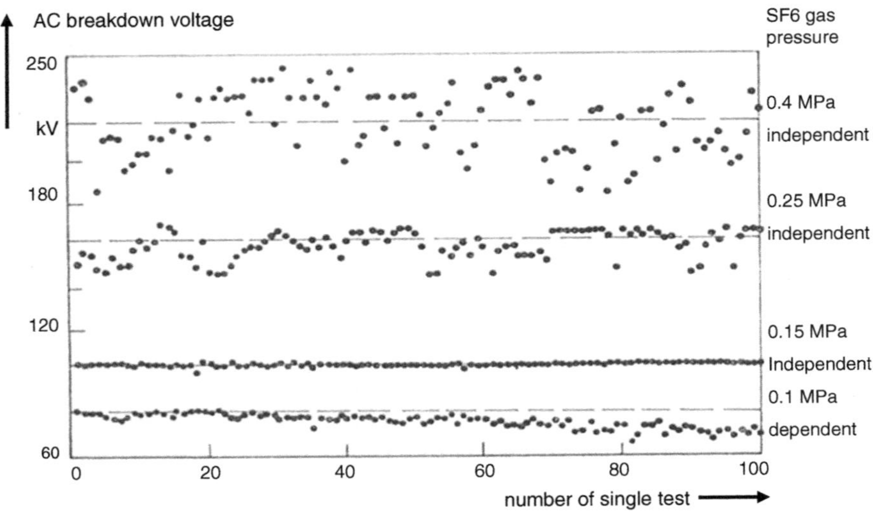

Fig. 2.28 Graphical check of independence of four test series in SF_6 gas

Each independent series shall be evaluated statistically, this means graphically represented and approximated by a theoretical distribution function. Both tasks can be connected when a so-called *probability grid* is used for the representation. A probability grid uses the inverse function of the considered theoretical distribution function on the ordinate. For each type of theoretical distributions, a probability grid can be constructed. Any empirical distribution of the same type as the grid appears as a straight line.

The following theoretical distribution functions are recommended for HV applications:

The *Gauss or normal distribution* is characterized by the parameters μ (estimated by the arithmetic mean value or the 50% quantile u_{50}) and the standard deviation σ [estimated by the mean square root of $(x_i - \mu)$ or the difference of quantiles $(x_{84} - x_{50}) = (x_{50} - x_{16})$]:

$$F(x; \mu; \sigma^2) = \frac{1}{\sqrt{2\pi}\sigma} \int_{-\infty}^{x} e^{-(z-\mu)^2/2\sigma^2} dz \tag{2.49}$$

The application of a certain distribution function should be based on its stochastic model: A normal-distributed random variable is the result of a large number of independent, randomly distributed influences when each of these makes only an insignificant contribution to the sum. This model is very well applicable to many random events, also to breakdown processes with partial discharges.

The *Gumbel or double exponential distribution* is—as the normal distribution—an unlimited function $(-\infty < x < +\infty)$ characterized by two parameters, its 63% quantile η and the dispersion measure γ [estimated by $\gamma = (x_{63} - x_{05})/3$]:

$$F(x; \eta; \gamma) = 1 - e^{-e^{\frac{x-\eta}{\gamma}}}. \tag{2.50}$$

Stochastic model: The double exponential distribution describes the distribution of realisations according to an extreme value, in case of HV tests it is the minimum of the electric strength. It is a mathematical description of the simple fact that "the breakdown of a slightly uniform electric field takes place at the weakest point". It can be well applied if there are slightly uniform insulations which show a quite high dispersion (Mosch and Hauschild 1979).

The *Weibull distribution* is also a distribution describing extreme values, but it is limited and characterized by three parameters, its 63% quantile $\eta = x_{63}$, the Weibull exponent δ as a measure of dispersion and the initial value x_0

$$F(x; \eta; \delta; x_0) = 1 - e^{-\left(\frac{x-x_0}{\eta}\right)^{\delta}} \quad x > x_0, \tag{2.51a}$$

$$F(x; \eta; \delta; x_0) = 0 \quad x \le x_0, \tag{2.51b}$$

$$\delta = 1.2898 / \log(x_{63}/x_{05}). \tag{2.51c}$$

The Weibull distribution is highly adaptable in its structure and therefore applicable for many problems (Cousineau 2009). For the case $x_0 = 0$ it is the ideal function for breakdown time investigation (two-parameter Weibull distribution), see e.g., Bernard (1989) and Tsuboi et al. (2010). In the case $x_0 > 0$, the initial value becomes an absolute meaning, e.g., as an ideal withstand voltage of the breakdown probability $p = 0$! Therefore consequences must be carefully considered when it is applied to breakdown voltage problems.

For all three theoretical distribution functions a *probability grid* can be constructed. Figure 2.29 shows the comparison of the different ordinates of these grids. It can be seen that in the region x_{15}–x_{85} the grids are very similar, but for very low and very high probabilities remarkable differences exist. This means that for estimation of withstand voltages the optimum selection of a theoretical distribution function for the adaptation of empirical data (test results) is very important. Additionally a sufficient sample size (number of realizations), e.g. $n \ge 50$, must be chosen, for evidence to low and high probabilities.

In HV tests the *empirical distribution function* is usually determined from a quite limited number of realizations, e.g., $10 \le n \le 100$. In that case it is recommended to arrange the realisations x_i according to increasing magnitude between x_{min} and x_{max} and to complete them by their relative, *cumulative frequencies*

$$h_{\Sigma i} = \sum_{m=1}^{i} \frac{h_m}{(n+1)} \tag{2.52}$$

where n is the total number of realizations and h_m is the absolute frequency of the mth voltage value (Eq. 2.43). Then the data are plotted as a "stair" in a suited probability grid. If the empirical (stair) function can be approximated by a straight line, the adaptation with the theoretical function of the grid is acceptable.

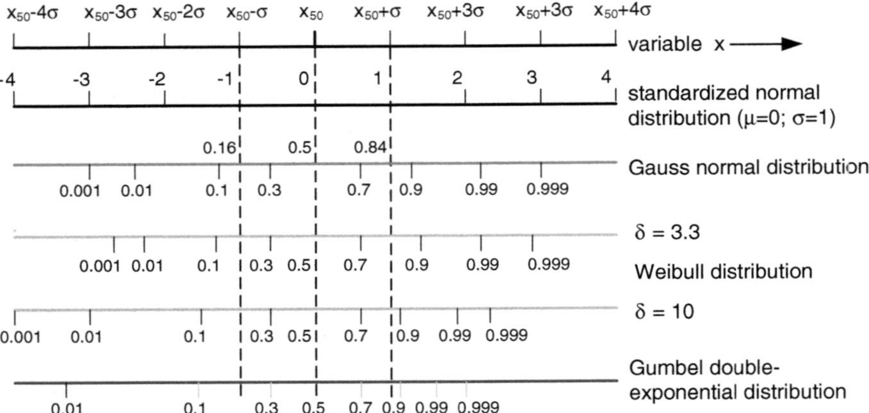

Fig. 2.29 Comparison of ordinates of probability grids with identical 50% quantiles (logarithmic abscissa for Weibull distribution)

Example Figure 2.30 shows, the stair function can be well adapted by a straight line in the Gauss grid of a normal distribution. This means a normal distribution describes the randomness of the performed test sufficiently. Its parameters can be estimated by quantiles: The mean value by $u_{50} = 953$ kV and the standard deviation by $s = u_{50} - u_{16} = 18.2$ kV.

A *confidence estimation* of the parameters can be performed using so-called test distributions, the *t*-distribution for confidence estimates of the mean value and the χ^2-distribution for the standard deviation. For details see e.g., Hauschild and Mosch (1992).

The *maximum likelihood method* delivers the most efficient estimation of parameters including their confidence limits. As the term "likelihood" is a synonym for "probability", the method delivers estimates of parameters of the selected distribution function, which are those of maximum probability for the given sample. The method has been introduced many years ago, but got its broad application with numerical calculations by personal computers (PC). It can be applied for all classes of HV breakdown tests and any type of theoretical distribution function (Carrara and Hauschild 1990; Yakov 1991; Vardeman 1994).

The mathematical calculations are based on the so-called *"likelihood function L"*. It is proportional to the probability p_R to obtain a description of the investigated sample (realizations x_i with $i = 1 \ldots$ n) using a distribution function, e.g., with the parameters δ_1 and δ_2. The n realizations of the sample are distributed to m levels of the random variable x_i (breakdown voltage or time). The likelihood function L is based on the probability p_R which is proportional to the product of the probabilities f_{Ri}:

$$L = Ap_R = A \prod_{i=1}^{n} f_{Ri}(x_i/\delta_1, \delta_2) = L(x_i/\delta_1, \delta_2) \qquad (2.53)$$

The factor "A" is for normalization only. The most likely *point estimates* of the parameters δ_1 and δ_2 are those which maximize L as shown in Fig. 2.31 (δ^*_1 and

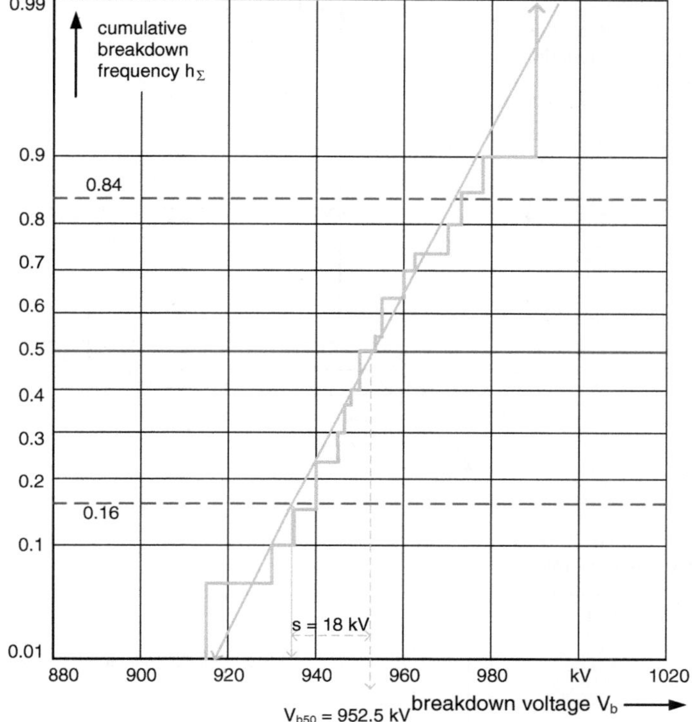

Fig. 2.30 Cumulative frequency distribution function on a Gauss grid

δ^*_2). They are calculated from the maximum conditions $dL/d\delta_1 = 0$ and $dL/d\delta_2 = 0$, usually performed with the logarithms where the same parameters indicate the maximum

$$d(\ln L)/d\delta_1 = 0 \quad \text{and} \quad d(\ln L)/d\delta_2 = 0. \tag{2.54}$$

The three-dimensional diagram shows the likelihood function (L normalized to its maximum) on the area of the two parameters. By help of a cross section through the "likelihood mountain" one can define the confidence limits of the parameters (δ_{1min}, δ_{1max}, δ_{2min}, δ_{2max}). Each parameter combination of the *confidence region* (Fig. 2.31) delivers one straight line on the probability grid (Fig. 2.32). The upper and lower border lines of the bundle of straight lines are considered as confidence limit for the whole distribution function. Figure 2.32 shows this schematically in the relevant probability grid used for the approximation. The maximum-likelihood method can also be applied to "*censored*" *test results*, e.g., when a life-time test is terminated after a certain time and only k of the n test objects have broken down. Then the likelihood function gets the form

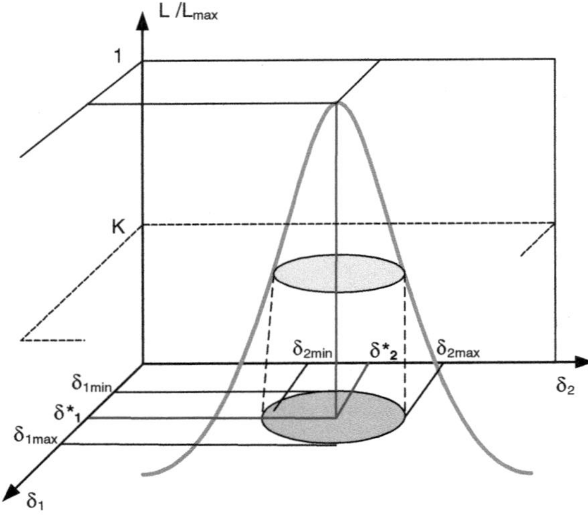

Fig. 2.31 Point and confidence estimation by the maximum likelihood function (schematically)

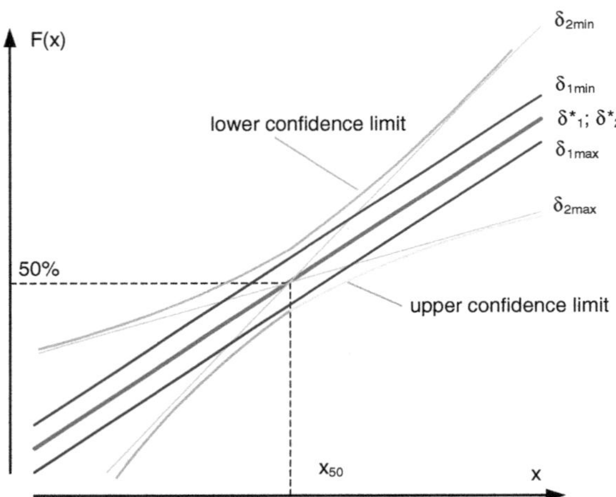

Fig. 2.32 Confidence limits of the distribution function derived from Fig. 2.31 (schematically)

$$L = A \cdot p_R = A \cdot \prod_{i=1}^{k} f_{Ri}(x_i/\delta_1, \delta_2) \cdot \prod_{i=k+1}^{n-k} (1 - f_{Ri}(x_i/\delta_1, \delta_2)). \qquad (2.55)$$

The same sample as shown in Fig. 2.30 is evaluated by a commercially available PC program of the ML method (Speck et al. 2009) under the assumption of a Weibull distribution (Fig. 2.33). The program delivers after *independence tests* a plot of the *cumulative frequency distribution* on a Weibull grid with logarithmic abscissa. The parameters are estimated as follows: initial value $v_0 = 750$ kV, 63% quantile $V_{b63} = (v_0 + x_{63}) = (750 + 211)$ kV and dispersion parameter $\delta = 8.4$. The

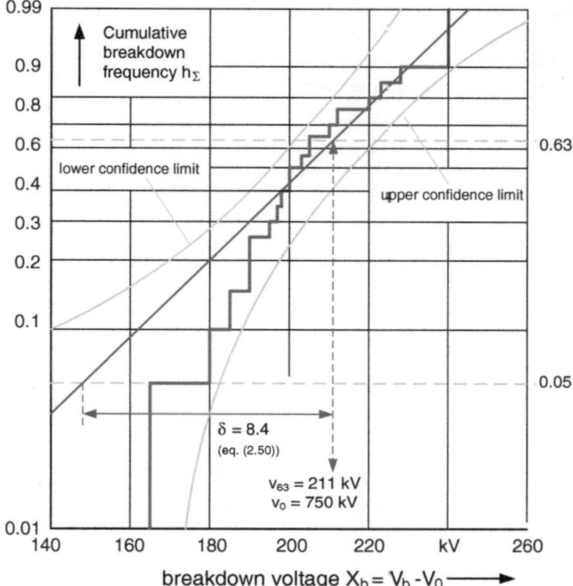

Fig. 2.33 Cumulative frequency function with 95% confidence limits on Weibull grid

evaluated lower 95% confidence limit of the cumulative frequency function should be taken for technical conclusions.

2.4.3   HV Tests Using the Multiple-Level Method

The *multiple-level method* (MLM) means the application of constant voltage tests (see Sect. 2.4.1) at several voltage levels (Fig. 2.34). For each level the test delivers an estimation of the breakdown probability including its confidence limits (Fig. 2.25). The relationship between stressing voltage and breakdown probability is not a distribution function in the statistical sense and therefore called *"performance function"* (sometimes also called *"reaction function"*). The performance function is not necessarily monotonically increasing, it can decrease (Fig. 2.35), e.g., in case of a change of the discharge mechanism depending on the height of the voltage. But it delivers exactly the information necessary for the reliability estimation and insulation coordination of a power system: The performance function supplies the probability of a breakdown in case of a certain overvoltage stress.

In most cases also the performance function shows a monotonic increase and can be mathematically described by a theoretical distribution function. Figure 2.36 shows the difference between the performance function $V(x)$ (simulated by a standardized normal distribution with $\mu = 0$ and $\sigma = 1$) and derived cumulative frequency functions $S_{\Delta x}(x)$ of different heights Δx of the voltage steps in the test. For statistical reasons, the principle relation is $S_{\Delta x}(x) > V(x)$. Both functions have a

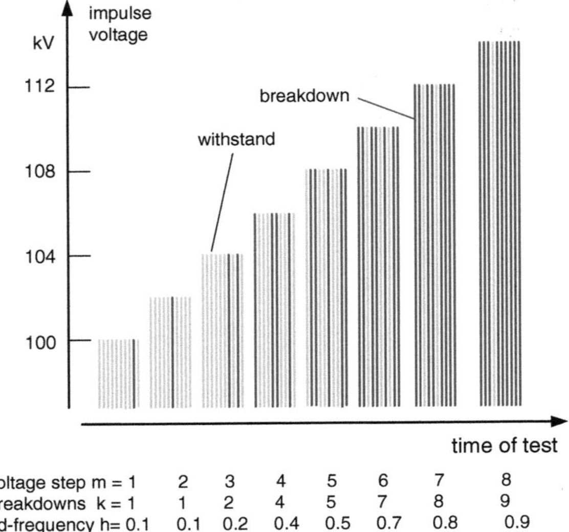

Fig. 2.34 Test procedure according to the multi-level method with $m = 8$ voltage steps and $n = 10$ impulses per step

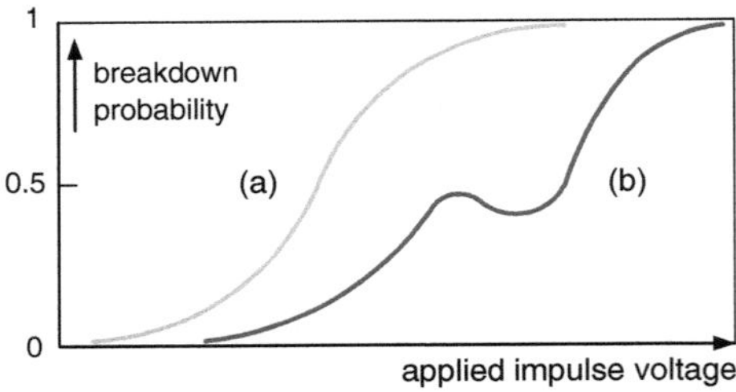

Fig. 2.35 Performance functions with monotonic (**a**) and non-monotonic (**b**) increase

different meaning: *cumulative frequency functions* consider the probability of breakdown *at all stresses up to a certain stress* value, performance functions do it *at a certain stress*. Cumulative frequency functions from stepwise increased voltages should be converted into *performance functions* (Hauschild and Mosch 1992).

A MLM test shall be performed at $m \geq 5$ voltage levels and $n \geq 10$ stresses per level. The number of stresses is not necessarily identical at all levels. If withstand voltages are considered, the number of stresses at low breakdown frequency might be higher. Then the *independence* of the outcomes of each level must be checked (see Table 2.8) and confidence estimations for the breakdown probability are determined (Fig. 2.25) and plotted in a probability grid.

Example From earlier experiments it can be expected that the performance function is monotonic increasing and can be approximated by a double exponential distribution function. Therefore, the point and confidence estimations are plotted in a Gumbel grid (Fig. 2.37). Because a straight line can be drawn through all confidence regions, the assumption of a double exponential or Gumbel distribution is confirmed. The small

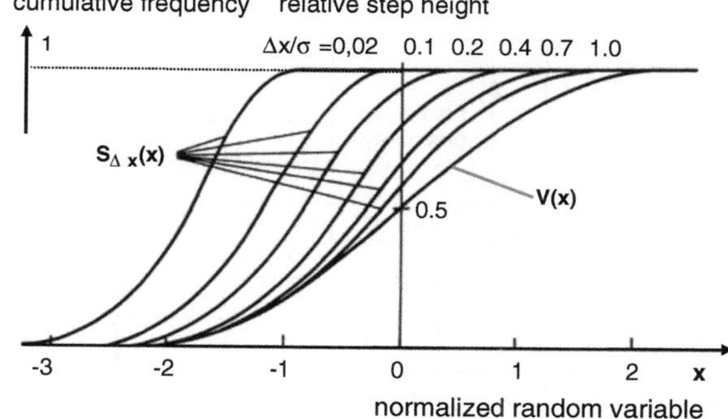

Fig. 2.36 Calculated cumulative frequency functions $S_{\Delta x}(x)$ determined at different step heights $\Delta x/\sigma$ based on one identical performance function $V(x)$

reduction of the relative breakdown frequency between 1083 and 1089 kV is not significant as it can be seen from the confidence limits. The parameters can be estimated from quantiles, $v_{63} = 1112$ kV and $\gamma = (v_{63} - v_{31}) = 12$ kV.

Also the maximum likelihood estimation can be applied when the performance function is approximated by a certain distribution function (parameters δ_1, δ_2). According to Fig. 2.34 there are $j = 1...m$ voltage levels and apply at each level n_j stresses. The probability of obtaining k_j breakdowns and $w_j = (n_j - k_j)$ withstands at the voltage u_j is expressed by the *binomial distribution* (Eq. 2.47) on the basis of the breakdown probabilities given by the performance function $V(v_j) = V(v_j/\delta_1, \delta_2)$. The corresponding *likelihood function* for all m voltage levels with n_j stresses is given by

$$L = \prod_{j=1}^{m} V\big(v_j/\delta_1, \delta_2\big)^{kj}\big(1 - V\big(v_j/\delta_1, \delta_2\big)\big)^{wj}. \tag{2.56}$$

Varying the parameters δ_1 and δ_2 the maximum of Eq. (2.56) is found as described above. One gets point and confidence estimates as well as a confidence region for the whole performance function which is also shown in Fig. 2.33.

For practical applications of the maximum likelihood method the application of suitable software is necessary. An optimum software package (Speck et al. 2009) contains all necessary steps of the HV test data evaluation, from several independence tests, tests for best fitting with a theoretical distribution function, representation of the empirical performance function (or the cumulative frequency distribution) on probability grid up to point and confidence estimations for the parameters and the whole performance (or distribution) function.

2.4.4 HV Tests for Selected Quantiles Using Up-and-Down Methods

A whole *performance function* is not always required, e.g., the withstand voltage of an insulation can be confirmed, when the test voltage value u_t is lower than the 10%

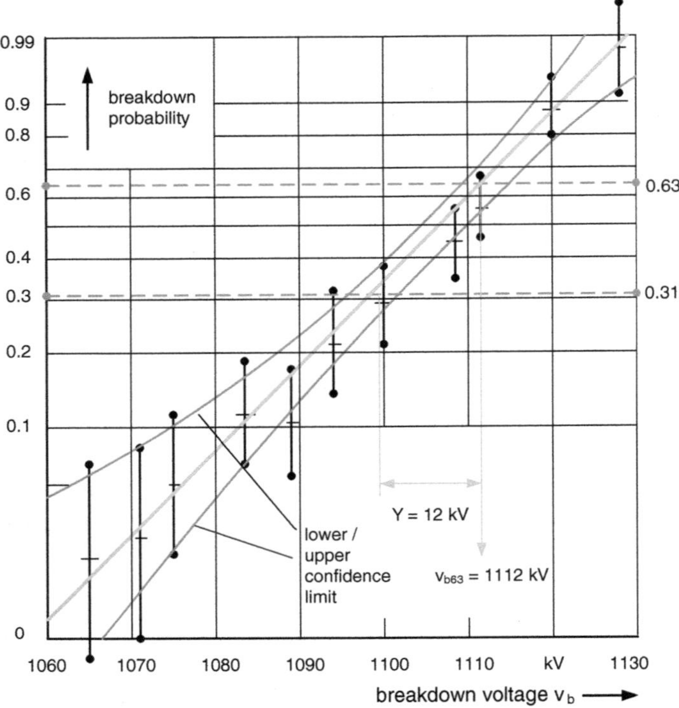

Fig. 2.37 Performance function with its confidence region according to a ML estimation and confidence limits of the single breakdown probabilities and *a*

quantile v_{10}. In that case it is sufficient to determine the value v_{10}, in other cases it might be sufficient to look for the quantiles v_{50} or v_{90}. The related up-and-down test method (UDM) which is based on the constant voltage tests (Sect. 3.4.1) has been introduced by Dixon and Mood (1948).

The method requires that the voltage is initially raised in fixed voltage steps Δv, from an initial value v_{00} at which certainly no breakdown occurs, until a breakdown occurs at a certain voltage (Fig. 2.38: v_1 is the first counted value). Now the voltage is reduced by Δv, if no breakdown occurs the voltage is increased in steps again until the next breakdown, otherwise in case of breakdown it is reduced by Δv. The procedure is repeated until a predetermined number $n \geq 20$ of voltage values v_1, v_2, ... v_n have been obtained. The mean value of these applied voltages is a first estimate for the *50% breakdown voltage* v_{50}:

$$v_{50}* = \frac{1}{n}\sum_{l=1}^{n} v_l. \tag{2.57}$$

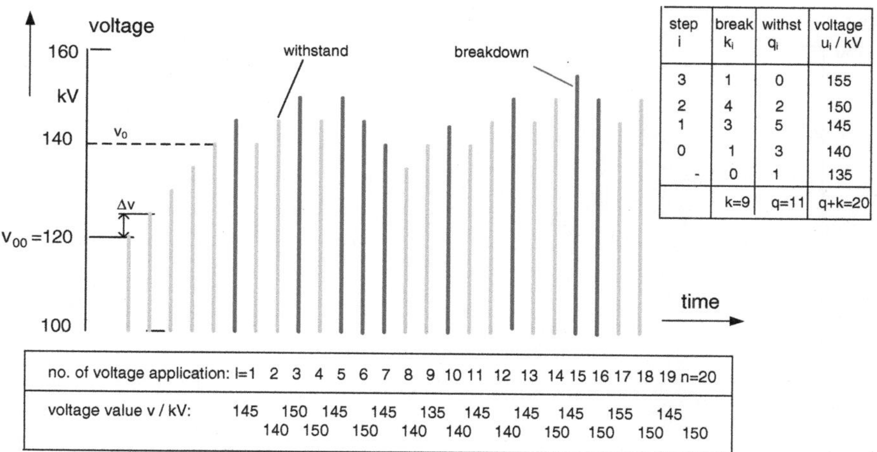

step i	break k_i	withst q_i	voltage u_i / kV
3	1	0	155
2	4	2	150
1	3	5	145
0	1	3	140
-	0	1	135
	k=9	q=11	q+k=20

Fig. 2.38 Up-and-down method (*UDM*) for the estimation of the 50% quantile v_{50}

A more detailed evaluation considers the influence of the step height Δv and uses the number of breakdowns k and the number of withstands q. The sum of the two complimentary events is identical with the number of voltage applications $n = k + q$ starting with the first breakdown. Additionally, the number of voltage levels or steps v_i (with $i = 0 \dots r$) is taken into account. It is counted $i = 0$ from the step of the lowest breakdown. On a certain voltage level v_i, there are k_i breakdowns. Then the 50% breakdown voltage can be estimated by

$$v_{50} = v_0 + \Delta v \left(\frac{\sum_{i=1}^{r} i \cdot k_i}{k} \pm \frac{1}{2} \right). \tag{2.58}$$

Example The test in Fig. 2.38 starts at $v_{00} = 120$ kV and has voltage steps of $\Delta v = 5$ kV. Including the first breakdown there are $l = 20$ voltage applications. The first estimate of v_{50}^* by Eq. (2.57) delivers $v_{50}^* = 145.5$ kV.

The lowest voltage level at which a breakdown occurs is $v_0 = 140$ kV. The number of breakdowns is $k = 9$, that of withstands is $q = 11$ and there are $i = 3$ voltage steps above the lowest breakdown voltage. With these data Eq. (2.58) delivers $v_{50} = 145.3$ kV. The difference between the two methods is very small, for most practical conclusions it can be neglected.

The UDM test is independent when the single voltage applications do not show a decreasing or increasing mean tendency. A UDM test according to the breakdown procedure starts at an initial voltage at which the breakdown is sure and goes down until the first withstand. It also delivers v_{50}, if the withstands are counted as the breakdowns above. Furthermore it should be mentioned that there are methods for the estimation of the standard deviation of the performance function, but the method cannot be recommended (see e.g., Hauschild and Mosch 1992). Confidence limits shall be calculated by the maximum likelihood function (see below) and not by the only roughly estimated dispersions.

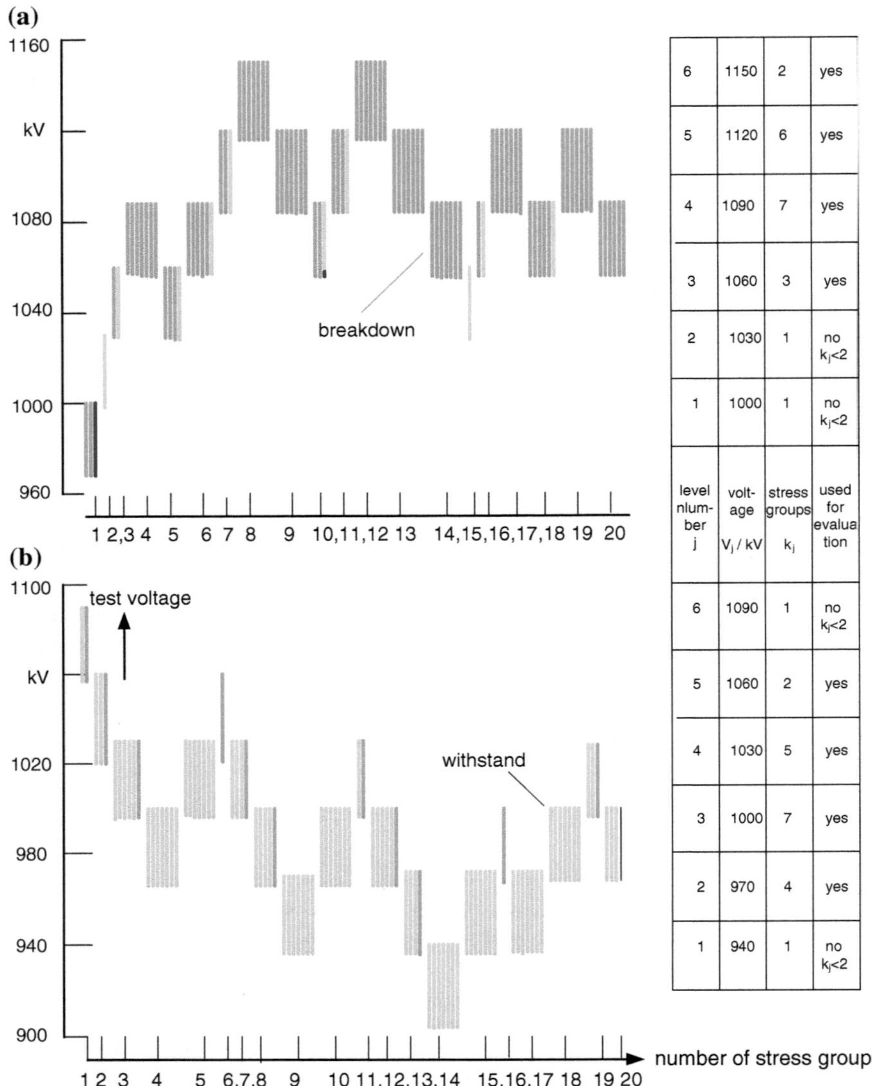

Fig. 2.39 Up-and-down test for the determination of the 90% quantile v_{90} (**a**), and respectively the 10% quantile v_{10} (**b**)

Carrara and Dellera (1972) proposed an *"extended up-and-down-method"* which applies series of stresses (Fig. 2.39) for the determination of pre-selected quantiles instead of single stresses (Fig. 2.38). For the determination of a certain quantile, a certain number of impulses in one series is necessary. Table 2.9 gives the relation between the order p of the quantile and the required number of impulses in one series. The *"withstand procedure"* starting with a withstand voltage v_{00} and raising

Table 2.9 Number of stresses n per UDM group for the estimation of the order p of quantiles

n	70	34	14	7	4	3	2	1	
p	0.01	0.02	0.05	0.10	0.15	0.20	0.30	0.50	(withstand procedure)
p	0.99	0.98	0.95	0.90	0.85	0.80	0.70	0.50	(breakdown procedure)

the voltage delivers the quantiles v_p of the order $p \leq 0.50$, the *"breakdown procedure"* starting with a breakdown voltage and decreasing voltage to withstands delivers the quantiles of the order $p \geq 0.50$.

Figure 2.39b shows the estimation of the statistical withstand voltage determined as the 10% breakdown voltage (quantile u_{10}) using the *withstand procedure* with $n = 7$ stresses per series. As soon as only withstands occur in a series, the voltage is increased by Δv to the next higher level. As soon as a breakdown occurs the voltage is decreased to the next lower level.

When the *breakdown procedure* is applied, the voltage is decreased when in a series only breakdowns occur and only increased when the first withstand appears. The expected *quantile* for this breakdown procedure is v_{90}. The evaluation of the point estimation of the quantile can be performed according to the simplified evaluation by Eq. (2.57).

Also the computer-aided maximum-likelihood method can be applied for the UDM tests provided related software is available (Speck 1987; Bachmann et al. 1991): The principle corresponds to the MLM (see Sect. 2.4.3). For each of the applied voltage levels the relative breakdown frequency including its confidence region is estimated (Fig. 2.40 for estimation of v_{10} with $n = 7$ stresses per series) and plotted in a suited probability grid (Fig. 2.40: Gauss grid). The maximum of the likelihood function delivers the expected quantile v_{10} including its confidence limits. The confidence region ($\varepsilon = 95\%$) of all quantiles is calculated and plotted as a violet line.

2.4.5 Statistical Treatment of Life-Time Tests

A *life-time test* is the stress of the insulation at a certain constant AC or DC voltage (or a series of impulses). The random variable is the breakdown time (or the number of impulses) which can be evaluated according to the PSM (see Sect. 2.4.2). When this test is performed at several voltage levels, the relationship between breakdown voltage and breakdown time—usually known as the *life-time characteristic* (LTC)—can be evaluated (Speck et al. 2009). It is described for a p-order quantile of the breakdown voltage v_p by

$$v_p = k_d t_p^{-1/n} \quad \text{or} \quad t_p = \left(\frac{k_d}{v_p}\right)^n , \tag{2.59}$$

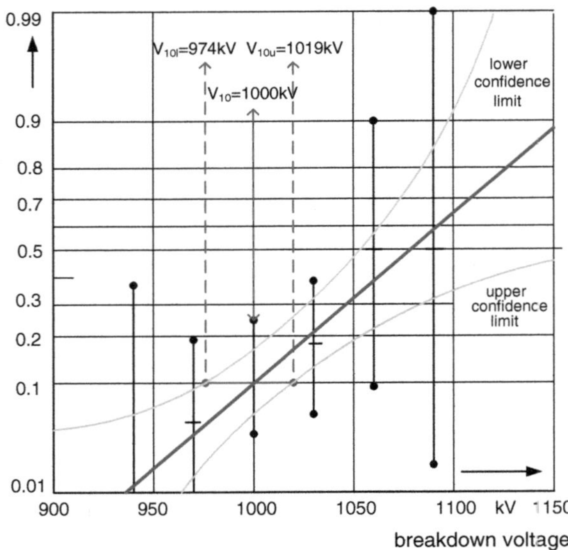

Fig. 2.40 ML evaluation of an extended UDM test for the determination of the performance function in the vicinity of the 10% quantile

where t_p—p-order quantile of breakdown time, n—life time exponent mainly characterizing the insulating material and k_d—a constant mainly characterising the field geometry. In a logarithmic grid the formula delivers a falling straight line which allows the estimation of the parameters n and k_d.

The breakdown time can be described by a two-parameter Weibull distribution (Eq. (2.51): $x_0 = t_0 = 0$) under the consideration of Eq. (2.59) and the relation that v_p is the applied voltage of the constant voltage test $v_p = v_t$:

$$F(t, v_t) = 1 - \exp\left(-\left(t\left(\frac{v_t}{k_d}\right)^n\right)^\delta\right).\tag{2.60}$$

Now the computer-aided maximum likelihood method (Speck 1987; Speck et al. 2009) is applied for the unknown triplet of the parameters k_d, n and δ. The maximum of the likelihood function delivers the best estimation for the triplet. In the usual way also confidence limits can be estimated. The life time-characteristic (Fig. 2.41) includes confidence limits now.

The method enables also the evaluation of censored life-time data, this means test objects are also considered, which have not yet broken down when the test has been terminated.

2.4.6 Standardized Withstand Voltage Tests

At a *standardized withstand voltage test*, the test object has to withstand a test voltage according to the insulation coordination (IEC 60071-1:2006) during an

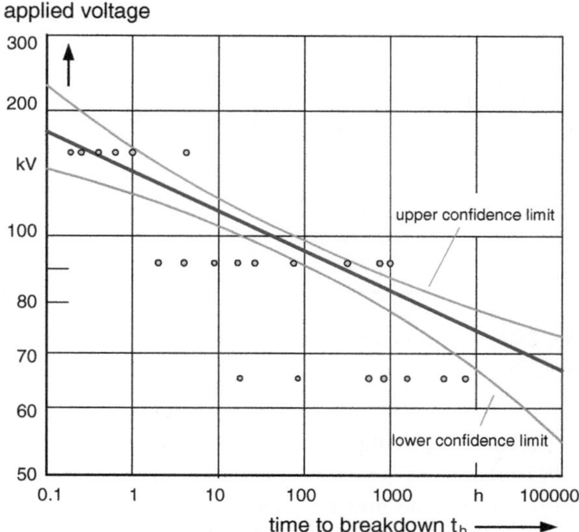

Fig. 2.41 Life-time characteristic including confidence intervals

agreed *test procedure*. In the following the procedures for type and routine tests are considered statistically in brief. The procedures have a long tradition, had been introduced without detailed statistical considerations but are connected with the remarkable experience of test field engineers. A simple change of the procedures would not be accepted and cannot be recommended. But it seems to be necessary that the statistical consequences of these procedures are understood.

The stochastic nature of electrical discharges causes, that a defective test object is not always rejected in a test. With a certain low probability it may pass the test and fail during the operation. This is called the risk of the user. But it may also happen that a test object without defects is rejected in the test. This is the risk of the manufacturer. Which risk is higher depends on the design and the quality of production of the object. If there is only a very small distance between real breakdown voltage and test voltage the risk of the user might be higher than that of the manufacturer. But with a sufficient safety margin the risk of both sides is acceptable.

For AC and DC test voltages (Fig. 2.42a, IEC 60060-1:2010) the voltage shall be rapidly increased up to 75% of the test voltage value. Then it shall be raised with about 2% of the test voltage value per second. When the test voltage value is reached it has to be maintained within $\pm 1\%$ for the test duration T_t, which is very often 1 min but sometimes longer, e.g. one hour. Then the voltage shall be decreased to 50% and switched off. Such a test is a kind of constant voltage test, the test voltage is a single stress and without knowing any details from the development, a statistical judgement of the single test is impossible. But if many test objects of the same type are tested, one should have a statistical evaluation of failure statistics for improvement of design and/or production.

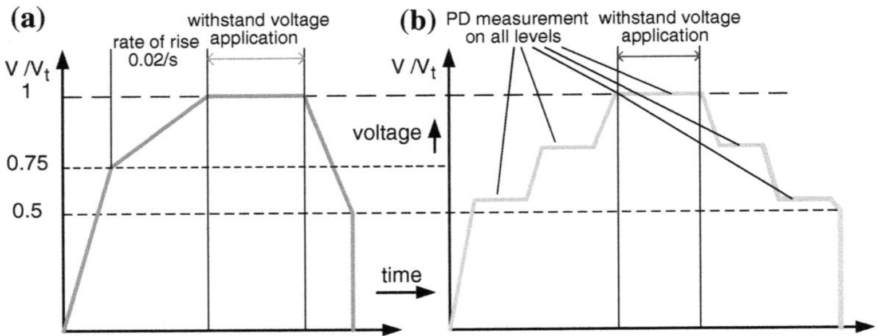

Fig. 2.42 Conventional (**a**), and PD monitored (**b**) withstand test procedure for AC and DC test voltages

AC and DC withstand tests are more and more completed by PD measurement ("*PD monitored withstand tests*"). Then a step test procedure has to be applied (Fig. 2.42b). The upwards and downwards steps should be at identical voltages to enable a comparison of the PD characteristics before and after the withstand test. The PD measurement should also be performed at the withstand test voltage for the specified test duration. Also the duration of the steps for PD measurement must be specified. For all steps, a duration $T \geq 1$ min is necessary. Which step voltage is considered for the withstand test is also a matter of specification. The combination of withstand and PD testing is the most efficient method for AC/DC testing today.

For *LI and SI withstand tests* several methods are recommended (IEC 60060-1:2010):

(A1) A withstand test of a self-restoring external insulations is passed, when it can be shown that the 10% quantile of the performance function is higher than the specified withstand voltage. The 10% quantile may be taken from a measured performance function or from an up-and-down test (see Sect. 2.4.4).

(A2) For such external insulations $n = 15$ test voltage impulses shall be applied and $k \leq 2$ breakdowns are allowed.

(B) For internal insulations $n = 3$ test voltage impulses may be applied and no breakdown is allowed.

By help of the *binomial distribution* (Eq. 2.47) the methods can be compared: Fig. 2.43 shows a diagram of the probability of passing the test depending on the *breakdown probability p* of the test object at the test voltage. The method A1 has a sharp criterion: When the breakdown probability reaches $p = 0.10$, the test object fails the test. The procedure A2 is not sharp, because at $p = 0.08$, 10% of the test objects fail, although according to procedure A1 they are considered as acceptable. But when $p = 0.30$—a too high breakdown probability—there is 15% probability of passing the test in case of method A1. The procedure B is still worse: A test object of the high breakdown probability $p = 0.30$ will pass the test with even 30% probability.

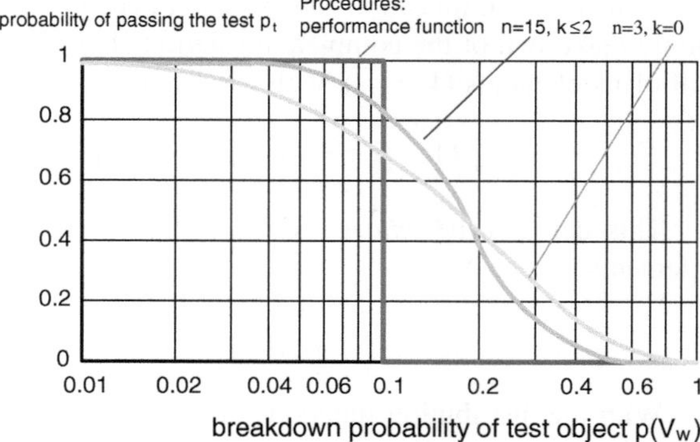

Fig. 2.43 Probability of passing a LI/SI voltage test depending on the breakdown probability of the test object for different test procedures

The example shows, no manufacturer shall design its products with a breakdown probability of 0.10 at the test voltage value. The breakdown probability for design should be $p < 0.01$!

2.4.7   The Enlargement Laws

If insulations are large and voltage sources are sufficiently powerful, discharges which may possibly produce breakdown can develop in parallel, both in space and time. The breakdown occurs in a random manner at the "weakest" of the parallel insulation elements. Examples of discrete elements are the many parallel insulators on overhead lines or in substations or the parallel windings in power transformers. Large insulation with continuous structure are e.g. the insulation of cables, GIS or GIL which may be subdivided into suited individual elements. HV experiments for development of equipment are usually made on such insulation elements. Their results must be transferred to the complete insulation of the equipment. Under the condition of statistical independence of the single elements, the enlargement can be done by the *enlargement law*. The book by Hauschild and Mosch (1992) contains many related details and references. Important relations for planning and evaluation of HV tests are given in the following.

2.4.7.1   Statistical Fundamentals

The enlargement may be in space (volume V, possibly reduced to area A or length l) or in time T. The general enlargement factor from a single reference element (V_1, T_1) to the enlarged insulation (V_n, T_n) is given by

$$n = \frac{V_n}{V_1} \cdot \frac{T_n}{T_1}; \quad 0 < n < \infty \tag{2.61}$$

With the multiplication law for non-dependent probabilities follows, that the non-breakdown of the complete insulation $(1 - p_n)$ requires the non-breakdown of all of its elements $(1 - p_1)_i$ with $i = 1, \ldots n$:

$$(1 - p_n) = (1 - p_1)_1 \cdot (1 - p_1)_2 \cdot \ldots \cdot (1 - p_1)_n \tag{2.62}$$

This delivers the *enlargement law* for the breakdown probability of identical elements

$$p_n = 1 - (1 - p_1)^n \tag{2.63}$$

With the distribution functions of the single elements $F_1(x)$ one gets that of the complete insulation $F_n(x)$

$$F_n(x) = 1 - (1 - F_1(x))^n \tag{2.64}$$

If the single elements have not an identical breakdown probability (or distribution function) the enlargement rule for discrete elements becomes

$$p_n = 1 - \prod_{i=1}^{n} (1 - p_{1i}) \tag{2.65}$$

respectively

$$F_n(x) = 1 - \prod_{i=1}^{n} (1 - F_{1i}(x)) \tag{2.66}$$

When the distribution function $F(x; \alpha; \beta)$ (x is the random variable, α and β are its parameters) of a suited reference element is known and differentially small insulation elements are considered, one gets the generalized enlargement law applicable for insulation with a continuous structure. For details see Hauschild and Mosch (1992) and Hauschild (1995) as well as Marzinotto and Mazzanti (2015).

2.4.7.2 Consequences of Enlargement

There would be no enlargement effect of the breakdown probability if there is no randomness of breakdown processes. For the enlargement of identical elements, Fig. 2.44 is the general evaluation of Eq. (2.63). It delivers the breakdown probability of the enlarged insulation p_n depending on the enlargement factor n and the breakdown probability of the reference element p_1.

Example After breakdown voltage measurements on newly designed cable samples of 10 m length it is evaluated, that the breakdown probability at test voltage is equal to 0.5% ($p_1 = 0.005$).

(1) The planned production and test length is 1000 m, corresponding to an enlargement factor of $n = 100$. Will the cable pass the routine test in factory and is the design good enough?

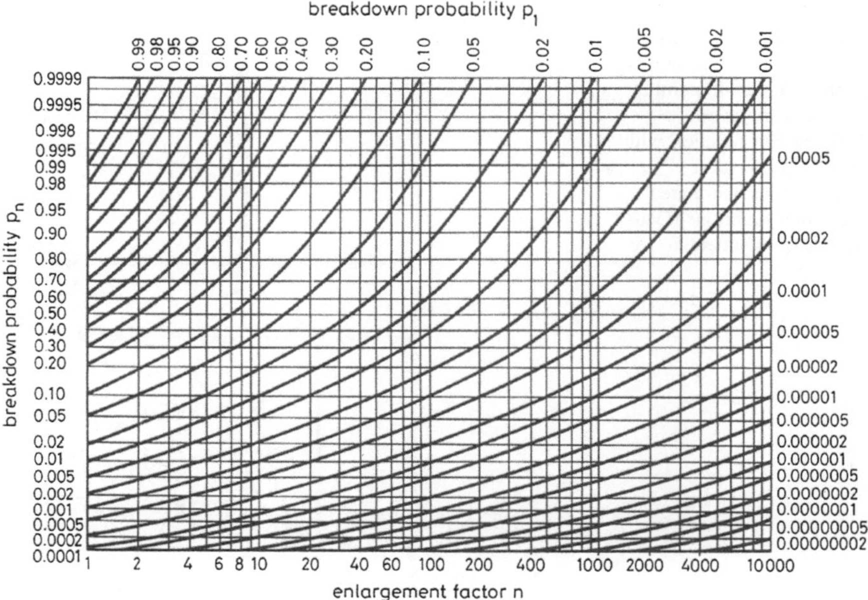

Fig. 2.44 Breakdown probability and enlargement factor for identical elements

Figure 2.44 delivers for $n = 100$ and $p_1 = 0.005$ at test voltage a breakdown probability of the long cable of 40% ($p_n = 0.40$) which is much too high!

(2) If it is assumed that a cable system of 10 km length shall be made with that cable and a commissioning test on site is agreed with the customer for the same test voltage as for the routine test in factory, the breakdown probability for $n = 1000$ becomes even 99.5%. The cable system cannot pass that test!

Conclusion: The design of the cable must be improved!

Figure 2.44 (now based on Eq. 2.67) can also be used to calculate the breakdown probability from measurements at complete large objects with p_n back to the very low breakdown probabilities p_1 of small samples:

$$p_1 = 1 - \sqrt[n]{1 - p_n} \tag{2.67}$$

Figure 2.45 schematically shows the results of enlargement as results of the statistical characteristics of the breakdown: If there is a single point distribution (a), this means no dispersion of results, then no enlargement effects can be expected.

In case of a distribution with an initial value as the Weibull distribution (Fig. 2.45b), The distribution converges for $n \rightarrow \infty$ to a single point distribution at the initial value x_0. If the variables of the *Weibull distribution function* are reduced by $y = (x - x_0)/\eta_1$ to its initial value x_0 and its 63%—quantile $\eta = x_{63} - x_0$, one gets parallel distribution functions for increasing enlargement factors and a

Fig. 2.45 Application of enlargement law to various types of distribution functions. **a** Single point distribution function. **b** Distribution function with lower limit (Weibull distribution). **c** Unlimited distribution functions (Gauss and double exponential distribution)

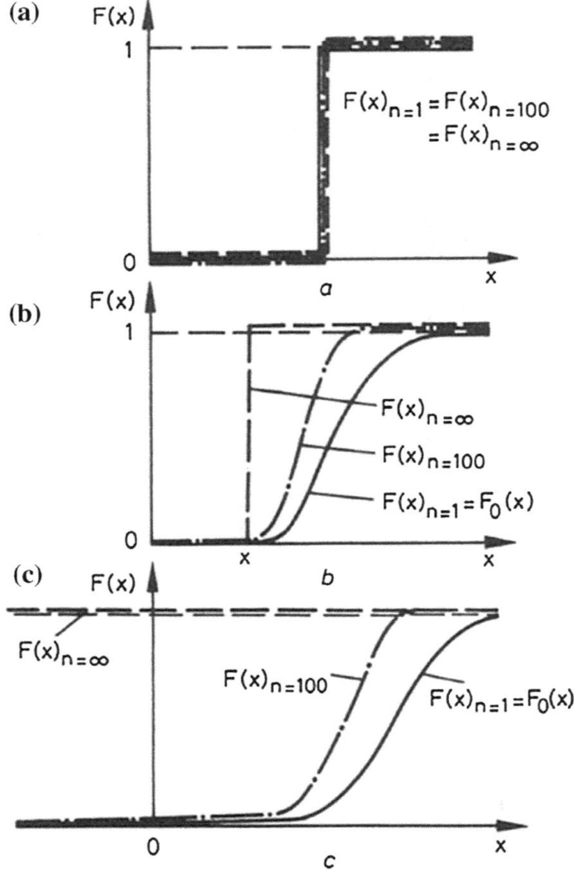

logarithmic y-scale (Fig. 2.46). The parallel straight lines indicate, the application of the enlargement law does not change the type of the distribution, a Weibull distribution remains a Weibull distribution. The limitation is caused by the initial value used for the reduction.

Unlimited distribution functions (Fig. 2.45c) converge for $n \to \infty$ against the breakdown probability of $p = 1$. This is physically non-realistic, therefore the enlargement law should be carefully handled.

In case of the *double exponential (or Gumbel) distribution function* the type of the distribution remains double-exponential (Fig. 2.47, parallel straight lines). Enlargement factors larger than $n = 10^3$ should only be used if the result remains physically logical.

In case of the *Gauss or normal distribution function*, the enlargement law changes its type (Fig. 2.48). The distribution function for enlarged insulation ($n > 1$) are not any longer in parallel to the Gauss distribution of a single reference

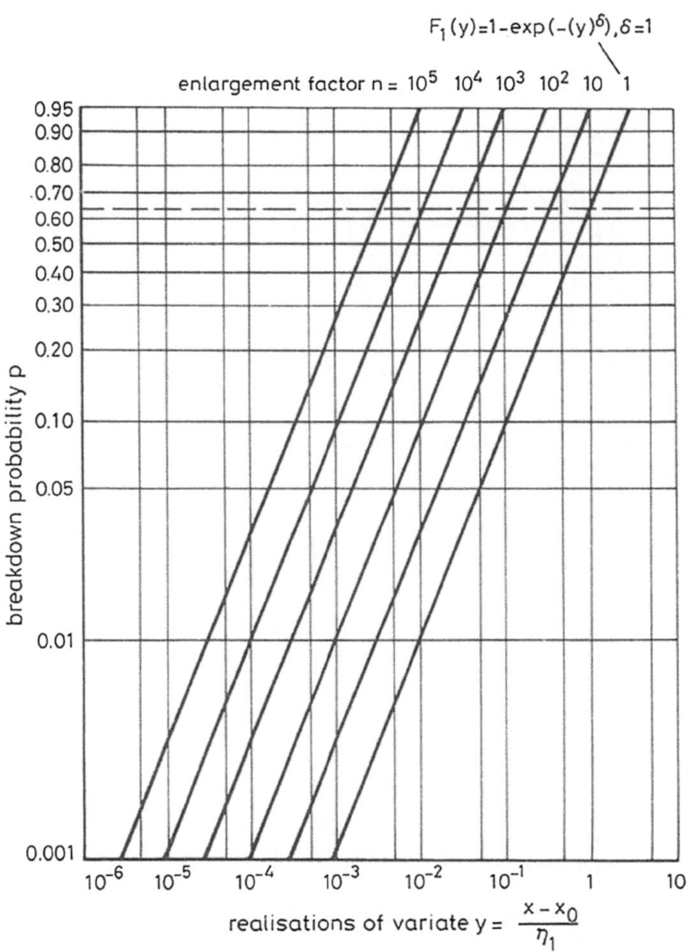

Fig. 2.46 Enlargement law for Weibull distribution function on Weibull probability paper

element (n = 1). It converges not so very fast as the double exponential function, but the careful use of the enlargement law is also recommended.

There are many confirmations for the applicability of the enlargement law for identical elements and for the *area effect*, because the pre-condition of statistically independent elements is often fulfilled, e.g. (Su et al. 2016; Rytöluoto et al. 2015).

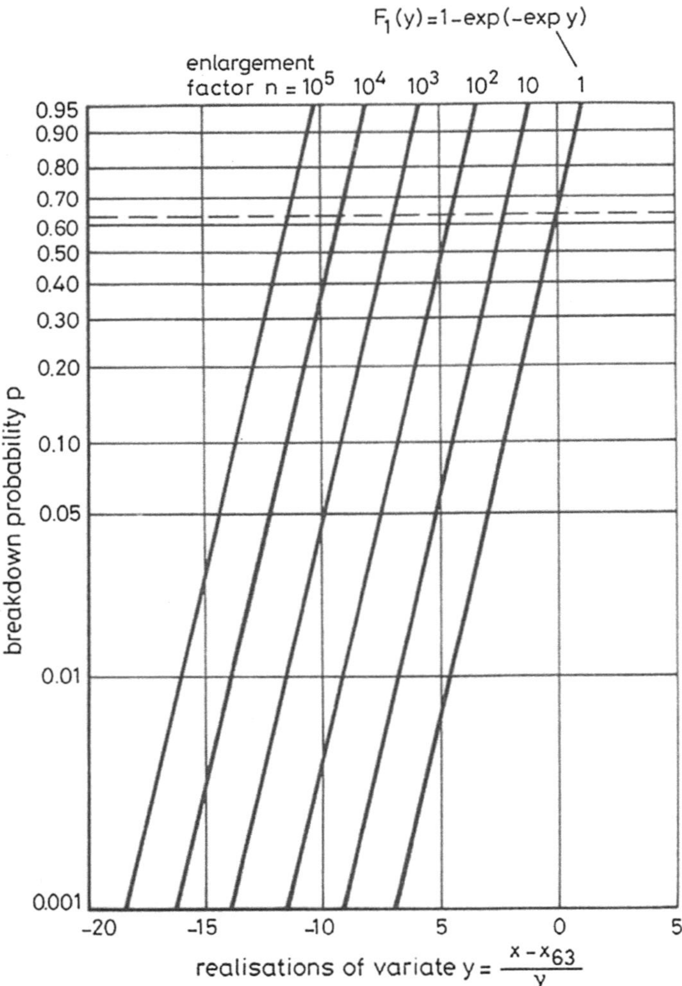

Fig. 2.47 Enlargement law for double exponential (Gumbel) distribution on Gumbel probability paper

For the *volume effect* it is more difficult to presume the independence of elements from each other, but sometimes possible (Zhao et al. 2015). The application of the enlargement law to the breakdown time (*time effect*) is often impossible.

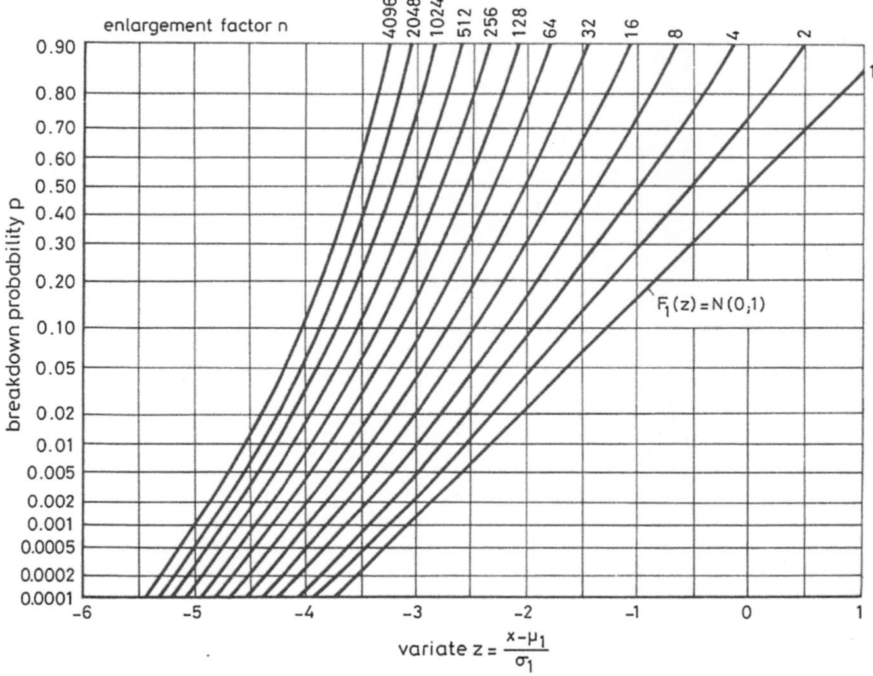

Fig. 2.48 Enlargement law for Gauss normal distribution on Gauss probability paper

The mathematically necessary assumption of the statistical independence of time elements, following each other, is not realistic because of aging processes, mainly in solid insulations.

Chapter 3
Tests with High Alternating Voltages

Abstract *HVAC test voltages* represent the stress of insulations by operational alternating voltages (50 or 60 Hz) and temporary over-voltages. For that reason they are the most important test voltages and applied for all kinds of withstand tests, lifetime tests and dielectric or partial discharges (PD) measurements. After the detailed description of *HVAC voltage generation*, the requirements of HVAC test voltages and the interaction between test system and test object are investigated. Measuring systems for HVAC test voltages are mainly based on capacitive voltage dividers and peak voltmeters, but for measurements during HVAC tests with an expected voltage drop, with harmonics or fast voltage changes, digital recorders become more and more necessary. The chapter is closed with a section on procedures for HVAC breakdown and withstand tests, dry, wet as well as pollution tests, long duration or lifetime testing. Examples for HVAC tests on cables, gas-insulated switchgear, power and instrument transformers are given.

3.1 Generation of HVAC Test Voltages

This section investigates the different possibilities for the HVAC test voltage generation. HVAC test voltages can be generated by test transformers or transformer cascades as well as by resonant circuits with reactors of tuneable inductance or of fixed inductance and by a power supply of variable frequency (frequency converters). The principles of voltage generation are described and recommendations for application are given. Also the special application of HVAC test voltage generation by power or voltage transformers under test (induced voltage tests) is considered.

© Springer Nature Switzerland AG 2019

W. Hauschild and E. Lemke, *High-Voltage Test and Measuring Techniques*,

https://doi.org/10.1007/978-3-319-97460-6_3

3.1.1 HVAC Test Systems Based on Test Transformers (ACT)

3.1.1.1 General Principle

The components of an HVAC test system based on *test transformers* (ACT, Fig. 3.1) are described as introduced for a general HV test circuit (Sect. 2.2; Fig. 2.5): The test transformer is the HV generator, the power supply is usually realized by one (for higher power several) switching cubicle(s), a regulating transformer and a compensation reactor and the voltage measuring system by a capacitive voltage divider with a peak voltmeter connected via a measuring cable. The control-and-measuring system completes the HVAC test system.

A *test transformer*—characterized by its

rated primary voltage (LV) V_{1m},
rated secondary voltage (HV) V_{2m},
rated current I_m and/or
rated test power S_m, both with a
rated duty cycle

—is a usual single-phase transformer with layer windings. The *transformer ratio* is the number of turns of the high-voltage winding w_2 over those of the low-voltage winding w_1. For the generation of a high voltage, a certain magnetic flux is necessary that requires a certain number of HV turns w_2. The number of the LV turns w_1 is then depending on the available (or selected) feeding voltage V_{1m}.

For basic considerations, the HV circuit can be simplified as given in Fig. 3.2a. The introduction of the *transformer ratio* w_2/w_1 for the feeding voltage V_1^* (Fig. 3.2b) delivers a simple R–L–C circuit where L_T represents the transformer stray inductance, R_T the active transformer losses and C the acting capacitance from test object, voltage divider and transformer.

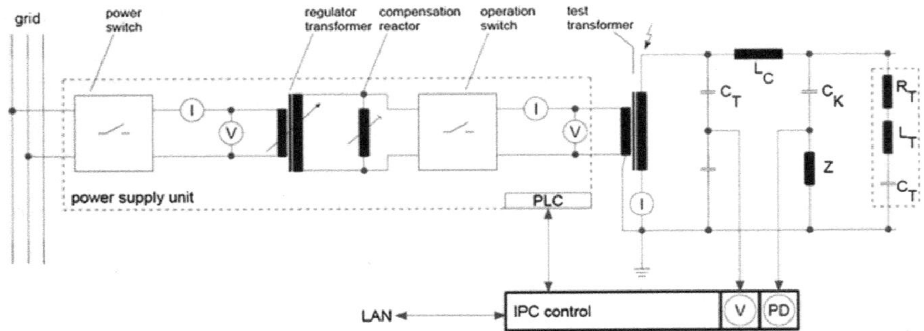

Fig. 3.1 Circuit diagram of an HVAC test system based on a test transformer (ACT)

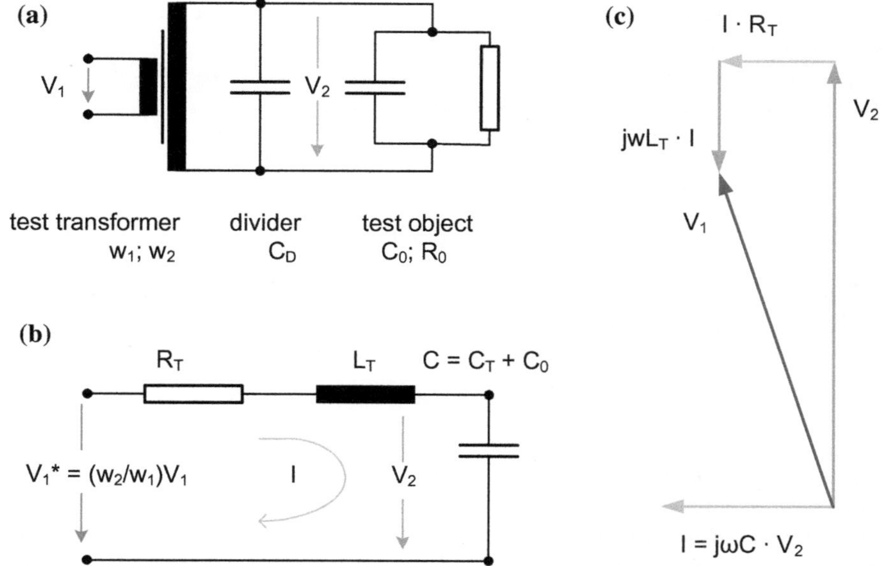

Fig. 3.2 HVAC test circuit, **a** simplified HV test circuit, **b** equivalent circuit, **c** phasor diagram

The voltage at the test object V_2 is shown in the phasor diagram (Fig. 3.2c). With the pre-given current I in the test circuit, the test voltage becomes

$$V_T = V_2 = \frac{I}{j\omega C}.$$

There is a resistive *voltage drop* on the transformer

$$V_R = I \cdot R_T.$$

The inductance of the transformer causes a reactive voltage drop

$$V_L = I \cdot j\omega L_T,$$

which partly compensates the capacitive test voltage $V_2 = V_T$. It can easily be seen that the transformed feeding voltage V_1^* is for a capacitive test object lower than expected from the test voltage V_2.

Note Correspondingly a feeding voltage V_1^* higher than expected from the test voltage V_2 would be required if the test object is inductive. In case of a resistive test object and low losses in the test transformer, $V_2 \approx V_1^*$ applies.

One can conclude that *the relation between the feeding primary voltage of a test transformer and the generated secondary test voltage depends on the parameters of the test object*. Therefore, the primary voltage cannot be used for the measurement of the test voltage (see Sect. 3.4).

If also the resistive losses in the test object are neglected ($V_R = 0$), the feeding and the test voltage would be almost in phase, one gets

$$V_2 = V_1^* + V_L \quad \text{and}$$
$$V_2 = \frac{V_1^*}{1 - \omega^2 C L_T}. \tag{3.1}$$

This delivers a second important conclusion: For $(1 - \omega^2 L_T C) \to 0$ one gets oscillations with the *natural frequency* of the test circuit

$$\omega_0 = \frac{1}{\sqrt{C \cdot L_T}} = 2\pi f_0 \tag{3.2}$$

and *resonance when the natural frequency f_0 is equal to the frequency f of the feeding voltage V_1^** (e.g. $f_0 = 50$ Hz). For the considered theoretical case without active losses ($R_T \to 0$), at resonance the test voltage would go ad infinitum ($V_2 \to \infty$). In practical test circuits with losses, it can reach quite high values and the circuit is out of control. *Consequently, resonance must be avoided and the ACT test circuit protected against over-voltages.*

> **Note** Such an *over-voltage protection* can be realized by an order from the control to reduce the test voltage and switch it off, e.g. for the case that the measured test voltage has reached 105% of its preselected value.

The *short-circuit voltage V_{kT}* of a transformer is the necessary primary voltage V_1 to drive the rated current I_m through the transformer when the secondary side is short-circuited ($V_2 = 0$). For the simplified equivalent circuit (Fig. 3.2b) without resistive losses ($R_T = 0$), one gets the short-circuit voltage

$$V_{kT}^* = \omega L_T I_m, \tag{3.3}$$

the relation of this short-circuit voltage to the rated secondary voltage $V_{2m} = I_m/\omega C$ of the transformer defines the *short-circuit impedance* (sometimes also "impedance voltage")

$$v_{kT} = \frac{V_{kT}^*}{V_{2m}} = \omega^2 C L_T. \tag{3.4}$$

For test transformers, the short-circuit impedance is an important parameter with values between 5 and 25%. A high short-circuit impedance causes voltage enhancements at capacitive load and—in case of heavy discharges or breakdown—current limitations that may impede the breakdown process. With the short-circuit impedance, one can estimate the capacitive enhancement of the test voltage based on Eqs. (3.1) and (3.4).

Example An HVAC test transformer is loaded with a capacitive test object (Fig. 3.2b) that requires its rated current at rated frequency. The enhanced voltage can be calculated by $V_2 = V_1^*/(1 - v_{kT})$. For an impedance voltage of $v_{kT} = 18\%$, one gets $V_2 = V_1^*/0.82 = 1.22\ V_1^*$. This means, the test voltage is enhanced by 22% compared with the transformer ratio. Vice versa the primary voltage for a certain secondary test voltage is lower.

A further increase in the total impedance voltage for complete HVAC test circuits is caused by the impedance voltage of the regulating transformer v_{kR} and impedances in the HVAC circuit (e.g. a feeding transformer v_{kF} between the mains and the regulator, limiting reactances or resistors on the LV or HV side). The short-circuit current can be calculated from the resulting impedance voltage which means the weighted sum of all impedance voltages. The weight is the power of the considered components. When only the test transformer (reactive test power S_T and impedance voltage v_{kT}), the regulator transformer (S_R, v_{kR}) and a feeding transformer (S_F, v_{kF}) are considered, the resulting impedance voltage at the test transformer is

$$v_k = v_{kT} + \frac{S_R}{S_T} \cdot v_{kR} + \frac{S_F}{S_T} \cdot v_{kF}. \tag{3.5}$$

The continuous *short-circuit current* I_k, which is important, e.g., for HVAC pollution testing (Sect. 3.2), follows from the rated current I_m with the resulting impedance voltage v_k

$$I_k = \frac{I_m}{v_k}. \tag{3.6}$$

Note This estimation of the short-circuit current describes the steady state conditions and neglects the transient currents after the instant of the breakdown. Furthermore, the impedance voltage is given as a dimension-less figure. When it is given in percent (%), Eq. 3.6 has to be modified accordingly.

In case of a breakdown of the test object, the short-circuit current causes a high mechanic stress to the windings of the test transformer. Due to this, some turns can be displaced and cause the damage of the transformer. Therefore, a stable mechanic design of the windings is important when a high short-circuit current is required, e.g. for wet and pollution tests.

A stable mechanic design is difficult because of the necessary high number of turns (sometimes several thousands) consisting of quite thin wires (diameter often in the order of 1 mm). This has to be realized under the strong HV insulation requirements. Therefore, the stable design is necessary, for example as follows: In a larger oil–paper-insulated test transformer, the whole winding is subdivided into winding elements (or "coils"), each based on a pressboard cylinder of different diameter, manufactured separately and—later on—coaxially arranged around a leg of the magnetic core. Such a coil consists of several layers of wires wrapped in insulating paper of some centimetres thickness (Fig. 3.3). The layers are connected together, and the coil is connected to the neighbouring coils. The mechanic stability depends strongly on a reliable design considering the shrinking of the paper, on the quality of making the windings and on a careful drying process.

Some test transformers and regulators are not resistant enough against short-circuit currents and require an impedance, often an inductance, in the HV circuit between test transformer and test object. This indicates certain problems of the mechanic stability of the windings that should not happen at a modern test transformer under normal test procedures in a test field.

In addition to the dielectric and the mechanic design, the thermal design is of great importance. Due to the thick layers of paper, the heat transfer from the windings to the oil is hampered. Therefore, cooling channels between the single layers are arranged to enable the heat convection to the streaming oil. Usually test transformers operate with natural cooling by oil streaming because of temperature differences inside the tank. In rare case also forced cooling is applied for further improvement of the heat transfer. The heated oil is cooled at the walls of the transformer by the surrounding air. Sometimes the walls are made from corrugated sheet or equipped with radiators or special heat exchangers.

Two types of oil–paper-insulated test transformers, with metal tank and with insulating cylinders, are in wide application. Each type has characteristic parameters and correspondingly its preferred application, both summarized in Table 3.1. In addition to these transformers with oil–paper-insulated windings, test transformer

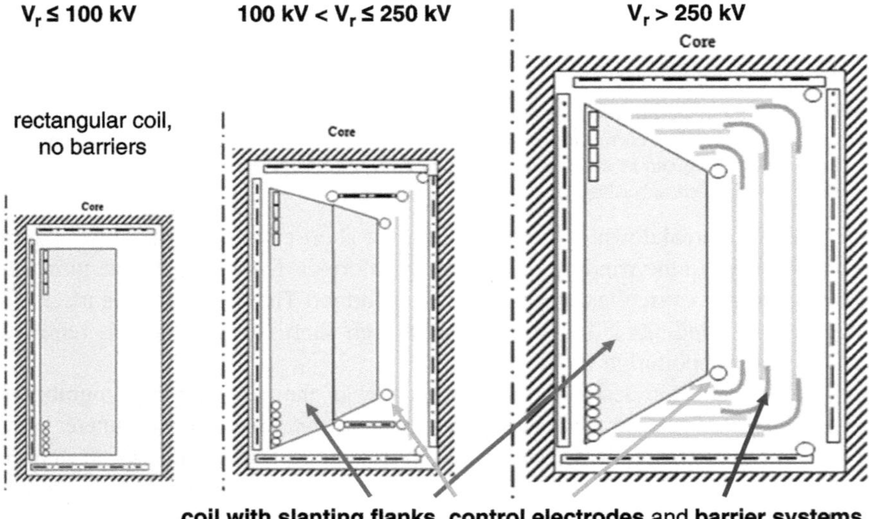

Fig. 3.3 Principle arrangement of layers, coils and windings of an oil–paper-insulated test transformer

Table 3.1 Parameters and application of different types of test transformers (single units)

Type and insulation characteristics	Tank type with oil–paper insulation	Cylinder type with oil–paper insulation	Metal-enclosed SF_6-impregnated foil insulation	Epoxy-resin insulation
Maximum rated voltage (kV)	1000	500	1000	100
Maximum rated current (A)	10	<2	<0.5	0.2
Maximum test power (kVA)	10,000	<1000	<500	<20
Maximum duty cycle	Continuous	Short time 10 h	Short time < 1 h	Short time < 2 h
Impedance voltage (%)	5–10	10–15	20	20
HV taps	Yes	No	No	Not important
Specific weight (kg/kVA)	10–15	8–12	8–10	15–20
Lab height demand	Low	High	Not important	Not important
Ground area demand	Average	Average	Small	Not important
For parallel connection provided	Yes	Conditioned	Yes	Not important
For cascades provided	Yes	Yes	No	Not important
Applicable for metal-enclosed circuits	Yes	No	Yes (preferred)	Conditioned
Outdoor application	Yes	No	No	No

with SF_6-impregnated foil insulation (for some mobile HVAC test systems) and epoxy resin cast (for small test systems of low power up to 100 kV) insulation are in use.

3.1.1.2   Tank-Type Test Transformers

Tank-type test transformers guarantee best cooling conditions because its active part is arranged in a metal (steel) tank (Fig. 3.4) that can additionally be equipped with "radiators" to increase the cooling surface of the tank. The active part can contain a single HV winding (Figs. 3.4 and 3.5) or consist of two HV windings in series as also used for cylinder type transformers (Fig. 3.8b). Tank-type transformers offer best

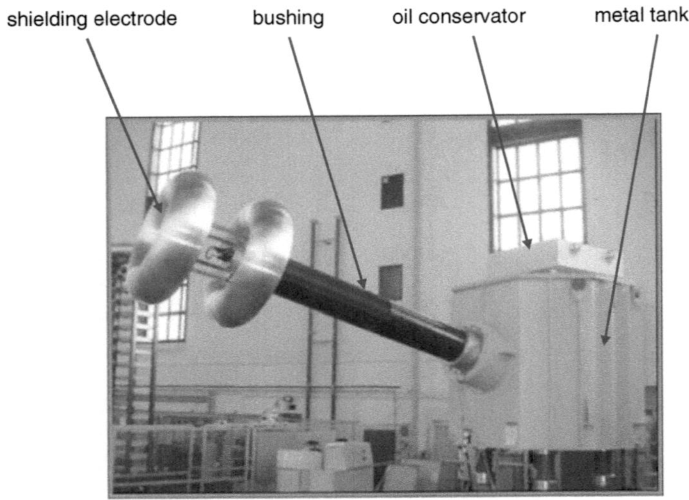

Fig. 3.4 Metal-tank test transformers transformer 600 kV, 2000 kVA with a single HV winding

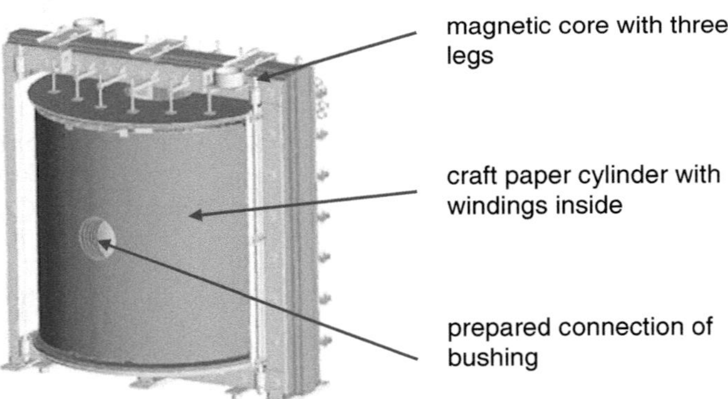

Fig. 3.5 Principle design of the active part of a metal tank test transformer (single HV winding and magnetic core)

cooling conditions and can be designed for the highest test currents, respectively, test powers, and enable continuous operation (Table 3.1). The magnetic core of the transformer has often three legs, and the inner LV and the outer HV winding are arranged around the central leg (Fig. 3.5). The core is on the same ground potential as the tank is, the lower end of the HV winding is usually also grounded via a bushing. At this bushing, the current in the HV test circuit can be measured. For an HVAC testing in ambient air, the transformer must be equipped with an oil-to-air bushing (Fig. 3.4), but metal tank transformers can also be connected by oil-to-SF_6

bushings to metal-enclosed test circuits that are mainly applied for testing of gas-insulated systems (GIS) and their components (Fig. 3.6; see also Sects. 3.2.3 and 10.4.1).

Because of the grounded tank, the tank-type test transformer does not require any clearance to neighbouring walls from the electrical point of view. In case of horizontal or diagonal bushings, other components (voltage divider and coupling capacitor) can be placed below or very near to the end electrode of the bushing. This enables a very compact and space-saving arrangement in a test laboratory, often the ground space is lower than that for a cylinder-type transformer (Fig. 3.7).

Metal tank transformers are well suited for outdoor operation. Therefore, this design should be applied for all open-air HV test fields. Furthermore, there is the opportunity to arrange the test transformer outside of a HV test hall and to have only its bushing inside of that hall. This is a further possibility to save the expensive space in the test laboratory.

Considering the possibility of AC test voltage generation by resonant circuits, tank-type transformers are mainly applied for wet and pollution testing (Sect. 3.2.4), for outdoor application and for metal-enclosed test circuits.

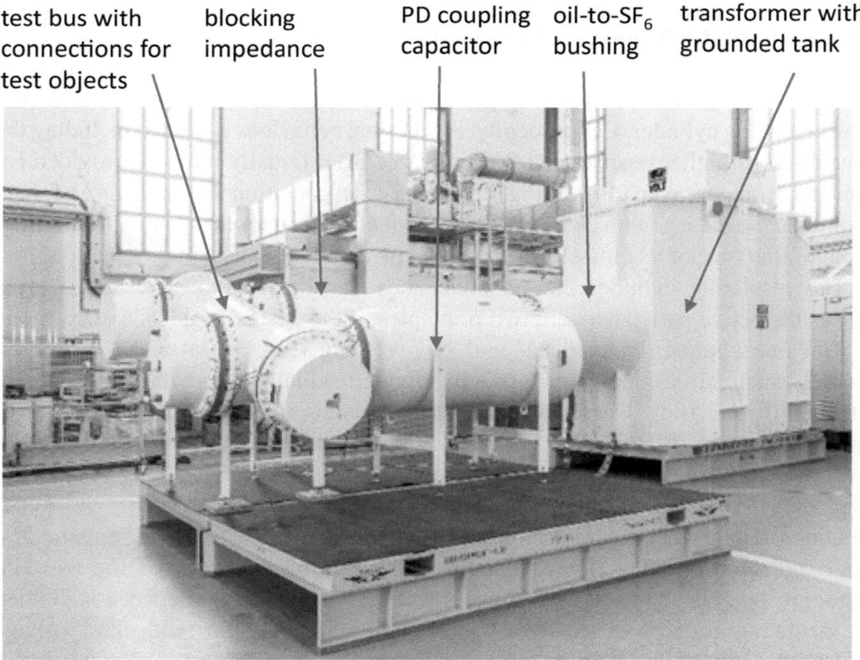

Fig. 3.6 Metal tank transformer 1000 kV/1000 kVA with oil-to-SF6 bushing to a SF$_6$-insulated test bus

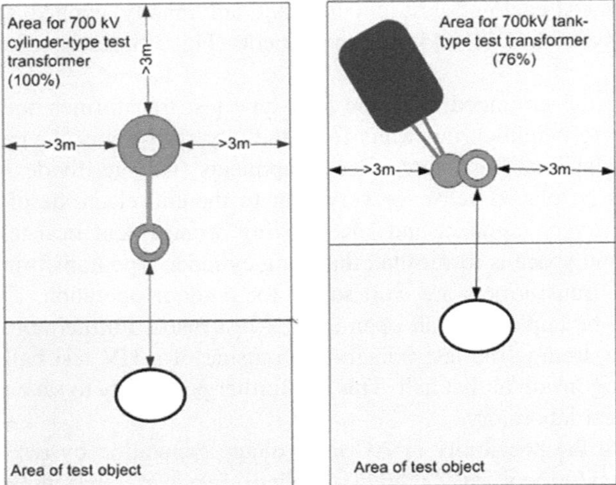

Fig. 3.7 Ground area demand of a cylinder-type test transformer compared with that of a tank-type test transformer (both 700 kV)

3.1.1.3 Cylinder-Type Test Transformers

Cylinder-type test transformers (Fig. 3.8a) do not require a bushing because their case is an insulating cylinder. Consequently, the thermal behaviour is limited including the generation of higher reactive test power (Table 3.1). Usually they are provided for short-term operation up to 10 h (Fig. 3.9), long-term operation would require a forced cooling of the oil, e.g. with external coolers connected by houses. In any case, the duty cycle of this type of test transformers should be carefully considered.

The insulating cylinder has on both sides metallic covers. The lower cover carries the active part. There are two principles of the design of the active part, one is with one common coaxial LV and HV winding (Fig. 3.10a) and the core on ground potential, the other one is with divided windings and the core on half potential (Fig. 3.10b). The latter requires an insulating support for the core, but it is better adapted to the geometry of the cylinder. The lower windings have the LV potential on its outer side and the HV "half" potential on its inner side. The core is connected to this "half" potential as well as a transfer winding for the upper windings. There, the exciter winding is next to the core, whereas the second HV "half"-potential winding is on the outer side. This means, between the two HV windings, one has the full high voltage, therefore, the oil gap in between is divided by insulating barriers to improve the insulation for a higher withstand voltage (Figs. 3.8b and 3.10b).

Cylinder-type transformers are well suited for transformer cascades up to 1500 kV (Fig. 3.11), but only recommended for indoor laboratories. They are well suited for multi-purpose test laboratories without special requirements, e.g. at universities or at test service suppliers.

Fig. 3.8 Cylinder-type test transformer 500 kV/500 kVA, **a** complete transformer, **b** active part with divided windings

Fig. 3.9 Available test duration of a cylinder-type test transformer depending on the required test current (T ambient temperature)

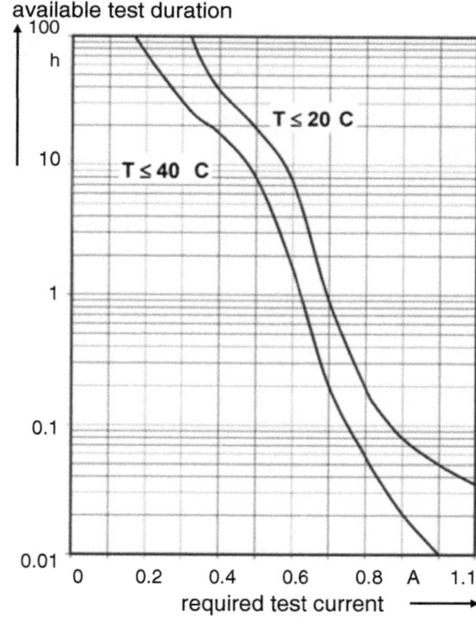

The principle of cylinder-type cascades is also used in HVDC modules (See Sect. 6.1.4; Fig. 6.9) as well as for cascaded HVDC voltage generators (Fig. 6.10).

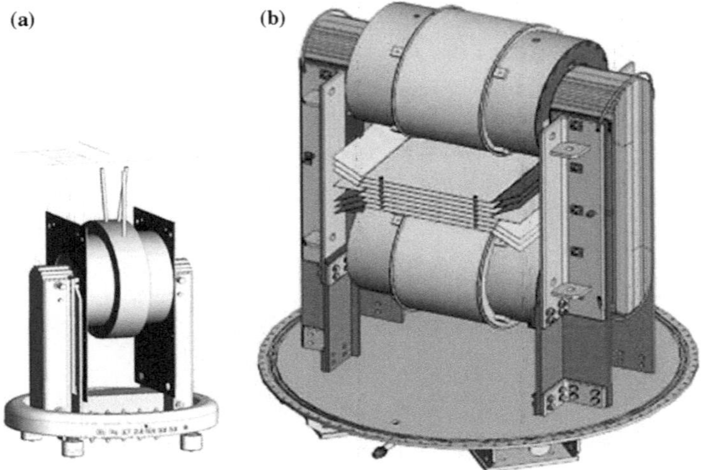

Fig. 3.10 Principle design of cylinder-type test transformers, **a** single winding and core on ground potential, **b** divided windings and core on half potential

Fig. 3.11 Transformer cascade 1000 kV/350 kVA of three cylinder-type transformers. Courtesy of TU Cottbus, Germany

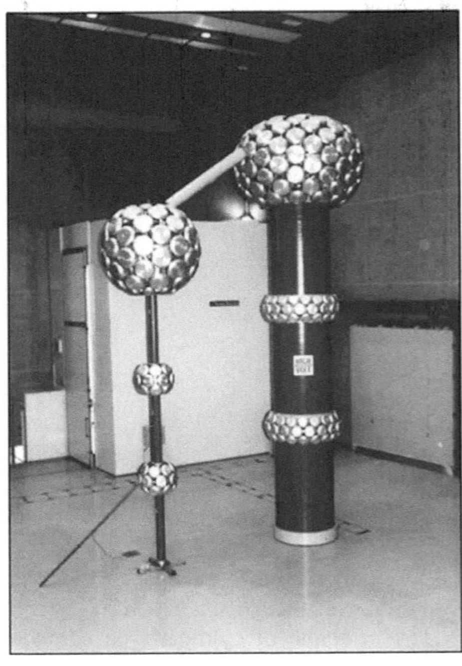

3.1.1.4   Test Transformers with SF6-Impregnated Foil or Solid Insulation

Test transformers with SF_6 -impregnated foil insulation (Fig. 3.12) are derived from instrument transformers of the same insulation type (Moeller 1975) and mainly used for GIS testing. In that case, they can directly be flanged to the GIS under test to establish a metal-enclosed, and therefore well-shielded HV test and partial discharges (PD) measuring circuit, also under on-site conditions.

The thermal behaviour of the SF_6-insulated foil insulation is poor compared with oil–paper insulation. Therefore, this type of test transformer has limited rated current and a very short duty cycle. Also the weight is relatively high due to the necessary magnetic core. For GIS testing, SF_6-insulated test transformers should be replaced by resonant circuits with SF_6-insulated reactors of much better weight-to-test power relation (Hauschild et al. 1997), especially for test voltages > 200 kV.

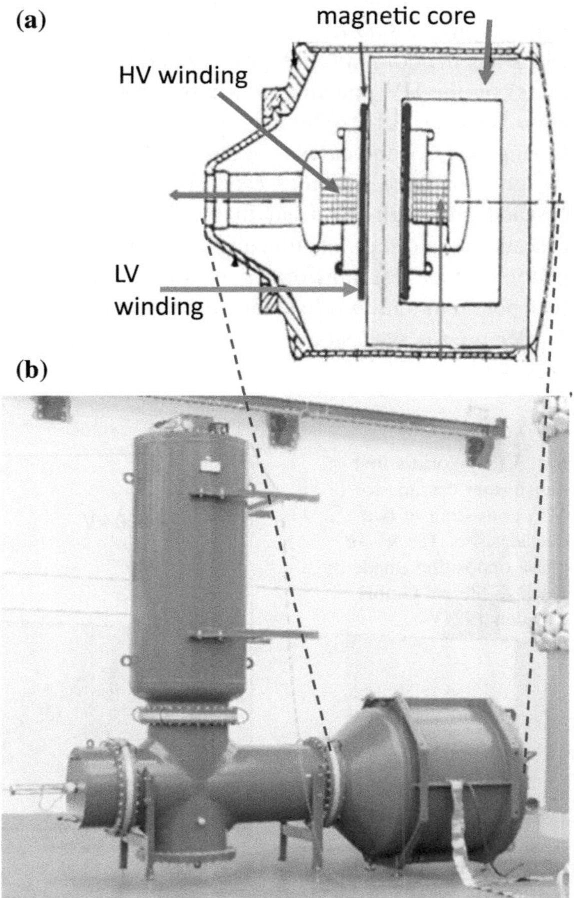

Fig. 3.12 Test transformer with SF_6-impregnated foil insulation, **a** cross section, **b** complete test system with transformer (right), blocking impedance (horizontal) coupling capacitor (vertical) and PD calibration coupling (left)

Test transformers with epoxy-resin insulation are only realized up to rated voltages of about 100 kV because of the necessary PD-free design. Because of the necessary thermal design, only quite low-rated currents for short-time operation are allowed (Table 3.1). Therefore, epoxy-resin-insulated test transformers can be used for limited HV test applications of low- and medium-voltage equipment, voltage generation in "testers" (e.g. for insulating oil), student's training and demonstrations.

3.1.1.5 Test Transformer Cascades

Test transformer cascades are applied for the generation of high test voltages that cannot be generated by single test transformers. The principle of transformer cascades is an early design of the HV test technique with contributions of W. Petersen with co-workers and A.-J. Fischer in 1915 (Fig. 3.13). It is useful only for oil–paper-insulated transformers that are considered in the following.

The application of a transformer cascade can be taken into consideration for voltages above 600 kV (Fig. 3.11). A test transformer used for transformer cascades has in addition to the primary and secondary winding a *transfer winding* (w_3 turns) on the HV potential with a transformer ratio to the primary winding of w_3: $w_1 = 1:1$ (Fig. 3.14). This winding feeds the primary (exciter) winding of the following stage of the cascade. Usually, the transformers of a cascade have identical design with three windings each. This enables its application on all stages in the cascade, although not used in the highest stage. Figure 3.15 shows the largest cascade transformer manufactured till now (TuR Dresden). This cascade is equipped with three test transformers of divided HV winding, magnetic core and tank on half potential and two bushings each of half-rated voltage (Fig. 3.16). This transformer type is well suited for outdoor cascades without limitations of the ground area demand.

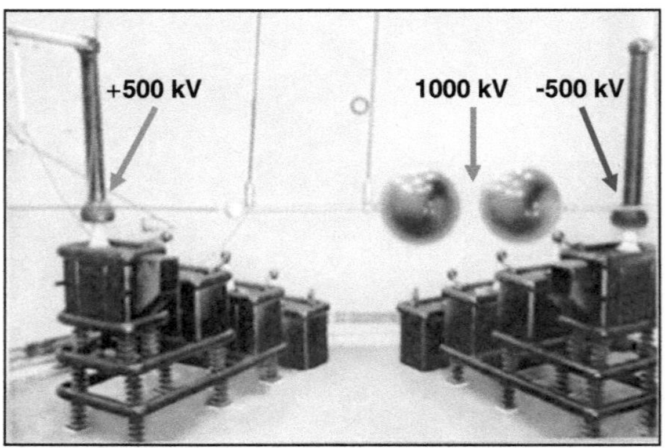

Fig. 3.13 World's first transformer cascade for 1 MV, consisting of two cascades 4 × 125 kV in phase opposition (made by Koch & Sterzel GmbH, Dresden 1921)

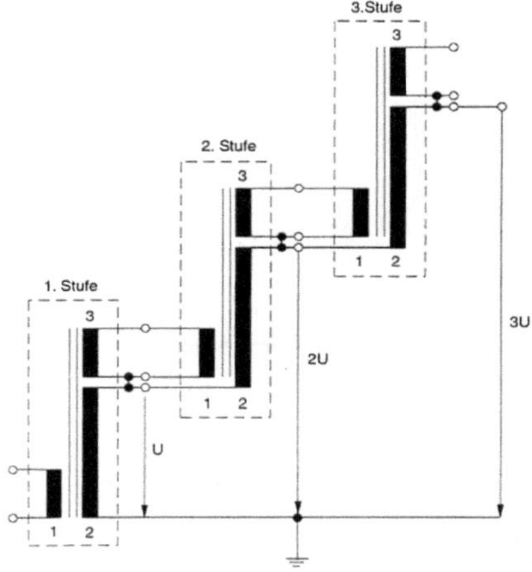

Fig. 3.14 Principle circuit of a three-stage transformer cascade

Fig. 3.15 World's largest transformer cascade for 3000 kV/4.2 A with capacitor banks for SI voltage generation

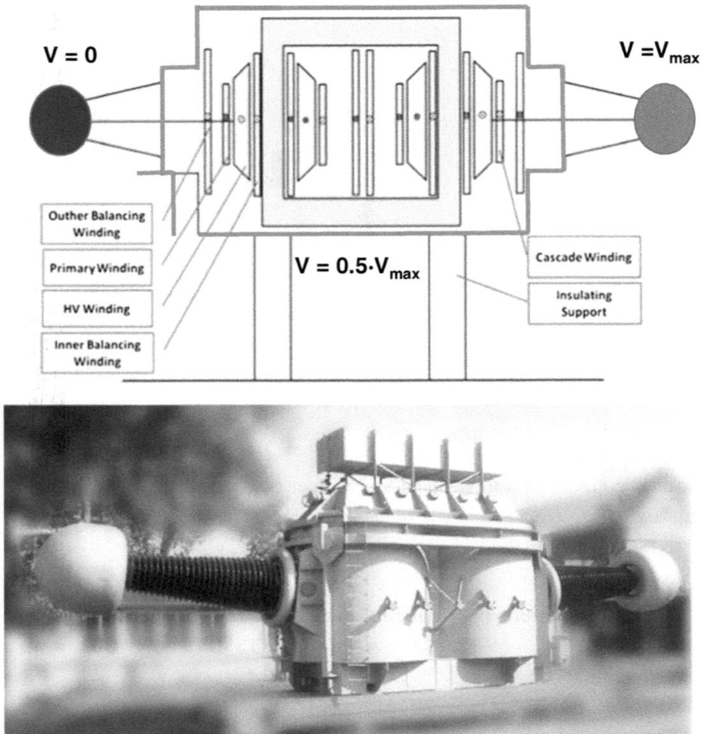

Fig. 3.16 Test transformer used for cascades with divided HV winding and tank on half potential

The *transformer cascade* is helpful for the generation of high voltages but one has to consider that all the test power must be supplied via the lower stages. This means the rated power of the cascade cannot be higher than the rated power of its lowest stage. Sometimes also the rated voltage of the cascade is a bit lower than the voltage of its components multiplied by the number of stages to meet non-linear voltage distributions in the cascade. For instance, the rated parameters of the single transformers of Fig. 3.16 are 1.2 MV and 14 MVA compared with 1 MV and 12.6 MVA when used in the cascade. To generate higher currents in transformer cascades, two transformers can be switched in parallel for the lowest stage.

Not only the available test current is reduced, the *short-circuit impedance* of the cascade transformer is increased according to the number of stages of the transformer cascade drastically. There are simplified calculations of the *short-circuit reactance* (Hylten-Cavallius 1988; Kind and Feser 1999). Kuffel et al. (2006) calculated the short-circuit reactance of the cascade X_{res} from the reactances of the single units (X_{HV} of the HV winding, X_{LV} for the LV exciter winding and X_{TV} for the transfer winding) and the number of stages n by

$$X_{\text{res}} = \sum_{i=1}^{n} \left(X_{\text{HV}i} + (n-i)^2 XL_{\text{LV}i} + (n+1-i)^2 + X_{\text{TV}i} \right), \qquad (3.7)$$

where it is presumed that all reactances are related to the same voltage, e.g. the HV output of the lowest transformer, and active losses, e.g. in the windings or in the core, are neglected. When three identical transformers ($n = 3$) are used for the cascade, one gets the resulting short-circuit reactance:

$$X_{\text{res}} = 3 \cdot X_{\text{HV}} + 5 \cdot X_{\text{LV}} + 14 \cdot X_{\text{TV}}. \qquad (3.8)$$

The formula indicates different voltage changes in the different stages and consequently non-linearities of the voltage distribution. The voltage distribution can be improved by compensation reactors on the stages. This remains without considerable influence on the short-circuit impedance. Test current reduction and increasing short-circuit impedance are the main reasons that cascades consist of no more than three stages usually.

To avoid the overload of the lower stages, the capacitive currents are compensated by reactors in parallel to the primary windings of the transformer and placed on the stages of the cascade. The *compensation reactors* are equipped with taps or even tuneable for the adaptation to the test object capacitance. In most cases, the reactors are separately arranged on the level of the stage. In some cases, fixed reactors inside of the tank of the test transformer are used, if only the capacitances of the test transformer itself shall be compensated. This can also be reached with a suited gap in the core of the transformer.

3.1.2   HVAC Test Systems Based on Resonant Circuits (ACR)

3.1.2.1   Principles of Resonant Circuits

If in the simplified test circuit based on a transformer (Fig. 3.2b), the inductance is modified to full compensation of the capacitive current of the capacitive load, one gets the phasor diagram *of a series resonant circuit* given in Fig. 3.17. This *"oscillating circuit"* is characterized by its *natural frequency* (Eq. 3.2)

$$f_0 = \frac{1}{2\pi\sqrt{L \cdot C}}. \qquad (3.9)$$

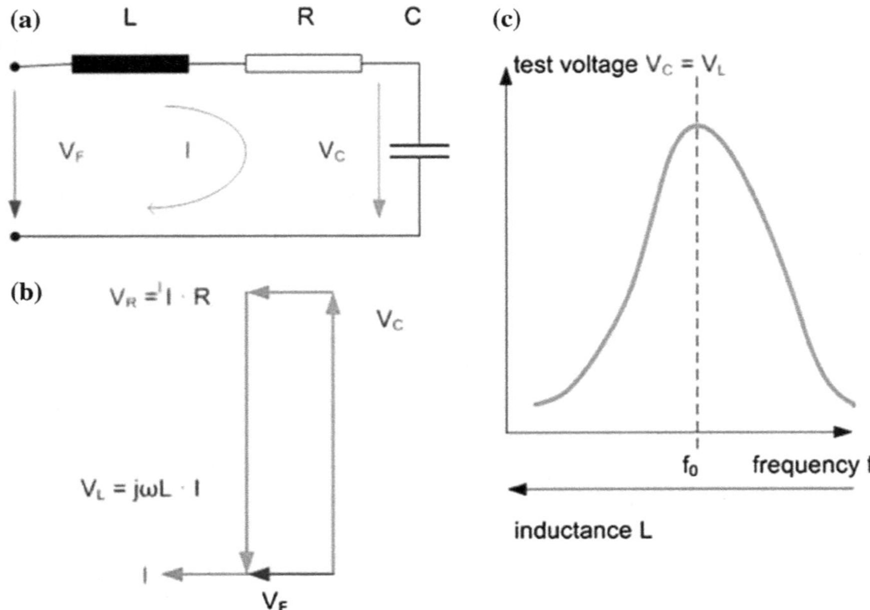

Fig. 3.17 HVAC resonant test circuit, **a** equivalent circuit, **b** phasor diagram, **c** resonant curve

When the resistive losses $P_R = V_R \cdot I$ are replaced by feeding a voltage of the natural frequency f_0 into this circuit, the system operates in series resonance. Then, the voltage is increased according to the so-called *quality factor Q*, which is the relation of the capacitive test power $S_C = V_C \cdot I$ to the resistive loss power P_R, and this is also identical with the relation between the test voltage V_C and the feeding voltage V_F (Figs. 3.17a and 3.18a):

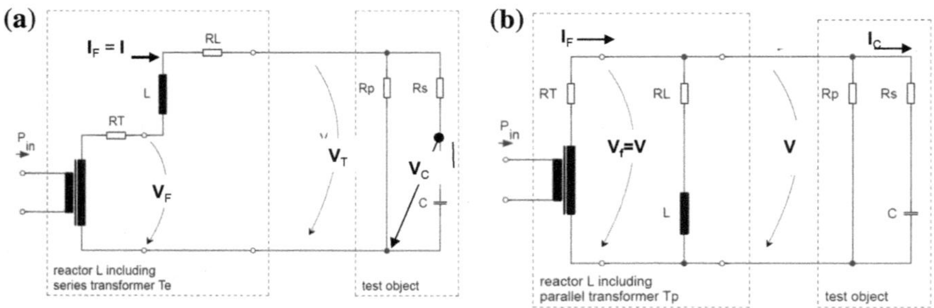

Fig. 3.18 Equivalent circuit of resonant circuits including exciter circuit, **a** series resonant circuit **b** parallel resonant circuit

$$Q = \frac{S_\mathrm{C}}{P_\mathrm{R}} = \frac{V_\mathrm{C}}{V_\mathrm{F}} = \frac{\frac{I}{\omega_0 C}}{I \cdot R} = \frac{1}{\omega_0 C \cdot R} = \sqrt{\frac{L}{C \cdot R^2}} \approx \frac{V_\mathrm{T}}{V_\mathrm{F}}. \tag{3.10}$$

There are two possibilities to reach resonance (Fig. 3.17c): One can adjust the inductance of the HV reactor until the natural frequency f_0 becomes identical to the frequency f_F of the feeding voltage (inductance-*tuned resonant circuit*: ACRL, see Sect. 3.1.2.2) or one can feed the circuit via a frequency converter with a voltage of the natural frequency of the test circuit determined before (*frequency-tuned resonant circuit*: ACRF, see Sect. 3.1.2.3). For both cases, one has to consider that the tuning ranges are limited, because the resonance condition $f_\mathrm{F} = f_0$ cannot be fulfilled for very low or very high capacitive load (there is no oscillation without a capacitive load). In case of a breakdown of the test object, the load capacitance changes drastically, the system goes out of resonance and the following "short-circuit current" is negligible. This means, in case of breakdown, the test object does not burn out.

The feeding into the resonant circuit is realized by an *"exciter transformer"* which adapts the voltage from the mains to the required output voltage $V_\mathrm{C} = Q \cdot V_\mathrm{F}$. The exciter transformer is designed according to the maximum necessary feeding power $P_\mathrm{F} = P_\mathrm{R}$ and voltage V_F, both following for the minimum assumed quality factor Q. In a series resonance circuit, the HV reactor is connected in series with the HV winding of the exciter transformer (Fig. 3.18a). Series resonant circuits are the most important application of the resonance principle.

The so-called *parallel resonant circuits* are applied for special tests at very huge capacitive test objects of relatively low voltage, e.g. capacitors for capacitor banks. In this case, the exciter transformer is switched in parallel to the HV reactor (Fig. 3.18b). This means that the voltage is fully controlled by the transformer. In case of resonance, the whole capacitive current is compensated by the reactor. Consequently, the *quality factor* becomes

$$Q_\mathrm{p} = \frac{S_\mathrm{c}}{P_\mathrm{R}} = \frac{I_\mathrm{C}}{I_\mathrm{F}} \approx \frac{I_\mathrm{T}}{I_\mathrm{F}}. \tag{3.11}$$

A parallel resonant circuit is a HV test transformer circuit that is fully compensated on the HV side. It is usual to combine the test transformer with a resonant reactor to one unit in one tank. In that case, the magnetic core is designed with gaps. Because of the HV transformer, the test voltage may be remarkably disturbed by harmonics whereas series resonant circuits supply a fine sine wave. In the following, only the more important *series resonant circuits* will be considered.

The complete test system may be understood as the parallel connection of the feeding circuit (with the quality factor Q_F) and the test object (with Q_T) (Fig. 3.18a). The combination of the two quality factors delivers the total quality factor of the test circuit:

$$Q = 1 \left/ \left(\frac{1}{Q_F} + \frac{1}{Q_T} \right) \right. = \frac{Q_F \cdot Q_T}{Q_F + Q_T}. \tag{3.12}$$

The quality factor is a related parameter, and therefore depending on the test condition. The value of Q_F may be considered as a fixed value for a fixed frequency (depending on the design between $Q_F = 50$ and 200). When, e.g., a long XLPE cable system is tested on site, the capacitive test power and Q_T are high, then the resulting quality factor is determined by Q_F. Vice versa in case of a type test on a short-cable sample using a water termination (see Sect. 3.6), the test object capacitance is low and the parallel resistance of the water termination causes additional resistive losses, then the resulting quality factor is determined by Q_T. Therefore, for the correct design of a resonant test system, the conditions of the later application should be well known.

Example A series resonant test system shall be selected for HVAC routine tests on XLPE cables with a length of 400 m at $V_T = 280$ kV/50 Hz. The total load capacitance is that of the cable (400 m $\cdot$ 260 pF/m $= 104$ nF) plus that of the capacitive divider, coupling capacitor and basic load (4 nF) $C_T = 108$ nF. The parallel water resistance of the termination has active losses of 22 kW (cooling power demand), which corresponds to a resistance R_p of 3.6 MΩ. Both parameters deliver the following parameters:

• Quality factor of the test object	$Q_T = 2\pi f \cdot C \cdot Rp = 121$
	(for parallel capacitance and resistance: $Q = I_c/I_r$!!)
• Quality factor of the power supply	$Q_F = 70$ (assumption)
• Total quality factor	$Q = 44.3$ (Eq. 3.12)
• Required test power	$S_T = 2\pi f \cdot C \cdot V_T^2 = 2650$ kVA
• Necessary exciter voltage	$V_F = V_T/Q \approx 6.3$ kV
• Necessary exciter power	$P_F = P_T = S_T/Q = 60$ kW

The *exciter transformer* should deliver an *"exciting" voltage*, which is well adapted to the required output test voltage and test power. The exciter transformer must be designed to deliver the active feeding power P_F that follows from the capacitive test power S_T and the quality factor Q by the equation $P_F = S_T/Q$.

If the exciter transformer would have only one voltage output and a test voltage should be generated, which is only 50% of the rated voltage, then only a quarter of the test power is available. The full test power can be used at the 50% test voltage level, when the exciter transformer has a tap for 50% output voltage. An optimum adaptation to both, output voltage and power, requires an exciter transformer with a well selected number of taps. Its rated voltage has to be selected according to the maximum voltage necessary for the HVAC test, and the test is controlled via a regulator transformer. As the above example shows, the voltages and powers necessary for exciting the oscillation circuit are low compared with those of the test object.

First ideas of HV resonant circuits came up by Charlton et al. (1939), a certain technical perfection has been reached for inductance-tuned (ACRL) test systems after 1960 (Reid 1974). Frequency-tuned (ACRF) test systems have been proposed by Zaengl and co-workers (Bernasconi et al. 1979; Zaengl et al. 1982; Schufft et al. 1995). In the following, the two basic types of series resonant circuits are described and compared.

3.1.2.2 Inductance-Tuned Resonant Circuits of Fixed Frequency (ACRL)

The inductance of a reactor can be changed when its magnetic core has a gap of an adjustable width (Fig. 3.19). The inductance of the reactor is proportional to the square of the number of the turns w^2 of the winding, the permeability μ_0 of the insulating material in the gap, the area A of the cross section of the magnetic core at the gap as well as inversely proportional to the width a of the gap (k is a factor of proportionality)

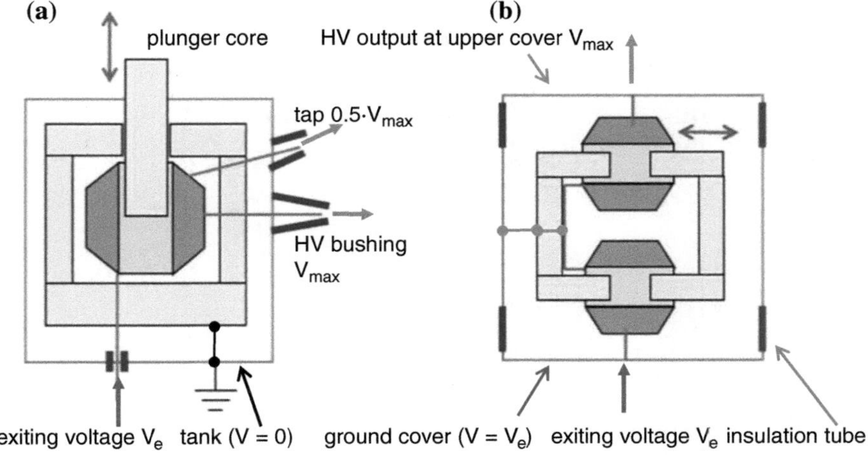

Fig. 3.19 Active part of a tuneable reactor, **a** shell-type core with one gap of width a, **b** two half-cores with two half gaps, each a/2

$$L = \frac{k \cdot \mu_0 \cdot w^2 \cdot A}{a}.$$

(3.13)

The maximum inductance L_{max} is connected with the minimum gap a_{min} and enables resonance (Eq. 3.9) with the minimum load capacitance C_{min}. Vice versa the maximum load C_{max} is related to the minimum inductance L_{min} and the maximum gap a_{max}. With increasing gap width, the magnetic stray flux and also the losses increase, too, and lead to a certain non-linearity of the inductance and a decrease of the quality factor. These—and also some technological limitations—cause the technical useable change of the gap up to $a_{max}/a_{min} \approx 20$. Based on Eqs. 3.9 and 3.13, the limits in a load-output voltage diagram (Fig. 3.20) are also $C_{max}/C_{min} \approx 20$:

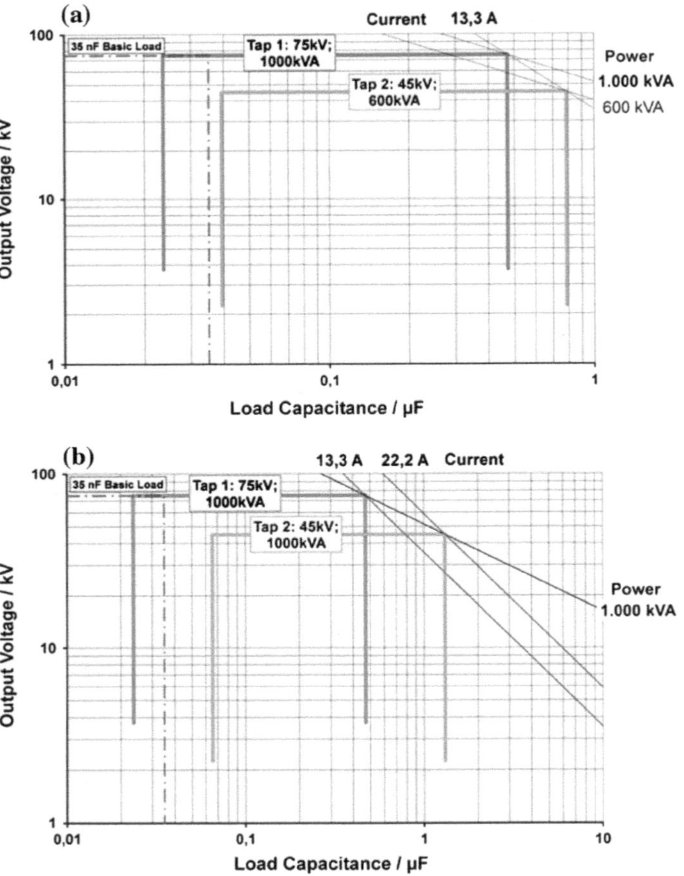

Fig. 3.20 Load diagram of an inductance-tuned resonant circuit, **a** design for a tap of identical (constant) current, **b** design for a tap of identical (constant) power

$$\frac{C_{max}}{C_{min}} = \frac{L_{max}}{L_{min}} \approx \frac{a_{max}}{a_{min}} \tag{3.14}$$

Example: Determine the minimum and maximum inductance of an ACRL test system for 50 Hz routine tests on cables with lengthes between 100 and 2000 m (capacitance: 220 nF/km)!

With f = 50 Hz and C_{min} = 22 nF and C_{max} = 440 nF, the Eq. (3.9) $L = 1/(C \cdot (2\pi f)^2)$ delivers L_{max} = 460 H (for testing the shortest cable) and L_{min} = 23 H (for the longest cable). This can be realized by a gap of the reactor core adjustable between 200 and 10 mm.

The precise adjustment of the gap—corresponding to the maximum of the resonant curve—is reached when the phase angle between test voltage and test current is maximized. A more robust criterium delivers the phase angle between feeding voltage at the input of the reactor and the test current. When this angle becomes zero the maximum of the resonant curve is reached. The higher the quality factor, the steeper is the resonant curve and the higher are the requirements to the accuracy of the gap control.

Usually a resonant circuit is equipped with a *basic load capacitor* $C_b \geq C_{min}$, which enables resonance and operation without a capacitive test object. This is important when the resonant test system itself is to be checked. The basic load capacitor is used as a voltage divider and/or a coupling capacitor for PD measurement or as a part of a HV filter.

Tank-type reactors: The design of a reactor depends also on the insulation and the thermal conditions. Often tank-type reactors are designed with taps for voltages V_i lower than their rated voltage V_m. Then, the winding can be designed for constant current I, which means it is made with one type of wire and the power at lower voltage is according to the ratio of the voltage V_i/V_m lower (Fig. 3.20a), the winding is made of identical sub-windings. It can also be designed for constant power. Then, the current ratio I_i/I_m must increase inversely to the voltage ratio V_i/V_m. Higher currents require thicker wires for the winding, the winding is made of different sub-windings.

For the tank-type design (Fig. 3.21), a *shell-type core* (Fig. 3.19a) is useful, because the very stable magnetic core of one tuneable central leg and three-fixed outer legs can overtake the high mechanic forces (Spiegelberg et al. 1993). The mechanic stability means less vibrations that results in a stable output voltage and less acoustic noise. The layer winding is designed similar to that described above for oil-insulated test transformers. As shown in Fig. 3.20, the winding of tank-type reactors is often designed with taps. If there is only one additional tap, a second bushing may be an economic solution. In case of three or more taps, the reactor is equipped with a no-load tap-changer. The advantages of tank-type reactors can be compared with those of tank-type test transformers (Table 3.1), but they are not used for cascading for higher voltages. Their main application is for *cable testing* up

(a)

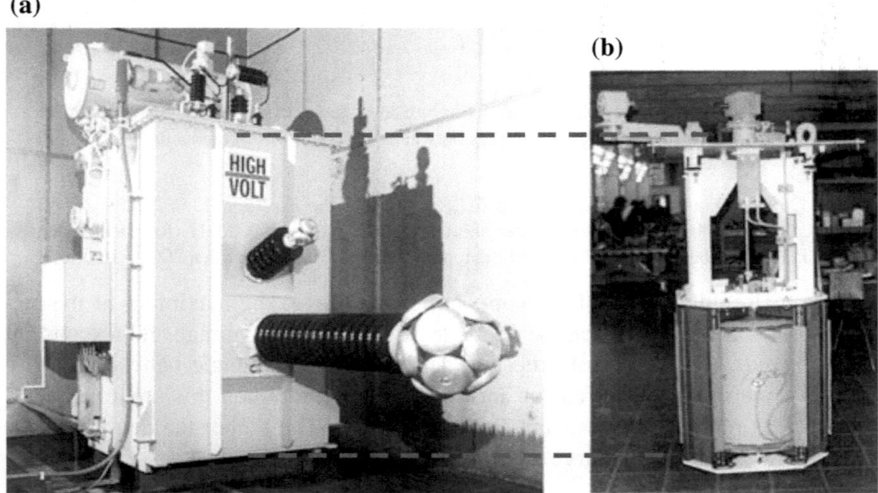

Fig. 3.21 Tank-type reactor (250–100 kV, 2000 kVA). **a** Reactor with two bushings, **b** active part with drive for core adjustment

to 400 kV including PD measurement. For that application with a Faraday cage, they can be arranged outside the cage, the tank is flanged to the grounded, metallic wall of the cage and only their bushing is arranged inside the test area.

Cylinder-type reactors: For the cylinder design of reactors (Fig. 3.22), a magnetic core of two half-cores (Fig. 3.19b) is applied. Depending on the design, both or only one half-core are adjusted by gears for the gap. The gap is designed for a certain distance—corresponding to the required inductance—which gives resonance with the capacitive object under test (Reid 1974). The winding is divided into two parts. The beginning of the lower part of the winding is connected with the grounded lower cover of the cylinder. The end of the upper part is connected with the upper cover on HV potential. The active part is behind the metallic middle part of the cylinder (grey on Fig. 3.22a) that is on half potential as the magnetic core itself. The metal cylinder controls the electric field inside for the active part in oil and outside for the surface of the reactor in air. The (blue) insulating cylinders overtake the role of the bushings of the reactor. The outer field strength distribution is controlled by the toroid electrodes. For a good utilization of the limited space inside the cylinder, the windings have a rectangular cross section (Fig. 3.23). Care is taken for the high mechanic stability of the wire layers by a special compact insulation.

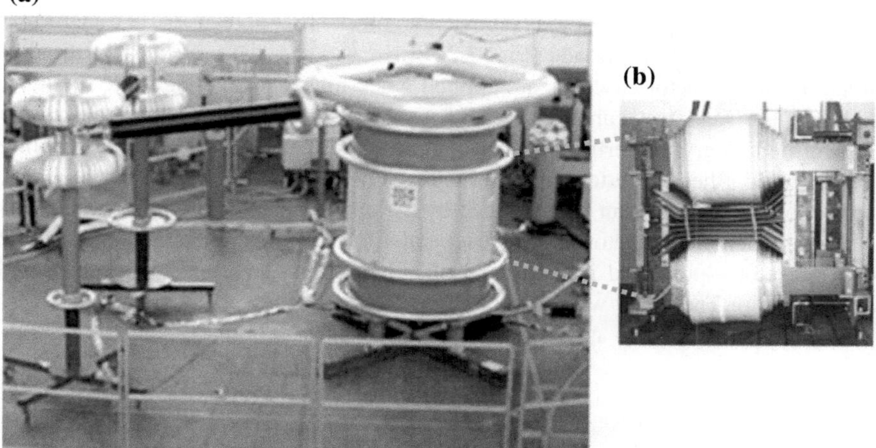

Fig. 3.22 Cylinder-type reactor (400 kV, 14 MVA). **a** Complete test reactor, **b** active part with two coils and one adjustable half-core

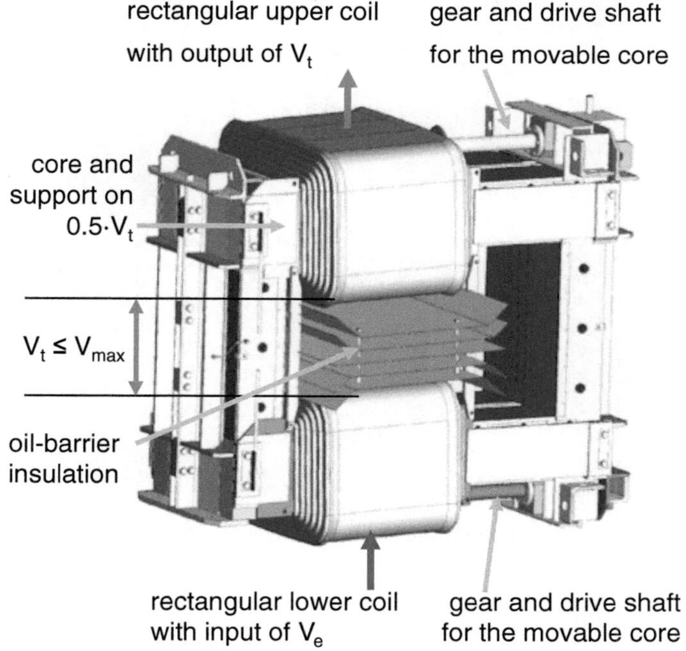

Fig. 3.23 Rectangular layer winding for a cylinder-type reactor

Reactor cascades: Cylinder-type reactors can easily be connected in series for higher voltages by stacking one above the other. In opposite to a transformer cascade such a reactor cascade (Fig. 3.24) increases the test power by each additional reactor. Reactor cascades are applied when higher test voltages are required, with not too high test currents for *GIS testing* and for *applied voltage tests* on transformers. Very powerful types of cylinder-type reactors with external cooling are also used when cable tests require voltages higher than the mentioned 400 kV.

Higher test currents can be generated by the parallel connection of reactors. This is simple when each reactor is placed separately. But the parallel connection of the single reactors—arranged in a cascade of one column—requires special bars and workmanship. The effort with up to three modules remains reasonable.

Fig. 3.24 Reactor cascade 1200 kV of three identical modules 400 kV each

3.1.2.3   Frequency-Tuned Resonant Circuits of Variable Frequency (ACRF)

An oscillating circuit with a fixed reactor of inductance L and of fixed capacitive test object C_0 has a fixed *natural frequency* f_0 according to Eq. 3.9:

$$f_0 = \frac{1}{2\pi\sqrt{L \cdot C_0}}.$$

It must be excited with that frequency for resonance. For a different test object, a different natural frequency appears. The tuning range of this resonant circuit depends on the acceptable frequency range. The maximum load capacitance C_{max} determines the minimum test frequency f_{min} for a pre-given inductance L of the reactor and vice versa. Usually, the standards for HVAC testing allow a certain frequency range and this determines the load range:

$$C_{max} = \frac{1}{(2\pi f_{min})^2 \cdot L} \quad \text{and} \quad C_{min} = \frac{1}{(2\pi f_{max})^2 \cdot L}. \tag{3.15}$$

The tuning range follows to

$$\frac{C_{max}}{C_{min}} = \left(\frac{f_{max}}{f_{min}}\right)^2, \tag{3.16}$$

and the maximum current to

$$I_{max} = 2\pi f_{min} \cdot C_{max} \cdot V_T = V_T \cdot \sqrt{\frac{C_{max}}{L}}. \tag{3.17}$$

From Eqs. 3.9, 3.15 and 3.16, the operation characteristic of an ACRF test circuit (Fig. 3.25) can be calculated. With increasing load from C_{min} to C_{max}, the natural frequency is decreasing from f_{max} to f_{min} (Fig. 3.25a). At the same time, the current is increasing with the capacitive load. It may happen that the required current I_{max} is higher than the current that results from the thermal design I_{max}^*. Then, the current must be limited to the technically acceptable value I_{max}^* by reduction of the test voltage in the relevant load range (Fig. 3.25b). The voltage can be generated in the range between minimum and maximum load, possibly with certain reductions for a necessary current limitation.

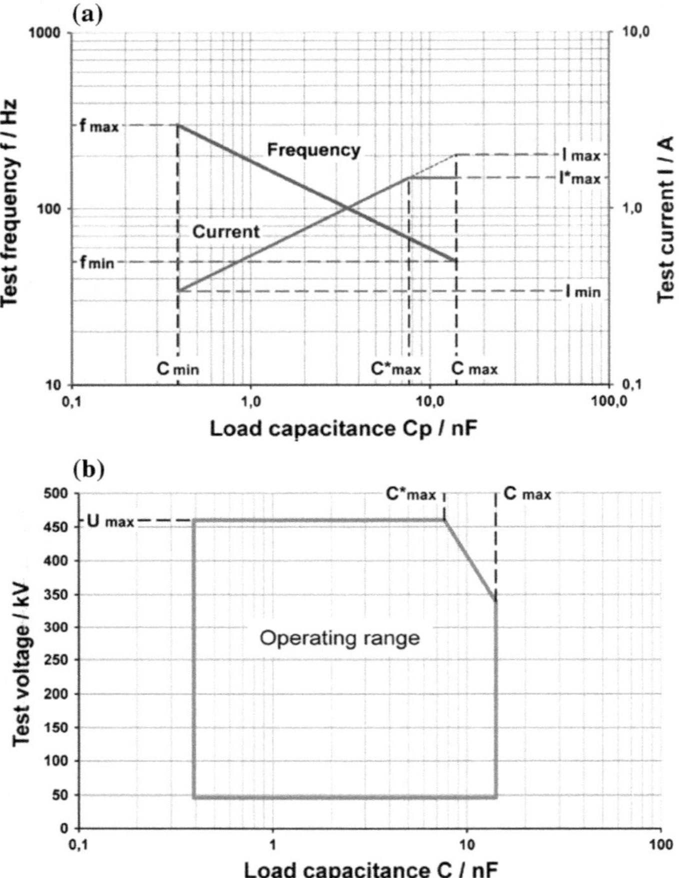

Fig. 3.25 Operation characteristic of a frequency-tuned resonant system. **a** Frequency and current versus load capacitance, **b** voltage operating range versus load capacitance

The maximum *capacitive test power* at f_{min} follows to $S_{\mathrm{T}} = V_{\mathrm{T}} \cdot I_{\mathrm{max}}$, but this is not a typical parameter because it depends on the test frequency. Therefore, it is usual for comparisons to use the 50 Hz (*or* 60 Hz) *equivalent power*

$$S_{50} = \frac{50}{f_{\mathrm{min}}} \cdot S_{T}. \tag{3.18}$$

The necessary *feeding power* that is equal to the loss power of the circuit can be calculated by the help of the quality factor from the test power S_{T}:

$$P_F = S_T/Q. \tag{3.19}$$

The *quality factor* of a frequency-tuned resonant test system is higher than that of an inductance-tuned system, because the design of a fixed reactor enables reduced stray fluxes, and consequently lower losses compared with a tuneable one. It can be assumed with values in the order $Q = 50$–150 (for ACRL systems it is about $Q = 20$–60).

It should be noted that the above mentioned feeding power is related to mains of a stiff network. If, for instance, on site a *Diesel motor-generator set* is used for the power supply, it is recommended to apply one with a rated power about three times higher than the calculated P_F.

Example: For on-site tests at cable systems, a test frequency range of 20–300 Hz is accepted. An HVAC resonant test system based on a reactor 160 kV/80A/16 H is available for that frequency range. Which maximum capacitance can be tested with this system at its rated voltage? Which basic load is necessary to check the test system without test object? Which length of cable can be tested when the specific capacitance of the cable is 200 nF/km? Which feeding power is required when a quality factor $Q = 100$ is assumed?

Maximum load by Eq. 3.15

$$C_{max} = \frac{1}{(2\pi \cdot 20\mathrm{s}^{-1})^2 16 \ \mathrm{VsA}^{-1}} = 3962 \ \mathrm{nF}$$

Basic load by Eq. 3.16

$$C_{min} = C_{max}\left(\frac{f_{min}}{f_{max}}\right)^2 = 3962 \cdot \frac{1}{225} = 18 \ \mathrm{nF}$$

At 160 kV, a capacitance of 18 nF is quite large and expensive. One can think about to use, e.g., a capacitance of that value, but only for 50 kV and check the performance of the test system at that reduced voltage.

Maximum current according to Eq. 3.17

$$I_{max} = 160 \times 10^3 \ \mathrm{V} \cdot \sqrt{\frac{3.962 \ \mathrm{AsV}^{-1}}{10^6 \cdot 16 \ \mathrm{VsA}^{-1}}} = 79.6 \ \mathrm{A}$$

This current is below the rated value, and there is no limitation due to the current. The testable length of the cable results $C_{Cable}/200$ nF/km = (3962 nF – 18 nF)/200 nF/km = 19.7 km.

Required feeding power by Eq. 3.19

$$P_F = \frac{I_{max} \cdot V_T}{Q} = \frac{79.6 \ \mathrm{A} \cdot 160 \times 10^3 \ \mathrm{V}}{100} \approx 130 \ \mathrm{kVA}$$

This power must be supplied from a stiff network. In case of the application of a mobile power supply, a 400 kVA motor-generator set should be used.

The necessary feeding power of variable frequency is generated by static *frequency converters* (Fig. 3.26). The three-phase power from the mains is first rectified to a direct voltage, and then a circuit of power transistors (IGBT's) or of SKIIP modules converts the DC voltage to a rectangular AC voltage (Figs. 3.26a, 3.27). The oscillation circuit acts as a HV filter and generates a fine sine wave. There is a phase shift of 90° between the rectangular voltage, which replaces the active power losses of the circuit and the test voltage at the capacitive test object. With respect to the high quality factors of ACRF test systems, the adjustment of the

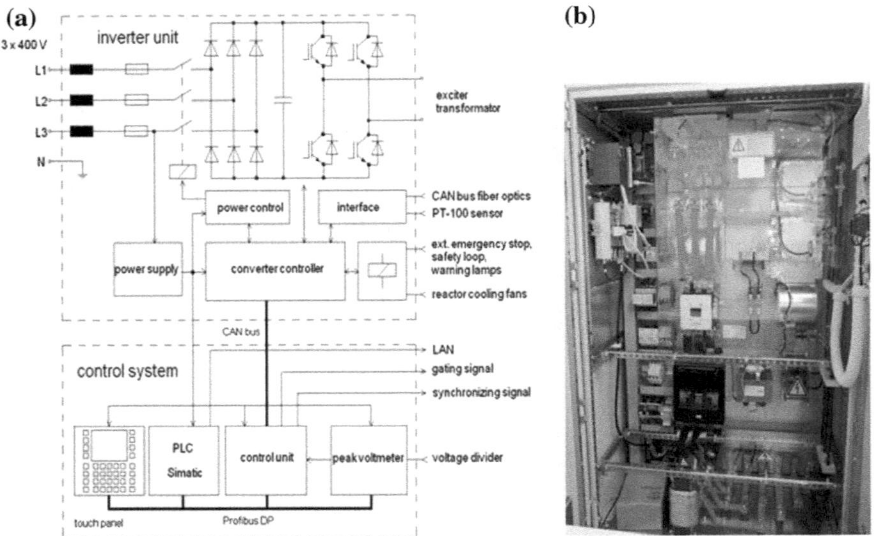

Fig. 3.26 Frequency converter for a frequency-tuned resonant test system. **a** Principle circuit diagram, **b** cubicle of the frequency converter

Fig. 3.27 Rectangular output voltage of the frequency converter for low (*above*) and high (*below*) sinusoidal test voltage on the capacitive test object

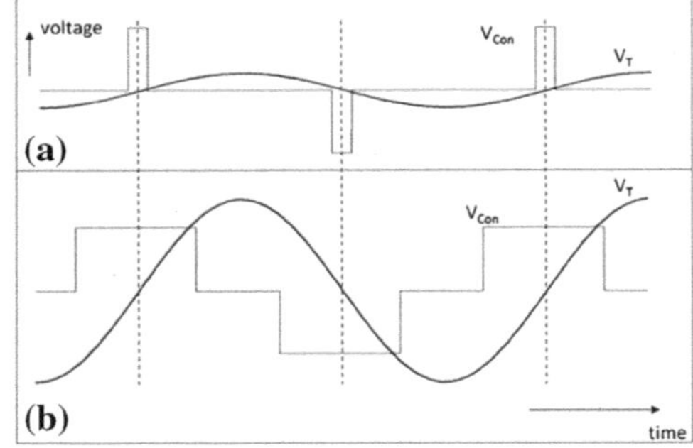

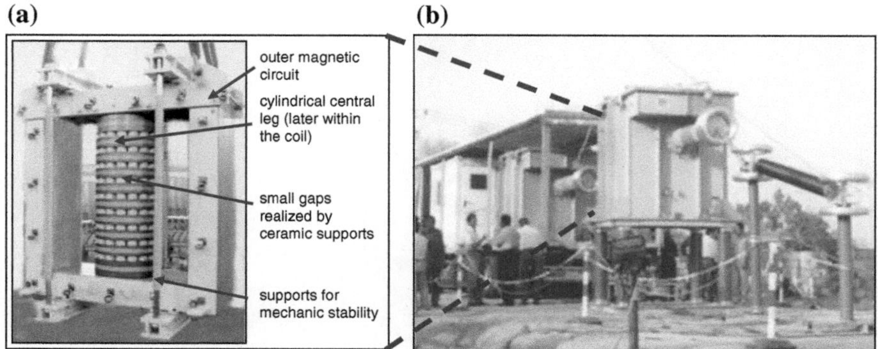

Fig. 3.28 Tank-type fixed reactor with oil–paper insulation. **a** Principle design of the magnetic core, **b** two reactors in series for test voltages up to 520 kV

frequency must be guaranteed with high precision (better than ± 0.1 Hz). The height of the voltage is determined by pulse-width modulation. The frequency converter—together with the control-and-measuring system—can be arranged in one cubicle (Fig. 3.26b) or—for lower power—in a movable desk (Fig. 3.28b, see also Fig. 10.21).

Tank-type fixed reactors: As tank-type test transformers (Fig. 3.5), tank-type fixed reactors (Fig. 3.28) are related to high test power (50 Hz equivalent power up to the order of 50 MVA). Their design with oil–paper insulation is derived from that of test transformers. In the central leg, the magnetic core has several small gaps of very low stray flux. Therefore, the fixed reactors have a higher quality factor than tuneable reactors. The tank of a single reactor is grounded, but they can be arranged in cascades when the higher stages are arranged on insulating support (Fig. 3.28b). The main application of these reactors is *on-site testing* of cable systems after laying or repair. But more and more they are also used for *routine testing* of the long submarine cables and for *pre-qualification tests* of newly developed cables (see Sects. 3.6 and 7.3; Hauschild et al. 1997, 2002; Schufft et al. 1999; Gockenbach and Hauschild 2000).

Cylinder-type fixed reactors: The design of these cylinder-type reactors (Fig. 3.29) is similar to that of the cylinder-type transformers (Fig. 3.10). They are for lower test power, but well-suited for reactor cascades by simple stacking one above the other. Their application is therefore for objects of higher required test voltages at lower required test power: Therefore, they are used for on-site testing of gas-insulated systems (GIS and GIL) and for *applied voltage tests* of power transformers, both in factories and on site (see Sects. 3.2.3, 3.2.5 and 10.4). Additionally small, lightweight reactors are realized with a bar core and a relatively small oil volume. The combinations of several reactors in series and/or in parallel (Fig. 3.29b) enable a good adaptation to different test objects (Kuffel and Zaengl 2006; Hauschild et al. 1997). Their low thermal capacity limits both, the rated current and the duty cycle remarkably.

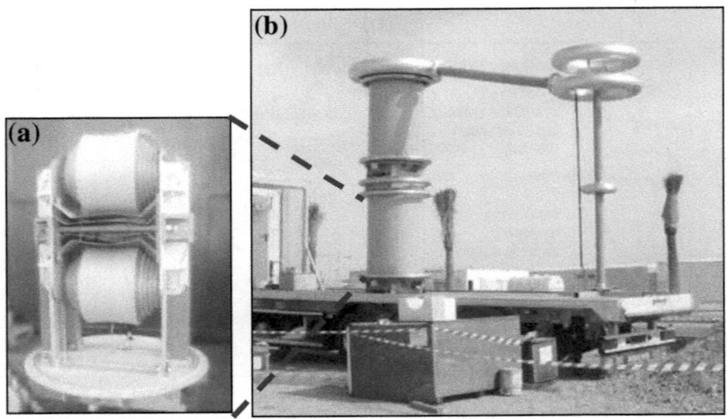

Fig. 3.29 Cylinder-type fixed reactor with oil–paper insulation. **a** Principle design of the active part, **b** two reactors 250/10 kV in series for a test voltage of 500 kV

Fixed reactors with SF₆ -impregnated foil insulation: This type of reactor is arranged in a non-magnetic metal enclosure (aluminium) that can be flanged directly to a gas-insulated test objects (Fig. 3.30). It is designed without iron core but with a divided winding of a very high number of turns. The single layers of each half winding are arranged on a conducting cylinder each. The SF_6 insulation of the gas gap between the two cylinders must be designed for the rated voltage of the reactor, whereas the voltage between the outer layers and the enclosure corresponds to the half of the full voltage. The thermal design is difficult because of the large number of turns and the control of the magnetic flux. This type of reactors is for rated voltages up to 750 kV per unit, test currents of few Amps and short-time operation. It can be recommended to clarify the necessary load-time characteristic of these reactors carefully. Their application is for testing GIS and their components (see Sects. 3.6 and 7.3). An improvement of both, rated voltage and duty cycle, is expected if the reactor is designed with a magnetic feedback (acting as a magnetic core, (Belinski et al. 2017) This allows also the application of a steel tank.

3.1.2.4   Comparison of ACRL and ACRF Test Systems

The comparison of the most important component in a resonant circuit, the *HVAC reactor*, shows that a *fixed reactor* is much simpler, more compact, more robust and consequently cheaper than one with tuneable inductance. The principle circuit diagram of both solutions (Fig. 3.31) shows that the *ACRF test system* has less components, also because of the cubicle that includes the switches, the frequency converter and the control.

But for most tests in factory, there is the disadvantage of the variable frequency. The frequency tolerance for HVAC test voltages between 45 and 65 Hz, accepted

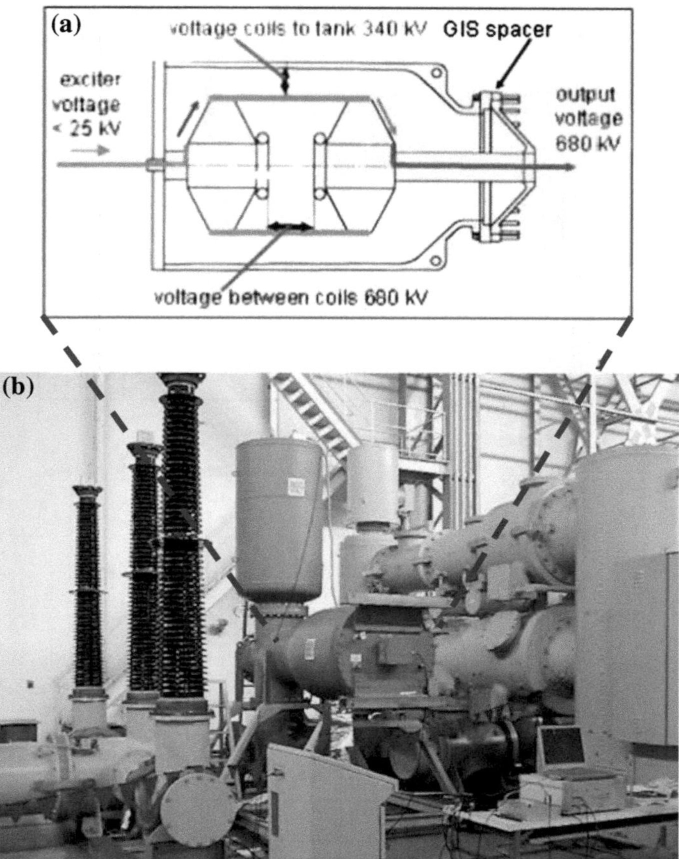

Fig. 3.30 ACRF test system for GIS routine tests. **a** Principle design of the SF$_6$-impregnated, foil-insulated reactor, **b** reactor (*blue, horizontally*), coupling capacitor (*blue, vertically*) and frequency converter (*small gray desk in the mid*)

by IEC 60060-1: 2010, is too small for a sufficient tuning range for the load. It corresponds only to a factor $(f_{max}/f_{min})^2 = 2$ between highest and lowest load.

Table 3.2 summarizes a comparison between the two principles for the application of on-site testing of HV and EHV cables. There is a *frequency range* of 20–300 Hz accepted (IEC 62067; IEC 60840) and used by the ACRF system, but not helpful for a system of fixed frequency of 50 or 60 Hz. The extended frequency range with the lower limit 20 Hz enables a 2.5 times higher test power (or a 2.5 times higher capacitance of the test object) of the ACRF system than the ACRL system at identical voltage. Because of the lower losses of the reactor, the *quality factor* of the ACRF system is about twice of that of the ACRL system. The frequency range enables an ACRF *load range* about ten times wider than that of an ACRL system. Both, wider frequency range and higher quality factor result in a

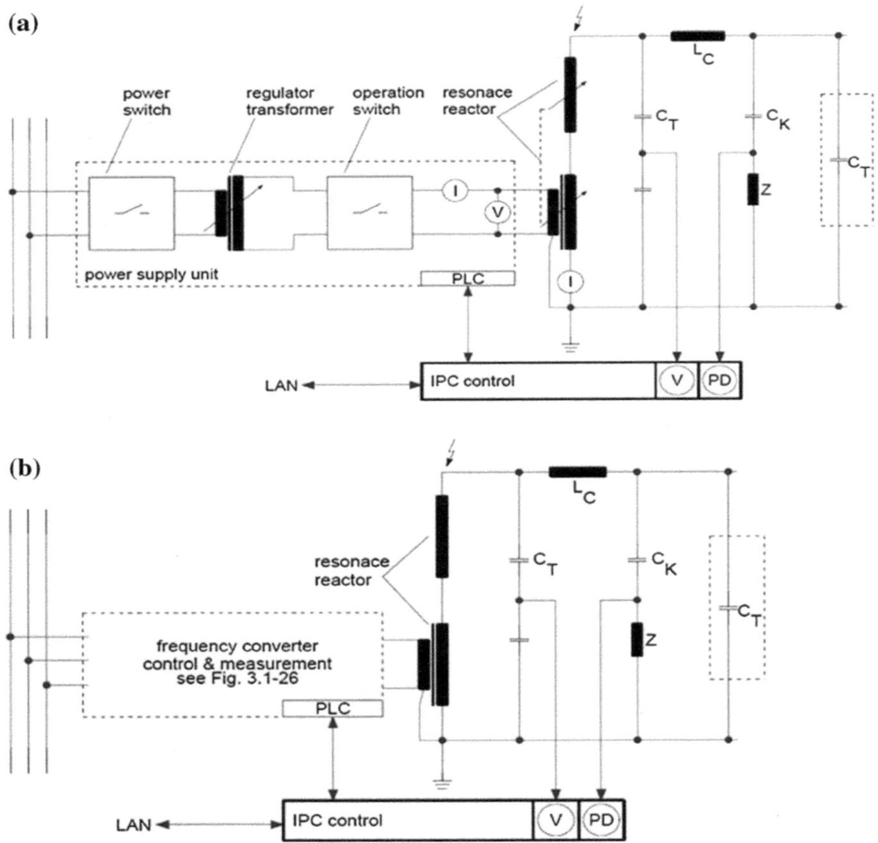

Fig. 3.31 HVAC test systems based on series resonant circuits. **a** Equivalent circuit of a complete ACRL test system, **b** equivalent circuit of a complete ACRF test system

feeding power that is about five times lower for ACRF than for ACRL test systems. The power supply of an ACRL system requires single- or two-phase operation. The *frequency converter* of an ACRF system uses a three-phase power supply that is a symmetric and lower load and simpler to handle on site. For rough transportation and operation conditions on site, the absence of moving parts is an advantage of ACRF systems, whereas the ACRL systems use regulator transformers and tuneable reactors. As also seen from Table 3.2 and Fig. 3.31, the number of components of an ACRF system is lower.

The comparison shows clearly that an ACRF test system is always the better economic solution if the frequency range is sufficient for an acceptable tuning range. This is the case for most on-site tests (see Chap. 7), but it can be applied also for some factory tests: The applied AC voltage test of power transformers (IEC 60076-3: 2012) shall be performed at a frequency of >40 Hz, when an upper limit

Table 3.2 Comparison between ACRL and ACRF test systems for cable-testing on site

Resonant circuit	ACRL (inductance-tuned)	ACRF (frequency-tuned)
Frequency	f_L = 50 Hz (60 Hz)	f_F = 20–300 Hz (Cables)
Maximum test power	$S_{Lmax} = 2\pi f \cdot CU^2$	$S_{Fmax} = 2.5\ S_{Lmax}$
Quality factor	Q_L = 40–60	Q_F = 80 to > 120
Load range	$C_{max}/C_{min} = L_{max}/L_{min} \approx 20$	$C_{max}/C_{min} = (f_{max}/f_{min})^2 \approx 225$
Feeding power	$P_{eL} = (2\pi f\, CU^2)_{qL}$	$P_{eF} = P_{eL} \cdot (f_F/f_L) \cdot (Q_L/Q_F)$
Power supply	Single or two phase	Three phase
Weight-to-power ratio (kg/kVA)	3–8	0.8–1.5
Components with moving parts	Tuneable reactor, regulator, transformer	None
Number of main components	Six components: Tuneable reactor, HV divider/PD coupler, excite transformer, switching cubicle, control-and-measuring rack	Four components: Fixed reactor, HV divider/PD coupler, exciter transformer, control-and-feeding unit

of 200 Hz is assumed, a sufficient load range of 1 to 25 can be reached. For super-long *HVAC submarine cables* (some 10 km long), no other solution than ACRF testing can be performed with acceptable economic effort. There are discussions for using the full frequency range 10–500 Hz given in IEC 60060-3 for on-site testing also for the factory tests (Karlstrand et al. 2005). A wider application of ACRF systems in factory tests requires the definition of acceptable frequency ranges. This can only be based on new research work on the influence of the frequency—e.g. in the range 10–100 Hz—on the discharge processes in the relevant insulations.

3.1.3   HVAC Test Systems for Induced Voltage Tests of Transformers (ACIT)

Above it has been shown that in a resonant test system, the capacitive test object, e.g. a HV cable, becomes a part of the HV generation circuit. Also in testing power or instrument transformers, the test object may become a part of the generation circuit. When it is energized on its low-voltage side, the test voltage on the HV side can be generated by this test object. This so-called *induced voltage test* (IEC 60076-3, Draft 2011) requires a feeding circuit that matches the low voltage to both, the required test voltage and the ratio of the transformer under test. This has to consider the non-linear magnetic behaviour of the transformer as a part of the test circuit.

Saturation effects of the iron core mean under test conditions that an increase of the test voltage causes an over-proportional increase in the test current (Fig. 3.32).

Fig. 3.32 Schematic
voltage-current characteristic
of a transformer under
induced voltage test

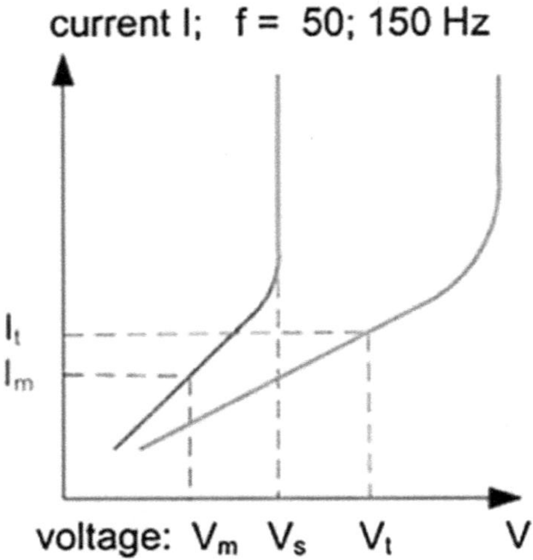

Finally, a test voltage remarkably higher than the operational voltage cannot be generated at operational frequency $f_m = 50$ or 60 Hz (Fig. 3.32). Because of the limitation of the magnetic flux Φ, the voltage can only be increased up to a certain value V_s when saturation becomes considerable. There is a frequency-depending relationship between voltage and magnetic flux:

$$V_S \sim f \cdot \Phi_S,$$

which shows that the "saturation" voltage V_s can only be increased for a certain "saturation" flux Φ_s when a higher test frequency is applied. Therefore, the induced voltage tests has to be performed at higher test frequencies f, usually higher than twice the operational frequency ($f \geq 2\,f_m$).

The transformer under test is not a simple load, its characteristic changes with the frequency (Fig. 3.33). The upper curve is the total test power demand S_t and the lower one the active power demand P_t. The difference between the two curves is the reactive power demand. At lower frequencies, the test object is an inductive load, at higher frequencies, it is a capacitive load for the feeding circuit. In between, there is the frequency of *self-compensation*, in Fig. 3.33 at $f \approx 160$ Hz. If this frequency is selected for a test, the required test power becomes a minimum.

In transformer testing not only the insulation of the windings must be tested at induced voltage, also the no-load and the short-circuit losses shall be measured with induced voltage of operational frequency. Therefore, the feeding system has to supply in minimum test voltages of two frequencies, operational frequency f_m and test frequency $f \geq 2\,f_m$. For that reason, a frequency converter for single-phase and three-phase operation is the basic element of the feeding circuit:

For a very long time *rotating frequency converters*, so-called *motor-generator (M/G) sets*, are applied and most transformer test fields are equipped with such sets (Fig. 3.34a). They consist of a motor of variable revolutions (traditionally a DC

Fig. 3.33 Example for the frequency–load characteristic of a power transformer under induced voltage test

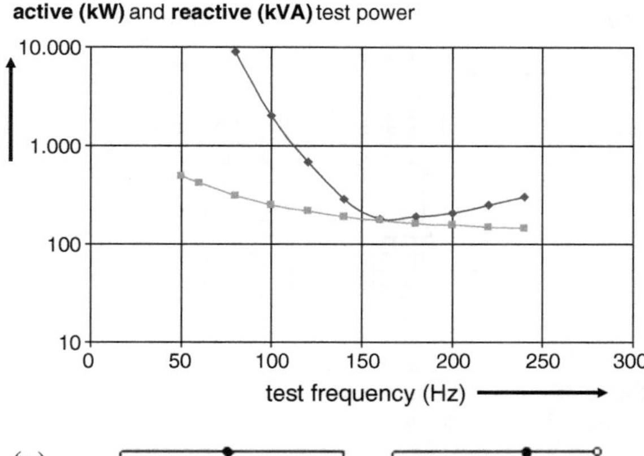

Fig. 3.34 Principles of frequency converters in circuits for induced voltage tests. **a** Rotating converter by M/G set, **b** static frequency converter (*SFC*)

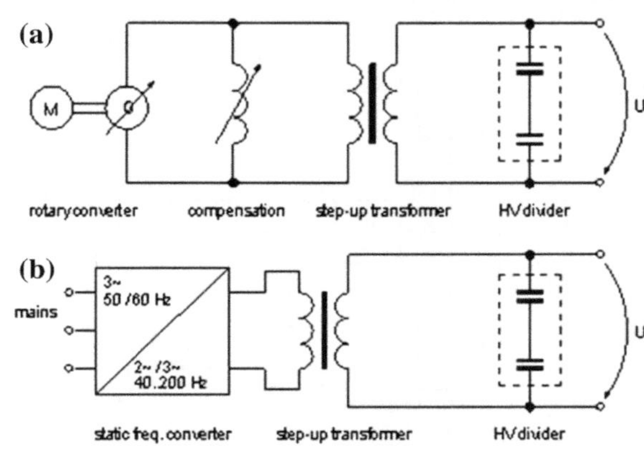

motor, nowadays a thyristor-controlled AC motor) that is mechanically coupled with a synchronous generator. The frequency is changed according to the revolutions of the motor, at the same time, the output voltage can be adjusted. In case of induced voltage tests (usually $f > 100$ Hz), the test object represents often a capacitive load that may cause dangerous over-voltages. To avoid such *self-excitations* ("*runaways*") exceeding the limits of the M/G set, sufficient inductive compensation must be applied. The output voltage of the M/G set is adapted to the necessary input voltage of the transformer under test by a matching (also called step up) transformer.

A modern alternative is the application of *static frequency converters* (SFC) based on power electronic modules (Fig. 3.34b; Kübler and Hauschild 2004; Hauschild et al. 2006; Martin and Leibfried 2006; Werle et al. 2006; Thiede and Martin 2007; Winter et al. 2007; Thiede et al. 2010). The converter itself (Fig. 3.35) uses a three-phase power supply. The voltage is rectified and then transferred into a single-phase or three-phase sine voltage by a unit of power transistors (IGBT) operating with pulse-width modulation. A sine-wave filter is very important for the suppression of the noise signals caused by the switching of the IGBT's. The SFC has a control for frequency and voltage height, which automatically adjusts the

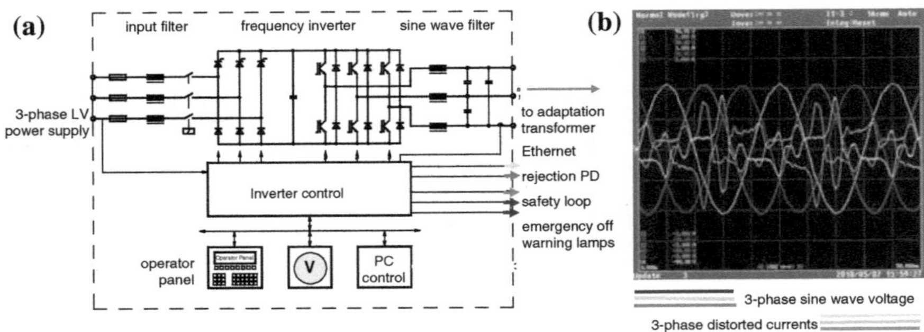

Fig. 3.35 Static frequency converter (*SFC*) for induced voltage tests (Thiede et al. 2010). **a** Principle circuit diagram, **b** sinusoidal voltages and disturbed currents

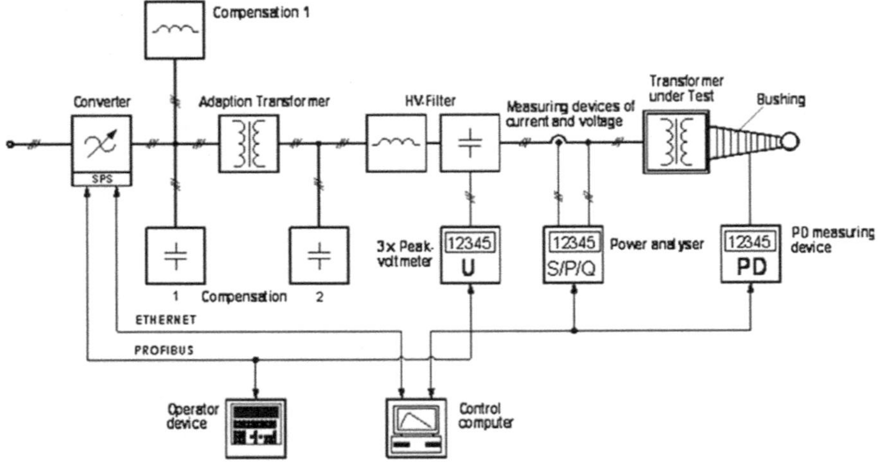

Fig. 3.36 Complete circuit for tests and measurements at induced voltage

converter to the above mentioned self-compensation. The *self-excitation* cannot happen. A SFC requires a matching transformer.

In opposite to a M/G set, a SFC is able to compensate also the capacitive power demand that may play a role in induced voltage tests at higher frequency (Fig. 3.33). In case of loss measurements at operational frequency of 50 or 60 Hz, for both principles an adjustable *capacitor bank* for compensation of the high inductive power demand is necessary. It is arranged after the matching transformer in the circuit for the induced voltage test (Fig. 3.36). Furthermore, *a medium-voltage filter (MVHF)* for the reduction of conducted noise signals is arranged between compensation unit and test object. Except for the two different frequency converters, the elements of the circuit may be identical (Fig. 3.37). Table 3.3 compares therefore only the two converters.

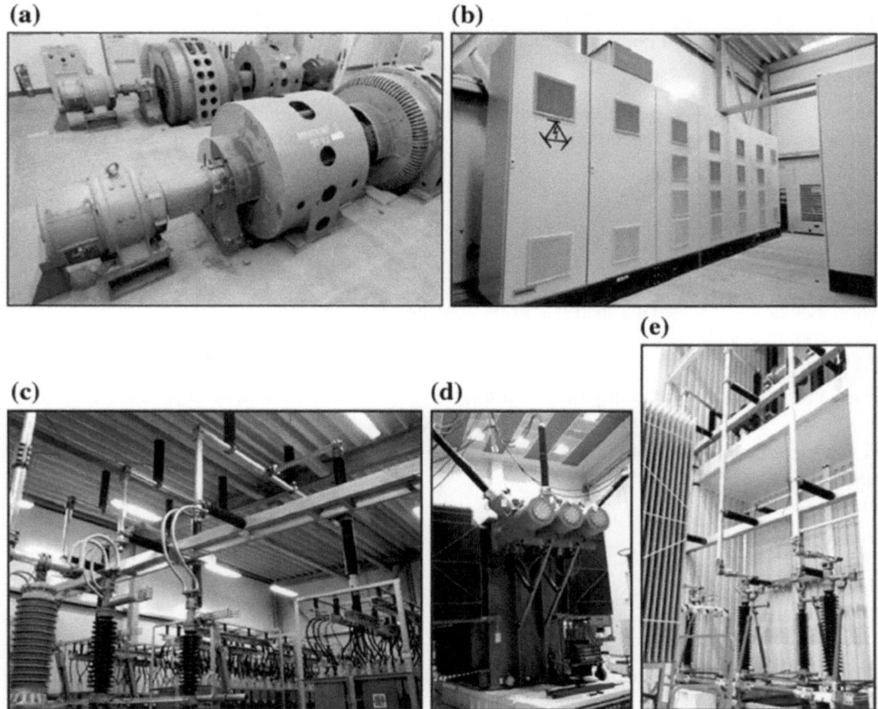

Fig. 3.37 Components of a circuit for induced voltage testing (Courtesy of Siemens Transformers Dresden). **a** Motor-generator (*M/G*) set, **b** static frequency converter (*SCF*), **c** top of matching transformer and capacitor bank, **d** and **e** transformer under test and medium-voltage (*MV*) filter

3.1.4   HVAC Test Systems of Variable Frequencies Based on Transformers (ACTF)

A frequency converter as shown in Fig. 3.35 can also be used as the power supply for a test transformer or a transformer cascade. ACTF test systems would enable *HVAC tests of pre-selected frequency* independent on the character of the test object (capacitive, inductive or resistive).

> **Note** Have in mind, the test frequency for frequency-tuned resonant circuits is not a fixed one and determined by the inductance of the reactor and the capacitance of the test object (Sect. 3.2.2). In such oscillating ACRF circuits only reactive test power can be generated. When the HVAC test voltage is generated by a transformer which is fed by a suited frequency converter (similar to that for transformer testing), both, active and reactive test power can be supplied.

ACTF test systems might be useful for research and development, but not yet for apparatus testing (except of that mentioned in 3.1.3) because most apparatus shall be tested with power frequency AC voltages (see Sect. 3.2.1).

Table 3.3 Comparison of motor-generator sets (M/G) and static frequency converters (SFC) without additional compensation elements

Characteristic	M/G	SFC
General characteristic	Set of electric machines with special parameters	Power electronic modules that can be combined, replaced or added
Max. available active power (MW)	5	8
Max. available inductive power (MVA)	30[a]	12[a]
Max. available capacitive power (MVA)	This load has to be avoided by additional reactors	12
Frequency values or range	Usually only two fixed frequencies, e.g. 50 and 150 Hz	Continuously adjustable, e.g. 40–200 Hz
Power electronic application	For control of the driving motor	For the whole converter
PD noise suppression necessary	Yes	Yes
PD characteristics described by basic noise level	<30 pC is usually reached	<10 pC can be reached
Optimum circuit harmonics (THD)	Standard value <5% fulfilled	Standard value <5% fulfilled
Dynamic behaviour	Limited, because determined by rotating machines	Excellent, because determined by short reaction of power electronics
Self-excitation	May happen, to avoid it adjustment of reactors at lower voltage is necessary	cannot happen
Weight and size	Large and heavy, M/G set requires foundations	Relatively lightweight and compact
Principle design, upgrade, repair	Conventional structure, upgrade of power is impossible	Modular design, upgrade of power or repair by additional or exchange of modules
Application	Traditionally for stationary application in factory testing	Application for both, factory and on-site testing

[a]The range of inductive power can be extended by a capacitor bank

3.2 Requirements to AC Test Voltages and Selection of HVAC Test Systems

3.2.1 Requirements to AC Test Voltages

The requirements for AC test voltages for *laboratory testing* are fixed in IEC 60060-1: 2010 and described here, those for *testing on site* are described in

Sect. 10.2.1.1 according to IEC 60060-3: 2006. For research purposes, each interesting type of alternating voltages may be used.

The AC test voltage shall be of approximately sinusoidal *wave shape* and its value is determined by the peak value because the highest stress determines usually the breakdown process (Fig. 3.38). To have the direct comparison with the usually applied rms value, the test voltage value is defined by

$$V_T = \frac{V_{peak}}{\sqrt{2}} = \frac{V_{peak+} + |V_{peak-}|}{2\sqrt{2}}. \tag{3.20}$$

If the peak values at the two polarities are different, the average value of both is applied as peak value. For some special test application, e.g. when heating processes influence the development of the breakdown, the *rms value* might be applied:

$$V_{rms} = \sqrt{\frac{1}{T} \int_0^T V(t)^2 dt}, \tag{3.21}$$

where T is the duration of a period. The two half waves should be symmetric within 2%

$$|\Delta V| = \left| \frac{V_{peak+} - |V_{peak-}|}{V_{peak+}} \right| < 0.02 \tag{3.22}$$

The tolerance of the *test voltage frequency* is 45–65 Hz according to IEC 60060-1: 2010.

Note This range is simply related to the "European" frequency of (50–5) Hz and the "American" frequency of (60 + 5) Hz. It is assumed that within the range (45–65) Hz, the

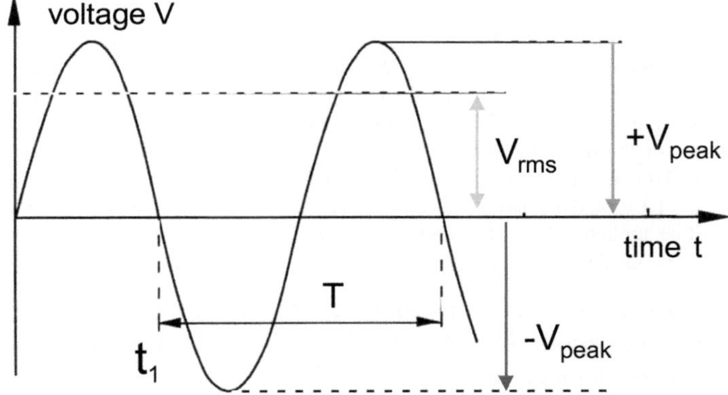

Fig. 3.38 Parameters of AC test voltages

influence of frequency on the breakdown process can be neglected for all types of insulations. Some IEC apparatus standards define tolerance ranges of frequency different from the previous. There are no physical investigations known which would deliver physical reasons for the selection of the frequency ranges.

For the deviation of the test voltage wave shape, IEC 60060-1: 2010 defines the *peak factor* as the relation between the peak and the rms value, which should be equal to $\sqrt{2}$ within $\pm 5\%$, this means

$$1.344 \leq \left| \frac{V_{peak}}{V_{rms}} \right| \leq 1.485.$$

This definition is only related to the peak voltage value, it does not consider the wave shape of the test voltage as a whole. This might be sufficient to withstand or breakdown tests, but PD are influenced by other instants of the test voltage, too. For such applications, the *total harmonic distortion (THD)* considers the peak values V_{npeak} of the superimposed *harmonics* on the basic frequency (peak value V_{1peak}):

$$\text{THD} = \frac{1}{V_{1peak}} \sqrt{\sum_{n=2}^{m} V_{n\,peak}^2}. \tag{3.23}$$

The IEC Technical Committee 42 had discussed to introduce an acceptable limit of 5% of the THD, but did not decide for its binding introduction with the last revision of IEC 60060-1 in 2010. One reason for that decision is that peak factor and THD are not proportional to each other. Table 3.4 (Engelmann 2007) shows an example: The voltage of a cascade transformer of a rated voltage $V_m = 1200$ kV was evaluated at very low output voltages and compared with a 5% limit of deviation of the peak factor, respectively, the THD. Whereas the output voltage of 2.8% related to V_m shows strong harmonics and both, peak factor and THD are too high to be accepted as an AC test voltage. The little voltage increase by 3% causes an improvement in the wave shape and acceptance by the peak factor, but none by the THD. The 24.1% output voltage shows a better sine wave and acceptance by both criteria.

Note The content of harmonics depends on all elements of the HVAC test circuit and consequently also from the position of the regulator. Some harmonics that come from the power supply might be amplified by resonances of certain parts of the test circuit. For very low positions of the regulator, a high content of harmonics is typical. Therefore, a peak factor (or a THD) within the mentioned limits of 5% is usually only specified for output voltages $V_t > 10\%$ of the rated voltage of an HVAC test system. When the HVAC test system shall be applied at lower test voltages, the wave shape has to be checked. Usually, it is sufficient to consider in Eq. 3.23 the harmonics of the numbers $n = 3, 5, 7$ and 9.

A partial discharge that causes a heavy current pulse may cause a drop ΔV of the test voltage for few periods. IEC 60060-1: 2010 allows $\Delta V/V_t < 20\%$. The standard did not specify the duration of that *voltage drop*. There is also no physical background for the value 20% [Compare the voltage drop for DC voltage (Sect. 6.2)!].

Table 3.4 Deviations from sinusoidal wave shape (Measurement on a 1200 kV cascade transformer by Engelmann 2008)

Voltage shape Output voltage $V_{peak}/V_T \cdot 100\%$ Parameter	2.8%	3.0%	24%
V_{peak}/kV	48.4	51.6	410
$V_t = V_{peak}/\sqrt{2}$/kV	32.2	36.5	290
V_{rms}/kV	30.8	37.4	287
$F = V_{peak}/V_{rms}$	1.571	1.380	1.413
Comparison: Peak factor $1.344 \leq F \leq 1.485$	Rejected	Accepted	Accepted
THD (%)	25	7	2.5
Comparison THD $\leq$ 5%	Rejected	Rejected	Accepted

The specified voltage drop should be understood as step in the right direction. In the following, a voltage drop $\Delta V < 10\%$ shall be considered as acceptable. A more detailed consideration of that is expected with the next revision of IEC 60060-1.

Last but not least, the *tolerance* of the test voltage value must be mentioned. For a test duration T_t up to one minute (1 min), the tolerance is $\pm 1\%$. For $T_t > 1$ min, it is $\pm 3\%$. The wider range for longer test durations is necessary because larger changes of the supply voltage are expected at longer test durations.

Note The tolerances mean if a withstand test is performed, e.g. at $V_t = 500$ kV for $Tt = 1$ min, the measured and displayed voltage should remain within 495–505 kV. If the test duration is 1 h, the test voltage should be within 485–515 kV. When one considers an accepted *uncertainty of the voltage measurement* (Sect. 2.3.4) of 3% additionally, the true, but unknown real stress during the 1 h test may lay approximately between 470 and 530 kV.

3.2.2 Test Systems for Multi-purpose Application

The selection of test systems is strongly dependent upon their use. HVAC test systems for multi-purpose application are for HV research work, for training in HV engineering, for calibration of measuring systems or for the HV testing of model insulations. Whereas, e.g., the HVAC test systems for routine testing of a certain group of products are selected according to the test parameters, for this special-purpose application, systems for multi-purpose application require detailed analyses of the later application. Usually, they are applied for dry tests on air insulations, tests on model insulation of gaseous, liquid or solid dielectrics. The objects have to be stressed up to few hundred kilovolts, in rare cases up to 1000 kV. Often a test current below 0.5 A is sufficient, but a quite sensitive PD measurement is mandatory today.

Under the mentioned conditions, HV test transformers (or transformer cascades) of rated voltage $V_m = 30$–1000 kV and rated current $I_m = 0.2$–0.5 A (in rare cases up to $I_m = 1$ A) are the ideal HV generators for multi-purpose test systems. Especially, cylinder-type test transformers are well suited. Their duty cycle should be some hours, only for long-term tests continuous operation is required. The control system should enable both, manual and computer-aided testing.

The HV circuit should be completed by a voltage measuring system based on a capacitive voltage divider with a peak voltmeter or—even better—with a transient recorder (IEC 61083-3 and -4, Draft 2011) for the measurement of all parameters mentioned above (Sect. 3.2.1). Furthermore, a PD measuring system, often also measuring systems for dielectric parameters, should be available.

For rated voltages up to 200 kV and low currents, modular HV test systems can be recommended, for voltages up to 500 kV, single cylinder-type transformers and for higher voltages, cascade transformers should be applied (Fig. 3.39).

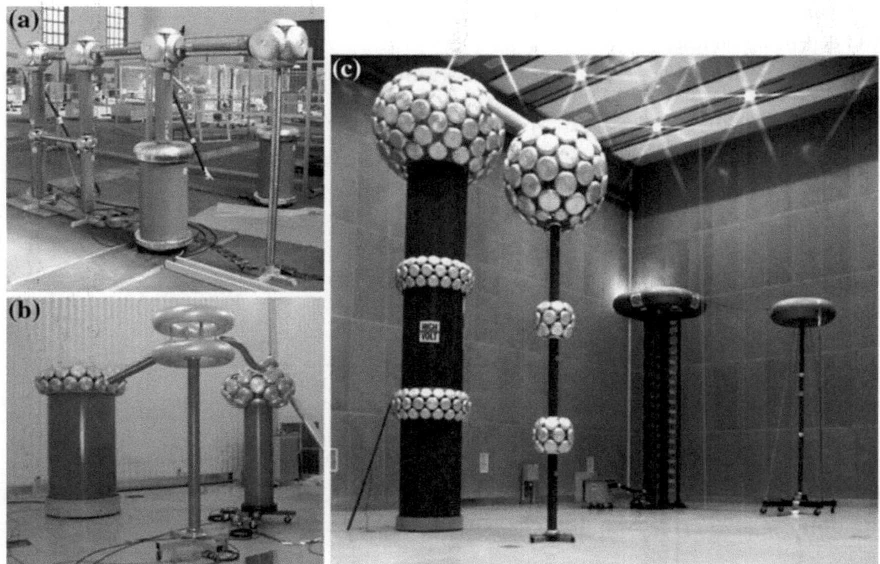

Fig. 3.39 Test transformers for multi-purpose application. **a** AC/DC voltage test circuit of a modular system, **b** test system with a 350 kV cylinder-type transformer, **c** 1000 kV test system with cylinder-type transformers (Courtesy of BTU Cottbus)

3.2.3 AC Resonant Test Systems (ACRL; ACRF) for Capacitive Test Objects

The capacitive load of an HVAC test system can cover a very wide range, from few 100 pF for the voltage divider and no-load conditions up to some 10 μF for submarine cables or capacitors. The limit of application between multi-purpose test systems based on test transformers and resonant test systems is not sharp, but should be characterized by the available current. Using the relation between total load capacitance C, test voltage V_t and test current I_t

$$I_t = \omega C \cdot V_t \leq 1 \text{ A} \qquad (3.24)$$

of a multi-purpose HVAC test system, one gets a limiting curve for the testable capacitance

$$C = \frac{I_t}{\omega V_t}. \qquad (3.25)$$

Figure 3.40 shows this relationship. For total load capacitances C above the curve of 1 A (depending on the situation possibly 0.5 A), the application of HVAC resonant test systems is better suited than that of a transformer-based system. One

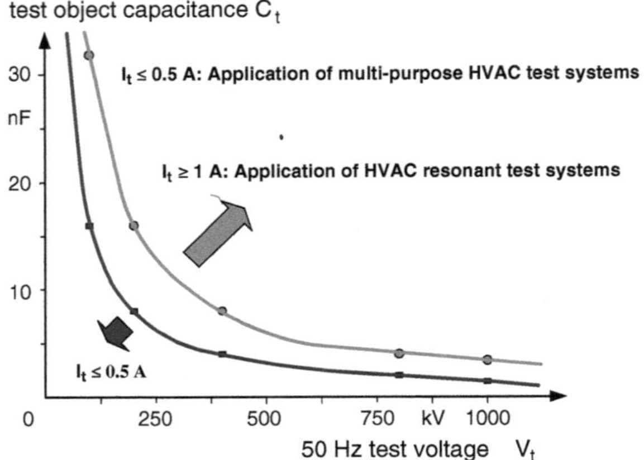

Fig. 3.40 Testable capacitive load depending on test voltage and rated current of the HVAC test system (lower limit for resonant test systems application)

has to consider that the total capacitance includes in addition to the test object also the voltage divider, a coupling capacitor and—for resonant circuits—a capacitive basic load.

Typical test objects in the range up to few 10 nF are components or complete bays of gas-insulated substations (GIS), power and instrument transformers, bushings and cable samples. In an HVAC routine test, a complete cable drum may have a capacitance of some 100 nF.

In case of *GIS testing*, the load of the test object can easily be estimated from the length of the GIS tubes and the specific capacitance $Ct \approx 60$ pF/m for a single-phase bus bar. Often, the HVAC test system is connected to a test bus, which can be arranged in the production hall (Fig. 3.41). The bus has several connection points for the GIS to be tested. Only one of the connection points is connected to the HVAC test system for a test. The others are separated and grounded, e.g. at a second, a GIS tested before is disassembled and at a third, a GIS is prepared for the next test. This solution does not require any separate test field, because the test bus and the test object have metal enclosures and are grounded. But the test bus increases the total load capacitance and the required test current.

When the *applied voltage test of a power transformer* is performed, the transformer acts as a capacitance in an approximate range of $C_t \approx 4$–30 nF, in very rare cases higher. Usually, cylinder-type resonant reactors are applied (Fig. 3.42). According to IEC 60076-3, the application of a fixed frequency of 50 or 60 Hz is not mandatory, moreover only a test frequency $f_t > 40$ Hz is required for the applied voltage test. This enables not only the application of ACRL test systems, but also of frequency-tuned ACRF systems. *ACRF test systems* are very economic solutions for that application.

The testing of power cables is the traditional application of resonant circuits. The relevant standards for liquid impregnated and extruded cables require a fixed test frequency of 50 or 60 Hz for the AC voltage withstand test and the PD measurement. Therefore, ACRL test systems are applied for land cables that are

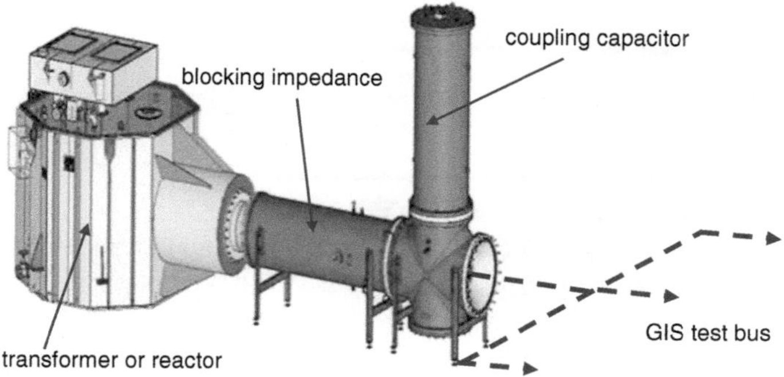

Fig. 3.41 AC test system for GIS testing with test bus

Fig. 3.42 ACRL test system during applied voltage test of a power transformer

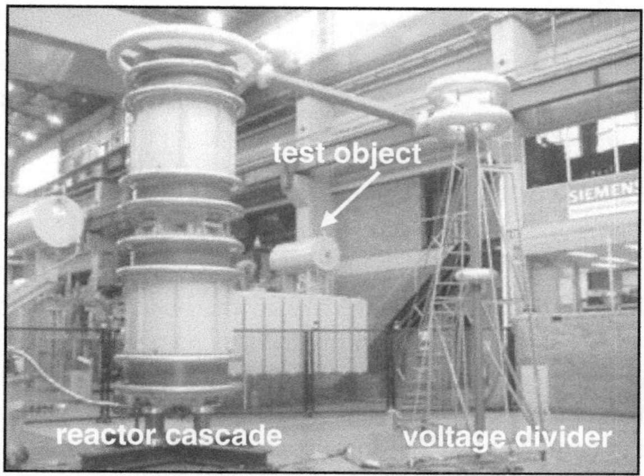

manufactured with lengths up to 1000 m per drum, for medium-voltage cables even longer. The specific capacitance—depending on the cable design (relation between conductor and cable shield) and the insulating material—is usually within 0.15 and 0.4 nF/m. The connection between the cable to be tested and the test system must be realized by a *cable end termination*. For medium (MV) voltage cables sliding test terminations of solid materials or simple oil-filled test terminations are applied. For HV and EHV cables water end terminations are used. Their design is an impressive example for the control of the field strength along the insulating surface of the end termination (Pietsch et al. 2003; Saya et al. 2016):

While the electric field within the cable is slightly non-uniform, the field of the cable ends is characterized by a locally high field strength causing a strongly non-uniform field strength distribution along the surface of the termination (Fig. 3.43a). The field strength distribution is improved when the insulated cable end is surrounded by deionized water of controlled conductivity. In the equivalent circuit diagram (Fig. 3.43b) the water is represented by the chain of resistors. The

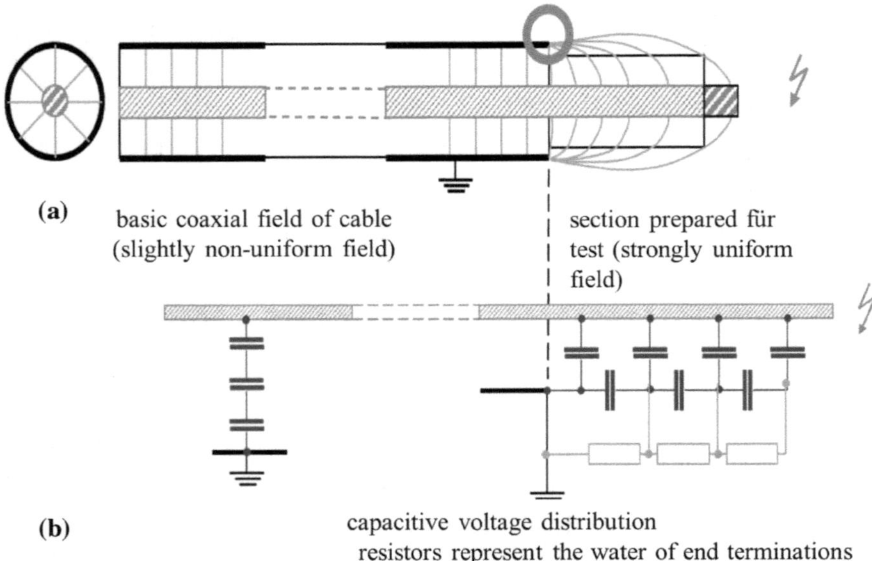

Fig. 3.43 Cable end prepared for testing (schematically) **a** electric field, **b** equivalent circuit diagram

higher the water conductivity σ, the better is the control of the electric field. But with increasing conductivity the loss increase also and the water is heated. To avoid an overheating, the water must be cooled and its conductivity shall be adapted to the test conditions by a water conditioning unit (Fig. 3.44a). There are two limits for the conductivity. The lower one is caused by a minimum conductivity required for the improvement of electric field. The upper one is caused by the available power of the HVAC test system to compensate the losses due to the heating of the water as well as for the power necessary for the cooling in the water conditioning unit. It determines the upper limit of the conductivity. The control algorithm for the conductivity results from the two opposite effects of the conductivity shown in Fig. 3.45. During a cable test the conductivity shall be maintained in the acceptance range (Pietsch et al. 2003).

A water termination reduces the quality factor of a resonance circuit. It has to be checked that the ACRL test system can sufficiently operate for short-cable samples connected with water terminations.

Cables require different HVAC tests, Table 3.5 mentions, e.g., those for EHV cables of 220–500 kV: HVAC test voltage is required for the voltage withstand tests, the partial discharge tests, the heating cycle test in combination with the high-current heating (Naderian et al. 2011), the LI and the SI voltage test with an HVAC repetition after the impulse voltage test. Especially, for heat cycle and pre-qualification tests (Bolza et al. 2002), HVAC test systems of high reliability are required. Therefore, test transformers have been applied, but now also resonant test

Fig. 3.44 Test of a HV cable
a cable termination system
with test arrangement,
b principle design of water
end terminations

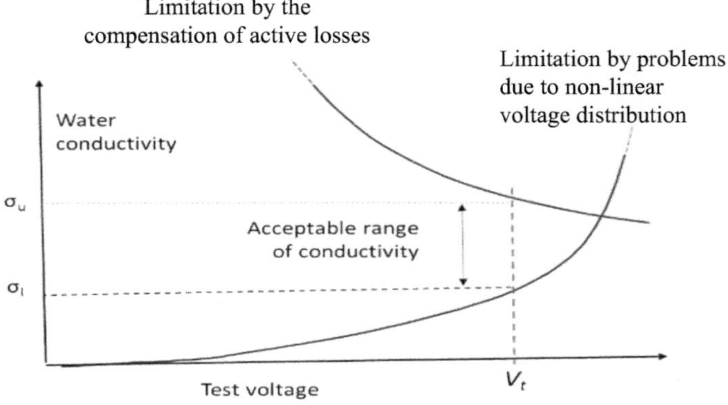

Fig. 3.45 Selection of the acceptable range of water conductivity

systems with tuneable inductance and even more with variable frequency are
applied for these tests.

Nowadays, *submarine cables* up to many ten kilometres length must be tested in
factory. The capacitive load may reach more than 10 μF, and an ACRL test system
becomes very huge and expensive. The way out is an ACRF test system (Fig. 3.46).

Table 3.5 Test voltages for extruded cables 220–500 kV (IEC 62067: 2001)

Rated voltage, V_r (kV)	Highest voltage for equipment, V_m (kV)	Conductor to ground voltage, V_0 (kV)	AC withstand voltage test		AC PD test voltage,	AC Heat cycle test,	LI voltage test (kV)	AC voltage after LI	SI voltage test (kV)
			Voltage, V_t (kV)	Duration, T_t (min)	1.5 V_0 (kV)	2 V_0 (kV)		2 V_0 (kV)	
220–230	245	127	318	30	190	254	1050	254	–
275–287	300	160	400	30	240	320	1050	320	850
330–345	362	190	420	60	285	380	1175	380	950
380–400	420	220	440	60	330	440	1425	440	1050
500	550	290	580	60	435	580	1550	580	1175

Fig. 3.46 Testable capacitive load on test voltage and rated current of the HVAC test system (limit between ACRL and ACRF test system application)

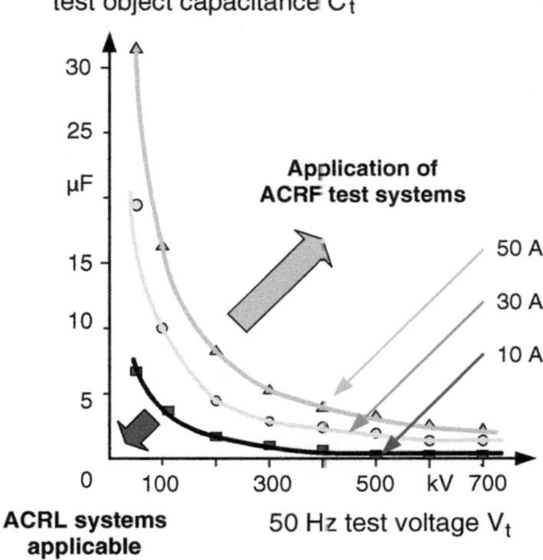

Up to a required test current of the order of few 10 A, ACRL test systems can be realized, but for higher currents, only ACRF test systems are recommended, also by the latest IEC draft on HVAC submarine cables (Karlstrand 2011). Figure 3.47 shows the rotating table for submarine cables and the ACRF test systems in a factory as well as the laying ship with its cable drums.

Fig. 3.47 ACRF test system for submarine cables (courtesy of ABB Karlskrona). **a** Rotating cable table, **b** ACRF test system of five reactors, **c** laying ship with cable abcard

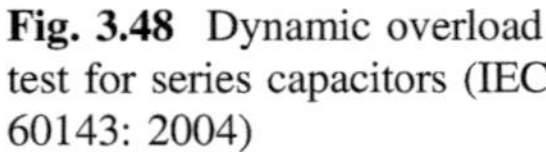

Fig. 3.48 Dynamic overload test for series capacitors (IEC 60143: 2004)

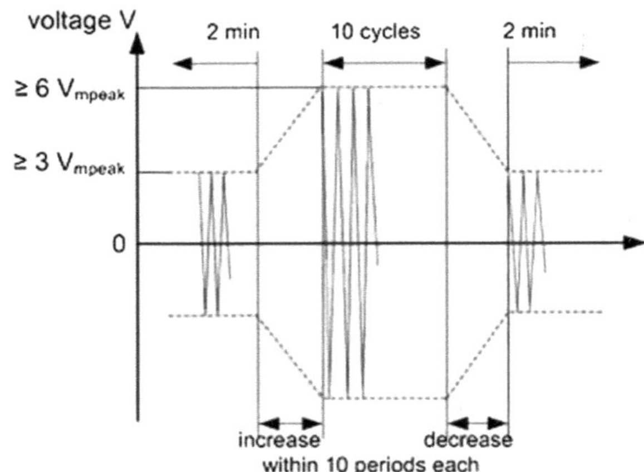

Fig. 3.49 Two ACRF test systems of 1000 and 600 kV for routine testing of GIS spacers. Courtesy of HYOSUNG Industries, Changwon

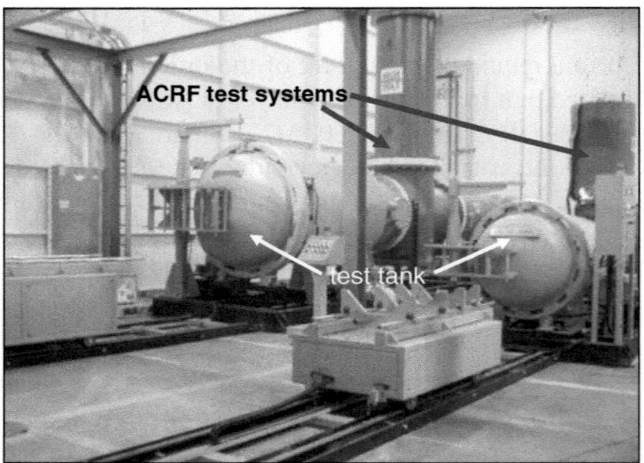

Capacitors with very high capacitances are also tested with resonant circuits. In case of the so-called cold duty test on series capacitors (IEC 60143-1: 2004), an additional over-voltage period (Fig. 3.48) requires a dynamic change of the test voltage within less than 10 cycles. For the better control of the test voltage, this test is performed with a parallel resonant circuit. This test voltage must be recorded using an adapted digital recorder (Draft IEC 61083-3: 2012).

ACRF test systems may also be optimum for special applications as shown in Fig. 3.49. Each spacer insulator for GIS is carefully PD tested after its production. This test is done at voltages up to 1000 kV and requires a metal-enclosed test system in the production hall. Because all test objects are identical, the inductance can be designed to guarantee a frequency range within 45–65 Hz. For that application, the ACRF test system is by far the most economic one.

3.2.4 HVAC Test Systems for Resistive Test Objects

A resistive or active load of an HVAC test system requires an *active test power* which can only be supplied by a *test transformer*. In this sense, also the above mentioned, so-called parallel resonant circuit is a transformer circuit. A mainly active characteristic of a test object is caused by one or by a combination of the following influences:

- a volume resistance of the tested insulation that is lower than the value of the capacitive impedance, e.g. a strongly aged oil–paper insulation;
- a high surface conductivity of the insulator under test, e.g. by wet and/or polluted surfaces due to *rain*, *pollution* and *humidity*;
- heavy pre-discharges at test objects in air, e.g. caused by leader discharges at very high AC voltages or by pre-arcs in pollution tests;
- continuous, quasi-stationary *corona discharges* on conductors, e.g. on overhead test lines or corona tests. The active power demand of the test object may be a quasi-stationary *leakage current*—e.g. for high volume resistance or for corona discharges. But for many other cases, the current impulse connected with a strong discharge requires a very fast active power supply by the test transformer. If the voltage source cannot supply the current fast enough, the voltage would drop down. This *voltage drop* can delay or even interrupt the development of pre-discharges to a breakdown, and this may lead to wrong, source-dependent test results. The acceptable voltage drop of 20% of the present IEC 60060-1:2010 is not related to pollution testing. The drop of 10% proposed in this book (See Sect. 3.2.1) might be better related to realistic test results, but a new value for the standard requires more investigation. Up to a certain degree, a too weak HVAC voltage source can be "supported" by a capacitor in parallel to the test object (Reichel 1977).

3.2.4.1 HVAC Test Systems for Artificial Pollution Tests

On a polluted insulator surface, the leakage current pulses have a typical shape as shown in Fig. 3.50 (Verma and Petrusch 1981). Many measured impulses are normalized to their peak currents and displayed versus their phase position. The currents start to increase at about 50° (el.), reach their peak at about 110° (el.) and extinguish at about 170° (el.). The current peak may reach values up to few Amperes (Rizk and Bourdage 1985). It causes a *voltage drop* over the internal impedances of the HV test system. Figure 3.51 shows a voltage drop of 20% at a peak current of 1.8 A. Such a voltage drop influences the measuring result remarkably, but in the relevant IEC Standard 60507 (1991), no measurement of the current or of the voltage drop is required. Several publications consider a voltage drop up to 10% as acceptable.

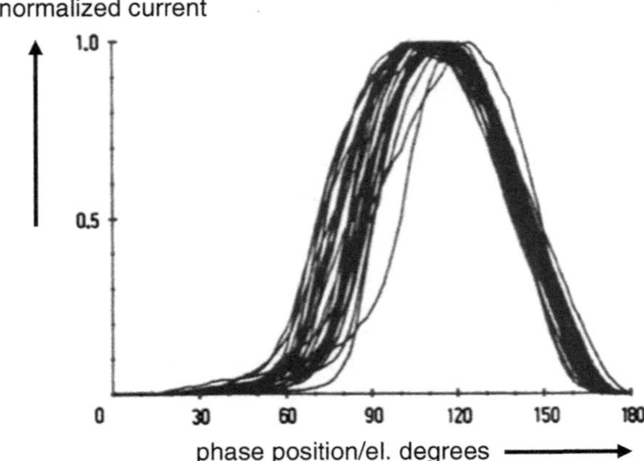

Fig. 3.50 Normalized current pulses in pollution tests depending on the phase position

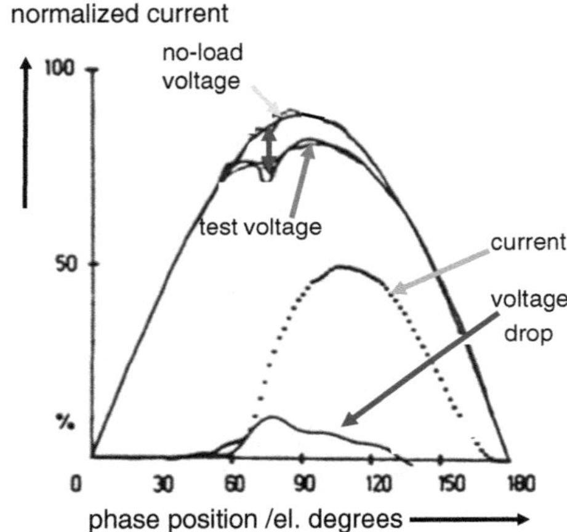

Fig. 3.51 Measured no-load voltage, leakage current, test voltage and voltage drop

In pollution testing (see Sect. 2.1.4), instead of direct measurement of the voltage drop, the rather old IEC Standards 60507 (1991) and the old IEC 60-1 (1989) define a set of requirements to the HVAC test system.

Note The actual IEC 60060-1 (2010) does not contain any requirement for pollution testing and contains only a hint to the latest edition of IEC 60507 that is at the moment the version of 1991. But consider that the voltage drop of 10% does not coincide with a voltage drop of 20% mentioned in IEC 60060-1 (2010). Pollution tests require a remarkably lower value.

These requirements are based on the voltage drop via the internal impedance of the HVAC voltage source that can be expressed by the short-circuit current. A high internal impedance means a low short-circuit current (see Sect. 3.1.1.1). To avoid a critical voltage drop, the *short-circuit current* must have a minimum value that varies with the test conditions. The test conditions can be expressed by the *specific creepage distance* (in mm/kV) of the insulator, which characterizes the type and shape of the insulator. An insulator with a certain specific creepage distance is selected according

Fig. 3.52 Required short-circuit current of the HVAC test system depending on the specific creeping distance (IEC 60507: 1991)

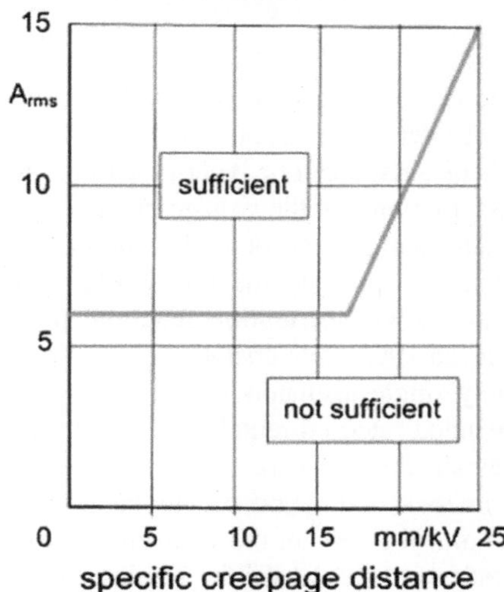

to the *pollution class*. Depending on the specific creeping distance, Fig. 3.52 shows the short-circuit current I_{scmin} in minimum required by IEC60507 (1991), to avoid the influence of the test system as the result of a pollution test.

A short-circuit current higher than I_{scmin} is necessary, but not sufficient requirement for the "suited" HVAC test system. It shall comply also with a

$$\text{resistance/reactance ratio} : \frac{R}{X} \geq 0.1, \qquad (3.26)$$

using the reactance according to Fig. 3.2b

$$X = \left(\frac{1 - \omega^2 LC}{\omega C}\right),$$

and with a

$$\text{capacitive current/short circuit current ratio: } 0.001 < (I_c/I_{sc}) < 0.1. \qquad (3.27)$$

The value of $I_{sc} = 6$ A is considered as an absolute minimum value for the short-circuit current of an HVAC test system for pollution testing. But for higher specific creeping distances, higher values are necessary according to Fig. 3.2w. When the test system does not reach the necessary short-circuit current I_{sc}, IEC 60507 allows to measure the highest leakage current pulse I_{hmax}. If the ratio between I_{sc} (as a rms value) and I_{hmax} (as a peak value) becomes

$$\frac{I_{sc}}{I_{hmax}} \geq 11, \tag{3.28}$$

the test result can be considered as independent from the test system.

Under the assumption of a uniform pollution of an insulator chain, the flashover characteristic can be determined on one insulator of the chain. This is based on the assumption that the voltage distribution along the chain is a resistive one. The withstand characteristic of the *insulator chain* is determined by multiplying the parameters (voltage values) by the number of insulators in the chain. In case of parallel insulators and known distribution function, the statistical enlargement rule (Hauschild and Mosch 1992) delivers the withstand parameters. The described principle of testing of only single insulators enables that the used powerful HVAC test systems (recommended rated current $I_m = 3\text{–}10$ A, recommended short-circuit current $I_{sc} > 20$ A) remain limited to rated voltages below 500 kV. Therefore, also the *pollution test chambers* and the other equipment for pollution testing are usually limited in size.

Salt fog test: In the withstand test procedure (IEC 60507: 1991) for the salt fog test (see Sect. 2.1.4), the insulator is subjected to a preconditioning process with a series of progressive voltage tests at the reference salinity of the water sprayed as a fog. Then, for the withstand test itself, the insulator is cleaned, pollution conditions are adjusted (specified salinity) and the test at the specified withstand voltage level is performed for the duration of 1 h. Repetitions of the test might be specified. Before each repetition, the insulator must be carefully cleaned. The insulator has passed the test, if no flashover occurs during all withstand procedures.

Solid layer test: In case of the solid layer method (IEC 600507: 1991), the withstand test procedure can be performed when the test voltage is applied after the wetting of the insulator under test has started (procedure A), or the test voltage is applied before the wetting starts (procedure B). For procedure A and specified pollution layer conditions, the voltage is applied for 15 min (or until a flashover occurs). For procedure B, the test voltage is maintained for 100 min (or until a flashover occurs). Both procedures must be repeated three times. The insulator complies with the specification if no flashover occurs during three repetitions. If only one flashover occurs and a fourth repetition remains without flashover, the insulator has also passed the test.

The IEC 60507: 1991 is related to ceramic insulators. At present, there is no standardized method for pollution testing of *composite and other polymeric insulators*. Therefore, the methods described above are often applied also for polymeric insulators. Gutman and Dernfalk (2010) and Gutman et al. (2014) presented modified solid layer methods for composite and polymeric insulators. Dong et al. (2012) found relationships between the shape of composite insulators and the pollution method. Further experiences on testing of polluted and iced conditions are published by Pigini et al. (2015), He and Gorur (2016) and Farzaneh (2014). These tests might be completed by tests related to the *hydrophobicity* (Bärsch et al. 1999; Schmuck et al. 2010). It can be assumed that HVAC test systems suited for pollution testing of ceramic insulators fulfil the requirements for pollution tests on polymeric insulators, too.

It can be expected for the future that during pollution tests both, the test voltage and the leakage current, will be measured and that a certain voltage drop, e.g. <10%, becomes a criterion for an acceptable test. Even now, it is recommended that these most important parameters are recorded during a pollution test using suited digital recorders.

3.2.4.2 HVAC Test Systems for Artificial Rain Tests

In artificial rain (or wet) tests (see Sect. 2.1.3), heavy streamer and even leader discharges connected with remarkable current pulses may occur. The peak currents do not reach a magnitude as the leakage current pulses of pollution tests, but multi-purpose test systems cannot guarantee that the related voltage drop remains within acceptable limits (<10%) and cannot influence the discharge processes. Therefore, more powerful HVAC test systems are applied for wet tests. Usually, an HVAC test system based on a transformer with a rated current of $I_m = 1$ A and a short-circuit current $I_{sc} > 10$ A (short-circuit impedance $v_k < 15\%$) is sufficient. A capacitor of about $C_p \approx 1$ to 3 nF in parallel to the test object is recommended additionally (e.g. related voltage divider).

When series resonant test systems shall be applied for wet tests, a parallel capacitor has to be applied with a capacitance even higher than mentioned above. In any case, the voltage should be recorded by a digital recorder to ensure that any voltage drop does not exceed 10%. The supervision of the voltage seems to be helpful in wet tests, too, and one can hope that for wet tests a physically based upper limit of the voltage drop should be agreed in a future standard.

3.2.5 HVAC Test Systems for Inductive Test Objects: Transformer Testing

Power and distribution transformers, instrument transformers and compensation reactors have quite complicated insulation systems and are for certain frequencies of the AC test voltage an inductive load for the voltage source.

As an example, power transformers shall be considered: Table 3.6 gives an overlook on the main HV tests on power transformers in the sequence of their application. The tests with AC voltage follow the tests with impulse voltage. This means, the AC tests, considered in the following, shall supply the final statement about the condition of the insulation.

In the AC "*applied voltage test*", the good condition of the insulation between the short-circuited HV line terminals and the grounded low-voltage side (winding, magnetic core, tank, etc.) shall be verified with an external HVAC source. The transformer is a capacitive test object and can be tested with a resonant circuit (see Sect. 3.2.3; Fig. 3.42). Because the relevant standard IEC 60076-3 does not require a fixed frequency, but allows a frequency range $f > 40$ Hz, frequency-tuned resonant test systems are very economic for applied voltage tests.

Table 3.6 Main HV tests on power transformers for different rated voltages according to IEC 60076-3

Rated voltage (V_m)	≤ 72.5 kV	72.5 kV $\leq U_m \leq$ 170 kV	≥ 170 kV
Full lightning impulse voltage (LI)	Type test	Routine test	Included in LIC test
Chopped lightning impulse (LIC)	Special test	Special test	Routine test
Switching impulse test (SI)	Not applicable	Special test	Routine test
Applied voltage test (AC-AV)	Routine test	Routine test	Routine test
Induced voltage withstand test [AC-(IVW)]	Routine test	Routine test	Replaced by SI and AC-IVPD tests
Induced voltage test with PD measurement (IVPD)	Special test	Special or routine test	Routine test

For AC induced voltage tests on the insulation of the line terminals and the connected windings, the transformer under test generates the HVAC test voltage by itself and must be excited with an appropriate voltage on its low-voltage side as described in detail in Sect. 3.1.3. This exciting voltage should be a three-phase voltage for three-phase transformers. Because of the saturation of the magnetic core (Fig. 3.32), the AC test voltage requires a frequency of more than twice of the operational frequency ($f_t \geq 2 \cdot f_m$). The voltage is traditionally generated by a *motor-generator set*, but nowadays more and more by *static frequency converters (SFC)*. It is stepped up to the necessary excitation voltage by a *step-up transformer*. The test system for exciting the power transformer is also used for the *heat run tests* and for the measurement of *no-load and short-circuit losses* at their operational frequency of 50 or 60 Hz. In those tests, the transformer is always an inductive load that needs always a well-adapted compensation. This is realized by a switchable capacitor bank on the low-voltage side of the transformer under test, for fine tuning sometimes additionally on the primary side of the step-up transformer (Figs. 3.36, 3.37c). In case of feeding via a SFC, the frequency-converter should be tuned to the self—compensation value (Fig. 3.33).

According to IEC 60076-3, power transformers shall be applied to an *"induced voltage withstand test (IVW)"* and after that to an *"induced voltage partial discharge measurement (IVPD)"* in both cases, at the same test frequency f_t. The duration of the IVW test depends on the relation between the rated frequency f_r of the transformer under test and the test frequency f_t:

$$t_t/s = 120 \cdot \frac{f_r}{f_t} \geq 15. \tag{3.29}$$

The IVW test shall commence at a voltage $\leq 0.33 U_t$, and the voltage shall be raised to the test voltage V_t and there maintained for the frequency-dependent test

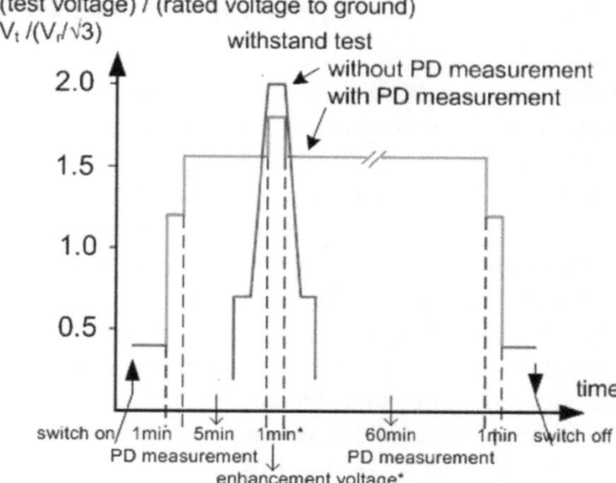

Fig. 3.53 Test cycles of an induced voltage test without (*blue*) and with PD measurement (*red*) on power transformers (Draft IEC 60076-3: 2011)

duration t_t. After that time, the voltage shall be reduced to $\leq 0.33\ U_t$ before switching off (Fig. 3.53, blue).

The IVPD test shall verify that the transformer is free of harmful PD. In the IVPD test, the voltage follows a step procedure (Fig. 3.53, red) with the highest (enhancement) level. The duration of the enhancement level is identical to the IVW test, only for power transformers with $V_m > 800$ kV, this duration is five times longer. At the three voltage levels before the enhancement level, a PD measurement is performed for a time long enough to have a stable result. At the enhancement level, no PD measurement is required in the standard, but should also be performed. Immediately after that, the PD level shall be recorded for 60 min on the specified PD measurement level. Then, the voltage is reduced to the same level as in the voltage increase and a PD measurement is made as before. The PD inception and extinction levels shall be determined. The IVPD test is passed, if no breakdown occurs during the whole cycle, if within the 60 min test none of the PD's recorded exceeds 250 pC, the PD level raises by not more than 50 pC or does not show any rising trend and if at the next lower voltage level not more than 100 pC are measured. The specified Picocoulombs may be modified according to contracts. For more details, see IEC 60076-3.

3.3 Procedures and Evaluation of HVAC Tests

The general statistical basis of this Sect. 3.3 has been described in Sect. 2.4 and more details are given by Hauschild and Mosch (1992), see also Yakov (1991), Carrara and Hauschild (1990) and Annex A of IEC 60060-1 (2010). In the following, procedures are considered that are relevant for tests with high alternating voltages.

3.3.1   HVAC Tests for Research and Development

Before a test starts, its clear statistical aim must be defined and transferred into an appropriate test procedure with well-defined parameters. This can be done based on similar earlier experiments, literature or own pretests and includes the clarification whether a complete *cumulative frequency function* (as an estimation of the *distribution function*) or only a certain *quantile* (for the estimation of a withstand voltage or an *assured breakdown voltage*) should be evaluated.

Sample size: The width of the *confidence region* of the estimate (quantile, distribution function) depends on the dispersion of the discharge process (expressed by its standard deviation) and the sample size (number of single tests). The higher the confidence requirements, the larger is the necessary sample size. Figure 3.54 shows the width of the confidence region (or interval) of the mean value ($V_{50\ upper} - V_{50\ lower}$)/$V_{50}$ depending on the variance $v = s/V_{50}$ and of the number of single tests. With a variance of $v = 6\%$ and $n = 10$ single tests, the confidence interval becomes 7%. To reduce that to 3%, one would require $n = 50$. The finally selected sample size should consider also the effort necessary for the test.

HVAC Test Procedures: The usual test procedure in an HVAC test is a *progressive stress test* (Fig. 2.26; Sect. 2.4.2). The voltage starts at an initial voltage V_0 and is continuously increased with a rate of rise dV_{peak}/dt until a breakdown occurs. The value of the breakdown voltage is a first realization V_{b1}. Then, the voltage is rapidly reduced to V_0, after a break Δt_p, the next single test follows. In total, n single tests are performed. The rate of rise may influence the results (e.g. by surface or space charges), therefore it is recommended to clarify the influence of the rate of rise on the measured breakdown voltages and to select a rate that enables independent results.

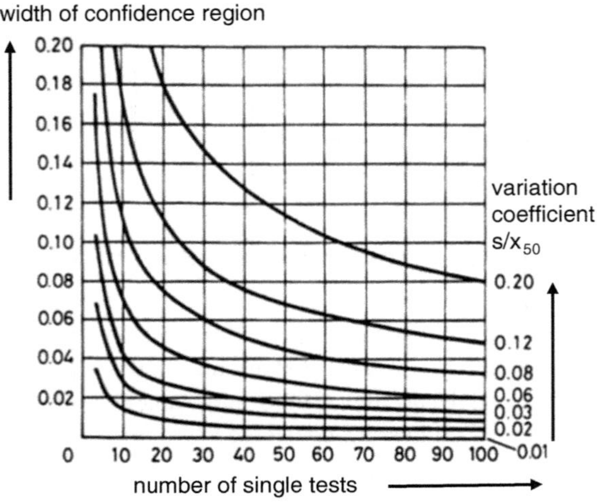

Fig. 3.54 Relative width of the confidence region depending on the variance of the measuring result and the sample size of the test (confidence level 95%)

More efficient than a continuous voltage might be a stepwise rise (Fig. 2.26 b). The initial voltage should be varied within limits lower than the step height to generate continuous breakdown voltages. Instead of the rate of voltage rise, the step height ΔV_s and the step duration Δts must be defined. This is better related to the procedures of withstand testing (see Sects. 2.4.6 and 3.3.2), especially when the step duration is equal to the duration of a withstand test (see Sect. 3.3.2). Under the assumption of independence—that the previous steps do not influence the following steps—one can calculate from the result of this test (cumulative frequency function) the *performance function* of single stresses that is directly related to withstand testing (Hauschild and Mosch 1992). There are actual applications of this principle, e.g. described by Tsuboi et al. (2010a, b, c).

Check of independence: When the test has been performed with the preselected parameters and n random breakdown voltages are available, their statistical independence must be checked. By means of a simple graphical method, the breakdown voltages (V_{b1} to V_{bn}) are graphically presented in the order of their occurrence. A visual assessment delivers an impression of the independence: If the realizations fluctuate in a random manner about the mean value, then there is no objection against the assumption of independence (Fig. 2.28: Upper three SF_6 pressures). When there is a falling or raising tendency, dependence must be assumed (Fig. 2.28 : Lower SF_6 pressure). Numerical *independence tests* are available and applied in related computer programs.

Approximation by a theoretical distribution function: Then the adaptation of the empirical cumulative frequency function by a theoretical distribution function is performed considering the correspondence between the physical model of the investigated breakdown process and the mathematical model of the applied distribution function, the good matching and the convenience of application (Hauschild and Mosch 1992). Table 3.7 gives an overview that theoretical distribution functions could be recommended for the approximation of breakdown test data: "Without PD" means that the insulation is slightly non-uniform, has no defects and the breakdown occurs immediately without stable PD. "With PD" means the insulation is strongly non-uniform or slightly non-uniform with defects and stable PD grow to the breakdown.

Note Not all experimental results can be approximated by one theoretical distribution function. Quite often so-called *mixed distributions* appear, e.g. in a slightly non-uniform field with defects (Fig. 3.55): When the breakdown process finds a starting electron at a defect, the breakdown voltage is quite low and shows a large dispersion (flat curve c). If the breakdown voltage is higher, the dispersion is low (steep curve b). For a mixed distribution, it is often sufficient to look for an approximation of the lower part of the mixed distribution. For more details, see Hauschild and Mosch (1992).

The application of a computer program of the *maximum-likelihood method* (MLM) delivers the best estimations and is expressly recommended. The MLM software is commercially available, and the evaluation can perform approximations by different distribution functions. In addition to the parameter of the selected distribution function, it delivers also the confidence limits of the distribution itself and those of their quantities (Fig. 3.56). Especially the lower confidence limits are

Table 3.7 Theoretical distribution functions recommended for the approximation of breakdown voltage data

Insulation and field	Air and gases		Liquids		Solids	
Distribution function	With PD	Without PD	With PD	Without PD	With PD	Without PD
Gauss (Eq. 2.48)	X		X		X	
Gumbel (Eq. 2.49)		X		X		
Weibull (Eq. 2.40)				X	X	X

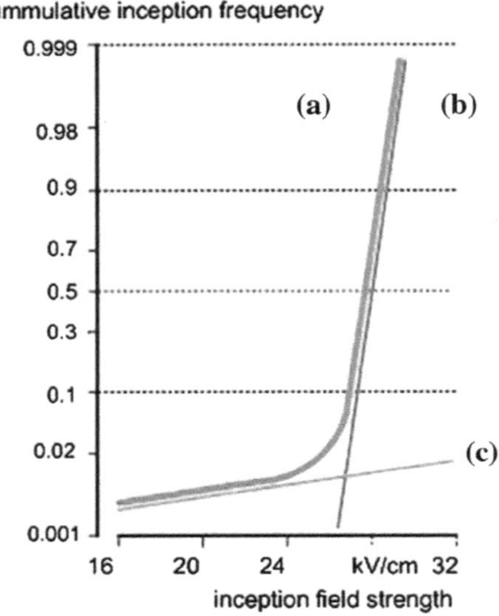

Fig. 3.55 Mixed distribution function of the breakdown voltage of a sphere-to-plane gap with a rough sphere surface (radius $r = 75$ cm; gap $d = 50$ cm)

important for the determination of design criteria of insulation, because they deliver "data on the save side".

Multi-level method: When an AC voltage of a certain height V and a certain duration Δt (e.g. 1 min) is defined as single stress, the multi-level method can be applied in a similar way as for impulse voltages. For details, see Sects. 2.4.3 and 7.3.1.1.

Lifetime tests: A certain number of samples of mainly solid insulation can be stressed at a constant voltage until the breakdown occurs. The random variable is the stressing time and the statistical evaluation is related to the breakdown time that can be well described by a *Weibull distribution* (Eq. 2.50a). If such tests are performed at different voltages, one can derive the lifetime characteristic (see Sect. 2.4.5) including its confidence limits using the MLM (Fig. 2.41). For each value of the breakdown time, one requires one single-test object (sample). Usually, a certain test time (e.g. 1000 h) is pre-given in such lifetime tests. The statistical evaluation allows also the consideration of *censored data*, this means samples that

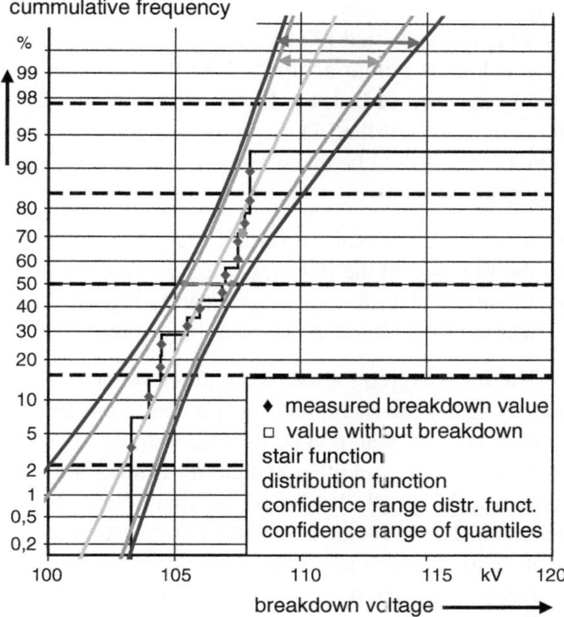

Fig. 3.56 Presentation of the results of a progressive stress test evaluated by the maximum-likelihood method and Gauss normal distribution function

did not break down within the pre-given test time (Speck et al. 2009). Their information is, e.g., that they withstood the test time without breakdown.

3.3.2 HVAC Quality Acceptance Tests and Diagnostic Tests

Usually acceptance tests shall verify the product quality—which consists of proper design, exact production of components and careful final assembling of the insulation of apparatus or systems—by withstand tests alone or withstand tests that include PD or dielectric measurement (*monitored withstand test*). In withstand tests, the products have to pass a certain procedure with the relevant test voltage values [(IEC 60071-1: 2006) and/or relevant apparatus standards] as, e.g., described in Sect. 2.4.6. If this is successfully done in type, routine and commissioning tests, the product can be handed over to the user for operation.

The classical withstand test procedure with continuously increasing voltages up to the test voltage value has been described in Sect. 2.4.6 and Fig. 2.42a. The test voltage value should be maintained on the test voltage level for a certain duration, usually of 1 min, quite often also of 1 h.

A monitored withstand test according to the step test procedure (Fig. 2.42b) delivers more information on the quality of the insulation than a simple withstand test. During the single steps and—for information—also during the withstand duration, PD—in special cases also the dissipation factor—shall be measured.

Usually, the acceptable PD level is not specified for the withstand voltage level, but for a certain specified PD measuring level after the withstand test. This level might have a longer duration (up to 1 h) than the steps before the withstand test. In a quality acceptance test, the PD level must be lower than a specified value, but also the comparison of the PD magnitude on identical voltage levels before and after the withstand test shall be taken into consideration. Consequently, the acceptance test is successful if

- there is no breakdown during the whole test procedure, and
- the PD level at the PD measuring voltage does not exceed the specified limit and
- PD levels should not be significantly higher on voltage levels after the withstand test than on identical voltages before the withstand test.

The first two requirements are given in the relevant apparatus standards. The last one is usually not mentioned in apparatus standards. It is additionally recommended to demonstrate that preceding test voltage stresses have not amplified PD phenomena at defects neither indicating a remaining deterioration, nor exceeding the specified PD level. It should be mentioned that in some traditional cases, the withstand test procedure and the PD measurement procedure are performed separately and not in one common cycle with the withstand test. But PD monitored withstand tests are recommended (Fig. 3.53).

For *condition assessment* of aged insulation also withstand tests are performed offline with suited mobile HVAC test systems. The aim of a diagnostic test is to estimate the remaining lifetime by suited stresses and measurements for detecting defects. The test voltage values can be lower than for a quality assurance test, they are often in the order of 80%. The selected measurements depend on the apparatus under test, the kind of its insulation and the experience of the test engineer. Very often, any kind of PD measurement is most suited, because PD—as the breakdown —are weak point phenomena.

For a *diagnostic test*, the PD- or a $\tan\delta$-monitored step procedure is advisable, but it should be adapted to the expected condition of the insulation (age, load, over-voltage stresses, climatic conditions, etc., Pietsch et al. 2012) Therefore, the performed procedure (e.g. selection and duration of steps, selection of measurands) should be left to the experience of the test engineer. Especially, the duration of PD measurements must often be related to the actual observations. For the evaluation, not only one PD parameter—e.g. the measured PD charge—should be used, but the full set of available PD characteristics (see Chap. 4).

3.4   HVAC Test Voltage Measurement

To measure high alternating test voltages, originally sphere gaps have been utilized, see Sect. 2.3.5. Even if these enable a direct measurement of high voltages up to the MV range, the procedure is extremely time-consuming. Another obstacle is the discontinuous procedure, i.e. the actual breakdown voltage of a test object cannot be determined due to the fact that the test object and the parallel connected sphere gap will never break down simultaneously. To overcome this crucial problem, the actual test voltage could be deduced from the primary voltage V_1 of the test transformer if this is multiplied with the turn ratio. This was a common practice at the beginning of the 1900s when HVAC voltage was increasingly used for power transmission. Such a simple approach, however, may cause severe measuring errors due to the inevitable voltage drop across the internal impedances of the test transformer, as can readily be deduced from the simplified equivalent circuit depicted in Fig. 3.57.

Here, a virtual voltage source V_{20} is connected in series with the inductance L_{12} and the series resistance R_{12} representing the effective impedances transformed to the HV side. Thus the secondary voltage without burden follows from the primary voltage V_1 if this is multiplied by the turn ratio:

$$V_{20} = V_1 \cdot \frac{w_2}{w_1}, \tag{3.30}$$

with w_1 and w_2—the turn numbers of the primary and secondary transformer coils, respectively, (see also Sect. 3.1.1.1). The cursor diagrams shown in Fig. 3.58, reveals that the actual test voltage V_{2T} appearing across the test object deviates strongly from the source voltage V_{20} without burden and depends furthermore on the kind of load, even at equal current magnitude.

Example Consider a test transformer having a turn ratio of $w_2/w_1 = 1000/1$, which is excited at frequency $f_e = 50$ Hz, so that without burden an assumed primary voltage of $V_1 = 100$ V causes a secondary voltage of $V_{20} = 100$ kV. The internal inductance and resistance shall be assumed as $L_{12} = 1000$ H and $R_{12} = 40$ kΩ, respectively. Moreover, each kind of test object connected to the transformer shall cause an equal peak current given by $I_R = I_L = I_C = 50$ mA. The peak values of the voltage appearing across the terminals of the test object are listed in Table 3.8. This reveals that at constant primary voltage the output voltage of the test transformer V_{2T} and thus the voltage applied to the test object decreases at both resistive and inductive burden, and increases at capacitive burden.

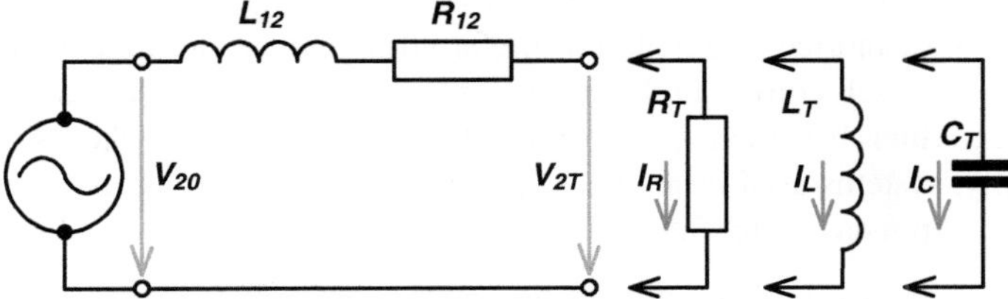

Fig. 3.57 Simplified equivalent circuit of a HV test transformer

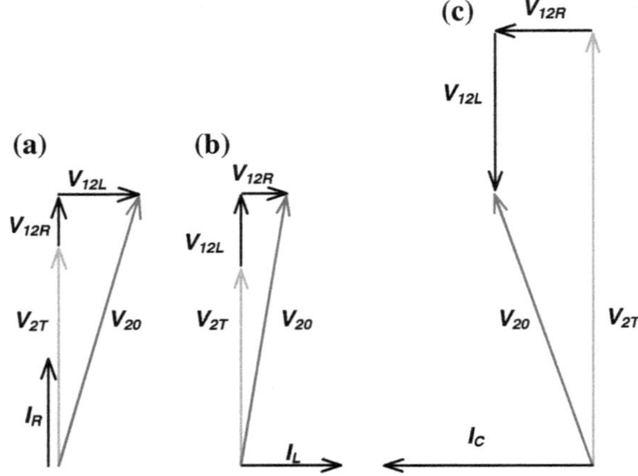

Fig. 3.58 Typical cursor diagrams deduced from the simplified equivalent circuit of a test transformer at various loads according to Fig. 3.57

Table 3.8 Variation in the secondary voltage at different load under constant source voltage $V_{20} = 100$ kV

Load	Resistive	Inductive	Capacitive
Test object	Polluted insulator	Instrument transformer	Power cable (length 10 m)
Value of circuit element	$R_T = 2$ MΩ	$L_T = 5350$ H	$C_T = 1.6$ nF
Test voltage generated	$V_{2T} \approx 97$ kV	$V_{2T} \approx 84$ kV	$V_{2T} = 110$ kV
Relative deviation (%)	−3	−16	+10

In this context it should be emphasized that the ratio between secondary and primary voltage is not only affected by the phase angle and peak value of the current flowing through the test object but also by higher harmonics due to core saturation and non-linear hysteresis effects. Therefore, it is not recommended to deduce the HVAC test voltage from the primary voltage.

Electrostatic voltmeters provide another option for the direct measurement of high alternating voltages, see Sect. 2.3.6. Such devices, however, can only be manufactured at reasonable expenditure for HVAC voltages up to approx. 100 kV. Thus, nowadays indirect methods are preferred to reduce the high voltage to an adequate low level conveniently measurable by means of classical indicating instruments, which became a common practice in the 1930s (Raske 1937).

Generally, systems used for indirect measurements of high voltages comprise the following components (Fig. 3.59):

- converting device (voltage divider and instrument transformer),
- transmission system (coaxial measuring cable or fiber optic link),
- measuring instrument (peak voltmeter or digital recorder).

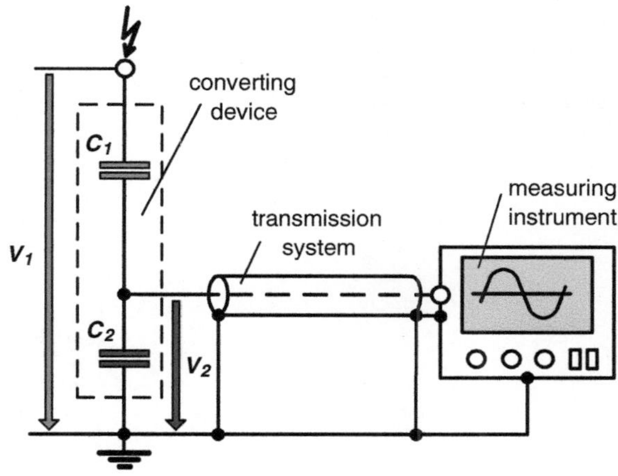

Fig. 3.59 Components of a system for indirect measurement of HVAC test voltages

3.4.1   Voltage Dividers

To reduce high alternating voltages down to an adequate low magnitude, resistive, capacitive or even inductive converting devices are commonly employed. However, inductive converting devices, such as instrument transformers, are only recommended for calibration purposes because these are very expensive, particularly if intended for HVAC measurements above 500 kV. Thus, in HV test laboratories commonly capacitive, resistive or even mixed voltage dividers are utilized, see Fig. 2.10. To avoid erroneous measurements, the top electrode as well as the grading electrodes of the divider column must be designed PD-free. Moreover, it has to be taken care that PD's are not ignited on account of dust and pollution deposited on the surface of the divider column, which is forced by a high air humidity. To keep the load of the HV test equipment as low as possible, the current through the divider should be limited below 10 mA, which is equivalent to a resistance/voltage ratio of 100 kΩ/kV. However, for resistive dividers this requirement can hardly be accomplished, in particular if designed for rated voltages exceeding 100 kV. This is due to the inevitable earth capacitances affecting the divider ratio, as will be treated more in detail in Sect. 7.4. Without going into more details it can be stated that capacitive *dividers* provide the most convenient solution for measuring high alternating voltages. As individual HV capacitors cannot be designed at appropriate expenditure for rated voltages above 300 kV, the HV arm is commonly composed of stacked capacitors, see Fig. 3.60.

The high voltage V_1 applied to the capacitive divider shown in Fig. 3.59 can be deduced from the measured low voltage V_2 as follows:

$$V_1 = V_2 \cdot (1 + C_2/C_1) \approx V_2 \cdot C_2/C_1. \tag{3.31}$$

However, this simple relation is not applicable in practice due to the impact of the inevitable *stray capacitances* between divider column and earth.

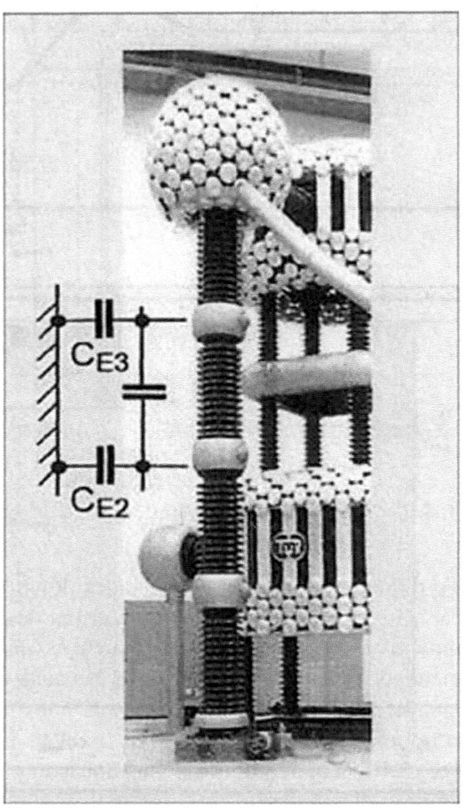

Fig. 3.60 Stacked capacitive voltage divider of a 1.2 MV test transformer composed of four units, each rated 300 kV. Courtesy of TU Dresden

Example Consider Fig. 3.61 where the HV arm is composed of two HV capacitors denoted C_{11} and C_{12}. For a simple estimation, it shall be assumed that each HV capacitor is screened by a metallic cylinder of length l and diameter d where each screen is connected to the bottom terminal of the capacitor. Under this condition the current and thus the voltage distribution is affected only by the earth capacitance C_e of the upper screen shielding the capacitor C_{11} arranged at a high $h = 1.2$ m above ground. Using the so-called antenna formula the earth capacitance can roughly be estimated based on the following approximation, which applies for vertical cylinders under the condition $l \gg d$ (Küpfmüller 1990)

$$C_e = \frac{2\pi\varepsilon l}{\ln\left\{\frac{2l}{d}\sqrt{\frac{4h+l}{4h+3l}}\right\}},\qquad(3.32)$$

with $\varepsilon = 8.86$ pF/m the dielectric permittivity of ambient air. For a quantitative estimation, the following practical values shall be introduced: $C_{11} = C_{12} = 200$ pF, $d = 0.2$ m, $l = 1$ m, $h = 1.2$ m. Inserting these values in Eq. (3.32), the earth capacitance of the upper HV capacitor becomes $C_e \approx 26$ pF, which is about 13% of C_{11}. Applying an AC voltage of $V_{12} = 100$ kV and frequency $f = 50$ Hz, this would cause the following capacitive current I_{12} through C_{12} and thus also through C_2:

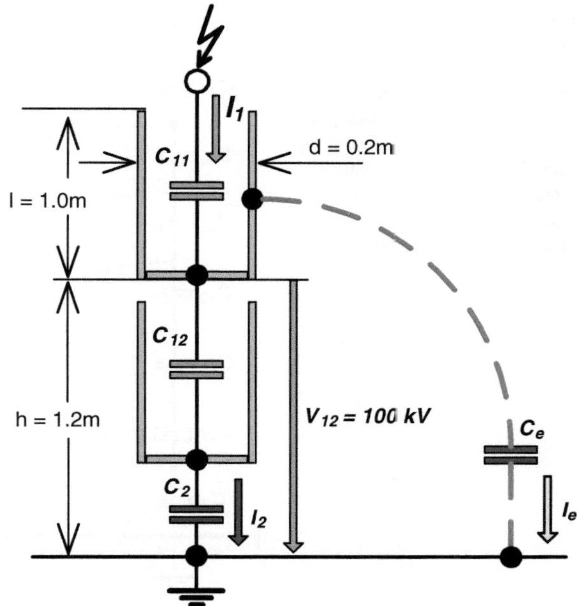

Fig. 3.61 Significant parameters utilized for the estimation of the earth capacitance C_e of a double-stage capacitive divider

$$I_{12} = I_2 = 2 \cdot \pi \cdot f \cdot V_{12} \cdot C_{12} = 6.28 \text{ mA}.$$

As the current through the stray capacitance amounts

$$I_e = 2 \cdot \pi \cdot f \cdot V_{12} \cdot C_e = 0.81 \text{ mA}.,$$

one gets the following total current through the upper HV capacitor C_{12}:

$$I_{11} = I_{12} + I_e = 7.09 \text{ mA}.$$

Thus the voltage appearing across the upper HV capacitor C_{12} attains

$$V_{11} = \frac{I_{11}}{2 \cdot \pi \cdot f \cdot C_{11}} = 113 \text{ kV}.$$

From this follows the total voltage appearing across the HV arm

$$V_1 = V_{11} + V_{12} = 113 \text{ kV} + 100 \text{ kV} = 213 \text{ kV},$$

Obviously, this is 6.5% greater than the theoretical value being $2 \cdot 100\text{kV} = 200\text{kV}$.

With other words: The inevitable earth capacitance of the divider column causes a fictive reduction of the capacitance of the HV arm. This becomes more pronounced at greater number n of stacked capacitors and even as lower the capacitance of each unit is. To estimate the potential distribution along the complete HV

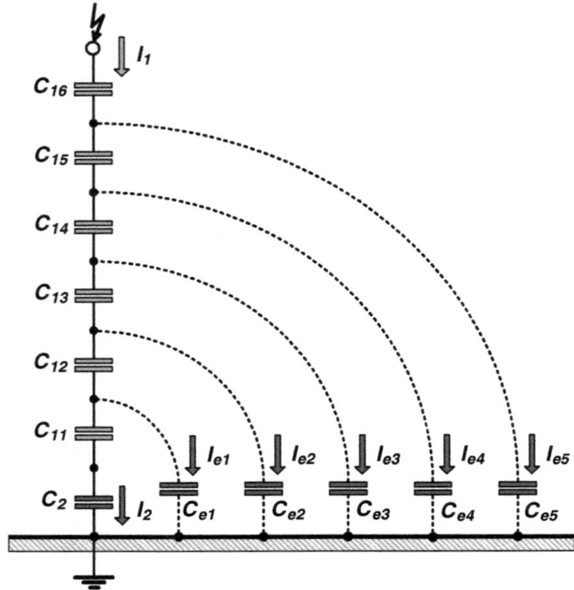

Fig. 3.62 Simplified equivalent circuit of a capacitive divider composed of six stacked HV capacitors, where only the impact of the earth capacitance is taken into account, while the proximity effect due to the surrounding walls and other metallic structures is neglected

divider column, the simplified equivalent circuit according to Fig. 3.62 is commonly utilized (Raske 1937; Hagenguth 1937; Elsner 1939).

Based on this, the effective capacitance C_1 of the HV arm composed of n stacked capacitors, each of capacitance C_{1n}, can roughly be approximated by

$$C_1 \approx \frac{C_{1n}}{n} - \frac{n \cdot C_e}{6}. \tag{3.33}$$

Theoretical and experimental studies revealed that the *earth capacitance* C_e of each capacitor depends only slightly on the high h above earth. Thus, Eq. (3.32) can be simplified as follows:

$$C_e = \frac{2 \cdot \pi \cdot \varepsilon \cdot l}{\ln\left(\frac{2 \cdot l}{d}\right)} \approx \left(56\frac{\text{pF}}{\text{m}}\right) \cdot \frac{l}{\ln\left(\frac{2 \cdot l}{d}\right)}. \tag{3.34}$$

Inserting this in Eq. (3.33) one gets

$$C_1 \approx \frac{C_{1n}}{n} - \left(9.3\frac{\text{pF}}{\text{m}}\right) \cdot \frac{n \cdot l}{\ln\left(\frac{2 \cdot l}{d}\right)}. \tag{3.35}$$

Example Consider a 2 MV divider consisting of $n = 10$ stacked capacitors, each of nominal capacitance $C_{1n} = 2000$ pF. The length and diameter of each unit shall be assumed as $l = 1$ m and $d = 0.2$ m, respectively. Inserting these values in Eqs. (3.34) and (3.35), one gets $C_e \approx 40$ pF and $C_1 \approx 160$ pF. That means, the virtual capacitance of the HV arm is approx. 20% lower than the theoretical one given by $C_u/n = 2000$ pF/10 $= 200$ pF.

In this context it must be emphasized by that the *effective divider capacitance* C_1 cannot be calculated at sufficient accuracy, even if C_1 is only slightly affected by the stray capacitances between divider column and top electrode. The only way is thus to determine the actual divider ratio experimentally, as treated in numerous technical papers and text books (Zaengl 1965; Kuffel and Zaengl 1984; Schon 2010, 2013) and as also specified in IEC 60060-2:2010. Basically, the Schering bridge or even a transformer ratio arm bridge could be utilized for this purpose. However, more convenient are comparative measurements using a *reference measuring system (RMS)*, see Sect. 2.3.3. A typical arrangement employed often for such reference measurements is shown in Fig. 3.63. Here, the standard capacitor providing the reference divider is located at the left (red unit), while the voltage divider under calibration is placed in the middle of the test area (blue unit). The AC voltage to be measured is generated by the 1000 kV test transformer cascade, which is located at the right-hand side.

Standard capacitors are composed of coaxial cylinder electrodes with guard rings at the LV side, as originally suggested by Schering and Vieweg in 1928. To achieve a high breakdown voltage at low gap distance, such standard capacitors are filled with compressed insulating gases, such as SF_6, which ensures an excellent stability of the capacitance at changing temperature. The construction principle is illustrated in Fig. 3.64, which underlines that the capacitance between the coaxial arranged electrodes are not affected by the earth capacitance, so that a proximity effect, typical for stacked capacitors, must not be taken into account.

As the capacitance of "long" coaxial cylinder electrodes is proportional to the cylinder length l and inversely proportional to the logarithm of the ratio between

Fig. 3.63 Photograph of a 800 kV HVAC reference measuring system (RMS) used to calibrate a capacitive 1000 kV AC voltage divider

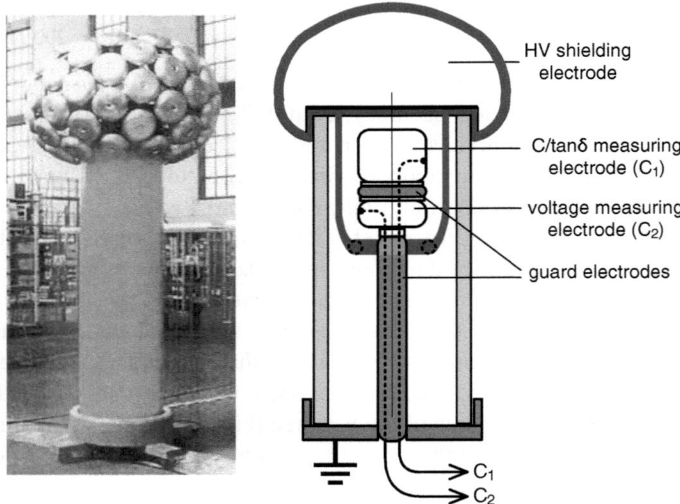

Fig. 3.64 SF_6-filled standard capacitor (rated voltage 800 kV, nominal capacitance 100 pF)

outer and inner conductor radius r_a/r_i, the following approach can be used to design compressed gas capacitors:

$$C_1 = \frac{2 \cdot \pi \cdot \varepsilon \cdot l}{\ln\left(\frac{r_a}{r_t}\right)} \approx \left(55.7 \frac{\text{pF}}{\text{m}}\right) \cdot \frac{l}{\ln\left(\frac{r_a}{r_t}\right)}, \tag{3.36}$$

Example For compressed gas capacitors of rated voltage $\geq$ 500 kV, the ratio of the outer and inner radii r_a and r_i is chosen such that the breakdown voltage approaches a maximum value, which is accomplished for ln $(r_a/r_i) = 1$ and thus $r_a/r_i = e = 2.718$. From Eq. 3.36 follows that the coaxial cylinder electrodes should have a length of 1.8 m to realize a capacitance of $C_1 = 100$ pF.

For rated voltages below 500 kV, the expenditure can significantly be reduced by choosing the ratio $r_a/r_i < e$. Assuming, for instance, $r_a/r_i = \sqrt{e}$, i.e. $\ln(\sqrt{e}) = 0.5$, the cylinder electrodes must have a length of only 0.9 m to achieve the desired capacitance of $C_1 = 100$ pF.

To determine the dynamic behaviour of HV dividers, it is a common practice to apply a sinusoidal voltage of known rms value and tunable frequency to the top electrode, and record the output voltage of the LV arm versus the frequency by means of a calibrated digital oscilloscope or even a spectrum analyzer. A simplified block diagram of such a circuit is depicted in Fig. 3.65.

To minimize the impact of ambient electromagnetic noises, which may appear at an extremely high divider ratio, and thus to improve the signal-to-noise (S/N) ratio, it seems reasonable to replace the original LV capacitance C_2 by a reference capacitor C_{2r} of capacitance substantially lower than C_2 to increase the divider ratio

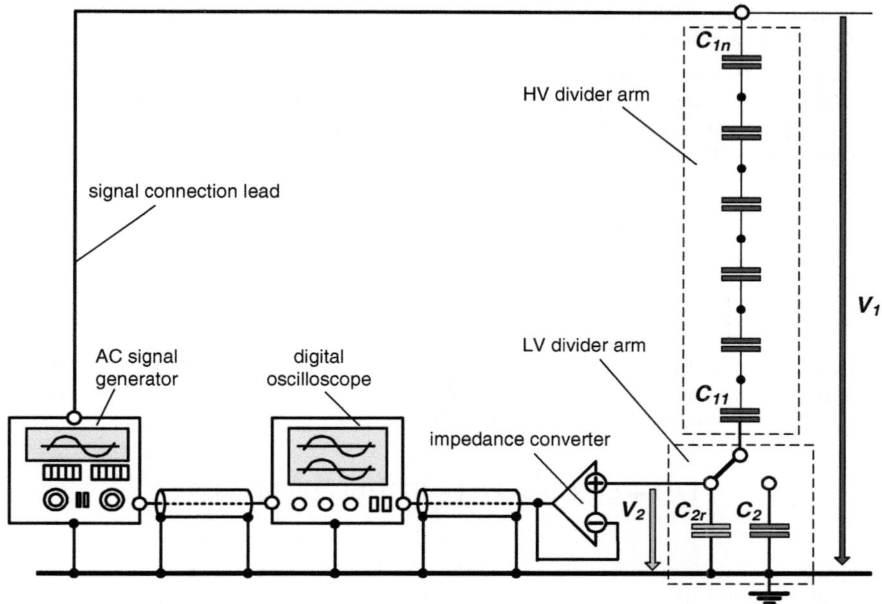

Fig. 3.65 Set-up for measuring the effective capacitance C_1 of a voltage divider

accordingly, for instance up to 1:1000. Under this condition the dynamic behaviour will not be affected significantly because this is governed by the performance of the HV arm. Applying, for instance, an AC peak voltage of 500 V to the top electrode, an output voltage of 500 mV across C_{2r} would appear, which can be measured at reasonable accuracy using a digital oscilloscope, particularly if this equipped with a built-in peak detector as well as an averaging tool. To avoid an increase of the lower limit frequency on account of the input impedance of the digitizer, which is usually in the order of 1 MΩ, the voltage across the reference capacitance C_{2r} must be captured by means of an impedance converter, where the input impedance should not be lower than 10 MΩ.

Example Consider a 2 MV capacitive voltage divider where the HV arm constitutes of $n = 5$ stacked HV capacitors, each of nominal capacitance $C_{11} = C_{12} = \cdots C_{15} = 1$ nF, and a LV capacitor of $C_2 = 877$ nF providing the LV arm. Neglecting the impact of the stray capacitance between HV divider column and earth, the theoretical capacitance of the HV arm would be $C_{1t} = 200$ pF. As mentioned above, for calibration purpose it seems reasonable to replace the capacitor $C_2 = 877$ nF by a reference capacitor of, for instance, $C_{2r} = 200$ nF, which increases the output voltage by almost 5-times. Applying, for instance, an AC peak voltage of $V_1 = 500$ V to the top electrode, the peak value of the output voltage appearing across C_{2r} would of attain $V_r \approx 0.5$ V, which can be measured at reasonable accuracy by means of nowadays available digital oscilloscopes. Provided, the input impedance of the impedance converter amounts 10 MΩ, one gets the following time constant for the LV arm: (200 pF) $\cdot$ (10 MΩ) = 2 s. This is equivalent to a lower limit frequency of about 0.8 Hz and thus capable of transferring the lowest nominal frequency of $f_{1n} = 20$ Hz at reasonable accuracy.

Tuning the frequency of the AC signal generator between $f_{1n} = 20$ Hz and $f_{2n} = 300$ Hz, it shall further be assumed that a constant voltage of $V_{2r} = 0.438$ V across C_{2r} is indicated by the built-in peak detector of the digital oscilloscope. From this follows for the "virtual" capacitance of the HV arm

$$C_1 = \frac{C_2}{\frac{V_1}{V_2} - 1} = \frac{200 \text{ nF}}{\frac{500}{0.438} - 1} = 172 \text{ pF.}$$

Obviously, this is 14% below the theoretical value given by $C_{1t} = 1000$ pF/5 = 200 pF. Replacing the reference capacitor $C_{2r} = 200$ nF again by the original capacitor $C_2 = 877$ nF, one gets the following "true" divider ratio:

$$1/(1 + C_2/C_1) = 1/(1 + 877/0.172) = 1/5099$$

Neglecting the earth capacitance, however, the following (incorrect) divider ratio would be obtained:

$$1/(1 + C_2/C_1) = 1/(1 + 877/0.200) = 1/4386.$$

In this context, it should be noted that the capacitance C_1 representing the HV arm is strongly affected by the clearance between HV divider column and the surrounding walls as well as other conducting structures in the vicinity. Thus, the geometrical configuration should not be changed after the actual capacitance C_1 of the HV arm has been determined. Moreover, it has to be taken into account that C_1 may also be affected by the temperature of the dielectric material used for HV capacitors, such as oil-impregnated paper. Due to the inevitable dielectric losses, the temperature rises at longer testing time and higher test level, which leads to an increase of the expanded measuring uncertainty. Thus, a measuring uncertainty below 3%, as specified in the relevant IEC publications, can hardly be achieved, particularly when the HV arm is composed of stacked oil-impregnated paper capacitors.

3.4.2 Measuring Instruments

The first approach for indicating the output voltage of a converting device by means of a reading instrument was proposed by Chubb and Fortescu in 1913. The basic circuit comprises a HV standard capacitor, connected in series with two back-to-back connected diodes. One of these diodes is connected directly to earth while the other one is connected via a moving coil instrument to earth in order integrate either the positive or negative half-cycle of the displacement current through the standard capacitor. In case of a pure sinusoidal HVAC voltage, the reading of the indicating instrument is direct proportional to both the rms and peak value. However, measuring HVAC voltages with superimposed harmonics, fatal measuring errors would appear.

Another approach to measure the peak value of HVAC voltages traces back to the year 1930, when Davis et al. used a capacitive voltage divider where the output voltage of the LV arm was rectified by means of a vacuum tube diode in series with a storage capacitor. To minimize the leakage current, an electrostatic voltmeter was utilized to indicate the peak voltage appearing across the storage capacitor. In the 1950s, when the first semiconducting rectifier diodes of high reverse resistance and very fast recovery time down to the ns-range were available, so-called sample and hold circuits have been employed for measuring the peak value of HVAC voltages, as illustrated in Fig. 3.66. The main benefit of such circuits is that the inevitable voltage drop across the rectifying diodes can almost completely be compensated.

Due to further achievements in microelectronics, the first *digital peak voltage meters* have been introduced in the 1970s. A typical block diagram and a photograph of such a device are shown in Fig. 3.67. Here, integrated microcomputers serve for an adjustment of the divider ratio and for controlling the entire measuring procedure. The high-resolution A/D converter ensures a measuring uncertainty as low as 0.5% for input voltages ranging between 10 and 1000 V and test frequencies ranging between 0 Hz (DC) and 1000 Hz.

Nowadays, digital recorders are almost exclusively employed to measure the output voltage of HVAC dividers. Using the advanced digital signal processing (DSP) and special software tools, these instruments are capable of measuring all HVAC test voltage quantities presented in Sect. 3.2. For more information in this respect see also Sect. 7.4. In principle, also conventional digital oscilloscopes equipped with a peak detector and averaging tools are capable of measuring HVAC test voltages at desired accuracy. To ensure the requirements according to IEC 60060-2:2010, however, only calibrated digital measuring instruments should be

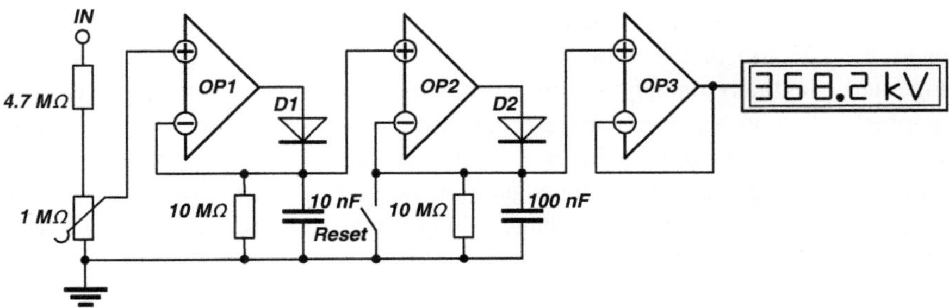

Fig. 3.66 Simplified circuit designed for the measurement of AC peak voltages where the inherent forward conduction voltage of the rectifier diodes is compensated by means of operational (OP) amplifiers

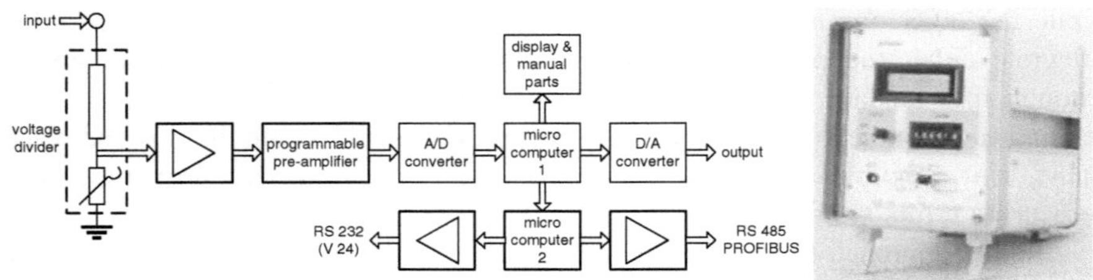

Fig. 3.67 Block diagram and photograph of a commercially available digital peak voltmeter

applied. Moreover, it has to be taken into account that the capacitance of the BNC measuring cable between divider and measuring instrument, which is in the order of 200 pF/m, may contribute to an increase of the capacitance C_2 of the LV arm and should thus not be replaced after the complete HVAC measuring system has been calibrated.

Another problem encountered with HVAC voltage measurements is related to the input resistance R_i of the measuring instrument. A rule of thumb is that the characteristic time constant $\tau_m = R_i \cdot C_2$ multiplied with the minimum test frequency f_{1min} to be measured should be equal or greater than 100. This condition is accomplished for

$$R_i \geq \frac{100}{C_2 \cdot f_{1min}}. \tag{3.37}$$

Example Consider a LV capacitance $C_2 = 1\ \mu F$ and a minimum test frequency of $f_{1min} = 20$ Hz. Inserting these values in Eq. (3.37) one gets $R_i \geq 5\ M\Omega$. As the input impedance of digital recorders amounts commonly 1 $M\Omega$, the output of the LV capacitor C_2 should be connected to the input of the digital recorder via a high-ohmic (5 $M\Omega$) resistive divider, as obvious from Fig. 3.67.

The *electromagnetic compatibility (EMC)* of the measuring system has also to be taken into account because fast transient voltages may be radiated from the test object in case of a breakdown. Under this condition the transient voltage appearing across the LV arm of the voltage divider might suddenly increase up to the kV range and eventually damage the measuring instrument. Thus the input has to be equipped with a proper over-voltage protection unit, which is commonly composed of discharge tubes in combination with spark gaps and very fast suppressor diodes. Moreover, the high-frequency impedance of the ground return should be kept as low as possible to prevent over-voltages along ground connection leads, as treated more in detail in Sect. 9.2.2.

3.4.3 Requirements for Approved Measuring Systems

As specified in IEC 60060-2:2010, HVAC measuring systems applied for quality assurance tests of HV equipment should have an expanded uncertainty $U_M \leq 3\%$, where U_M shall be evaluated with a coverage probability of 95%. For more details in this respect, see also Sect. 2.3.4. Traditionally, the peak voltage Vp is measured where the test voltage is defined in IEC 60060-1:2010 as $V_p/\sqrt{2}$. For sinusoidal test voltages, this quantity equals the rms value. In practice, however, superimposed harmonics may appear (see Sect. 3.2.1) so that the term $V_p/\sqrt{2}$ may substantially deviate from the rms value being of interest. In such cases, particularly when thermal processes play a role, as in the case of pollution tests, not only $V_p/\sqrt{2}$ should be measured but also the true rms value.

As HVAC test systems operate not only at the nominal power frequency, such as 50 or 60 Hz, but also at variable frequencies, as desired, for instance, for induced voltage tests of power transformers and instrument transformers as well as for on-site testing of long-power cable lines (Hauschild et al. 2002), a frequency range larger than 50/60 Hz is of interest. So, the normalized amplitude A_m/A_0 representing the scale factor shall be constant within 1% between the lowest nominal frequency f_{1n} and the highest nominal frequency f_{2n}. An example for this is shown in Fig. 3.68 that refers to $f_{n1} = 20$ Hz and $f_{n2} = 300$ Hz, respectively. For this specific case, the normalized amplitude limit shall be within the red and blue lines plotted in Fig. 3.68. To ensure an acceptable measuring uncertainty even in case of harmonic distortions of the test voltage in the higher frequency range, the amplitude–frequency response should be investigated for a nominal frequency range between $f_{n2} \leq f \leq 7f_{n2}$, as indicated by the shaded area in in Fig. 3.68. This is also of relevance for measuring composite test voltages, i.e. if transient voltages are superimposed on the HVAC test voltage, as treated in Chap. 8.

In this context it seems important to note that *type tests* and *routine tests* of the major components of measuring systems, such as the voltage divider and the measuring instrument as well as the transmission system if other than the BNC measuring cable (for instance, fiber optic link) have to be performed by the manufacturer, see Table 3.9, whereas *performance tests* (see Sect. 2.3.3) for the qualification of complete HVAC measuring systems must be performed under the users responsibility, see Table 3.10.

Additionally to the performance tests also *performance checks* are required using either the comparison method with a second approved measuring system or even sphere gaps according to IEC 60052. The assigned scale factor can be accepted if the deviation between the comparative measurements is less than 3%. Otherwise, the scale factor has to be determined repeating the *calibration procedure*, as described above and in Sect. 2.3. For the clarification of too large deviations it is highly recommended to check separately the scale factor of the individual components, such as the voltage divider and the measuring device including the signal

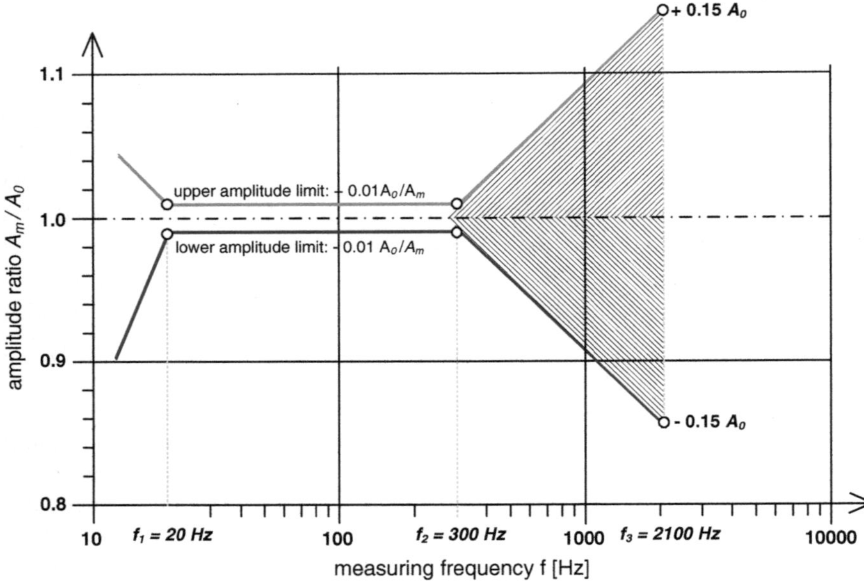

Fig. 3.68 Normalized limits for the amplitude–frequency response of an AC voltage measuring system to ensure a measuring uncertainty $\leq 3\%$ for the nominal frequency range 20 Hz $\leq f_{1n}$ $\leq$ 300 Hz

Table 3.9 Survey on tests recommended in IEC 60060-2:2010 for approved HVAC measuring systems to be performed on each component by the manufacturer

	Type test	Routine test
Linearity		X
Dynamic behaviour	X	
Short-term stability		X
Long-term stability	X	
Ambient temperature effect	X	
Proximity effect	X	
	(if applicable)	
Software effect	X	
	(if applicable)	
Dry withstand test on converting device	X	X
		(if applicable)
Wet or polluted withstand test on converting device	X	
	(if applicable)	
Scale factor of converting device	X	X
Scale factor of transmission system other than cable	X	X
Scale factor of measuring instrument	X	X
Repetition rate (recommended)		

Table 3.10 Survey on tests recommended in IEC 60060-2 for approved HVAC measuring systems to be performed by the user

	Performance test	Performance check
Scale factor calibration	X	
Scale factor check		X
Linearity	X	
	(if applicable)	
Dynamic behaviour	X	
Long-term stability	X	
	(if applicable)	
Proximity effect	X	
	(if applicable)	
Repetition rate (recommended)	Annually, at least every five years	According to the stability, at least annually

transmission link if other than cable. For this, a calibrated AC voltage source of expanded uncertainty less than 1% should be employed. The deviation between the comparative measurements should not exceed 1%. Otherwise, it seems necessary to determine the actual value of the assigned scale factor.

Chapter 4
Partial Discharge Measurement

Abstract This chapter is devoted to the measurement of partial discharges (PD) originating in weak spots in the insulation of HV apparatus and their components. As most HV equipment used to generate, transmit and distribute electric power are energized by high voltage alternating current (HVAC), this chapter focuses primarily on *PD measurements* under alternating voltage. However, specific problems which arise for PD tests under DC and impulse voltage will also briefly be considered. Sensitive PD measurements are often interfered by electromagnetic noises in the measuring surroundings. Therefore advanced features developed in the past to cancel disturbing noises will also be highlighted in this chapter as well as in the relevant sub-sections of Sects. 10.3 and 10.4. For a better understanding of the principles, procedures and instrumentation required for the measurement of electrical PD quantities, first some fundamentals of the PD occurrence will be presented, where also the PD quantities recommended in IEC 60270:2000 for the insulation condition assessment are considered. In the following section, the existing PD models proposed to estimate the PD charge transfer will critically be reviewed. Thereafter specific aspects of pulse charge measurements, as specified in IEC 60270:2000, are considered, which include the major components required to design PD measuring systems used to acquire and display the captured PD data and even the procedures employed to calibrate PD measuring systems. The following sections are dealing with the localization of PD faults, the visualization of PD events, and features developed to reduce or cancel electromagnetic noises interfering sensitive PD measurements. Finally, so-called non-conventional PD detection methods will briefly be highlighted, such as the measurement of electromagnetic PD transients up to the ultra-high frequency range as well as the detection of ultrasonic signals emitted from PD sources.

4.1 Fundamentals

4.1.1 PD Occurrence

Partial discharges (PD) are often caused by imperfections in dielectric materials, such as gaseous inclusions in liquid and solid dielectrics as well as sharp edges in ambient air, which cause a local field enhancement, so that the intrinsic field

© Springer Nature Switzerland AG 2019

W. Hauschild and E. Lemke, *High-Voltage Test and Measuring Techniques*,

https://doi.org/10.1007/978-3-319-97460-6_4

strength may be exceeded. As a consequence, self-sustaining avalanche-discharge may be ignited, where the electrons liberated from neutral gas molecules on account of collision and photo ionization are moving at extremely high drift velocity. This is associated with a very fast displacement current characterized by time parameters in the nanosecond range, which induces a pulse charge at the electrodes. This can thus be detected by means of a coupling device connected to the terminals of the test object or even via field sensors to capture the associated electromagnetic transients radiated from the test object. As defined in the relevant standard IEC 60270:2000, partial discharges are

> localized electrical discharges that only partially bridge the insulation between conductors and which can or cannot occur adjacent to a conductor. Partial discharges are in general a consequence of local electrical stress concentrations in the insulation or on the surface of the insulation. Generally such discharges appear as pulses having durations of much less than 1 μs.

The first experimental study of PD signatures dates back to the year 1777 when Lichtenberg discovered dust figures like stars and circles. These appeared on the surface of an amber cake after it has been hidden by spark discharges up to 40 cm in length (Lichtenberg 1777, 1778). In the middle of the nineteenth century, besides the dust-figure technique initially used by Lichtenberg also photographs became a valuable tool to study PD phenomena, such as surface and interfacial discharges (Blake 1870; Toepler 1898; Müller 1927). At the beginning of the last century, when high alternating voltage was increasingly employed for long-distance power transmission links, it became known that partial discharges can be considered as a precursor for an ultimate breakdown, particularly when occurring in gaseous inclusions embedded in solid dielectrics. Since that time various tools, such as optical, chemical, acoustical and electrical methods, were increasingly used to detect partial discharges. Since the 1960s, the electrical PD measurement has become a widely established procedure for quality assurance tests of HV apparatus and their components, as will be discussed in more detail in the following sections.

Basically, partial discharges are caused by the ionization of gas molecules. Thus PD events occur not only in ambient air but also in gaseous inclusions in solid dielectrics or in gas bubbles and even in water vapour in liquid dielectrics. Therefore it is widely accepted that the discharge processes occurring in gaseous inclusions are comparable with those appearing in ambient air, such as Townsend and streamer discharges as well as leader discharges, which have extensively been investigated since the late 19th and early 20th centuries and published in numerous technical papers and text-books (Paschen 1889, Townsend 1915, 1925; Schumann 1923; Meek and Craggs 1953; Loeb 1956; Raether 1964; Park and Cones 1963; Devins 1984).

Photographs of typical discharge filaments observed in ambient air, insulating oil, and in PMMA are exemplarily shown in Fig. 4.1 (Lemke 1967; Hauschild 1970; Pilling 1976).

As already mentioned above, the formation of single electron avalanches occurs within the nanosecond range due to the extremely fast drift velocity of the electrons.

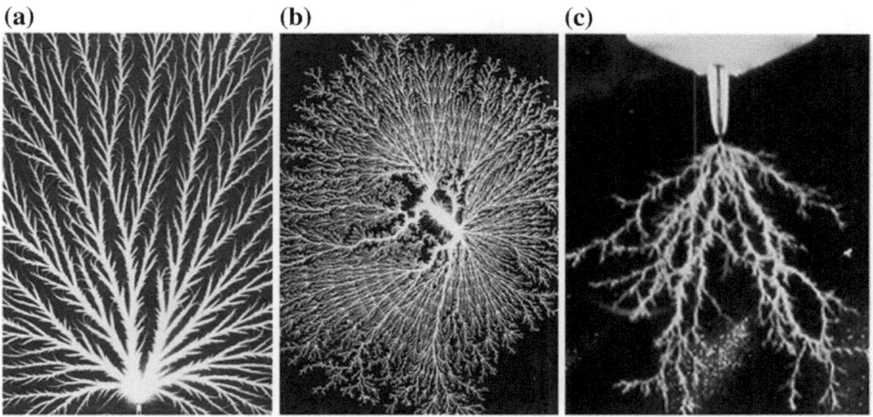

Fig. 4.1 Photographs of discharge channels. **a** Streamer discharge in air (Lemke 1967). **b** Leader discharge in oil (Hauschild 1970). **c** Electrical trees in PMMA (Pilling 1976)

This is associated with a very fast rising displacement current and induces thus a pulse charge at the electrodes of the test object, which occurs also within the nanosecond range, as has been estimated theoretically by Raether in 1964 and by Bailey in 1966 and confirmed experimentally by Fujimoto and Boggs in 1981 as well as by Boggs and Stone in 1982 using the first available 1 GHz oscilloscope (Fig. 4.2). As the origin of PD defects in the insulation of power equipment is usually not accessible, the true shape of PD current pulses cannot be measured when decoupled from the terminals of the test object. This is because the frequency content of the PD signal is dramatically reduced due to the inevitable attenuation and dispersion of the PD signal when transferred from the *PD source* to the terminals of the test object.

As an example, consider the oscilloscopic screenshot shown in Fig. 4.3. Here, an artificial PD pulse created by a PD calibrator was injected in a power cable of 16 m in length. The distance between injection point and near cable end, where the oscilloscope was connected, was chosen as 4m. Thus the first (direct) pulse traveled 4 m, while the second pulse reflected at the remote cable end traveled in total $(16 - 4)$m + 16 m = 28 m. Considering Fig. 4.3b, where the bandwidth of the oscilloscope was set to 200 MHz, the peak value of the second (reflected) pulse is substantially lower than the peak value of the first arriving (direct) pulse. Reducing the bandwidth down to 20 MHz, however, the peak value of the first pulse appears also substantially reduced, while the peak value of the second pulse is only marginally affected. With other words: the peak values of both pulses become invariant when the *measuring frequency* is chosen as low as possible, which is in principle accomplished by a low-pass filtering and thus by a so-called quasi-integration. Under this condition the peak values become proportional to the time integral of the recorded pulses, which can be considered as the background for measuring the PD quantity *"apparent charge"* (Kind 1963; Schon 1986; Zaengl and Osvath 1986). Based on this approach it is recommended in IEC 60270:2000 as well as in the

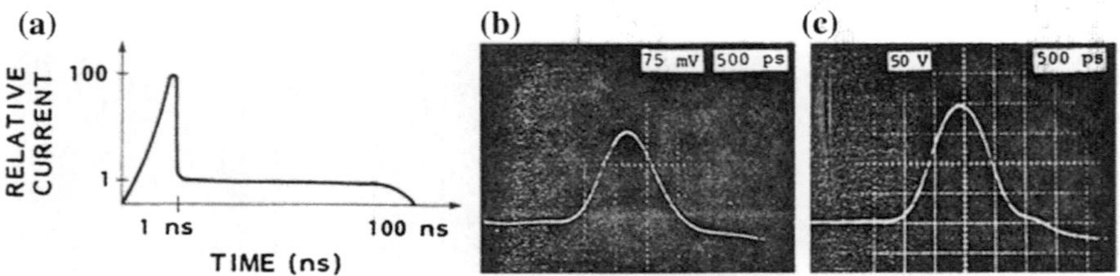

Fig. 4.2 Time parameters of PD current pulses. **a** Theoretical pulse shape calculated by Bailey for small voids embedded in solid dielectrics. **b** Pulse shapes measured by Boggs and Stone for a sharp point (left) and a floating particle in SF6 (right)

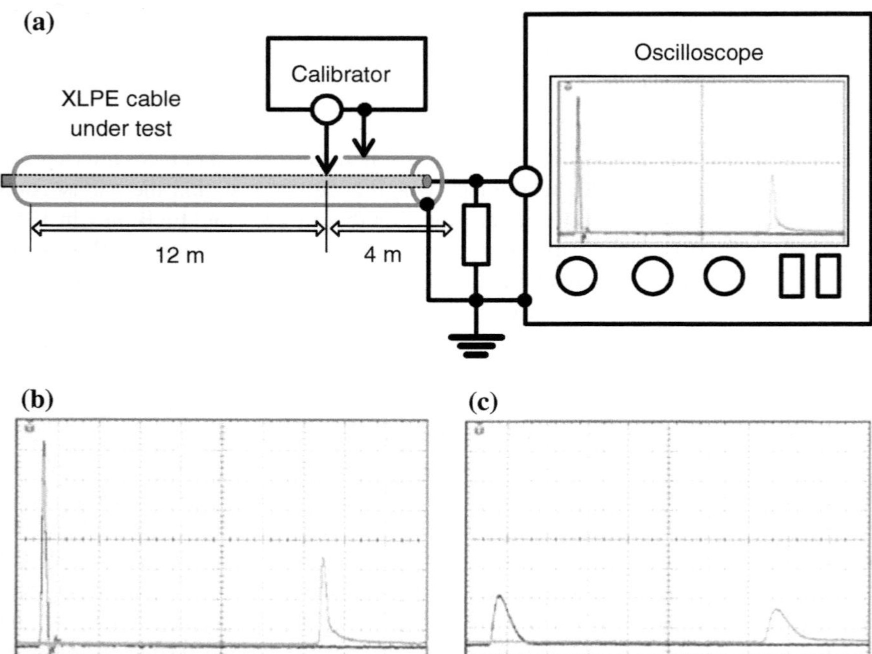

Fig. 4.3 Influence of the measuring frequency on the attenuation of a PD pulse travelling along a XLPE power cable of 16 m in length. **a** Experimental set-up. **b** Oscilloscopic record gained at overall bandwidth of 200 MHz. **c** Oscilloscopic record gained at overall bandwidth of 20 MHz

associated Amendment published in 2015, that the upper limit frequency of the entire measuring system should be chosen below 1 MHz to accomplish the desired quasi-integration of the captured PD signal.

4.1.2 PD Quantities

As mentioned above, to evaluate the PD severity of HV apparatus in service, chemical, optical, acoustical and electrical PD detection methods are commonly employed. In the following, however, only PD quantities gained in the course of electrical PD measurements will be considered, as defined in IEC 60270:2000 as well as in the Amendment to this standard published in 2015.

1. *Apparent charge*

The apparent charge, which can be considered as the main PD quantity, is defined in the relevant IEC 60270:2000 as

> that charge which, if injected within a very short time between the terminals of the test object in a specified test circuit, would give the same reading on the measuring instrument as the PD current itself. The apparent charge is usually expressed in picocoulombs (pC).

Moreover it is noted in this standard, that

> the apparent charge is not equal to the amount of charge locally involved at the site of the discharge, which cannot be measured directly.

For more information in this respect see the Sects. 4.3 and 4.4, respectively.

2. *PD inception and extinction voltage*

The inception voltage V_i and the extinction voltage V_e are usually of interest for HV equipment or its components, which failed the type or development test or even the quality assurance test after manufacturing. Moreover it is a common practice to determine V_i and V_e of HV apparatus, which cannot be designed fully PD-free under operation voltage, such as, for instance, the stator bar insulation of rotating machines. As a practical example, consider the PC screenshots shown in Fig. 4.4, which refer to development tests of a turbo-generator. The voltage profile applied is shown in Fig. 4.4a. Considering Fig. 4.4b, the PD level recorded under rising AC test voltage (red trace) deviates only marginally from that recorded under decreasing test voltage (green trace). Such a low hysteresis is commonly a signature for a "healthy" insulation, which is also manifested by the comparatively low difference between the inception voltage V_i and the extinction voltage V_e. Different to this, harmful PD defects can often be recognized by a clear hysteresis and an extinction voltage being significantly lower than the inception voltage, as exemplarily shown in Fig. 4.4c.

As known, PD pulses are only detectable when the apparent charge exceeds the background noise level, which should thus be kept as low as possible to prevent erroneous test results. For a better understanding consider Fig. 4.5, which refers to a PD test of a defective XLPE cable termination, rated voltage 36 kV. Under laboratory conditions (noise level < 0.5 pC) an inception voltage of 12 kV_{peak} was measured. Assuming, for instance, a background noise level of 5 pC, the measurable inception voltage would increase up to 25 kV_{peak}, as indicated in Fig. 4.5.

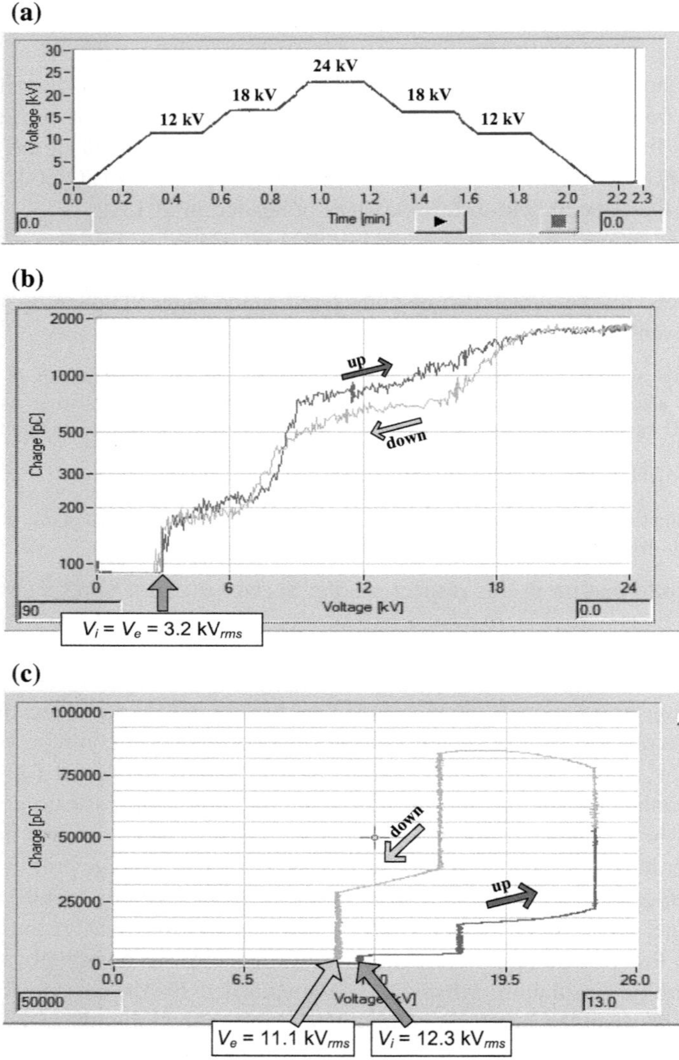

Fig. 4.4 PC screenshots gained in the course of development tests of turbo-generators, rated voltage 24 kV. (**a**) Profile of the applied 50 Hz Ac test voltage (**b**) PD level of a "healthy" stator-bar under raising (red trace) and decreasing (green trace) test voltage (**c**) PD level of a "faulty" stator-bar

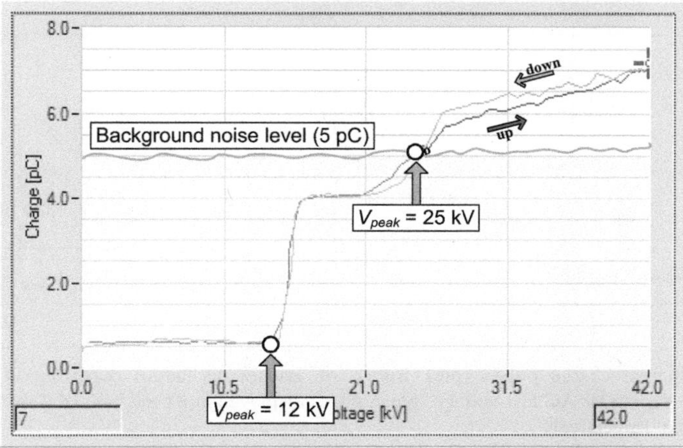

Fig. 4.5 Impact of the background noise level on the determination of the inception and extinction voltage (explanation in the text)

3. *Phase-angle*

The determination of the phase-angle Φ_i according to IEC 60270:2000 is based on the following equation:

$$\Phi_i = \frac{\Delta t_i}{T_c} \cdot 360^0, \tag{4.1}$$

with Δt_i the time span elapsing between the negative-to-positive crossing of the applied AC test voltage and that instant at which the PD pulse is ignited, and T_c the cycle duration of the applied AC test voltage.

Performing practical PD tests, however, the actual phase-angle of each individual PD pulse is usually of minor interest, while the qualitative distribution of the PD pulse trains correlated to the phase-angle of the applied AC test voltage is often helpful to identify harmful PD sources. For a better understanding consider the oscilloscopic screenshots shown in Fig. 4.6, which refer to so-called Trichel discharges. Due to the comparatively high pulse repetition rate, the individual charge pulses appeared only time-resolved when the initially chosen time scale of 4 ms/div (Fig. 4.6a) was set to 0.4 ms/div (Fig. 4.6b). As can be seen, the Trichel pulses are randomly distributed around the negative peak region of the applied AC test voltage, i.e. the phase-angle is scattering around 270°.

As another example, consider the oscilloscopic records shown in Fig. 4.7, which refer to surface and cavity discharges. Here the PD pulses are randomly distributed around the zero-crossing of the AC test voltage, i.e. the phase-angle scatters around 0 and 180°, respectively. As can be seen, the PD signature of cavity discharges is qualitatively comparable to that of surface discharges. However, the pulse magnitudes of surface discharges are substantially greater than those of cavity discharges, which offers the opportunity to discriminate between both kinds of discharges.

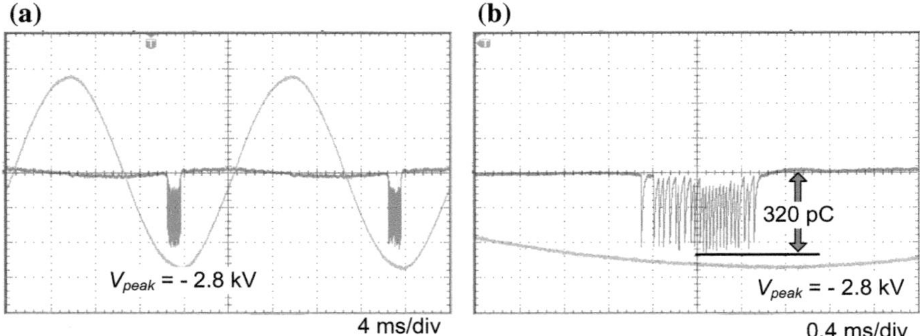

Fig. 4.6 Apparent charge pulses (pink traces) of Trichel discharges occurring in a 3.5 mm needle-plane gap under AC test voltage (green trace) displayed at a time base of 4 ms/div (**a**) and 0.4 ms/div (**b**), respectively

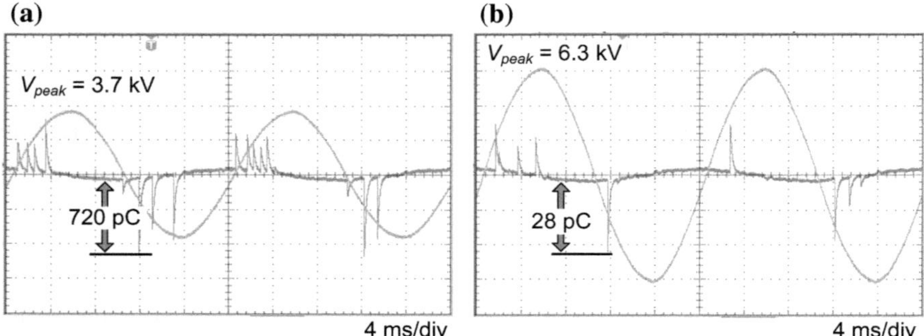

Fig. 4.7 Oscilloscopic screenshots of apparent charge pulses (pink traces) gained for surface discharges (**a**) and cavity discharges (**b**) under power frequency test voltage (green traces)

4. *Pulse repetition rate*

The determination of the PD pulse repetition rate n according to IEC 60270:2000 is based on the following equation:

$$n = \frac{N_p}{\Delta t} \tag{4.2}$$

with N_p the total number of PD pulses and Δt the recording time interval. In practice only those PD pulses are evaluated, which exceed either a specified magnitude or occur within a specified range of magnitude.

As a practical example consider the oscilloscopic screenshots shown Fig. 4.8, which were gained in the course of on-site PD tests of 110 kV power transformers. Rising the power frequency (50 Hz) test voltage up to only V_{rms} = 19 kV, a single

PD pulse appeared in each half-cycle, where the apparent charge exceeded 60 nC in magnitude, see Fig. 4.8a. Here a recording time of $\Delta t = 50$ ms was chosen, which covers five half-cycles. Thus the pulse repetition amounts $n = 5/(50\ \text{ms}) = 100\ \text{s}^{-1}$. However, increasing the test voltage level further, at least up to the highest phase-to-earth-voltage being $V_{rms} = 71$ kV, the magnitude of the apparent charge pulses remained nearly constant, while the PD pulse number increased up to seven per half-cycle, i.e. the repetition rate attained $n = 35/(50\ \text{ms}) = 700\ \text{s}^{-1}$.

Note Based on further experimental studies it has been found that the extremely high magnitudes of the charge pulses displayed in Fig. 4.8 are the result of spark discharges originating in a transformer bushing on account of a poor grounding of the bushing tap terminal.

5. *Accumulated apparent charge*

The accumulated apparent charge q_n is defined in the Amendment to IEC 60270:2000 and can be determined using the following expression:

$$q_n = /q_1/ + /q_2/ + /q_3/ + \cdots + /q_i/, \tag{4.3}$$

with $/q_1/$, $/q_2/$, $/q_3/$ ... $/q_i/$ the absolute values of the apparent charge, and i the number of PD pulses captured within the time interval Δt_r.

Under practical condition only PD pulses are considered, which exceed either a specified threshold level or occur within specified apparent charge limits. For a better understanding consider Fig. 4.9, which refers to the PD behavior of a defective XLPE cable termination of 24 kV rated voltage. Rising the AC voltage up to a test level of $V_{rms} = 12.8$ kV, the maximum apparent charge of the recorded PD pulse trains approached almost 500 pC (pink trace), where the accumulated apparent charge exceeded 7000 pC per cycle (green trace).

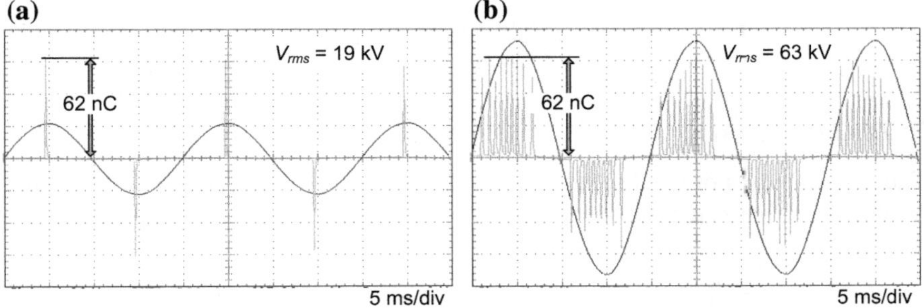

Fig. 4.8 Oscilloscopic screenshot showing the PD signatures of a defective 110 kV transformer bushing at inception voltage $V_{irms} = 19$ kV (**a**) and at highest phase-to-earth voltage being $V_{rms} = 71$ kV (**b**)

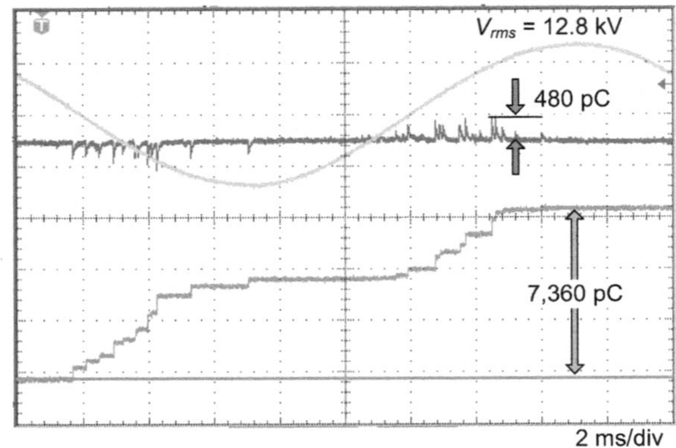

Fig. 4.9 Oscilloscopic screenshot gained for a defective XLPE cable termination (rated voltage 24 kV) showing besides the apparent charge pulses (pink trace) also the accumulated apparent charge (green trace), where the recording time covers both half-cycles of the applied power frequency (50 Hz) test voltage (blue trace)

6. *Average discharge current*

The determination of the average discharge current I_n according to IEC 60270:2000 is based on the following equation:

$$I_n = \frac{1}{T_{ref}} \cdot \{/q_1/ + /q_2/ + /q_3/ + \cdots + /q_i/\} = \frac{q_n}{T_{ref}}, \tag{4.4}$$

with T_{ref} the chosen reference time interval and i the number of the captured PD pulses. As a quantitative example consider the above presented Fig. 4.9. Inserting the accumulated charge being 7360 pC in Eq. (4.4) one gets the following average discharge current:

$$I_n = \frac{7360\,pC}{20\,ms} \approx 0.37\,\mu A.$$

7. *Discharge power*

The determination of the discharge power P_n according to IEC 60270:2000 is based on the following equation:

$$P_n = \frac{1}{T_{ref}} \cdot \{/q_1 \cdot v_1/ + /q_2 \cdot v_2/ + /q_3 \cdot v_3/ + \cdots + /q_i \cdot v_i/\}, \tag{4.5}$$

with $v_1, v_2, v_3 \ldots v_i$ the values of the AC test voltage at instants when the PD pulses of $q_1, q_2, q_3 \ldots q_i$ in magnitude are captured. Basically it has to be taken into account that the phase-angle of the measured AC test voltage is equal to that applied to the terminals of the test object. To determine the discharge power P_n at reasonable measuring accuracy in the real-time mode needs the use of computerized PD measuring systems.

8. *Quadratic rate*

The determination of the quadratic rate D_n according to IEC 60270:2000 is based on the following equation:

$$D_n = \frac{1}{T_{ref}} \cdot \{q_1^2 + q_2^2 + q_3^2 \cdots + q_i^2\}. \tag{4.6}$$

To determine this PD quantity in the real-time mode at reasonable measuring accuracy, the use of computerized PD measuring systems is inevitable. This seldom used quantity was introduced to underline the PD severity of high PD pulse magnitudes.

4.2 PD Models

As known, the electromagnetic transients due to PD events are only detectable at the terminals of the test object. Therefore it seems of interest if this *external pulse charge* is correlated with the *internal pulse charge* flowing through the PD defect. To analyse the PD charge transfer, instead of realistic dielectric imperfections shown in Fig. 4.10 (Kreuger 1989), simple shapes of gaseous inclusions are commonly considered, such as spherical, elliptical and cylindrical cavities. Generally, it can be distinguished between the *network-based PD model* represented by a capacitive equivalent circuit, and the *dipole-based PD model* represented by bipolar space charges deposited at the cavity walls.

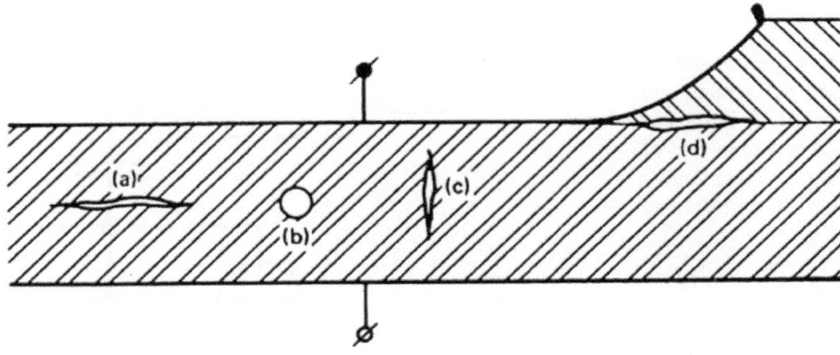

Fig. 4.10 Typical sizes of gaseous inclusions in solid dielectrics according to Kreuger (1989) **a** Flat cavity perpendicular to the electric field, **b** Spherical cavity, **c** Flat cavity aligned with the electric field, **d** Interfacial cavity

4.2.1 Network-Based PD Model

The origin of the network-based PD model traces back to the year 1928, when Burstyn analyzed theoretically the breakdown sequences per half-cycle of a spark gap connected via a small capacitance to an AC test voltage source. The main aim behind this approach was to underline that the comparatively high loss factor of mass-impregnated power cables is the consequence of spark discharges in gaseous inclusions embedded between the laminated insulation (Burstyn 1928). This simple approach has later been investigated experimentally by Gemant and Philippoff using the network proposed by Burstyn, as shown in Fig. 4.11 (Gemant and Philippoff 1932). To measure the breakdown sequences per half-cycle, the spark gap F was connected in parallel to the horizontal deflection plates of a high-voltage oscilloscope. To change the effective capacitance C_2, which is being discharged via the spark gap, additional capacitors were connected in parallel. The capacitance C_1 and resistance R connected in series, see Fig. 4.11, served for a limitation of the transient current through the spark gap F. Typical oscilloscopic records gained for two different test levels are shown in Fig. 4.12. The studies revealed that the measured breakdown sequence depending on both the test voltage level and the capacitance C_2 being discharged was in reasonable agreement with theoretical estimations.

To calculate not only the breakdown sequences depending on the test voltage level and the capacitance being discharged but also to deduce the detectable *external charge* from the *internal charge* flowing through the gas-filled cavity, the circuit shown in Fig. 4.11, has later been modified by Whitehead (1951) and Kreuger (1964), as depicted in Fig. 4.13. In this circuit, C_a represents the capacitance of the bulk dielectric between the electrodes of the test object, while C_b and C_c represent the series connection of the fictive capacitance of the "healthy" dielectric column with the intuitively assumed cavity capacitance, which is being periodically discharged via the parallel connected spark gap F_c. Due to the three characteristic capacitances the equivalent circuit shown in Fig. 4.13 is traditionally referred to as *abc-model*.

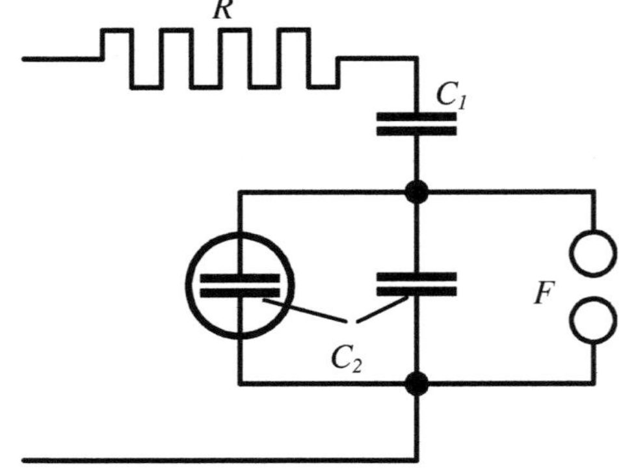

Fig. 4.11 Experimental set-up used by Gemant and Philippoff (1932) to measure the breakdown sequence of the spark gap depending on the applied AC test voltage level

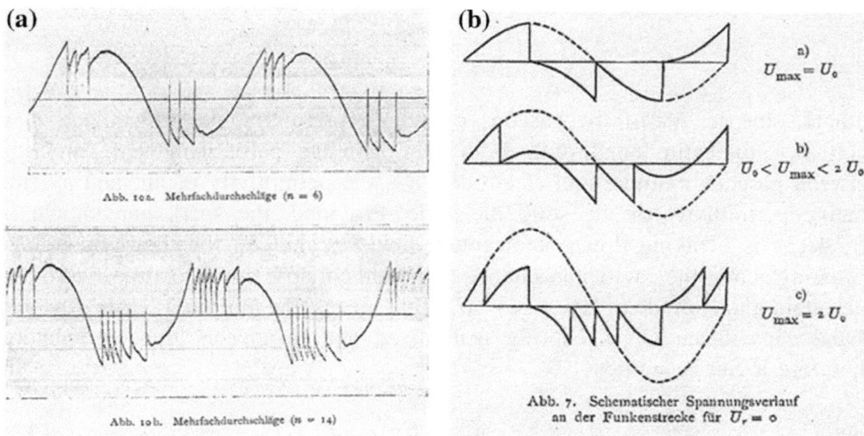

Fig. 4.12 Oscilloscopic records taken by Gemant and Philippoff (1932) (a) and comparison with theoretical estimations (b)

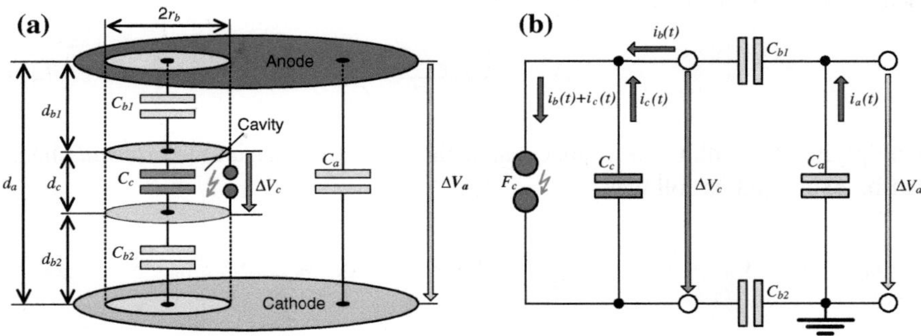

Fig. 4.13 Network-based PD model. a Plane-parallel electrodes with capacitive circuit elements. b Capacitive equivalent circuit (abc-model)

To analyse the pulse charge transfer from the PD source to the terminals of the test object, a cylinder-like column of diameter $2r_b$ is commonly considered, as illustrated in Fig. 4.13a. To simplify the following treatment the radius of the "healthy" dielectric column is assumed as equal to that of the gas-filled cylindrical cavity. Under this condition the equivalent capacitances C_b and C_c can simply be expressed by

$$C_b = \varepsilon_0 \cdot \varepsilon_r \cdot \frac{\pi \cdot r_b^2}{d_b},$$

(4.7a)

$$C_c = \varepsilon_0 \cdot \frac{\pi \cdot r_b^2}{d_c}, \tag{4.7b}$$

with d_c—the length of the gaseous cylinder representing the PD defect, $d_b = d_{b1} + d_{b2}$—the entire length of the both solid cylinders (partial lengths d_{b1} and d_{b2}) between gaseous cylinder and electrodes, ε_0—the permittivity of air, and ε_r—the relative permittivity of the solid dielectric. Provided, the spark gap shown in Fig. 4.13a is breaking down at inception field strength E_i, the transient voltage appearing across the cavity capacitance C_c would collapse down to almost zero and thus attain a magnitude of $\Delta V_c \approx E_i \cdot d_c$. That means, the *internal charge* q_c stored in the capacitance C_c is entirely neutralized and disappears thus completely. Therefore it can be written:

$$q_c \approx \Delta V_c \cdot C_c \approx E_i \cdot \varepsilon_0 \cdot \pi \cdot r_b^2. \tag{4.8}$$

As indicated in Fig. 4.13b, the above mentioned voltage collapse ΔV_c appears also across the series connection of the equivalent capacitances $C_{b1} - C_{b2} - C_a$. Assuming for simplification the relation $C_{b1} = C_{b2} = 2\,C_b$, one gets the following measurable voltage step, which appears across C_a:

$$\Delta V_a = \Delta V_c \cdot \frac{C_b}{C_a + C_b} \tag{4.9}$$

Multiplying this with the test object capacitance C_a, the measurable *external* charge can be expressed as follows:

$$q_a = \Delta V_a \cdot C_a = \Delta V_c \cdot \frac{C_b \cdot C_a}{C_b + C_a} \approx \Delta V_c \cdot C_b \approx E_i \cdot \varepsilon_0 \cdot \varepsilon_r \cdot \pi \cdot r_b^2 \cdot \frac{d_c}{d_b} \tag{4.10a}$$

$$q_a \approx q_c \cdot \frac{d_c}{d_b} \tag{4.10b}$$

For technical insulation, the inequality $d_c \ll d_{b1} + d_{b2} = d_b \approx d_a$ is generally satisfied and the relative permittivity ε_r of the used solid dielectrics is commonly below 5. Under this condition follows from Eq. (10):

$$\frac{q_a}{q_c} \approx \varepsilon_r \frac{d_c}{d_a} \ll 1. \tag{4.11}$$

Due to this relation, the *external* charge detectable in the connection leads of the test object is referred to as *apparent* charge and it is noted in IEC 60270:2000 that the apparent charge is not equal to the amount of that charge involved at the site of the discharge, which cannot be measured directly.

4.2.2 *Dipole-Based PD Model*

The classical apparent charge concept based on the capacitive equivalent circuit according to Fig. 4.13 has been rejected by Pedersen and his co-workers (Pedersen 1986, 1987; Pedersen et al. 1991, 1995). In particular they criticized that the voltage collapsing across the intuitively assumed cavity capacitance is not the result of a spark discharge but caused by the formation of an electrical dipole, which is established by bipolar space charges deposited at the cavity walls. As these bipolar charge carriers are released from neutral gas molecules in the course of collision and photo ionization processes, the number of electrons and negative ions deposited at the anode-side dielectric boundary equals that number of positive ions deposited at the cathode-side dielectric boundary (Fig. 4.14). As the Poissian field between the space charges of opposite polarity opposes the Laplacian (electrostatic) field resulting from the applied test voltage, the ionization of gas molecules will suddenly be quenched. The time interval required for the separation of the bipolar charge carriers lies usually in the nanosecond range, which is thus also representative for the time parameters of the associated current pulse.

To estimate the transient PD current through the equivalent capacitance C_a and thus through the test object shown in Fig. 4.14, the classical continuity equation applies. That means the conductive current $i_c(t)$ through the gaseous inclusion, represented in Fig. 4.14b by the capacitance C_c, equals the displacement current $i_b(t)$ through the solid dielectric column represented by the capacitances C_{b1} and C_{b2}, and continues thus also through the test object capacitance C_a, which causes a voltage jump ΔV_a. That means, the *external* PD charge q_a detectable at the terminals of the test sample equals the *internal* PD charge q_c, which is given by the time-integral of the transient current flowing through the gaseous inclusion on account of the motion of the bipolar charge carriers. Obviously, this is in contrast to the classical *apparent charge* concept deduced from the previously considered network-based PD model.

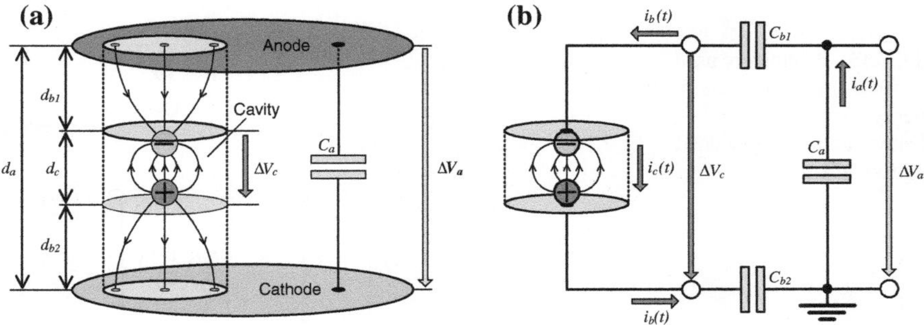

Fig. 4.14 Dipole model according to Pedersen (1986, 1987). **a** Space charge field due to bipolar point charges deposited at the cavity wall. **b** Circuit elements

To estimate the PD charge transfer quantitatively, Pedersen proposed a field theoretical approach, which is based on the Maxwell equations (Pedersen 1986). However, this alternative concept has been ignored in the past, while the traditional *abc*-model is promoted also nowadays (Achillides et al. 2008, 2013, 2017). The reason for that is, apparently, that a PD charge transfer cannot easily be explained when Pedersens concept is adopted, because this needs an excellent knowledge of the field theory based on the Maxwell equations (Maxwell 1873). However, the analysis of the PD charge transfer can substantially be simplified when instead of spherical, ellipsoidal or cylindrical cavities, as frequently investigated and published in numerous technical papers, the establishment of a dipole moment under quasi-uniform field conditions is considered, i.e. the Laplacian field between the electrodes is assumed as constant for the time span required for the separation of the charge carriers (Lemke 2012, 2016). Therefore, to estimate the PD charge transfer, only the change of the Poissian field caused by the separation of the charge carriers of opposite polarity has to be analyzed.

For a better understanding, consider first the motion of a single electron and the associated positive ion, where the propagation to the electrodes shall be hampered by two dielectric layers arranged at spacing d_c, see Fig. 4.15. Provided, at inception field strength E_i only a single electron carrying the elementary charge—e is liberated from a neutral molecule, which occurs at position $x = x_i$, this electron will be attracted by the anode due to the Coulomb force $F = -e \cdot E_i$. Thus, after travelling the maximum possible distance x_i, the field energy transferred to the electron becomes

$$W_e = F \int_{x_i}^{o} \mathrm{d}x = -e \cdot E_i(0 - x_i) = e \cdot E_i \cdot x_i. \tag{4.12}$$

In an analogue manner one gets the following expression for the energy, which is being transferred from the field to the associated positive ion carrying the elementary charge $+e$:

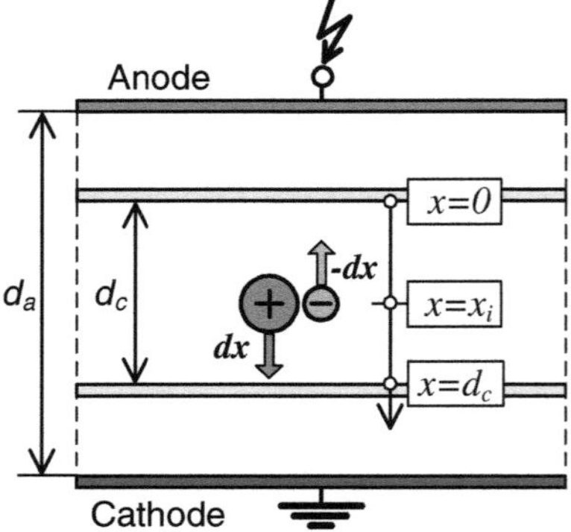

Fig. 4.15 Parameters used for the estimation of the field energy transferred to an electron and the associated positive ion

$$W_p = F \int_{x_i}^{o} \mathrm{d}x = e \cdot E_i \cdot (d_c - x_i). \tag{4.13}$$

Combining the Eqs. (4.12) and (4.13) the entire field energy transferred to both the electron and the positive ion after approaching the solid dielectric boundaries, becomes:

$$W_t = W_e + W_p = e \cdot E_i \cdot d_c. \tag{4.14}$$

Obviously, this expression is independent on the actual site x_i where the electron was released from a neutral molecule. Thus the maximum field energy transferred to an electron avalanche created by the ionization of n_i neutral gas molecules can simply be expressed by

$$W_a = e \cdot n_i \cdot E_i \cdot d_c. \tag{4.15}$$

If one imagines that just before the ignition of a PD event, which occurs at the inception voltage V_i, the test sample is disconnected from the HV test supply, the energy transferred to the moving charge carriers is delivered exclusively from the field energy stored in the test object capacitance C_a, see Fig. 4.14b. For simplification it shall be assumed that at instant t_d all charge carrier approached the dielectric boundaries. Under this condition the classical energy balance theorem can be written as follows:

$$W_a = V_i \int_{o}^{t_d} i_a(t) \cdot \mathrm{d}t = V_i \cdot q_a = e \cdot n_i \cdot d_c \cdot E_i = P_m \cdot E_i, \tag{4.16}$$

with $P_m = e \cdot n_i \cdot d_c$—the dipole moment created by the bipolar charge carriers, and E_i—the inception field strength. Separating the charge q_a delivered from the test object capacitance C_a one gets

$$q_a = e \cdot n_i \cdot d_c \cdot \frac{E_i}{V_i} = P_m \cdot \frac{E_i}{V_i}. \tag{4.17}$$

In this context it seems worth to notice that a similar approach can also be deduced using the concept of image charges (Shockley 1938; Kapcov 1955; Frommhold 1956). This has often been employed by many researchers to account for the displacement current associated with the motion of charge carriers between plane-parallel electrodes (Meek and Craggs 1953; Raether 1964).

Generally it is supposed that the insulation deterioration caused by PD events is mainly governed by the field energy transferred to the charge carriers. Without going into further details it can thus be stated based on Eqs. (4.15) and (4.16) that the measurable *external PD charge* q_a represents a reasonable quantity to assess the *PD severity*. So the density of the dielectric flux lines entering the electrodes and thus the danger for an ultimate breakdown increases at increasing cavity length d_c, see Fig. 4.16. This is not reflected by the previously discussed network-based PD model, where the internal charge decreases at increasing cavity length, because this leads to a decrease of the cavity capacitance C_c.

Note: With reference to the previously considered network-based PD model according to Fig. 4.14 it is noteworthy that the *internal PD charge* q_c could be measured under the condition when all negative and positive charge carriers would be able to cross the entire distance between the anode and cathode. As can readily be shown under this condition the *internal charge* q_c becomes equal to the measurable *external PD charge* q_a. Based on the Eqs. (4.10b) and (4.17) it can thus be written:

$$q_a \approx q_c \cdot \frac{d_c}{d_b} = e \cdot n_i \cdot \frac{d_c \cdot E_i}{V_i} = e \cdot n_i \cdot \frac{d_c \cdot E_i}{d_b \cdot E_i} = e \cdot n_i \cdot \frac{d_c}{d_b}$$

$$q_c \approx e \cdot n_i.$$

That means the "fictive" *internal PD charge* q_c deduced from the network-based PD model equals that charge amount, which is carried by either the positive ions alone or even by the negative electrons alone. However, this is in contrast to the physics of gas discharges due to the fact that in case of a cavity discharge the charge amount carried by the positive ions is equal but of opposite polarity to that charge amount carried by the electrons and negative ions. Due to the very short distance between the bipolar space charges deposited at anode-side and cathode-side dielectric boundary, a great amount of the the positive space charge is compensated by the negative space charge, as can also be deduced from Fig. 4.16a. Consequently, the net charge attains only a low fraction of the unipolar charge carried by the positive ions alone, which is given by $e \cdot n_i$, with n_i—the number of ionized gas molecules.

Even if Eq. (4.16) applies to quasi-homogenous field conditions, this is in principle also applicable for technical electrode configurations, provided the cavity

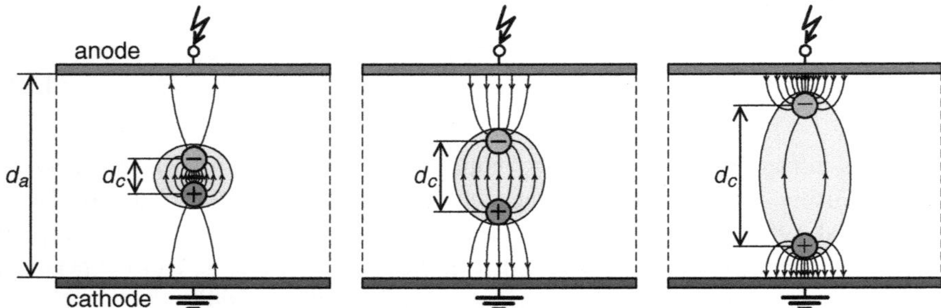

Fig. 4.16 Distribution of the dielectric flux density depending on the distance d_c between the positive and negative space charge

length d_c in field direction is substantially lower than the electrode spacing (Lemke 2013). As an example, consider Fig. 4.17, which refers to a spherical cavity embedded in the bulk dielectric between coaxial cylinder electrodes, as representative for power cables.

To simplify the following treatment, a virgin (space-charge-free) spherical cavity of radius r_c embedded in the bulk dielectric between plane electrodes of distance d_a shall be considered. This seems reasonable because the charge carriers affecting the field distribution are created just after the instant when the inception field strength E_i is exceeded. Assuming the inequality $r_c \ll d_a$, which is generally satisfied for technical insulation, the so-called field enhancement factor k_ε can be approximated as follows (Schwaiger 1925):

$$k_\varepsilon = \frac{3\varepsilon_r}{1 + 2\varepsilon_r}. \tag{4.18}$$

Considering a polyethylene-insulated cable, where the relative dielectric permittivity amounts $\varepsilon_r = 2.3$, one gets $k_\varepsilon \approx 1.2$. Inserting this in Eq. (4.17), the detectable pulse charge can be estimated using the following approach:

$$q_a \approx P_m \cdot \frac{1.2}{r_c\left[\ln\left(\frac{r_a}{r_i}\right)\right]}. \tag{4.19}$$

Determining the dipole moment P_m quantitatively, however, is a challenge due to the fact that the number n_i of ionized molecules is randomly distributed over an extremely wide range. Thus it seems reasonable to investigate only the worst case, which arises when streamer-like discharges are ignited. For this specific case the dipole moment can be expressed by the following semi-empirical approach (Lemke 2013):

$$P_m \approx (270 \text{ pC/mm}) \cdot d_c^2 \quad \text{for } 0.1 \text{ mm} < d_c < 2 \text{ mm}. \tag{4.20}$$

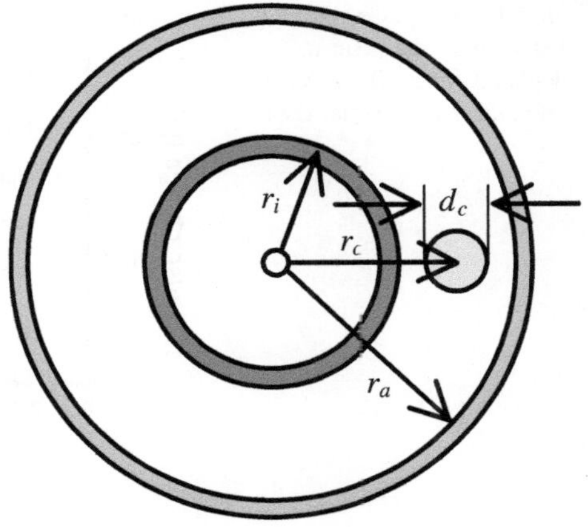

Fig. 4.17 Parameters used for analyzing the PD charge transfer in extruded power cables

Introducing this in Eq. (4.19) and inserting the following assumed geometrical parameters r_i = 8.5 mm, r_a = 14 mm and r_c = 10 mm, which are representative for a 20-kV polyethylene-insulated power cable, one gets

$$q_a \approx (67 \text{ pC}) \cdot \left(\frac{d_c}{d_o}\right)^2.$$

(4.21)

with d_0 = 1 mm—a reference cavity diameter. The curve q_a versus d_c following from this approach is plotted in Fig. 4.18. For comparison purposes, experimental data are also plotted in this figure. In this context it should be noted that some experimental data refer to configurations other than the coaxial cylinder electrodes, as has been investigated here. Nevertheless, the calculated curve meets the measured data quite well, where a better agreement cannot be expected due to the inherent large scattering of the PD pulse magnitudes.

With respect to quality assurance tests of extruded power cables, the pulse charge q_a versus the *PD inception voltage* V_i is the most interesting function. To solve Eq. 4.17, the ratio E_1/V_i must be known. An appropriate approach to determine the inception voltage E_i versus the cavity diameter d_c can be deduced from the Schumann curves (Schumann 1923), which reads:

$$E_i \approx E_0\left(1 + \sqrt{\frac{d_r}{d_c}}\right) \quad \text{for } 0.1 \text{ mm} < d_c < 2 \text{ mm},$$

(4.22)

with E_0 = 2.47 kV/mm—the static inception field strength of ambient air under atmospheric reference conditions and d_r = 0.82 mm a reference cavity diameter. Inserting these values in Eq. (4.17), one gets the following approach to determine the inception voltage of a virgin spherical cavity embedded in the bulk dielectric between coaxial cylinder electrodes:

Fig. 4.18 PD pulse charge q_a versus cavity diameter d_c calculated for a 20 kV XLPE cable and experimental data

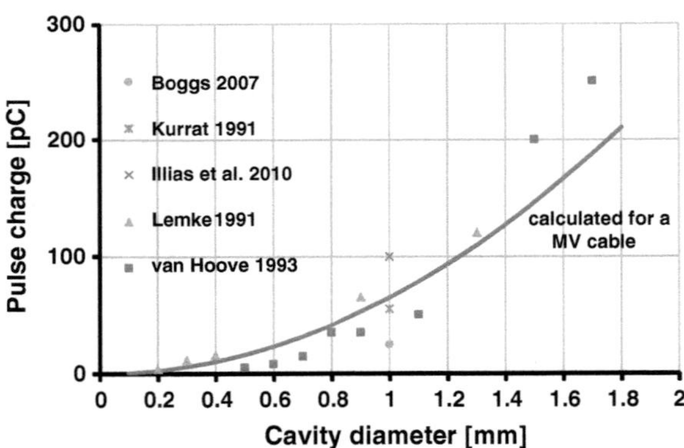

$$V_i = \frac{E_0}{k_\varepsilon} \cdot r_c \cdot \left[\ln\left(\frac{r_a}{r_i}\right)\right] \cdot \left(1 + \sqrt{\frac{d_r}{d_c}}\right) \approx V_i \approx (10.5 \text{ kV}) \cdot \left(1 + \sqrt{\frac{0.82 \text{ mm}}{d_c}}\right).$$

$$(4.23)$$

Combining the Eqs. (4.21) and (4.23), the detectable PD pulse charge q_a versus the inception voltage V_i has been calculated for and plotted in Fig. 4.19. In this context it seems worth to notice that in IEC 60502:1997 a maximum test voltage level of 1.73 V_0 is recommended for medium voltage cables with extruded insulation, where V_0 represents the *rms* value of the phase-to-earth voltage. For the here considered 20 kV cable this is equivalent to a peak voltage of 28 kV. In Fig. 4.19a this value is indicated by a circle, which yields that the cable would pass the PD test only when the *PD level* remains below 5 pC. This result is in satisfying agreement with the recommendations of IEC 60502:1997, which specifies that *the magnitude of the discharge at* 1.73 V_0 *shall not exceed* 10 *pC*. Moreover, it can be deduced from Eq. (4.23) that a PD inception voltage above 28 kV can only be guaranteed for a cavity diameter $d_c < 0.3$ mm.

Even if the calculated curves plotted in the Figs. 4.18 and 4.19 are in satisfying agreement with practical experience, it must be emphasized here that the quantitative values are only approximations. This is because Eq. (4.17) refers to a virgin cavity filled with ambient air under atmospheric normal conditions. However, for technical insulation, neither the cavity size nor the gas pressure are known. Moreover, the dipole moment established by subsequent PD events may strongly be affected by space charges deposited at the cavity walls on account of preceding PD events. This might also be the reason for the large scattering of the pulse charge magnitudes, as commonly encountered in practice.

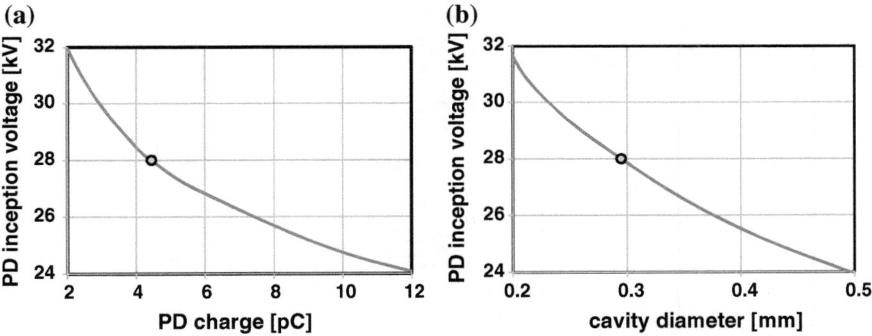

Fig. 4.19 PD inception voltage versus PD charge (**a**) and cavity diameter (**b**) calculated for an XLPE medium voltage cable

4.3   PD Pulse Charge Measurement

4.3.1   Decoupling of PD Signals

As has been discussed already in Sect. 4.1, the frequency content of *PD current pulses* is substantially reduced when the electromagnetic transients are travelling from the PD site to the terminals of the test object, see Fig. 4.3. Different to this, the time integral of the transient PD current and thus the pulse charge is more or less invariant. For a better understanding, consider Fig. 4.20, which refers to a gas-filled cavity embedded in the bulk dielectric between plane-parallel electrodes, as has already been investigated in Sect. 4.2.

As the PD transients are characterized by time parameters down to the nanosecond range and thus by a frequency spectrum up to 100 MHz and even more, the impedance of the HV inductance between HV test supply and test object becomes very high, so that for the duration of each PD event the test object can be considered as disconnected from the HV test supply. Under this condition the charge q_c flowing through the PD defect is provided only by the test object capacitance C_a, which is associated with a fast voltage collapse ΔV_a across the test object. Following the dipole-based PD model, the *pulse charge q_c* flowing through the PD defect is equal to that charge amount flowing through the capacitance C_a of the test object. Therefore it can be written:

$$q_c = q_a = \Delta V_a \cdot C_a. \tag{4.24}$$

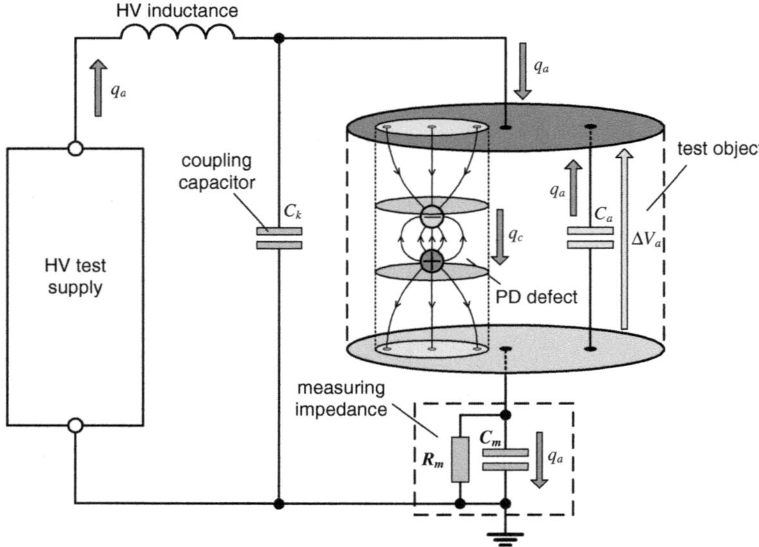

Fig. 4.20 Principle of pulse charge measurement

At instant when the PD process is quenched and the charge carriers are deposited at the cavity walls, the PD current decays to zero so that the frequency content and thus the effective impedance of the HV connection leads decreases accordingly. Therefore the test object capacitance C_a will be recharged again by the HV test supply, i.e. the former voltage step ΔV_a appearing across C_a is compensated, which appears at inverted polarity. That means the time integral of the current recharging C_a and thus the apparent charge can also be assessed using Eq. (4.24), even if the actual shape of the recharging current is very different from that of the original PD current. This offers the opportunity to measure the PD pulse charge by means of a *measuring impedance*, if connecting the low voltage electrode of the test object to earth potential (Fig. 4.20). Additionally the terminals of the test object should be shunted by a HV capacitance C_k to ensure that the entire current recharging C_a is flowing through the *measuring impedance*.

For a direct measurement of the pulse charge, the measuring impedance could be equipped with a measuring capacitance C_m. Under this condition, the magnitude of the voltage jump appearing across C_m is direct proportional to the pulse charge to be measured. At alternating test voltages, however, the capacitive load current flowing through the test object and thus through the *measuring impedance* could substantially exceed the signal level caused by the PD events, as exemplarily shown in Fig. 4.21a. To overcome this crucial problem, the measuring capacitance C_m is commonly shunted by a measuring resistor R_m, which reduces the superimposed AC current accordingly, see Fig. 4.21b and c. Practical experience revealed, however, that a measuring impedance equipped with an RC network shown in Fig. 4.20 is only applicable for low-capacitive test samples used for fundamental PD studies, such as point-to-plane gaps, but not for PD tests of HV equipment and their components.

Even if the highest measuring sensitivity is achieved by connecting the measuring impedance between low voltage side of the test object and ground potential, this approach is not applicable in general. On one hand the ground connection lead of the test object can often not be interrupted and, on the other hand, the measuring impedance must be able to carry the entire capacitive load current through the test object, as discussed above, and additionally the fast transient current that would occur in case of an unexpected breakdown. To overcome this crucial problems, the

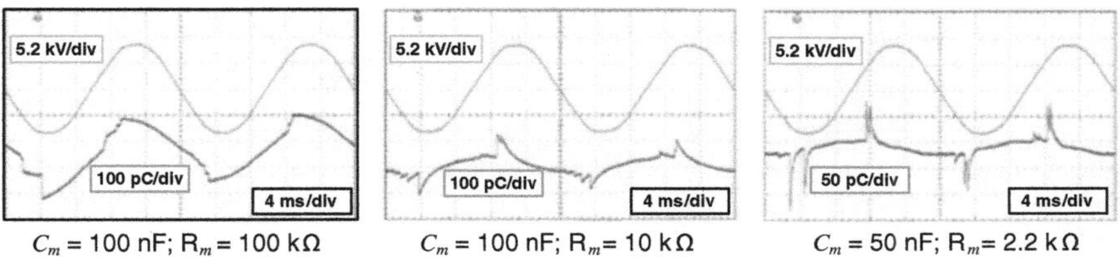

Fig. 4.21 Records of typical voltage signals caused by PD events (pink trace), which were captured from an RC measuring impedance connected in series with a point-to-plane gap subject to AC test voltage (green trace)

measuring impedance is usually connected in series with the coupling capacitor C_k, as illustrated in Fig. 4.22. Here, the measuring resistor R_m is shunted by an inductance L_s, which carries the entire alternating load current trough the coupling capacitor. The over-voltage protection unit O_p is required to suppress fast over-voltages due to unexpected breakdowns and prevents thus a damage of the measuring impedance and the measuring instrument, as well. As the PD coupling unit shown in Fig. 4.22 provides a high-pass filter, it has to be taken care that the lower limit frequency f_1 is chosen as low as possible, preferably below 100 kHz.

Example To design a coupling device according to Fig. 4.22, which is intended for induced voltage tests of power transformers, the following parameters shall be assumed:

$$\text{Lower limit frequency:}\quad f_1 = 50\ \text{kHz}$$
$$\text{Effective measuring impedance:}\quad R_m = 1\ \text{k}\Omega$$
$$\text{Maximum applied test voltage level:}\quad V_a = 200\ \text{kV}$$
$$\text{Maximum test frequency:}\quad f_{ac} = 400\ \text{Hz}$$

As the lower limit frequency is given by

$$f_1 = \frac{1}{2\pi \cdot C_k \cdot R_m} = 50\ \text{kHz}$$

the minimum capacitance of the coupling capacitor can be determined as follows:

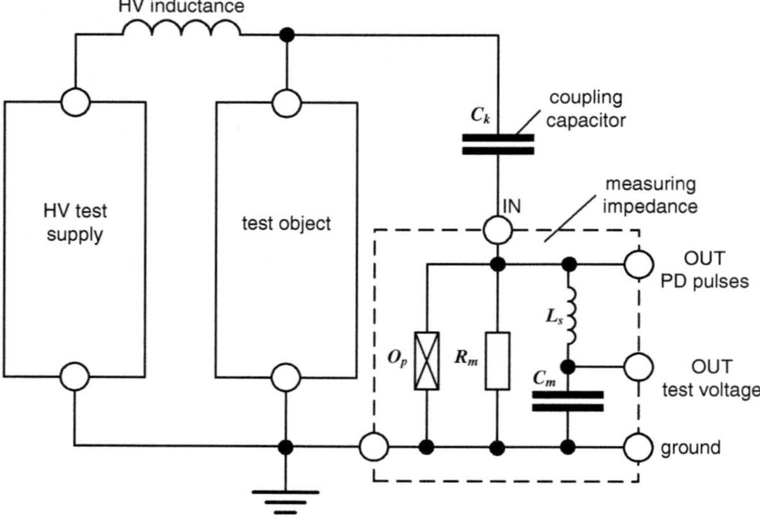

Fig. 4.22 PD measuring circuit, where the measuring impedance is connected in series with the coupling capacitor

$$C_k = \frac{1}{2\pi \cdot f_1 \cdot R_m} \approx 3.2 \text{ nF}.$$

Inserting the maximum exciting frequency being f_{ac} = 400 Hz, one gets the following capacitive current through the coupling capacitor:

$$I_c = 2\pi \cdot f_{ac} \cdot C_k \cdot V_{ac} = 1.6 \text{ A}.$$

As this current flows through the measuring resistor R_m = 1 kΩ, a voltage drop as high as 1600 V would appear. Of course, this is dangerous for the operator and could also damage the connected PD measuring system. Thus, to reduce this high voltage drop, R_m should be shunted by an inductance L_s (Fig. 4.22). However, applying this option it has to be taken care that the lower limit frequency may significantly exceed the previously mentioned lower limit frequency, which should be chosen below 100 kHz. This condition is accomplished for

$$2\pi \cdot f_m \cdot L_s \geq 5R_m; \quad L_s \geq 16 \text{ mH}$$

For the here considered maximum frequency f_{ac} = 400 Hz of the applied test voltage the inductive impedance attains

$$Z_l = 2\pi \cdot f_{ac} \cdot L_s \approx 40 \ \Omega.$$

Under this condition, the above mentioned load current of $I_c = 1.6$ A flowing through the coupling capacitor of $C_k = 3.2$ nF causes a comparative low voltage drop across L_s, which attains 64 V. This value can further be reduced using a high-pass filter of higher order.

To display the PD pulse in a phase-resolved matter by means of an oscilloscope or even a computer-based PD measuring system, the PD coupling unit could further be configured using an additional measuring capacitor C_m, as illustrated in Fig. 4.22. Due to the very different frequency spectra of the PD pulses and the AC test voltage, these both signals can simply be discriminated. Considering Fig. 4.22, these appear at the outputs "PD pulses" and "test voltage". If, for instance, the above-introduced maximum test voltage level of V_{ac} = 200 kV should be attenuated down to 50 V, a divider ratio of 1:4000 would be required. Using a coupling capacitor of C_k = 3.2 nF, a low-voltage measuring capacitance of C_m = 12.8 μF would be required.

4.3.2   PD Measuring Circuits According to IEC 60270

The basic *PD measuring circuits* recommended in IEC 60270:2000 are shown in Fig. 4.23. To avoid any danger for the operator due to hazardous over-voltages in case of an insulation breakdown, the measuring impedance should always be placed inside the HV test area. Moreover, it has to be taken care that the HV connection leads are designed PD-free up to the highest applied test voltage level.

The grounding leads used for the current return should be kept as short as possible and made of Cu or AL foil of approx. 100 mm in width to minimize the inevitable inductance in the higher-frequency range and thus to prevent electromagnetic interferences. For more information in this respect see Sects. 4.5.2 and 9.2.2.

The most commonly employed coupling mode is the use of a coupling capacitor in series with a measuring impedance (Fig. 4.24), as has already been discussed previously based on Fig. 4.22.

Electromagnetic noises interfering sensitive PD measurements can also be eliminated to a certain extend using the so-called balanced PD bridge according to Fig. 4.23c. Here, the adjustable measuring impedances Z_{m1} and Z_{m2} are installed in the ground connection leads of both the test object and the coupling capacitor providing the measuring branch and the reference branch, respectively. Adjusting Z_{m1} and Z_{m2} accordingly to balance the bridge, common mode noises appearing at the high-voltage terminals are more or less suppressed by the differential amplifier. Thus only the PD signal originating in the test object appears at the output of the differential amplifier and is thus measured by the PD instrument. To ensure a high common mode rejection, the bridge should be designed as symmetrical as possible. Thus it is advisable to use instead of the coupling capacitor a complementary PD-free test object as a reference. Despite of the benefits of the balanced bridge for noise suppression, this approach is not generally employed in practice because the design is a challenge due to the fact that both branches must have an equivalent frequency response over the full bandwidth used for the PD signal processing. Moreover it has to be taken into account that the time-delay of the interfering signal traveling along the complementary PD-free test object is equal to that signal traveling along the test object under investigation.

Another option frequently used for PD tests of power transformers is the so-called *bushing tap coupling* mode, as illustrated in Fig. 4.25. Of course, this coupling mode is only applicable for capacitive-graded bushings equipped with a bushing tap intended for $C/\tan \delta$ measurements. Here the HV bushing capacitance C_1 provides in principle the coupling capacitor C_k shown in Fig. 4.23a.

4.3.3 PD Signal Processing

As discussed in Sect. 4.2.2, the time integral of the transient PD current flowing through the connection leads of the test object and thus the measurable pulse charge is correlated to that charge amount flowing through the PD defect. Moreover, the pulse charge is more or less invariant even if the PD current pulse may substantially be distorted when travelling from the PD source to the terminals of the test object. Thus the transient voltage appearing across a resistive measuring impedance is commonly integrated to measure the so-called apparent charge. This is conveniently achieved by a *band-pass filtering*, commonly referred to as quasi-integration. For

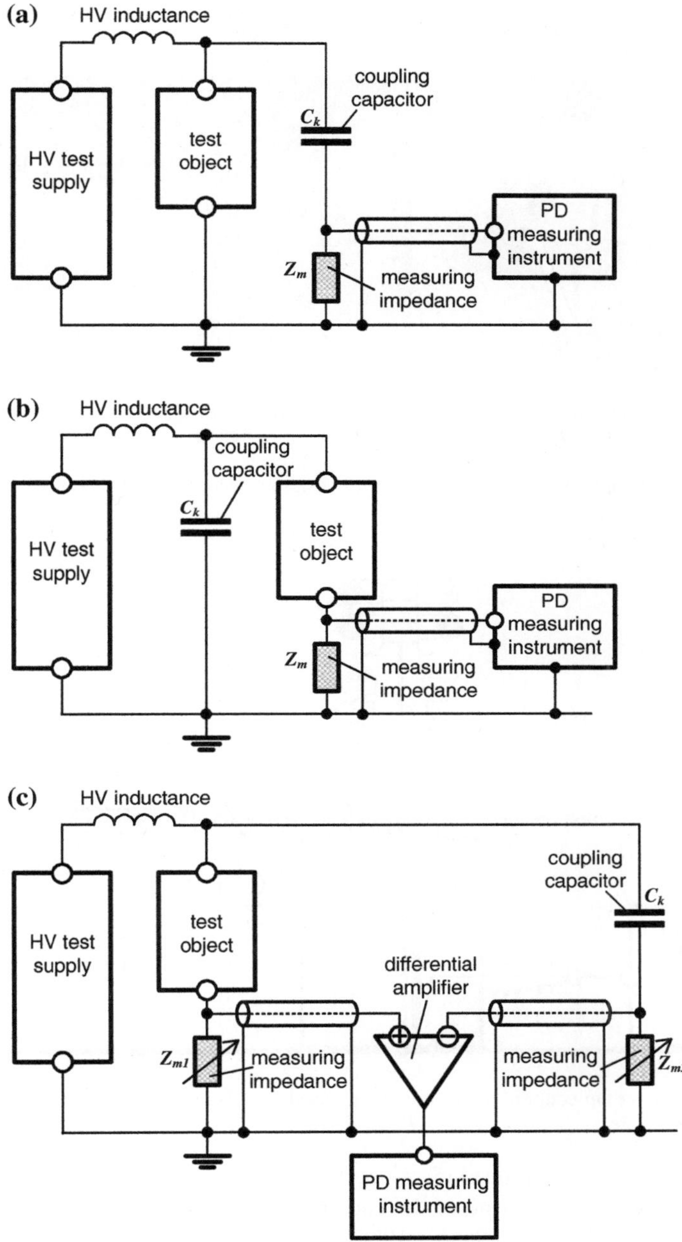

Fig. 4.23 Basic PD measuring circuits recommended in IEC 60270:2000. **a** Measuring impedance in series with a coupling capacitor used for grounded test objects. **b** Test object grounded via the measuring impedance. **c** Bridge circuit recommended for noise reduction

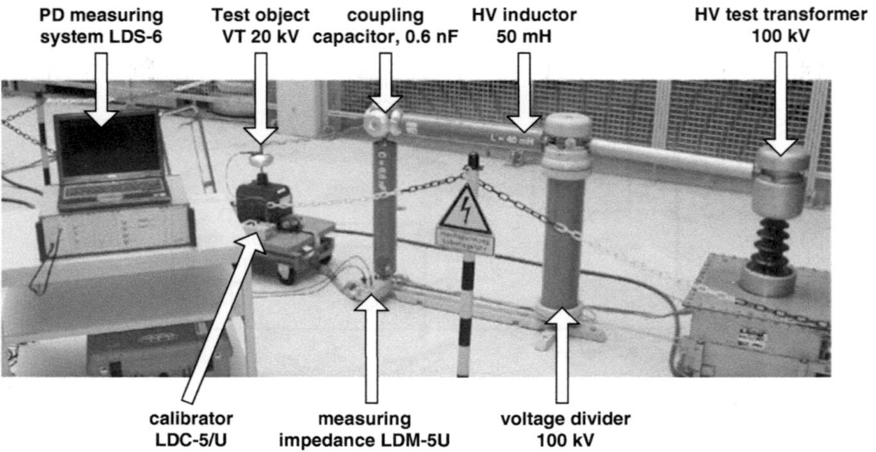

Fig. 4.24 Photograph of a PD measuring circuit designed according to IEC 60270. Courtesy of Doble Lemke

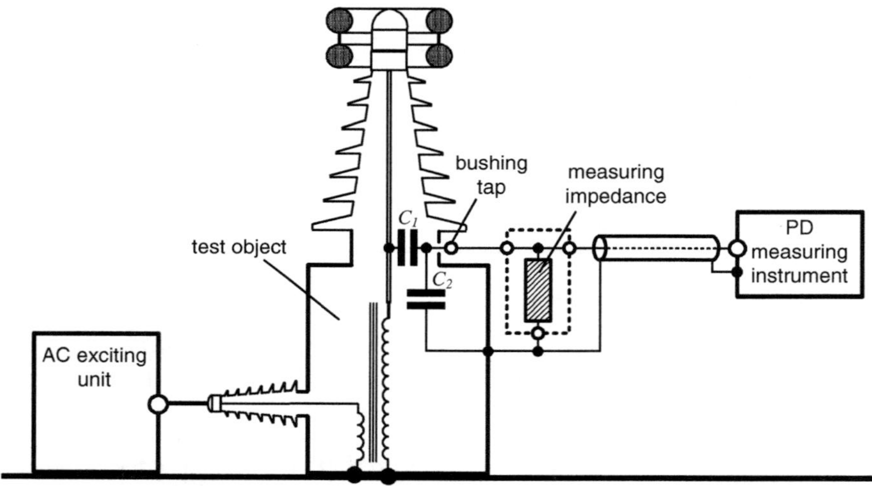

Fig. 4.25 Bushing tap coupling mode commonly used for PD tests of power transformers

this purpose the signal processing is performed in a frequency range where the amplitude–frequency spectrum of the captured PD pulses is nearly constant (Kind 1963; Kuffel et al. 2006; Schon 1986; Zaengl and Osvath 1986; König and Narayana 1993; Lemke 1997), as illustrated in Fig. 4.26. Practical experiences revealed that this requirement is accomplished for most test objects when the upper cut-off frequency is limited below 1 MHz, as has also been recommend in the Amendment to IEC 60270:2000, published in 2015.

Depending on the bandwidth $\Delta f = f_2 - f_1$ used for PD signal processing it is generally distinguished between *wide-band* and *narrow-band instruments*. For wide-band PD instruments the following frequency parameters are recommended in the Amendment to IEC 60270:2000:

$$\text{Lower limit frequency:} \quad 30 \text{ kHz} \leq f_1 \leq 100 \text{ kHz}$$
$$\text{Upper limit frequency:} \quad 130 \text{ kHz} \leq f_2 \leq 1000 \text{ kHz}$$
$$\text{Bandwidth:} \quad 100 \text{ kHz} \leq \Delta f \leq 970 \text{ kHz}$$

Note According to IEC 60270: 2000, the term "wide-band" refers to the band-pass filter characteristics of the PD processing unit characterized by a bandwidth $\Delta f = f_2 - f_1$, which is substantially greater than the lower limit frequency f_1. In this context, it must be emphasized that this term is not correlated with the frequency spectrum of real PD current pulses, which covers often a frequency range up to the GHz range, see Sect. 4.1.

The PD pulse response of a band-pass filter having the *upper* and *lower limit frequency* of $f_2 = 320 \text{kHz}$ and $f_1 = 40$ kHz, respectively, is shown in Fig. 4.27a. Based on the network theory it can be stated that the peak value of the output pulse is proportional to the time integral of the input pulse where the duration of the output pulse is substantially greater than that of the input pulse.

In this context it should be noted that the integration performance is only governed by the *upper limit frequency f2* and not by the lower limit frequency f_1. Thus also narrow-band filters could in principle be used to accomplish a quasi-integration. For this purpose the bandwidth Δf must be chosen substantially lower than the center frequency $f_0 = (f_2 - f_1)/2$. As a typical measuring example, consider Fig. 4.27b, which refers to the PD pulse response of a narrow-band amplifier characterized by a center frequency of $f_0 = 780$ kHz and a bandwidth of $\Delta f = 9$ kHz. Here the maximum peak-to-peak value of the oscillating response is direct proportional to the pulse charge injected in the the narrow-band amplifier, which follows also from the classical network theory (Schon 1986). In IEC

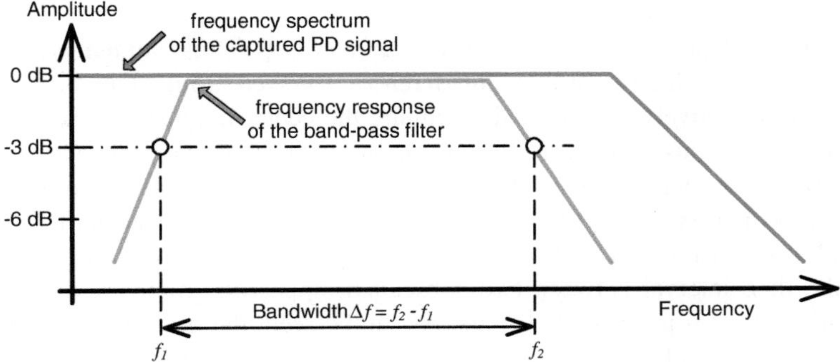

Fig. 4.26 Frequency spectrum of PD pulses in comparison to the frequency band recommended for PD pulse charge measurements according to IEC 60270:2000

(a)

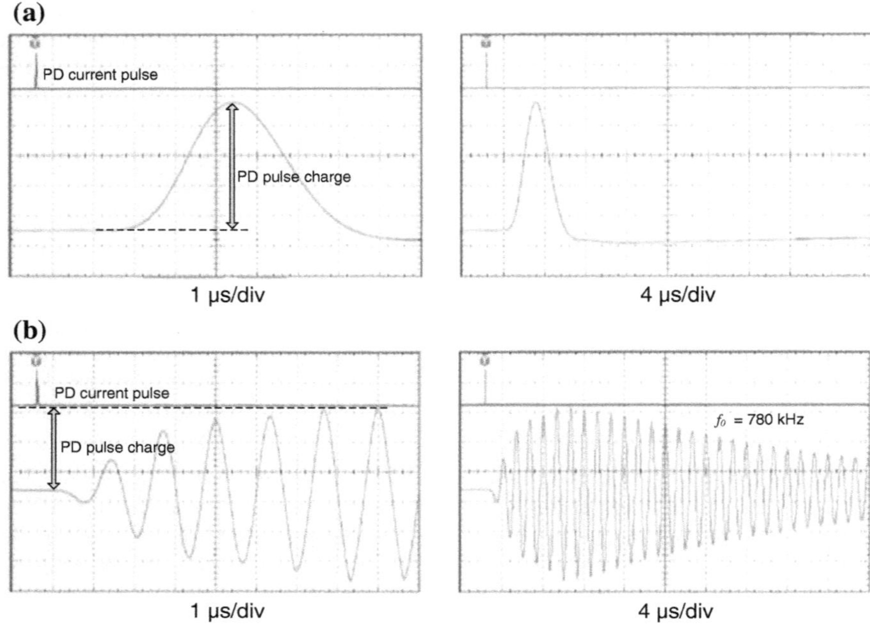

(b)

Fig. 4.27 PD pulse responses of a wide-band (**a**) and a narrow-band (**b**) processing unit. **a** Lower limit frequency $f_1 = 40$ kHz and upper limit frequency $f_2 = 320$ kHz. **b** Mid-band frequency $f_0 = 780$ kHz and bandwidth $\Delta f = 9$ kHz

60270:2000 the following frequency parameters are recommended for narrow-band PD instruments:

Center frequency : $50 \, \text{kHz} \leq f_0 \leq 1000 \, \text{kHz}$

Bandwidth : $9 \, \text{kHz} \leq \Delta f \leq 30 \, kHz$

The main advantage of *narrow-band amplifiers* is the noise immunity, i.e. continuous appearing high-frequency interferences received from radio broadcasting stations can effectively be canceled by tuning the center frequency f_0 accordingly, see Sect. 4.5. In this context it must be emphasized, however, that fatal superposition errors may appear due to the comparatively long duration of the oscillating response, which exceeds often several hundreds of µs. As an example consider Fig. 4.28, which refers to the double-pulse response of a narrow-band amplifier. Here the time interval between two subsequent appearing impulses was reduced from originally 1.4 µs down to 0.7 µs. As can be seen, the peak-to-peak values of the oscillating signal decreases substantially under this condition. Such a behavior is typical for PD pulses decoupled from power cables due to their reflection at the cable ends. Moreover, superposition errors may also occur on account of oscillations excited by PD events in the windings of rotating machines and power transformers. This is the reason why narrow-band amplifiers are in general not recommended for measuring the apparent charge of PD pulses in terms of pC.

4.3.4 PD Measuring Instruments

4.3.4.1 General

Schering bridges in combination with oscilloscopes can be considered as the first instruments used for the electrical PD detection. However, the measuring sensitivity was comparatively low. This was substantially increased in the 1920s when the first super-heterodyne receivers equipped with narrow-band amplifiers were available (Armann and Starr 1936; Dennhardt 1935; Koske 1938; Lloyd and Starr 1928; Müller 1934; Schering 1919). To ensure comparable and reproducible PD measurements, the requirements for such instruments were first specified in the USA and North America in 1940, when the standard *"Methods for Measuring Radio Noise"* was published by the *"National Electrical Manufactures Association NEMA"*. This standard was later revised by the *NEMA Publication 107 "Methods of Measurement of Radio Influence Voltage (RIV) of High-Voltage Apparatus"* and issued in 1964. An equivalent standard for RIV measurements of HV apparatus was also edited in Europe by the *"Comité International Spécial des Perturbation Radioélectrique (CISPR)"*, published in 1961.

Radio interference voltages (RIV) are commonly measured in terms of μV and weighted according to the acoustical noise impression of the human ear. Therefore it cannot be expected that these are correlated to the apparent charge of PD pulses measured in terms of pC, as can readily be proven experimentally (Harrold and Dakin 1973; Vaillancourt et al. 1981). Moreover, fatal superposition errors might appear at high PD pulse repetition rate or even in case of reflections and oscillations excited by the fast PD transients in cables and inductive components, as has been

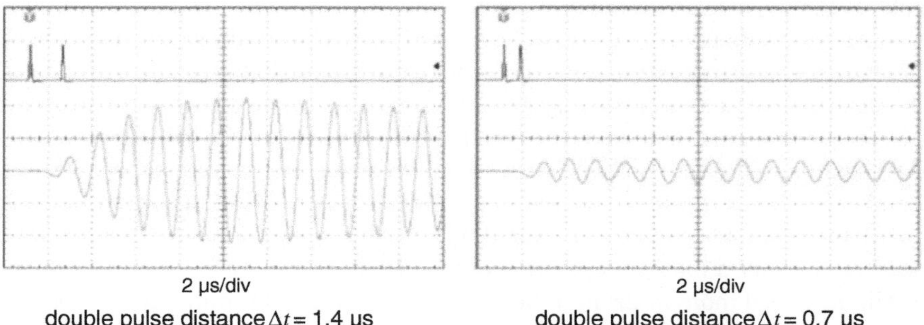

Fig. 4.28 Impact of the double-pulse distance on the oscillation magnitude of a narrow-band PD processing unit

discussed previously. As a consequence, the *"International Technical Commission (IEC), Technical Committee No. 42: High-Voltage Testing and Measuring Technique"* decided the edition of a separate standard on PD measurements. The first edition of *"IEC Publication 270"* appeared in 1968, where besides the definition of the *PD quantity "apparent charge"* as well as the *inception* and *extinction voltage*, several other PD quantities were introduced, such as the repetition rate as well as the power of consecutive PD pulses. Additionally, rules for calibrating PD measuring circuits were specified, and guidelines were attached which supported the identification of typical PD defects under AC test voltage based on oscilloscopic records using either the elliptical or the linear time base to record the characteristic PD pulse trains in a phase-resolved manner.

The second edition of IEC Publication 270, published in 1981, contained more details on the calibration procedure. Additionally, the PD quantity *"largest repeatedly occurring PD charge"* was specified. Based on this standard, the electrical PD measurement became an indispensable tool for tracing dielectric imperfections in HV apparatus, which might be caused by design failures as well as by a poor assembling work. Therefore, the measurement of partial discharges was increasingly requested with respect to increased quality requirements, which was also forced by the enhancement of the design field strength and, last but not least, by demands concerning the enlargement of the lifetime of HV equipment.

The following treatment is based on the third edition of IEC 60270 published in 2000, which can be considered as an extensive revision of the second edition. The specification covers besides the traditional analogue PD signal processing also the advanced digital acquisition of the captured PD pulses. Moreover, a section has been added, which refers to maintaining the characteristics of PD measuring systems and the associated calibrators, as will be considered more in detail in Sect. 4.3.7.

4.3.4.2 Analogue PD Instruments

A simplified block diagram of analogue PD instruments is shown in Fig. 4.29. The input of the device is commonly equipped with an attenuator to adjust the desired measuring sensitivity as well as a fast over-voltage protection unit to avoid a damage of the instrument in case of an unexpected breakdown of the test object. The desired integration of the captured PD signal is commonly performed by the use of a band-pass amplifier, as mentioned previously. As an alternative, a wide-band amplifier in combination with an electronic integrator can also be used (Lemke 1969), as illustrated in Fig. 4.29. This concept offers the opportunity to record the true shape of the captured PD pulses, which enables flight-of-time measurements used for the localization of the PD site, for instance, in power cables, as will described more in detail in Sect. 4.4. Another benefit of this concept is that pulse-shaped noises can effectively be rejected using various features for gating and windowing, as will be presented in Sect. 4.5.

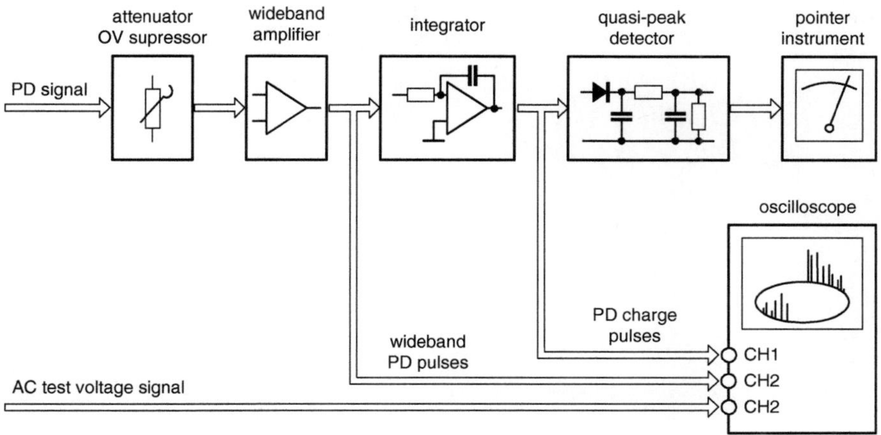

Fig. 4.29 Simplified block diagram of an analogue PD instrument

The quasi-peak detector in combination with the reading instrument is intended
for measuring the *"largest repeatedly occurring PD magnitude"* according to IEC
60270:2000. This unit averages the reading, which is particularly useful to weight
randomly distributed PD pulse magnitudes, see Fig. 4.30. Another benefit of this
unit is that stochastically appearing noise pulses of comparatively low repetition
rate are either not indicated or their magnitude is substantially reduced. As specified
in IEC 60270:2000, the charging and discharging time constant of the quasi-peak
detector should be chosen as $\tau_1 \leq 1$ ms and $\tau_2 \approx 440$ ms, respectively, to accomplish the pulse train response according to Fig. 4.31. In this context it should be
emphasized that this approach ensures more or less reproducible test results under
power frequency (50/60 Hz) AC voltages, but not when the test frequency is
changed, as usual for quality assurance tests of power transformers and instrument
transformers, where the test frequency is occasionally increased up to 400 Hz. This
has also to be taken into account when resonance test sets are used for on-site PD
tests of power cables, where the test frequency is often varied between 20 Hz and
300 Hz (Rethmeier et al. 2012).

To evaluate PD test results, it is highly recommended to display the *phase-
resolved PD patterns (PRPDP)*, which supports the identification of potential PD
defects and enables often the discrimination of disturbing noises from real PD
events. For this purpose, either the built-in oscilloscope or even an external connected multichannel oscilloscope as well as a computerized measuring system can
be used.

As discussed previously, narrow-band instruments are commonly not capable of
measuring the apparent charge of PD pulses in terms of pC. Nevertheless, such
devices, which are commonly designed to measure radio interference voltages
(RIV) as well as to investigate the electromagnetic capability (EMC) of electronic
devices, are nowadays also widely employed for PD measurements, particularly for

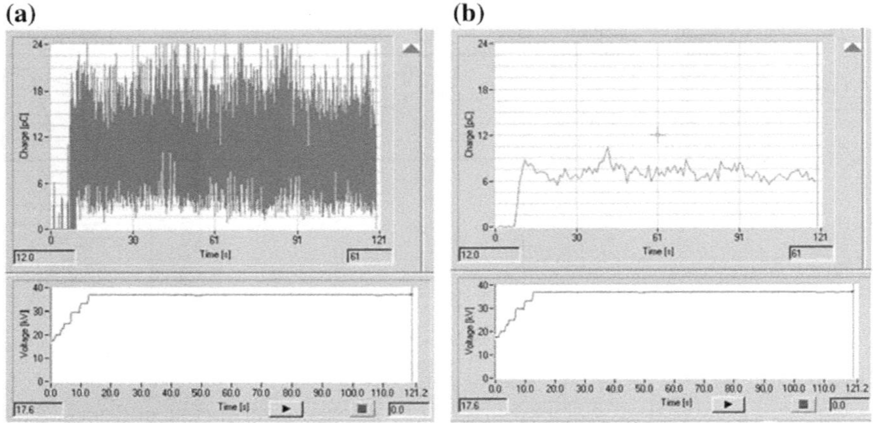

Fig. 4.30 Screenshots gained by means of a computerized PD measuring system showing the PD level of a MV power cable termination under AC (50 Hz) voltage (test level 38 kV, recording time 120 s). **a** Peak values of each individual charge pulse. **b** Largest repeatedly occurring PD magnitude weighted according to IEC 60270:2000

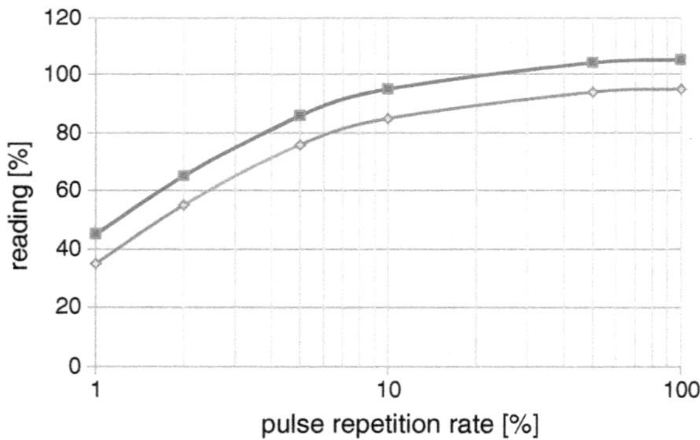

Fig. 4.31 Pulse train response specified in IEC 60270:2000 to measure the largest repeatedly occurring PD magnitude

PD diagnosis tests of HV apparatus under on-site condition due to their excellent noise immunity. Even if such instruments measure the captured PD signal in terms of μV and not in terms of pC, there is no doubt that the PD inception and extinction voltage as well as the change and the trend of the PD activity can well be determined by means of narrow-band instruments.

4.3.4.3   Digital PD Instruments

Due to the recent achievements in micro-electronics and in particular in digital signal processing (DSP), the traditional analogue PD instruments are nowadays increasingly replaced by advanced digital PD measuring systems. The first concept of a computerized PD measuring instrument has been presented by Tanaka and Okamoto (1978). After that time, various solutions for computer-based PD measuring systems have been proposed, for instance, by Kranz (1982), Haller and Gulski (1984), Okamoto and Tanaka (1986), van Brunt (1991), Gulski (1991), Kranz and Krump (1992), Fruth and Gross (1994), Shim (2000), Lemke et al. (2002), Plath et al. (2002).

Currently, two basic measuring principles are frequently employed. The first one is illustrated in Fig. 4.32a, which performes an *analogue pre-processing* of the

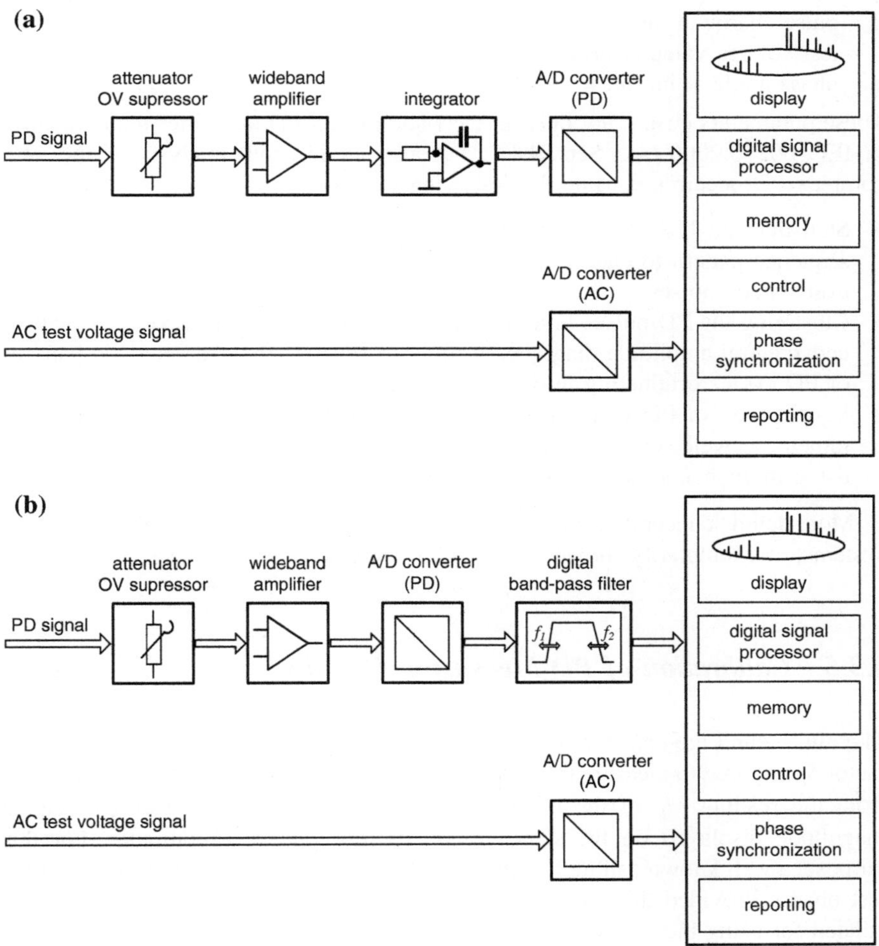

Fig. 4.32 Block diagram of digital PD instruments. **a** Analogue pre-processing of the PD signal followed by an A/D conversion of the apparent charge pulses. **b** Direct A/D conversion of the wide-band amplified PD pulses

captured PD pulses to establish the apparent charge pulses using a band-pass amplifier, as treated previously. Thereafter a *digital signal processing* is performed, where the A/D conversion requires a comparatively low sampling rate. Another option is the use of a very fast A/D converter to digitize the short PD pulses captured from the test object. The band-pass filtering required for the quasi-integration of the PD pulses is commonly performed by an adjustable digital filter and a numerical integrator. The main advantage of digital PD measuring systems is the ability to acquire, store and visualize the following characteristic PD quantities:

t_i instant of PD occurrence
q_i pulse charge at instant t_i
u_i test voltage value at instant t_i
ϕ_i phase angle at instant t_i

These parameters ensure not only an evaluation of all PD quantities recommended in IEC 60270:2000, see Sect. 4.1.2, but also an in-depth analysis of the very complex PD occurrence using the following features:

- Statistical analysis based on phase-resolved 2D and 3D patterns and pulse sequence pattern to classify and identify PD sources as well as to cancel electromagnetic noises.
- Clustering the PD pulses in homogenous families, based on waveform analysis and spectral amplitude diagrams in order to separate the characteristic patterns of PD events originating in different dielectric imperfections.
- Localization of PD faults in power cables and GIS using time-domain reflectometry as well as in the windings of rotating machines and power transformers using multichannel techniques.

More details concerning the capabilities of computerized (digital) PD measuring systems, as exemplarily shown in Fig. 4.33, are discussed in the Sects. 4.4–4.8.

4.3.5 *Calibration of PD Measuring Circuits*

The aim behind the *calibration* of PD measuring circuits is to determine the scale factor S_f required to measure the apparent charge q_a of PD pulses, which is deduced from the reading M_p of the PD measuring instrument or even from the signal magnitude displayed on the screen of an oscilloscope or a computer. For this purpose, a well known calibrating charge q_0 is injected between the terminals of the test object. Provided this causes the reading M_0, the apparent charge can be calculated for using the following relation:

Fig. 4.33 View of a computerized PD measuring system. *Courtesy* of Doble Lemke

$$q_a = \frac{q_0}{M_0} \cdot M_p = S_f \cdot M_p. \tag{4.25}$$

The calibration procedure is based on the fact that each PD event, which transfers a pulse charge from the PD site to the test object capacitance C_a, causes a transient voltage step ΔV_a across C_a detectable between the terminals of the test object, see Sect. 4.2. An equivalent response appears when a calibrating charge is injected between the terminals of the test object, as illustrated in Fig. 4.34 (Lemke et al. 1996; Lukas et al. 1997).

Example Assuming a calibrating charge of $q_0 = 20$ pC, which is injected between the terminals of the test object and causes a deflection of 5.4 divisions on the display of an oscilloscope, which is connected to the output of the PD instrument. Thus, the scale factor becomes $S_f = 20$ pC/ 5.4 div = 3.7 pC/ div. Performing an actual PD test, where a recorded PD pulse causes a maximum deflection of 8.6 div, the apparent charge amounts $q_a = (3.7$ pC/ div) $\times$ 8.6 div $\approx$ 32 pC.

In practice the capacitances C_{01} and C_{02} shown in Fig. 4.34 are substituted by only a single calibrating capacitor C_0, which is commonly connected between the internal pulse generator of the calibrator and the HV terminal of the test object in order to inject the calibrating charge, see Fig. 4.35.

As a measuring example, consider the oscilloscopic records shown in Fig. 4.36b. Here a step voltage (CH1), which is commonly caused by a PD event, was injected

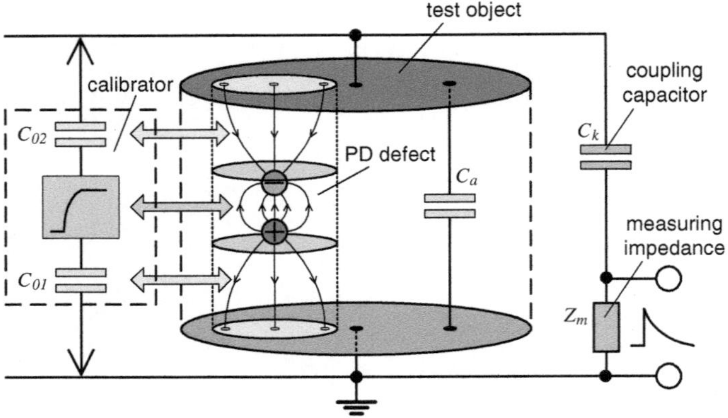

Fig. 4.34 Principle of the calibration procedure used to determine the scale factor of PD measuring systems

Fig. 4.35 PD Calibrator connected to a 20 kV instrument transformer

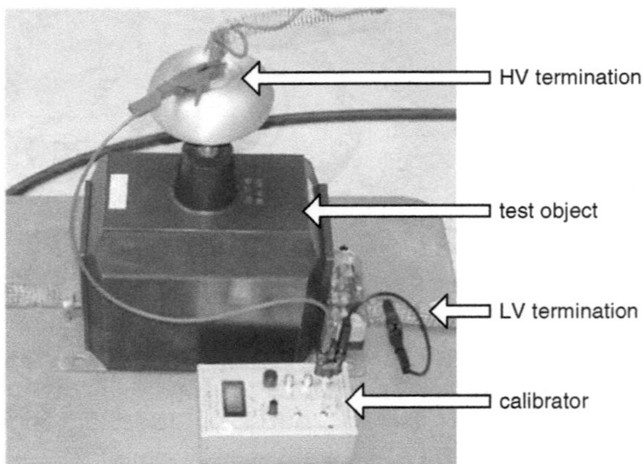

at the HV site of a 110-kV transformer bushing, as illustrated in Fig. 4.36a. This signal appeared differentiated when decoupled from the bushing tap terminal (CH2), which is due to the high-pass filter characteristics following from the series connection of the HV bushing (capacitance being close to 200 pF) and the measuring impedance (equipped with a 50 Ω measuring resistor). Under this condition the characteristic time constant attains nearly 10 ns. This signal was integrated (CH3) by means of a PD measuring instrument to reproduce a signal, which is proportional to the magnitude of the injected step pulse, and could in principle also be caused by a real PD event originating in a transformer under test.

To ensure reproducible PD test results, the step voltage shape of PD calibrators is specified in the Amendment to IEC 60270:2000 as follows (Fig. 4.37):

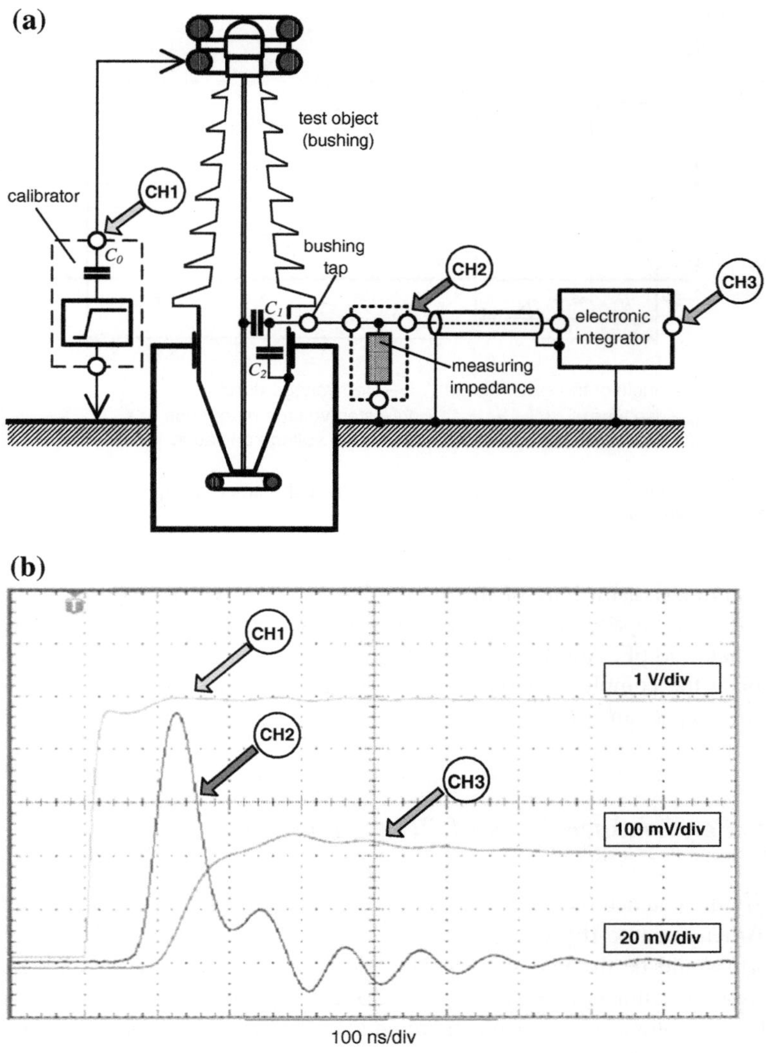

Fig. 4.36 Set-up to demonstrate the step pulse response of a transformer bushing (**a**) and characteristic signals injected at the HV terminal (CH1) and decoupled from the bushing tap (CH2) as well as from the output "apparent charge" of the connected PD measuring instrument (CH3) (**b**)

rise time:	$t_r \leq 60$ ns
time to steady state:	$t_s \leq 200$ ns
step voltage duration:	$t_d \geq 5$ μs
absolute voltage deviation:	$\Delta V \leq 0.03\ V_o$

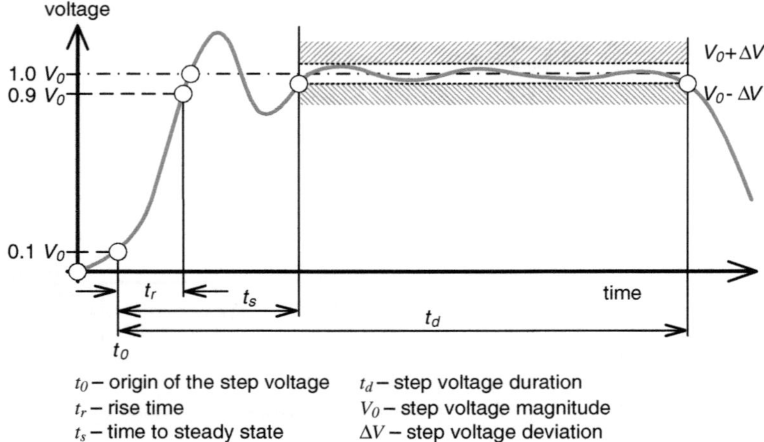

Fig. 4.37 Step voltage parameters specified for PD calibrators in the Amendment to IEC 60270:2000, published in 2015

To minimize distortions of the step voltage shape, which may occur at comparatively high capacitive load, the calibrating capacitor C_0 should be chosen not greater than 200 pF. Additionally, the condition $C_0 \leq 0.1\,C_a$ should be satisfied to accomplish the complete charge transfer to the test object capacitance C_a via the calibrating capacitance C_0.

4.3.6   Performance Tests of PD Calibrators

To verify the characteristic time and voltage parameters of PD calibrators specified in the Amendment to IEC 60270:2000, a performance test is required. The simplest approach would be the injection of the charge q_0 created by the calibrator into a measuring capacitor C_m, see Fig. 4.38a. To accomplish the entire charge transfer from the calibrator to C_m, which is required to apply the relation $q_0 = C_m \cdot \Delta V_m$, the condition $C_m \geq 100\,C_0$ must be satisfied. If, for instance, the calibrator is equipped with a capacitor of $C_0 = 100$ pF, the measuring capacitor should be chosen as high as $C_m \geq 10$ nF. Applying a conventional foil capacitor to guarantee a high-temperature stability, however, the front of the step pulse may be distorted drastically on account of the inherent inductive component, as exemplarily shown in Fig. 4.38b. This record refers to a calibrating charge of $q_0 = 120$ pC, which was injected via a 100 pF series capacitor into a 20 nF measuring capacitor. To overcome this crucial problem, an appropriate damping resistor R_d should be connected between calibrator and measuring capacitor, which leads to a response comparable to that shown in Fig. 4.38c. Another option to reduce the effective inductance is the use of numerous measuring capacitors connected in parallel.

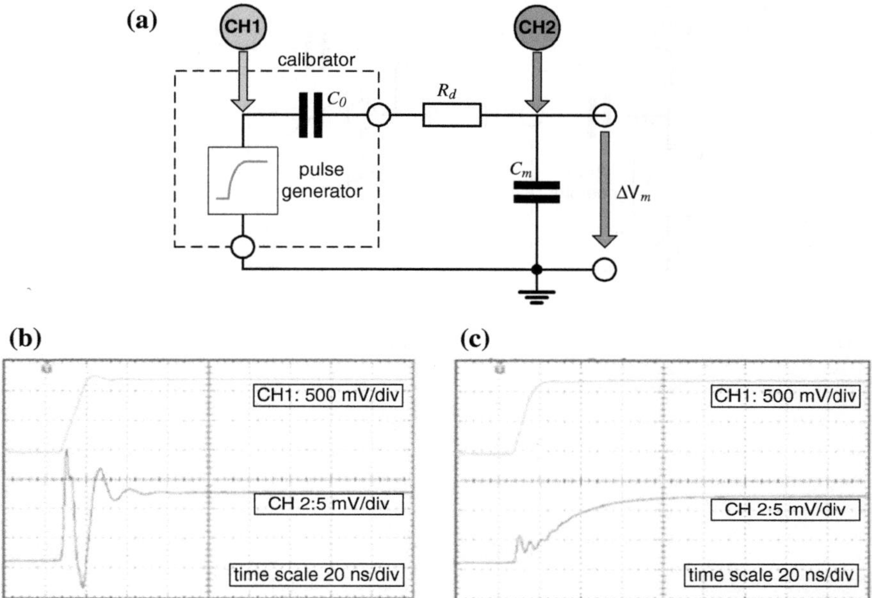

Fig. 4.38 Set-up for measuring the calibrating charge (**a**) and typical oscilloscopic records gained by the use of damping resistors by $R_d = 50\ \Omega$ (**b**) and $R_d = 390\ \Omega$ (**c**), respectively

The main obstacle of the calibrating circuit shown in Fig. 4.38a is the fact that calibrating charges below about 50 pC can hardly be measured by means of conventional digital oscilloscopes, because a calibrating charge of $q_0 = 50$ pC injected in a 10 nF measuring capacitor causes a voltage jump as low as $\Delta V_m = 5$ mV, i.e. a change in voltage as low as 0.2 mV must be recognized to achieve an appropriate measuring accuracy. To overcome this crucial problem, an electronic integrator could be employed to enhance the measuring sensitivity accordingly (Lemke 1996). A schematic diagram of such a circuit is shown in Fig. 4.39, which ensures the complete charge transfer from the calibrator to the capacitor C_m, even if the capacitance of C_m is reduced 100 times, i.e. from originally 10 nF down to about 100 pF, where the latter equals the value of the calibrating capacitor C_0. As obvious from the oscilloscopic records shown in Fig. 4.39, which refers to a calibrating charge of approx. 3 pC, a sufficient high-voltage step of approx. 30 mV is achieved, which can conviently be measured at desired accuracy, in particular when the oscilloscope is equipped with an averaging tool. In this context it should also be noted that despite the comparatively low integrating capacitance of only 100 pF the decay time constant of the measured signal is sufficiently high (Fig. 4.39c), so that the recorded voltage signal CH2 decreases only marginally during the recording time, which is chosen as 5 μs and corresponds thus to the minimum duration t_d of the step voltage response specified in the Amendment to IEC 60270:2000.

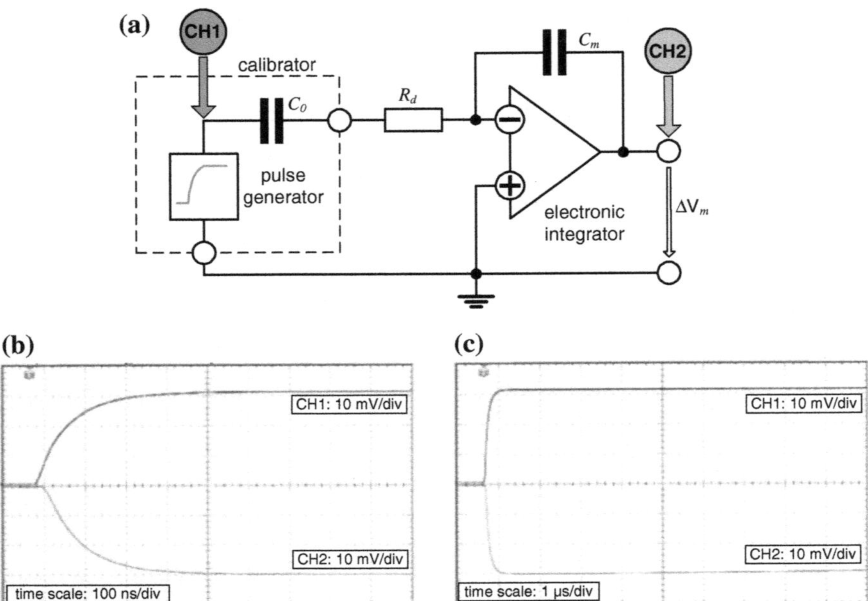

Fig. 4.39 Principle of an electronic integrator for measuring the charge of PD calibrators (**a**) and typical signals recorded at time scale of 100 ns/div (**b**) and 1 µs (**c**), respectively

Another option recommended in IEC 60270:2000 is the injection of the calibrating charge into a measuring resistor R_m, as illustrated in Fig. 4.40a. Obviously, the series connection of C_0 and R_m represents a high-pass filter, so that the time-dependent voltage $v_m(t)$ appearing across R_m must be integrated again to determine the charge q_0 created by the calibrator. For this purpose, a digital oscilloscope equipped with a feature for numerical integration could be be employed, where the A/D converter should have a vertical resolution not lower than 10 bits at 50 MS/s sampling rate to achieve an adequate resolution of the input signal. Additionally, the analogue bandwidth should be not lower than 50 MHz. Generally, digital oscilloscopes calibrated in compliance with IEC 61083-1:2002 should only be employed.

4.3.7   Maintaining the Characteristics of PD Measuring Systems

The third edition of IEC 60270:2000 recommends the following three levels for maintaining the characteristics of PD measuring facilities composed of the coupling device, the PD measuring instrument, and the PD calibrator including the necessary connection leads:

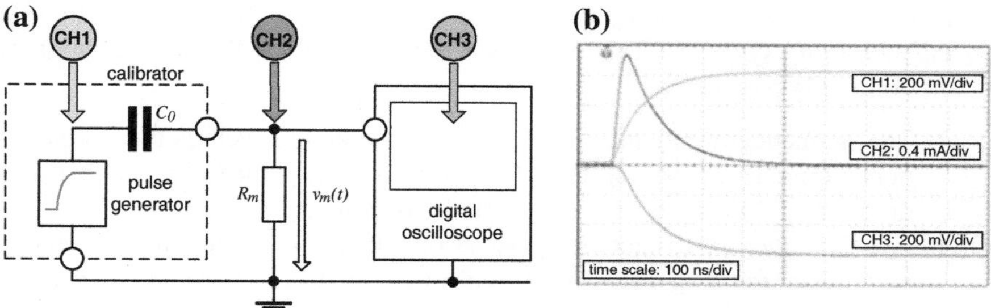

Fig. 4.40 Set-up recommended for a numerical integration of the calibrating charge q_0 injected in a measuring resistor of $R_m = 50\ \Omega$ (**a**) and typical oscilloscopic records (**b**) gained for a calibrating charge of $q_0 = 150$ pC

1. The routine calibration of the complete PD measuring system connected to the HV test circuit. This should be performed just prior a PD test, where the calibration provides the scale factor S_f of the entire measuring system to be used in the actual PD test. Nowadays, this procedure is mainly used to adjust the reading of the PD measuring instrument to obtain a direct reading of the PD magnitude, i.e. S_f should satisfy preferable values (e.g. 1, 2, 5, 10, 20…). For this routine calibration, there are no major changes as compared to the IEC 270 edited in 1981.
2. The determination of the specified characteristics of the complete PD measuring system should be performed at least once a year or after major repair.
3. The calibration of the PD calibrator itself, as presented above.

In general, manufacturers of PD measuring devices have to provide the necessary guidelines required for the verification of the specified technical parameters. Independent on such guidelines, the current (third) edition of IEC 60270:2000 recommends additional test procedures, where the results have to be recorded in a "Record of Performance (RoP)" to be established and maintained by the user. The RoP should include the following information:

- Nominal characteristics (identification; operation conditions, measuring range, supply voltage)
- Type test results
- Routine test results
- Performance test results (date and time)
- Performance check results (date and time; result: passed/failed: if failed: action taken)

Verifications of PD measuring systems and PD calibrators shall be performed once as acceptance tests. Performance tests should be performed annually or after any major repair, but at least every 5 years. Performance checks have to be performed at least once a year. To maintain the characteristics of *PD measuring instruments*, the following tests should be performed:

(a) *Type tests* are to be done by the manufacturer and shall be performed for one
PD measuring system of a series and shall at least include the determination of
the following parameters:

- The frequency-dependent *transfer impedance* $Z(f)$ as well as the lower and
  upper limit frequencies f_1 and f_2 over a frequency range in which it has
  dropped to 20 dB from the peak band-pass value;
- The *scale factor* k to calibrating pulses of at least three different pulse charge
  magnitudes ranging between 10 and 100% of the full reading at a pulse
  repetition rate n around 100 s^{-1}. In order to prove the linearity of the PD
  measuring instrument, the variation of k shall be less than 5%;
- The *pulse resolution time* T_r by applying calibration pulses of constant
  magnitude but decreasing time interval between consecutive pulses;
- *The pulse train response* for pulse repetition rates N ranging between 1 s^{-1}
  and $>100 \text{ s}^{-1}$.

(b) *Routine tests* are to be done by the manufacturer and shall include all tests
required in a performance test as listed below. Routine tests shall be performed
for each measuring system of a series. If the test results are not available from
the manufacturer, the required tests shall be arranged by the user.

(c) *Performance tests* shall include the determination of the following parameters:

- The frequency-dependent transfer impedance $Z(f)$ as well as the lower and
  upper limit frequencies f_1 and f_2 over a frequency range in which it has
  dropped down to 20 dB from the peak band-pass value.
- The linearity of the scale factor k to be verified between 50% of the lowest
  and 200% of the highest specified PD magnitude. Using calibrating pulses
  of adjustable magnitude having a repetition rate of approximately
  $n = 100 \text{ s}^{-1}$, the scale factor k shall vary not more than 5%.

(d) *Performance checks* shall include the determination of the transfer impedance Z
(f) at one frequency selected in the band-pass range in order to verify that the
value deviates not more than 10% from that one recorded in the performance
test.

To maintain the characteristics of *PD calibrators*, the following tests should be
performed:

(a) *Type tests* are to be done by the manufacturer and shall be performed for one
PD calibrator of a series. Type tests shall include at least all tests required in a
performance test. If results of type tests are not available from the manufacturer,
the required tests for verification the technical parameters of PD calibrators
shall be arranged by the user.

(b) *Routine tests* are to be done by the manufacturer and shall include all tests
required in a performance test. Routine tests are to be performed by the manufacturer for each measuring system of a series. If the test results are not
available from the manufacturer, the required tests shall be arranged by the
user.

(c) *Performance tests* shall include the determination of the following parameters:

  - The actual magnitude of the pulse charge q_0 for all nominal settings, where a measuring uncertainty within 5% or 1 pC, whichever is greater, is acceptable.
  - Rise time t_r of the voltage step U_0, where a measuring uncertainty within 10% is acceptable.
  - Pulse repetition frequency N, where a measuring uncertainty within 1% is acceptable.

(d) *Performance checks* include the determination of the actual magnitude of the calibrating charge q_0 for all nominal settings, where a measuring uncertainty within 5% or 1 pC, whichever is greater, is accepted.

4.3.8   PD Test Procedure

The main aim behind PD tests according to IEC 60270:2000 is to prove the integrity of insulation systems of HV apparatus and their components. The test procedures applied for quality assurance tests after manufacturing and repair as well as the test voltage levels and the limits of tolerated pulse charge magnitudes deduced from long-term experience are specified for each kind of HV apparatus by the relevant Technical Committee. As the test procedures vary for different HV equipment, only one typical application shall be presented in the following, which refers to a PD test of a single-phase power transformer under *induced voltage* based on the test circuit sketched in Fig. 4.41.

The PD test procedure commonly applied can generally be divided into the following steps:

(a) Configuration of the HV test circuit:

To minimize the impact of electromagnetic noises, the transformer under test should well be grounded. Moreover, the use of a *low-pass filter* at the LV site is highly recommended. Using the bushing tap coupling mode, the measuring impedance should be located as close to the test object and connected to the PD measuring instrument not only via the measuring cable but additionally via a parallel connected ground connection lead made of Cu or Al foil. Occasionally shielding electrodes should be arranged on the top electrodes of the bushings to prevent corona discharges, see Figs. 4.41 and 4.42. Moreover, it has to be taken care that the bushing is cleaned and dried to prevent surface discharges. To record the phase-resolved PD patterns, additionally to the PD signal an AC voltage should be recorded as reference, which could be taken from the measuring impedance connected to the bushing tap, as reported in Sect. 4.3.1 and illustrated in Fig. 4.22.

(b) Adjustment of the *measuring frequency range*:

For tall and inductive test objects, such as power transformers, the frequency content of the detectable PD signal travelling from the PD site to the bushing tap is

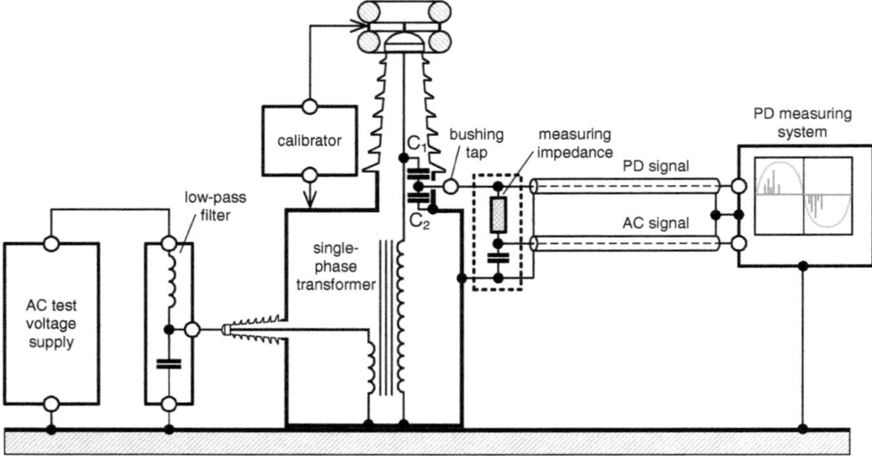

Fig. 4.41 Set-up applied for a PD tests of a power transformer under induced voltage using the bushing tap coupling mode

Fig. 4.42 Shielding electrodes assembled at the *top* of the bushings of a 500 kV single-phase transformer intended to prevent disturbing corona discharges

drastically reduced. Thus, the upper limit frequency f_2 should be selected preferably below 300 kHz, while the lower limit frequency f_1 should be adjusted below 100 kHz to ensure a PD signal processing in a wide-band range. Reducing f_1 substantially below 100 kHz, however, serious disturbances might appear due a possible iron core saturation as well as other core-related noises and even harmonics superimposed on the exciting AC voltage.

(c) *PD calibration*:

The main objective of the calibration procedure is to determine the ratio between calibrating charge q_0 injected into the top electrode of the bushing and the reading M_0 of the PD instrument, which provides the scale factor S_f, as pointed out in Sect. 4.3.5. The ground connection lead between calibrator and transformer tank should be of low inductance and should thus be kept as short as possible. To prove the linearity of the PD measuring system, calibrating charges of different magnitudes should be injected. If, for instance, the full reading of the PD instrument is obtained for a calibrating charge of 200 pC, it is recommended to reduce thereafter the calibrating charge down to 100 pC and at least to 50 pC. Under this condition the deviation of the reading from that to be expected for a linear operating instrument should be below 10% of the full reading. After the calibration procedure has been finished it has to be taken care that the calibrator is removed from the test object to avoid any damage when the HV test voltage is switched on.

(d) Actual PD test under HVAC voltage:

The test voltage profile to be applied for quality assurance tests is specified in the relevant apparatus standards. A typical measuring example is shown in Fig. 4.43, which refers to the IEEE standard C57.113: 2010. First, the induced AC test voltage level has to be raised up to approx. 50% of the rated voltage V_1 to determine the "energized background noise level" in terms of pC. This should not exceed the 50% value of the specified apparent charge, which is being accepted. Thereafter, the test voltage has to be raised up to the specified 1 h test value V_2 and held constant for few minutes to verify whether there are any PD problems. If not, the test voltage has

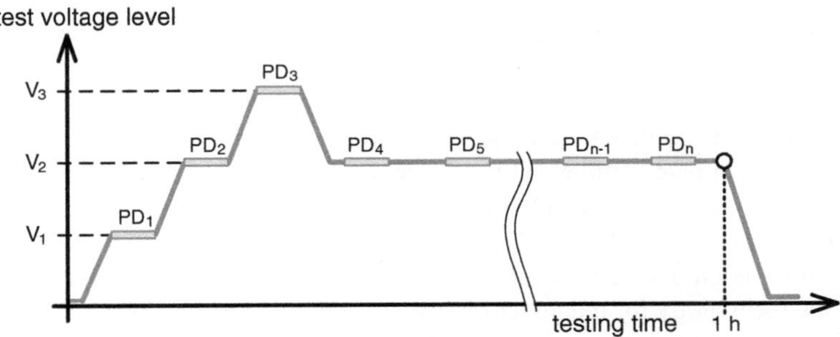

Fig. 4.43 Profile of the HVAC test voltage recommended in the IEEE Standard C57.113 for PD testing of liquid-filled power transformers and shunt reactors

then to be raised up to the enhancement (withstand) level V_3 and held constant for 7200 cycles in order to observe the PD trend.

> **Note** As HVAC test voltages substantially higher than the rated voltage are applied, the exciting frequency must be enhanced accordingly, which can conveniently be realized by the use of HVAC test sets of variable test frequency. For this reason, the duration of the PD test period at enhancement voltage level is expressed in terms of cycles. If, for instance, a test frequency of 120 Hz is applied, 7,200 cycles are equivalent to a test period of 60 s. For more details in this respect see Sect. 3.2.5.

Finally, the applied HVAC test voltage is reduced down to the 1 h test voltage level V_2. Under this condition, the apparent charge is recorded at subsequent five-minute intervals, where the recording time can be limited to about 1 min at each interval.

(e) Evaluation of PD test results:

As stated in the above mentioned IEEE Standard C57.113, the transformer has passed the PD test when the mean value of the maximum PD pulse magnitudes expressed in terms of pC and recorded during the 1 h test interval

- is below a specified value in terms of pC,
- is within a specified tolerance band,
- does not exhibit any steadily rising trend, and
- does not suddenly increase during the last 20 min of the 1 h test period.

In this context, it has to be taken into account that during the above mentioned 5-min test intervals sporadic noises may be encountered, for instance, by the switching of cranes. Provided, the test results do not comply with the specified limits, the PD-tested transformer should not warrant immediate rejection but lead to consultation between purchaser and manufacturer to decide further actions.

4.4 PD Fault Localization

To assess the PD severity, besides the apparent charge and the PD pulse repetition rate also the origin of the PD source should be known. This is particularly of importance for high-polymeric power cables, where the insulation may irreversibly be deteriorated by PD events having a magnitude of only few pC. Therefore, the localization of PD failures became a well established method since the 1970s, when high-polymeric power cables where increasingly used in distribution networks (Eager and Bahder 1967, 1969; Lemke 1975, 1979; Beinert 1977; Kadry et al. 1977; Beyer and Borsi 1977). The main benefit of the *PD fault localization* in power cables is on one hand, that the reason for typical PD defects can be clarified so that the technology for manufacturing such cables can be improved. On the other hand, the entire cable length must not be replaced in case of a single PD failure, but rather that short section containing the PD defect. In this context it should be noted that in case of routine tests most of the recognized PD failures appear at the cable ends, mainly due to a poor assembling of the stress cones required for a field grading.

Using the so-called time-domain reflectometry for the PD fault localization, first the *travelling wave velocity* v_c has to be determined. As a measuring example consider Fig. 4.44a. Here, a calibrating pulse was injected in that cable end where the coupling unit is connected, which is usually referred to as *near* cable end. As electrically long power cables behave like an electromagnetic waveguide, the pulse injected at the near end travels first at wave velocity v_c towards the *far* or *remote* cable end, where the signal is reflected, and thereafter towards the near cable *end*. Consequently, a second pulse is also detectable at the near cable end after a time span t_c required for travelling twice the entire cable length, i.e. $2 \cdot l_c$. Hence the travelling wave velocity can simply be accounted for using the following relation

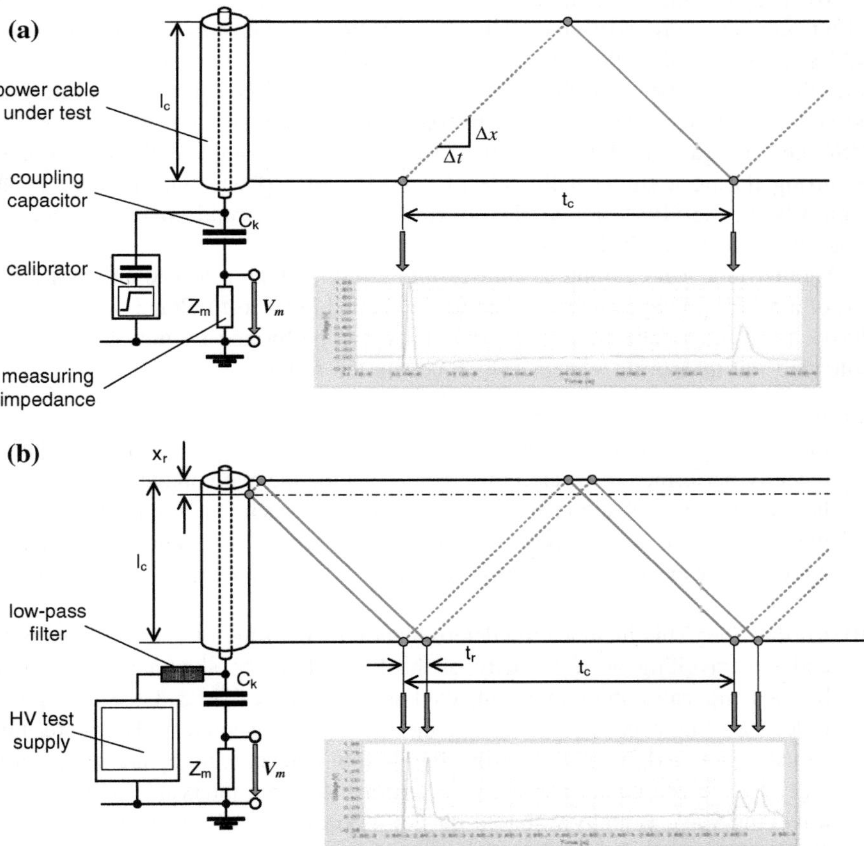

Fig. 4.44 Principle of PD fault localization in electrically long power cables. **a** Determination of the travelling wave velocity. **b** Using the time-domain reflectometry (*TDR*) to determine the PD site

$$v_c = \frac{2l_c}{t_c'} \tag{4.26}$$

Provided, a PD fault is located at distance x_n from the near cable end, the PD pulse would travel from the site of origin into both directions, i.e. directly towards the near cable end within the time span t_n and also towards the remote cable end within the time span t_r, where it is reflected, see 4.44b. Thus, the reflected pulse travels twice the distance x_r between PD site and remote cable end and thereafter the distance x_n, which is equal to that travelled by the direct PD pulse. Based on this it can be written:

$$x_r = 0.5 \cdot v_c \cdot t_r. \tag{4.27}$$

$$l_c - x_r = 0.5 \cdot v_c(t_c - t_r). \tag{4.28}$$

Nowadays available computerized PD measuring systems are mostly equipped with features for time-domain reflectometry (TDR) to localize PD failures in power cables (Lemke et al. 1996, 2001). The main challenge is, however, to measure the time difference t_r between the direct and the reflected PD pulse as accurately as possible, which requires an A/D conversion at sampling rate not lower than 100 MS/s and a signal resolution of 10 bit. To record the complete PD data stream occurring during a single half-cycle of a 50-Hz test voltage, the memory depth should be in the GByte range. The overall bandwidth should cover a frequency range between about 50 kHz and 20 MHz.

Another practical example is shown in Fig. 4.45, which refers to an on-site PD test of a XLPE power cable subjected to a so-called DAC test voltage (Lemke et al. 2001). Here the localization of the PD faults was performed by means of a computerized PD measuring system, which covers the following steps:

(a) *Inserting the cable data*: Besides the fundamental cable parameters (manufacturer, type and insulation of the cable, rated voltage, operation voltage, recently performed tests, etc.) this should include the test voltage parameters to be applied (test voltage levels, number of shots at each test voltage level), also the cable length and the positions of the accessories (joints and terminations), which are especially of interest to localize the PD defects as accurate as possible.

(b) *Calibration*: This includes the determination of both the measuring sensitivity and the travelling wave velocity of the PD pulses. As shown in Fig. 4.45a, besides the calibrating pulse injected at the near cable end, several pulse reflections might occur, where only the first one is of interest. Therefore, this signal is zoomed, as obvious from Fig. 4.45b, and the cursors are set accordingly by the computer software to determine the time interval t_c and thus the travelling wave velocity v_c based on Eq. 4.26 as precisely as possible.

(c) *PD measurement*: Recording of the consecutive PD pulses occurring within a pre-selected time interval (Fig. 4.45c) and evaluation the of the PD pulse magnitudes. For this purpose each PD pulse magnitude is initially indicated in

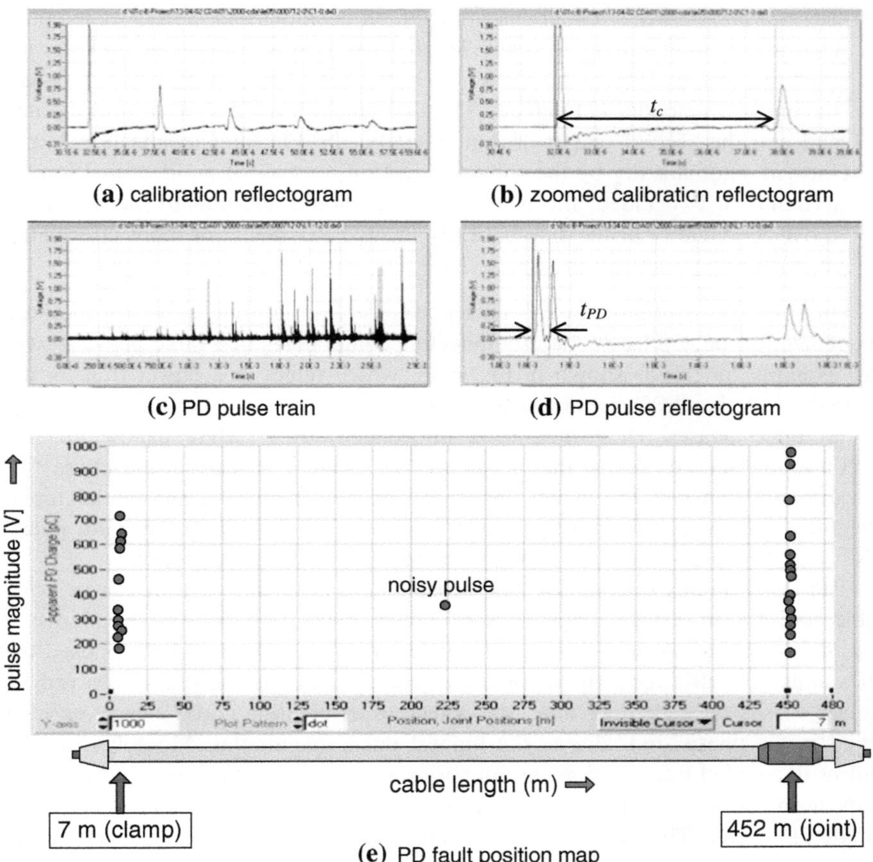

Fig. 4.45 Screenshots of a computer-based PD fault localization system (explanation in the text)

terms of Volts, and based on this the pulse charge appearing at each test voltage application is calculated for by the software and stored in the computer memory to perform a statistical analysis of the data stream.

(d) *PD fault localization*: To apply the time-domain reflectometry (TDR), only those PD pulses showing typical reflections within the time interval $t_r \leq 2\,t_c$ are extracted. Thereafter, these pulses are zoomed and the cursors are set accordingly to measure the time interval between each direct and the associated reflected pulse, see Fig. 4.45d. Based on this, the distance between PD source and either the near or the remote cable end is determined. This procedure is repeated several times in order to perform an averaging and thus to enhance the measuring accuracy. Occasionally, a digital filtering of the captured signal may be performed to minimize the impact of radio interference voltages on the test results.

(e) *PD mapping*: Displaying all determined fault positions and the associated pulse charge magnitudes along the cable length. A typical measuring example for this is shown in Fig. 4.45e, which refers to a 20 kV XLPE cable of 480 m in length having two potential PD defects. The first one is located at 3 m from the near end and the second one at 452 m. Usually, the PD fault localization is performed automatically by the software. Only if a repair of an identified joint or termination is decided, the manual feature should additionally be applied to prove the validity of the automatically located PD sites. For more information in this respect see also the Sects. 7.1.3 and 10.2.2.2.

For some HV equipment other than power cables, the time of arrival measurement can also be applied to identify defective components of a three-phase system. An example for this is shown in Fig. 4.46, which refers to a power transformer. Here the wide-band PD signal was decoupled simultaneously from the bushing taps of all three phases using measuring impedances of 20 MHz bandwidth. The oscilloscopic records reveal that a potential PD source appears in the phase "T", because the transient PD signal appeared first and showed a substantial higher magnitude if compared with those signals decoupled from the other both phases "R" and "S", respectively. This was confirmed by a visual inspection.

The synchronous three-phase PD measurement has also been proven as a feasible tool for in-service diagnostics of power cable terminations, as shown in Fig. 4.47. Recording only the time-dependent PD level, it could be supposed that PDs appear in all three terminations. However, displaying the phase-resolved PD pulses versus the recording time, the defective termination could clearly be identified in the phase "red" due to the highest pulse magnitudes. Moreover, the PD pulses decoupled from all three terminations appeared at phase angles equal for all three terminations, which means that the captured signal was radiated from only one single PD source. This conclusion was finally confirmed by a visual inspection after the identified termination has been de-assembled. Replacing this cable termination by a new one, no any critical PD level was recognized.

Another approach to distinguish between different PD sources is the presentation of typical clusters in a 3-phase amplitude relation diagram, which is based on a

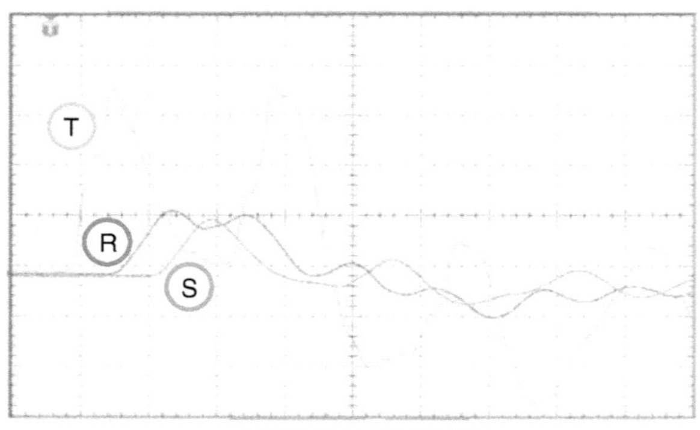

Fig. 4.46 PD signals decoupled simultaneously from the three phases (R, S, T) of a power transformer using the bushing tap coupling mode

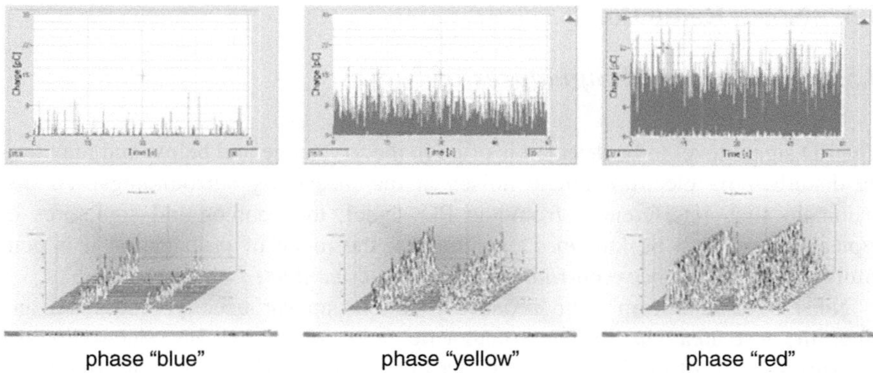

Fig. 4.47 Characteristic PD signatures captured from the cable terminations connected to a three-phase gas-insulated switchgear

synchronous multichannel PD measurement (Emanuel et al. 2002). An enhancement of this method is the presentation of so-called three-center-frequency relation diagrams where three different frequencies selected from the complete spectrum of a single PD pulse are evaluated and displayed on the computer screen. This feature provides not only valuable information on the discharge nature itself but can also be used to localize the origin of PD defects (Rethmeier 2009). For more details in this respect see Sect. 4.6.

Another promising tool proposed for the localization of potential PD defects in HV equipment is the so-called pulse waveform analysis. This is based on the extraction of a set of PD pulse parameters, such as the rise time and the decay time as well as the PD pulse width (Montenari 2009). Displaying the characteristic clusters like star diagrams, multiple PD failures can also be recognized, as presented also in Sect. 4.6.

In this context, it should be noted that besides the above-described electrical methods, also the acoustic emission (AE) technique is widely used, in particular to localize PD defects in metal-encapsulated HV apparatus, such as gas-insulated switchgears (GIS) and gas-insulated lines (GIL) as well as large power transformers. The combination of both the electrical and acoustic method can also be very effective, for instance, to enhance the signal-to-noise ratio. For more information in this respect see Sect. 4.8.

Even if the localization of PD faults is nowadays performed by means of advanced computerized PD measuring systems, it should not be overlooked that commercially available digital oscilloscopes can also conveniently be employed for this purpose. In this context it should also be emphasized that a great deal of practical experience is required to decide if a HV equipment showing a high PD activity should really be taken out of order or even kept in service and PD monitored permanently to recognize a sudden increase of the PD activity and thus to prevent an unexpected breakdown. For more information in this respect see Sect. 10.3.

4.5 Noise Reduction

4.5.1 Sources and Signatures of Noises

The PD signal level to be detected is often in the mV range and below and may thus be disturbed by electromagnetic noises in the measuring surroundings. To discriminate such interferences from the PD signal, the sources and signatures of typical noises must be known. Depending on the mode of propagation it is generally distinguished between *radiated noises* and *conducted noises*.

Noises radiated from radio broadcast stations appear usually modulated and enter the test area via the electromagnetic field, where the HV electrodes and measuring loops of the PD test circuit act like antennas. Moreover, high-frequency transients associated with corona discharges igniting in the vicinity of the test area at sharp edges and protrusions on the surface of HV electrodes can also be classified as radiated noises.

Electromagnetic noises may also originate in the mains or even in the HV test facility itself du to switching processes, which appear thus stochastically or even periodically. These are transmitted via conductors to the PD test area and hence classified as conductive noises. This refers also to a sparking between metallic parts on floating potential, for instance, due to a poor grounding. Screenshots of typical conductive noises, often encountered in PD test laboratories, are displayed in Figs. 4.48 and 4.49.

4.5.2 Noise Reduction Tools

To minimize the impact of radiated noises it is a common practice to erect electromagnetically well-*shielded test laboratories*, as described in Sect. 9.2.2, where the fundamental laws of HF technology have to be taken into consideration. Particularly wire loops acting as inductive antennas should be kept as low as possible in cross-section to minimize the induction of interferring voltages on account of radiated noises. Moreover, the ground connection leads should also be of low inductance which is best accomplished by using Cu or Al foil.

If PD test laboratories are not carefully shielded against radiated electromagnetic noises, it could be helpful under certain conditions to use the *balanced bridge circuit* according to Fig. 4.23c. Practical experiences revealed that for comparatively small test objects, such as instrument transformers and bushings, the signal-to-noise ratio can be enhanced by a factor up to 10. However, for tall test objects, such as power transformers, this method is commonly not effective.

An option of the bridge method is the *pulse polarity discrimination*, originally proposed by Black (1975), where a balance procedure is not required. As the PD pulses decoupled from both bridge branches appear at opposite polarity, these can conveniently be discriminated from radiated electromagnetic noises because these

(a)

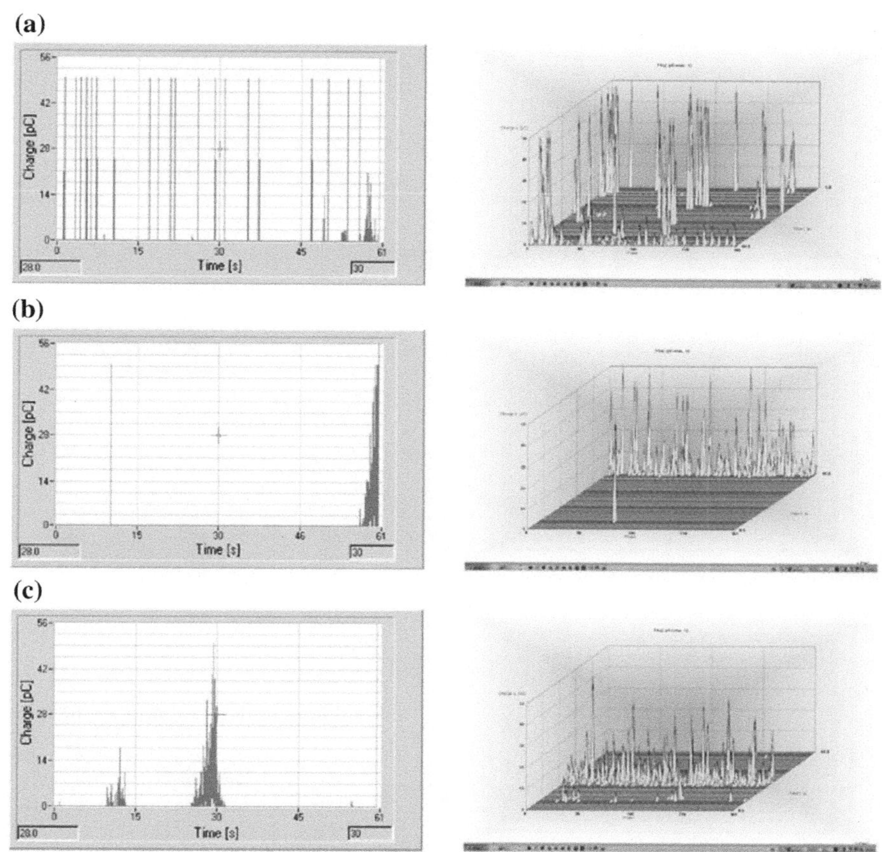

(b)

(c)

Fig. 4.48 Signatures of stochastically appearing pulse-shaped noises. **a** Maintenance work (drilling worker), **b** Car starting nearby, **c** Switching of a crane in the test lab

appear unipolar. For tall test objects, however, this method is commonly not applicable because the PD pulses transmitted via both bridge branches might excite oscillations, so that the true polarity is lost.

From a theoretical point of view, the *signal-to-noise ratio* can substantially be enhanced when the PD signal processing is performed in a higher frequency range, i.e. at frequencies substantially greater than the 1 MHz limit, as specified in the Amendment to IEC 60270:20060270. This concept has originally been adopted by Lemke (1968), where the PD pulses where first amplified by means of a 20 MHz wide-band amplifier and thereafter integrated by means of an electronic integrator in order to establish the pulse charge. A comparison of this alternative concept with the classical method is shown in Fig. 4.50, where the oscilloscopic records refer to 20 pC calibrating pulses (CH1) injected in a 20 kV XLPE cable. As can be seen, the charge pulse gained on account of a low-pass filtering ($f_2 = 600$ kHz) cannot clearly be identified (CH2, green trace), while the charge pulse gained by an initial

(a)

(b)

(c)

(d)

(e)

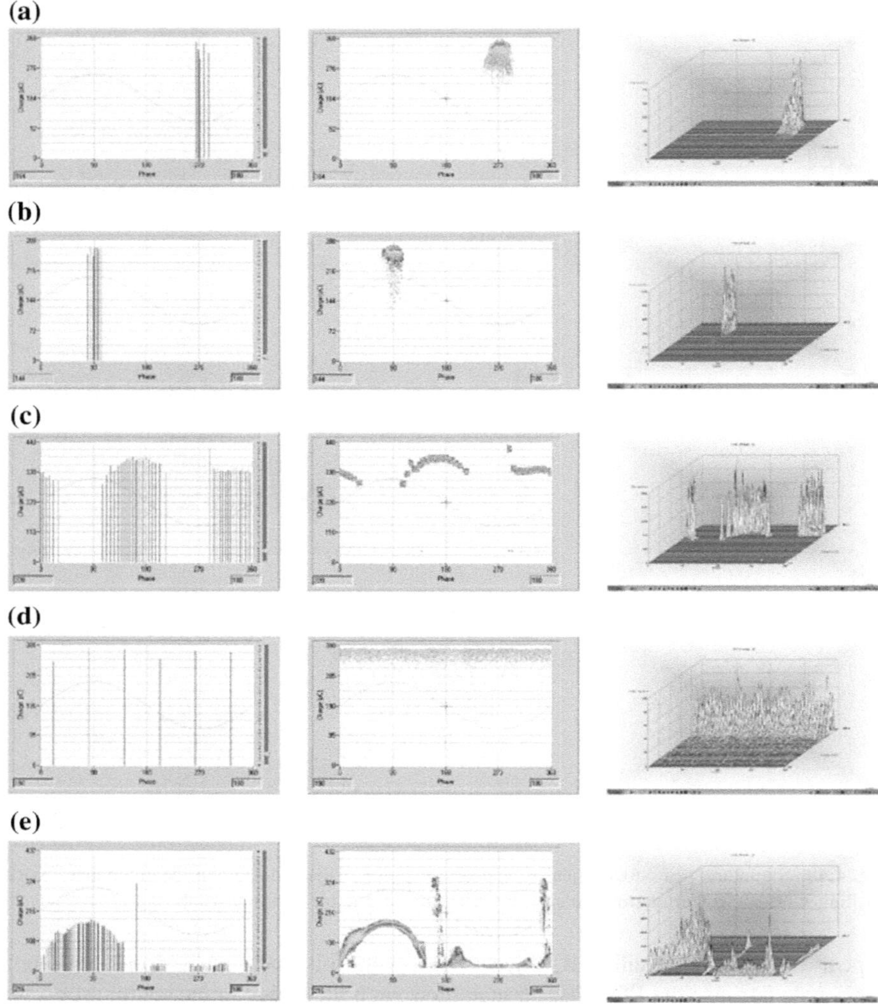

Fig. 4.49 Signatures of periodically appearing pulse-shaped noises. **a** Protrusion at the surface of a HV shielding electrode. **b** Sharp edge of a metallic structure on ground potential. **c** Sparking between metallic parts on floating potential. **d** Frequency converter feeding a resonant test set of variable frequency. **e** Defective xenon lamp in the control room

wide-band amplification ($f_2 = 20$ MHz) and a following electronic integration is not significantly interfered (CH3, pink trace).

The basic concept of the above mentioned non-conventional PD measuring instrument equipped with various tools for noise rejection is illustrated in Fig. 4.51 (Lemke 1979; Hauschild et al. 1981; Lemke 1981), where the operation principle covers the following features:

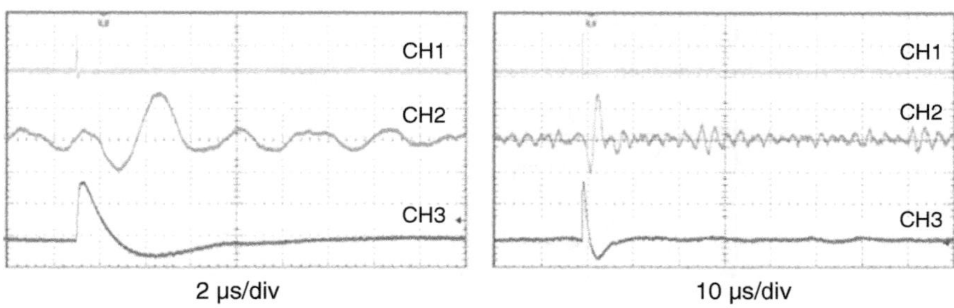

Fig. 4.50 Oscilloscopic screenshots gained for a 20-pC calibrating pulse injected in a XLPE cable: CH1: Input signal captured from a wide-band measuring impedance; CH2: Signal after conventional band-pass filtering; CH3: Signal after non-conventional wide-band amplification ($f_2 = 20$ MHz) followed by an electronic integration

- Wide-band amplification of the PD signal captured from the measuring impedance, where a bandwidth up to about 20 MHz seems reasonable.
- *Automatic gating* of pulse-shaped noises appearing periodically and even stochastically. To receive the noisy signals for triggering the gating unit, the rod or loop antennas should be installed as close as possible to the supposed noise sources.
- Canceling of radio interference voltages. This is achieved by adjusting the threshold level for passing the RIV rejection unit slightly above the noise level, which is controlled automatically.
- Electronic integration of the de-noised PD signal to measure the pulse charge of the wide-band amplified PD signal.

Typical oscilloscopic records illustrating the operation principles of various noise reduction tools are presented in Fig. 4.52. The screenshots displayed in Figs. 4.52a, b refer to the rejection of radio frequency (RF) noises, as already discussed previously based on Fig. 4.50. In this context it should be mentioned that the above mentioned noise canceling method, which is based on a wide-band pre-amplification of the captured PD pulses followed by an electronic integration, was also used for the design of a hand-held, battery-powered PD probe intended for on-site PD diagnostics of HV apparatus in service (Lemke 1985, 1988, 1991). Pulse-shaped noises appearing periodically or even stochastically can also effectively be canceled using a windowing and gating unit, which is triggered by the noisy pulses itself captured by means of an antenna, see Figs. 4.52c, d.

Measuring errors due to the superposition of reflected on direct pulses as consequence of PD events in power cables can also be prevented, as illustrated in Figs. 4.52e, f. Here the superimposed reflected pulse recorded in Fig. 4.46e is canceled by means of an electronic reflection suppressor, as recommended in IEC 60885-3:2003. For this purpose a gating unit is triggered by the direct (first arriving) PD pulse, which closes the gate and prevents thus a passing of the reflected (second) PD pulse. This tool operates reliable for double pulses having distances greater than 0.2 µs (Lemke 1979, 1981).

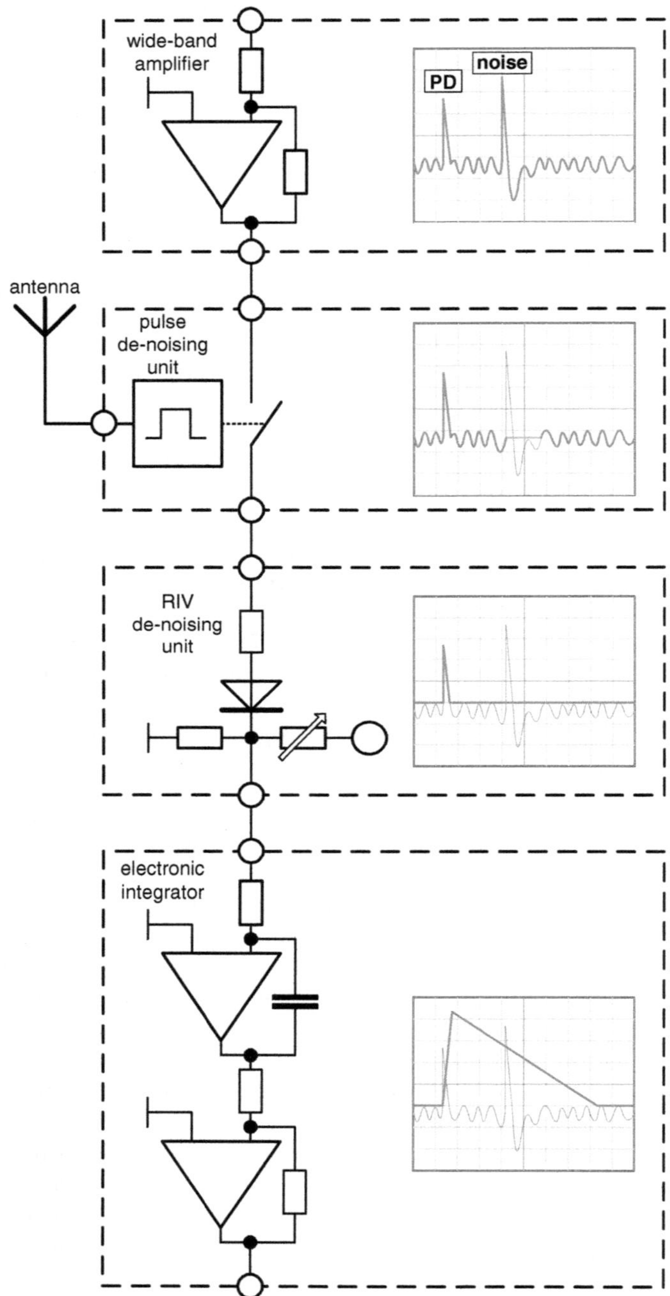

Fig. 4.51 Block diagram of a non-conventional wide-band PD measuring system equipped with various de-noising tools, as described in the text

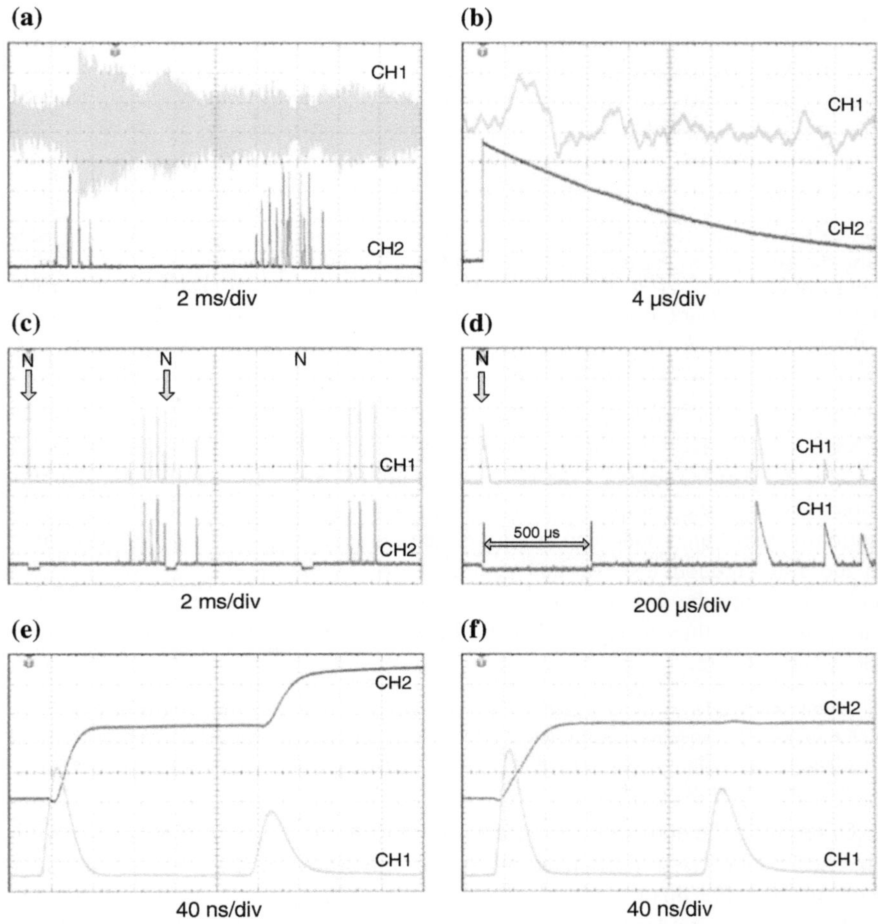

Fig. 4.52 Oscilloscopic records gained for a wide-band PD measuring system equipped with an electronic integrator and various tools for de-noising interfered PD signals (description in the text.)

Even if today's digital PD measuring systems are commonly equipped with various *de-noising software* tools, it should not be overlooked that the "windowing" and "gating" hardware, originally developed for the traditional analogue PD signal processing, is also applicable for digital PD instruments, where the noise-canceling is performed just prior the analog/digital conversion (Lemke 1996; Lemke and Strehl 1999). A practical example for this is shown in Fig. 4.53, which refers to PD tests of a defective instrument transformer of 110 kV rated voltage. Applying an AC test voltage of variable frequency, the phase-resolved PD patterns yields the superposition of two characteristic noise signatures, see Fig. 4.53a. Here the randomly distributed dots, which are scattering in magnitude and phase angle, are caused by pulse-shaped noises. As can be seen, these are not correlated with the

(a)   **(b)**   **(c)**

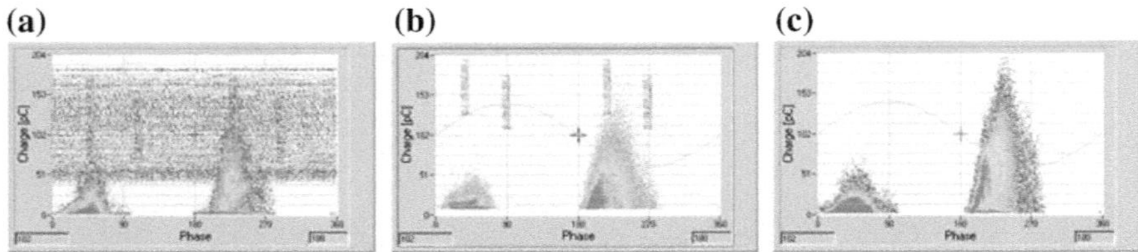

Fig. 4.53 Phase-resolved PD pattern of a defective 110-kV instrument transformer interfered by pulse-shaped noises originating in the mains (60 Hz) and in the frequency converter (92 Hz) of the applied resonant test set. **a** PD signatures without noise canceling. **b** Gating of noisy pulses originating in the mains. **c** Additional gating of the noisy pulses originating in the IGBTs of the frequency converter

applied 92-Hz test frequency, but apparently with the frequency of the mains, which is being 60 Hz. To cancel this noisy signal, an inductive sensor has been attached close to the power supply of the HV test facility to trigger the gating unit, which is implemented in the computerized PD measuring system. Under this condition, the randomly distributed pulses could entirely be eliminated, see Fig. 4.53b. Moreover, an additional inductive sensor has been installed close to the frequency converter in order to capture the switching pulses from the IGBTs used for triggering a second gating unit and thus to reject also the heavy noises originating in the frequency converter, see Fig. 4.53c.

> **Note** In practical tests using ACRF test systems (see Sect. 3.2.3) the noisy pulses caused by the frequency converter appear often at stable phase angle and can thus simply by identified Under this condition it is not absolutely necessary to perform a noise pulse gating, which might lead to an information loss on the PD occurrence

Different to the above presented measuring examples, the noisy pulses shown in Fig. 4.53 b can also be eliminated off-line, i.e. after the actual PD measurement has been finished. For this purpose the software tool "*windowing*" has been developed. As an example consider Fig. 4.54 which refers to the classical 2D display mode and the more sophisticated 3D display mode, as well. In this context it should be noted that most of the nowadays available digital PD measuring systems are equipped with software packages to discriminate pulse-shaped noises from real PD events, while hardware tools also suitable for this purpose is only rarely implemented in computerized PD instruments.

Due to the recent advancements in digital signal processing, very promising *denoising* software tools have been developed, mostly based on a *cluster separation* approach. An example for this is the establishment of *star diagrams* using synchronous three-phase PD measurements (Plath 2002; Kaufhold et al. 2006), as exemplarily shown in Fig. 4.55, which refers to a PD test of a power transformer. Here, external noises due to corona discharges show almost equal magnitudes, which were detected in all three phases and indicated by the three blue clusters in Fig. 4.55. However, PD pulses originating in the "Yellow" phase created a different colored cluster. Using the classical phase-resolved PD pattern (PRPDP)

(a)

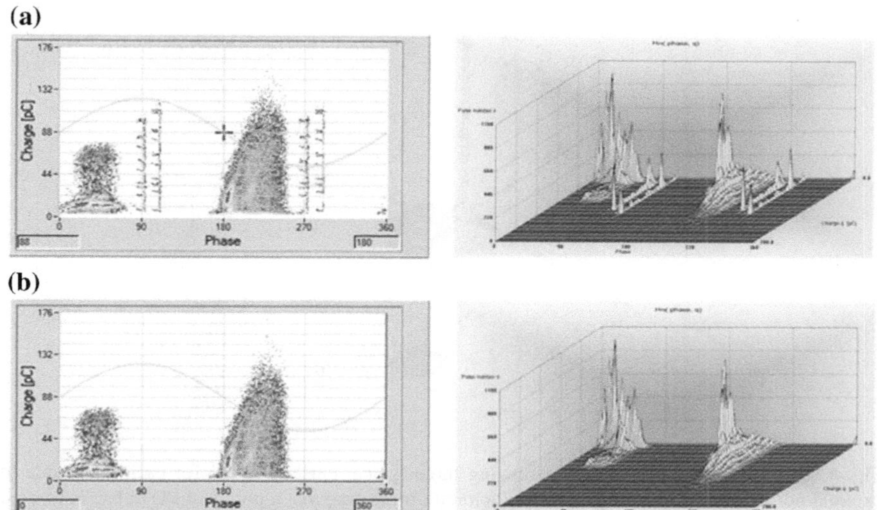

(b)

Fig. 4.54 Screenshots of phase-resolved PD patterns and synchronously appearing pulse noises originating in the frequency converter (**a**) and de-noising the PD patterns by gating (**b**)

recognition, a potential PD defect was identified in this phase. In principle, the vectors obtained by the three-phase PD measurement could also be added to establish a three-phase amplitude relation diagram. Under this condition, only one single *noise cluster* would be established. This is located close to the center point, whereas PDs in the single phase appear outside the center point, which leads to the conclusion that disturbing cross-talking phenomena can be neglected.

Another approach based on the cluster separation is the decomposition of the acquired PD pulse waveforms. For this purpose characteristic PD pulse parameters are used in either the frequency or the time domain, such as the rise and decay time as well as the pulse width (Cavallini et al. 2002; Rethmeier et al. 2008). A typical measuring example is illustrated in Fig. 4.56, which refers to a defective power cable termination, where the PD measurement was carried out in a poor-screened test laboratory. After acquiring the complete PD data stream captured from the test object, the strongly interfered phase-resolved PD pattern shown in Fig. 4.56a was obtained. Analysing the pulse shapes in the time domain according to the test series shown in Fig. 4.56b, the characteristic PD patterns could clearly be separated from the noise signatures, see Fig. 4.56c. In principle, this approach is also applicable for multi-source PD separation (Plath 2005).

In this context it should be emphasized that advanced de-noising tools can successfully be applied only by experienced test engineers, who must be familiar not only with the fundamentals of PD measurements but also with the operation principle as well as the capabilities and obstacles of the de-noising tools adopted. Even if sophisticated de-noising software is often implemented in advanced computerized PD measuring systems, it should not be overlooked that electromagnetic

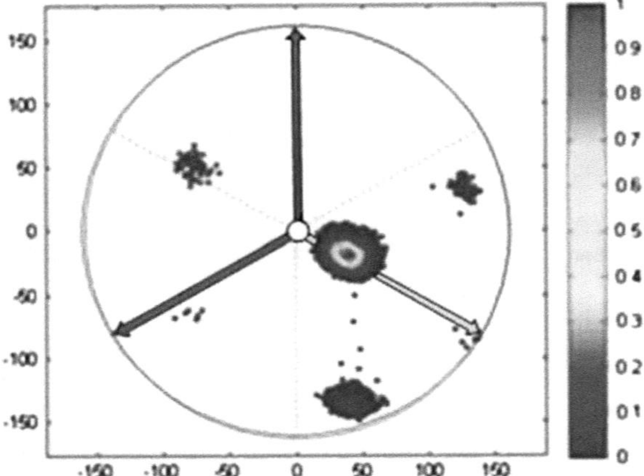

Fig. 4.55 Three-phase star diagram showing three typical clusters (blue) for each phase due to external noises as well as a single cluster (colored), indicating that a potential PD defect is located in phase "Yellow" of the investigated power transformer (Plath 2002)

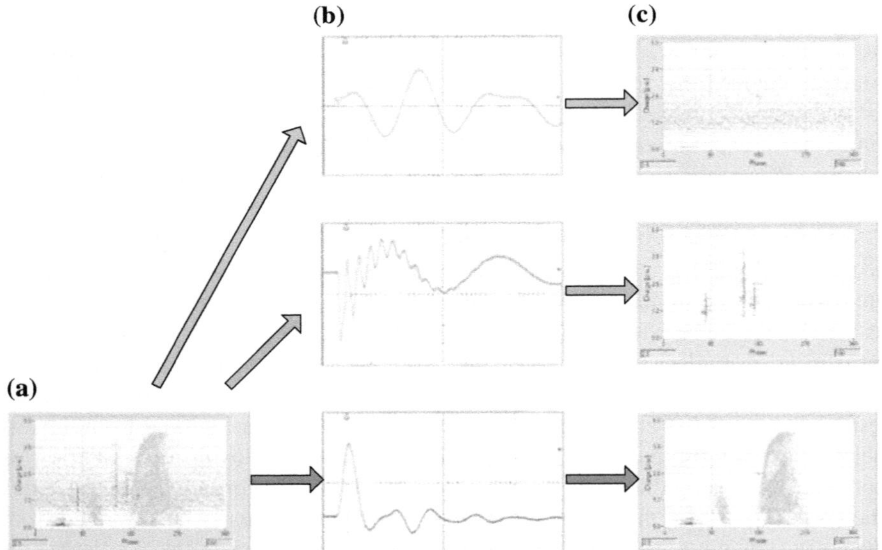

Fig. 4.56 De-noising of PD signals using cluster separation according to Cavallini and Montenari (2007). **a** Acquired data stream captured during a testing time of 10 min. **b** Waveform analysis. **c** Cluster separation

interferences can often simply be discriminated from PD pulses by means of multichannel digital oscilloscopes and thus conveniently be canceled using classical analogue features, such as windowing and gating, as discussed previously.

4.6   Visualization of PD Events

The main aim behind the visualization of phase-resolved PD patterns (PRPDP) by means of oscilloscopes or computerized PD measuring systems is the recognition and identification of typical PD sources. The first tool applied for this purpose was the electron-beam tube, also known as Braun tube, which has been employed by Lloyd and Starr already in 1928. Connecting the AC test voltage signal to the horizontal deflection plates and bridging the vertical deflection plates by a measuring capacitance, so-called Lissajous figures could be recorded, as exemplarily shown in Fig. 4.57a. In this context, it seems worth to notice that this circuit represents in principle an integrating bridge, which enables the measurement of the power losses of discharges based on the so-called parallelogram method, which has first been used by Dakin and Malinaric in 1960.

When the first PD detectors were available (Arman and Starr 1936; Mole 1954), the classical Lissajous figure technique has also been modified in order to display the *pulse charge trains* superimposed on an elliptical loop, where the time base covers a single AC cycle, see Fig. 4.57b. Later the use of a linear time-base became a common practice in order to record the pase-resolved PD patterns (PRPDP), as exemplarily shown in Fig. 4.58. This presentation refers to a needle-plane test sample, which is often used for fundamental PD studies, in particular to classify typical PD defects.

Further essential steps in the evaluation of PD events where achieved in 1969, when Bartnikas and Levi presented a *pulse-height analyser* for PD rate measurements, and in 1978, when Tanaka and Okamoto presented the first

(a)   **(b)**

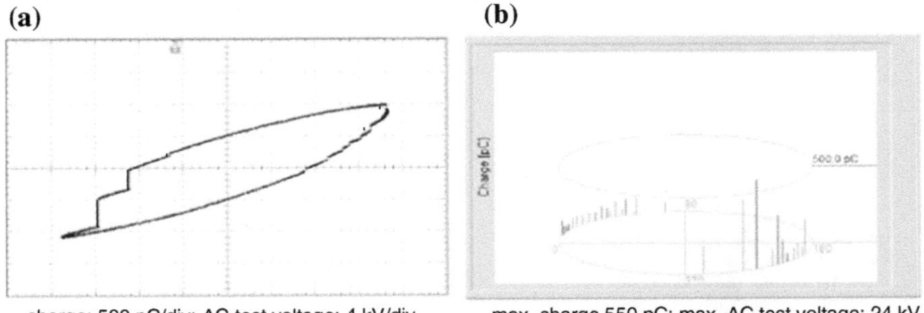

charge: 500 pC/div; AC test voltage: 4 kV/div   max. charge 550 pC; max. AC test voltage: 24 kV

Fig. 4.57 So-called Lissajous figure technique used since the 1960 s to display the charge of PD pulses occurring within a single cycle of the applied AC test voltage. **a** Voltage appearing across the measuring capacitance of an integrating bridge. **b** Traditional elliptical display mode

minicomputer-based PD measuring system. At the very beginning of computerized PD measurements, PD pulse trains appearing within individual AC cycles have often been displayed like waterfall diagrams, see Fig. 4.59.

Today's digital PD measuring systems acquire and store the vector $[q_i; u_i; t_i; \varphi_i]$ for each captured PD pulse. Here are:

q_i the pulse charge of the individual PD current pulse
u_i the instantaneous value of the applied test voltage
t_i the instant of PD occurrence
φ_i the phase angle at instant of PD occurrence

Based on this, the cumulative (integral) *phase-resolved PD pattern* is commonly displayed, as illustrated in Fig. 4.60. Since the 1980s, this display mode is widely used for the so-called phase-resolved PD pattern recognition (Kranz and Krump 1988; Ward 1992; Fruth and Gross 1994). As the PD data stream acquired in the real-time mode can completely be stored in the computer memory, these can be recalled again and visualized in a so-called replay-mode like real PD events, even after the actual PD test has been finished, as exemplarily shown in Fig. 4.61 (Lemke et al. 1996; Lemke and Strehl 1999).

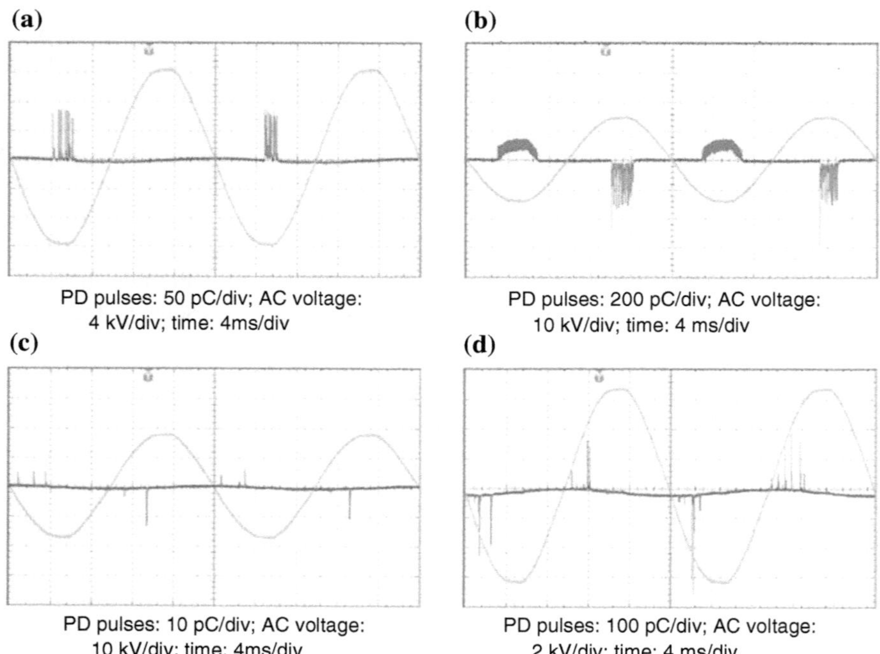

Fig. 4.58 Oscilloscopic screenshots of phase-resolved PD patterns gained for needle-to-plane test samples under power frequency (50 Hz) AC voltage. **a** Discharges in air at inception voltage. **b** Discharges in air at test level substantially above inception voltage. **c** Cavity discharges in XLPE at inception voltage. **d** Surface discharges at inception voltage

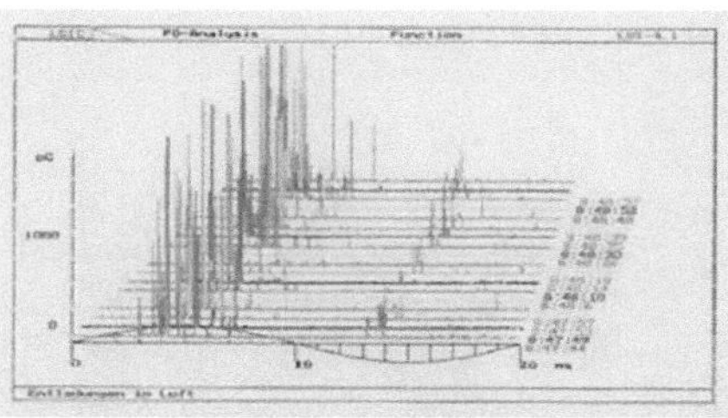
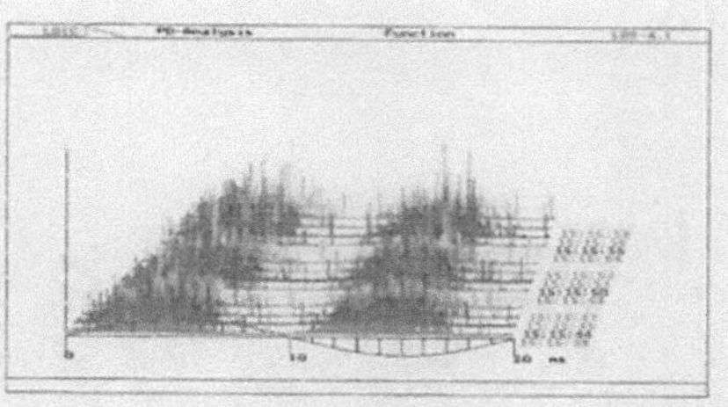

XLPE cable termination   XLPE cable joint

Fig. 4.59 PD signatures of defective cable accessories displayed like waterfall diagrams

Using the replay-mode besides the traditional two-dimensional graphs shown in Fig. 4.61, the 3D presentation is frequently used, which is in particular useful to discriminate electromagnetic interferences from real PD events, as has already been discussed based on the Figs. 4.48 and 4.49. Moreover, the phase-resolved PD patterns can be displayed in a traditional manner using either the elliptical or linear time base, as shown in Fig. 4.62.

Another approach sometimes adopted for *PD pattern recognition* is the analysing of PD pulse sequences, as proposed by Hoof and Patsch (1994). The algorithm is based on the evaluation of the voltage differences measured between three subsequent PD pulses, as illustrated in Fig. 4.63a. That means, the voltage difference ΔV_{n-1} between the reference pulse "n" and the previous occurring pulse "$n-1$", as well as the associated voltage difference ΔV_n between the reference pulse "n" and the following pulse "$n+1$", is plotted in a graph displaying the value pairs ΔV_n versus ΔV_{n-1}. A measuring example for this is shown in Fig. 4.63b which refers to slot discharges detected in a rotating machine.

Fig. 4.60 Principle for displaying cumulative phase-resolved PD patterns

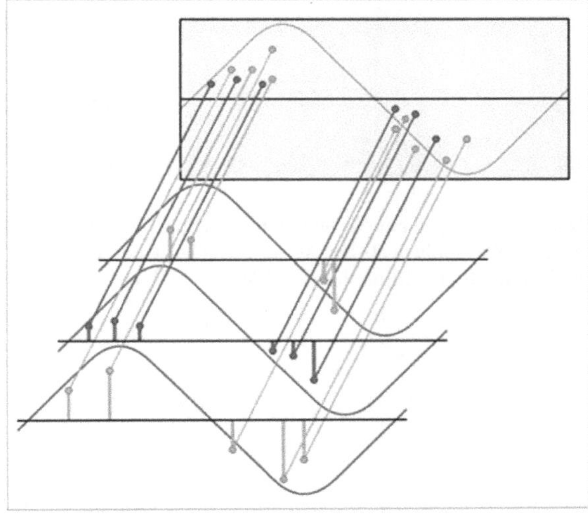

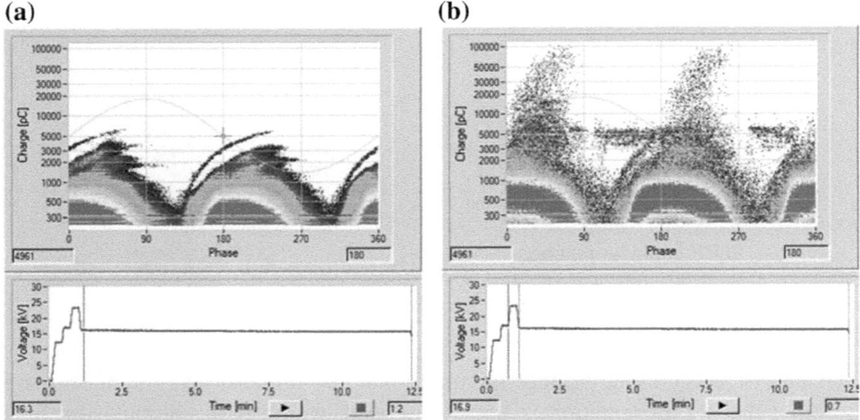

Fig. 4.61 Using the replay-mode to visualize phase-resolved PD patterns originally acquired and stored in the computer memory in the course of development tests of stator bars of a hydro-generator visualized by the use of the replay-mode. **a** Long-term PD test under AC voltage (test level 18 kV, test duration 10 min, which equals 30,000 AC cycles). **b** Short-term PD test under enhanced AC voltage (test level 24 kV, test duration 2 min, which equals 6000 cycles)

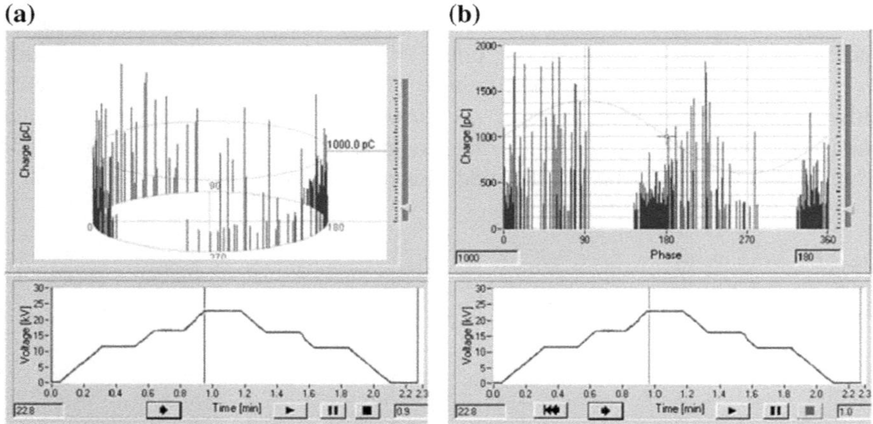

Fig. 4.62 Using the replay-mode to visualize phase-resolved PD pulses occurring within a single cycle of the applied 50-Hz AC test voltage (see cursor position obvious from the test voltage profile). **a** Elliptical time base. **b** Linear time base

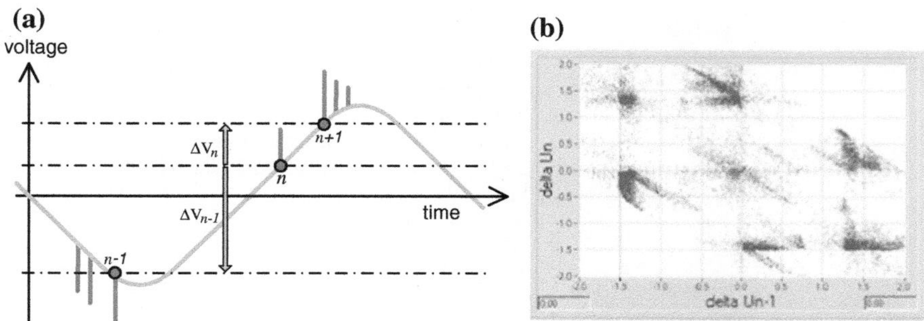

Fig. 4.63 Principle of PD pulse sequence analysis (**a**) and measuring example (**b**) gained for cavity discharges in a power cable termination

To classify typical discharge sources, Tanaka and Okamoto proposed (1986) the establishment of PD fingerprints based on the following statistical operators:

- standard deviation,
- skewness,
- kurtosis and
- cross-correlation.

Later the feasibility of such statistical operators to establish PD fingerprints and to identify and classify typical PD sources, has been confirmed by Gulski and his co-workers (Gulski 1991). To create characteristic PD fingerprints, the following statistical parameters shown in Fig. 4.64 are commonly displayed:

H_n (*phi*): Number of PD pulses occurring within each phase window versus the phase angle,
H_q (*phi*)$_{peak}$: Peak values of PD pulses occurring within each phase window versus the phase angle,
H_q (*phi*)$_{mean}$: Mean values of PD pulses occurring within each phase window versus the phase angle. This quantity is deduced from the total charge amount within each phase window divided by the pulse number occurring in this phase window.

Based on PD fingerprints established for various types of failures and stored in the computer memory, these can be compared with each other, where the test conditions have also to be taken into consideration. Practical experiences revealed that graphs according to Fig. 4.65 provide a valuable tool for the insulation condition assessment and thus for maintenance decisions.

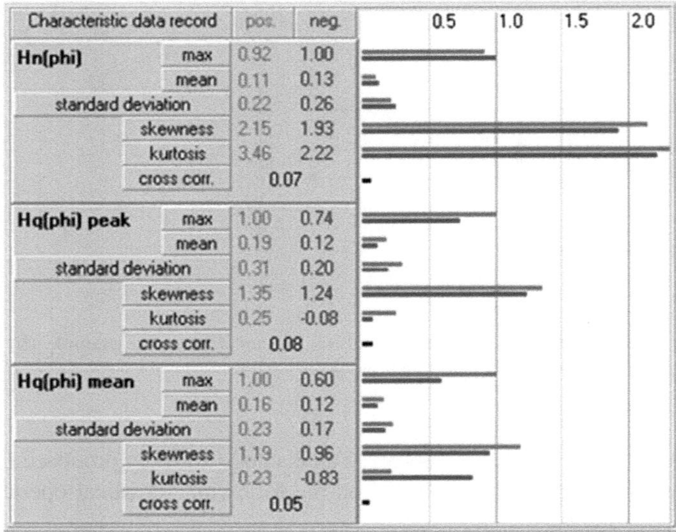

Fig. 4.64 PD fingerprint of a discharge source recognized in a HV cable termination in service

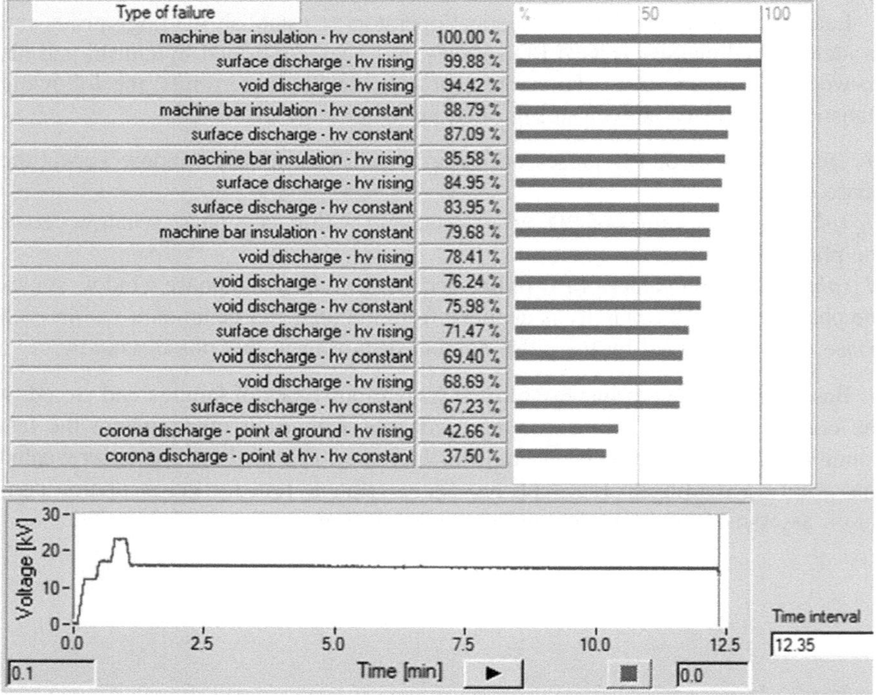

Fig. 4.65 Comparison of the PD fingerprint gained for stator bar insulation with earlier established PD fingerprints gained for various test samples

4.7 PD Detection in the VHF/UHF Range

4.7.1 General

Performing PD measurements in compliance with IEC 60270:2000, the upper cut-off frequency has to be limited below 1 MHz, as recommended in the Amendment to this standard and discussed more in detail in Sect. 4.3. However, under this condition the signal magnitude is substantially attenuated, so that well shielded test laboratories are required to perform sensitive PD tests. Obviously, the signal-to-noise ratio can significantly be enhanced when the PD signal is captured in the VHF/UHF range. This non-conventional method has first been employed for quality assurance tests of gas-insulated substations by Fujimoto and Boggs (1981). Later the benefits of this approach has successfully been proven for on-site PD diagnosis tests of HV/EHV cable accessories (Pommerenke et al. 1995) and even for PD diagnostics of power transformers under on-site condition (Judd et al. 2002). Moreover, the capability and limits of the non-conventional PD detection in the VHF/UHF range has extensively been investigated by various CIGRE Working Groups. The main issues of these studies are summarized in the Technical Brochures No. 444 (2010) and No. 502 (2012). Based on these publications, the IEC 62478:2015 provides valuable recommendations for the design of VHF/UHF PD measuring systems including the sensors required to capture the PD transients radiated from the test object.

Due to the extremely wide frequency spectrum of PD pulses, which covers the ranges of radio frequency (RF: 3–30 MHz), very high frequency (VHF: 30–300 MHz), and ultra-high frequency (UHF: 300–3000 MHz), various kinds of PD couplers have been developed in the past, which are commonly classified as capacitive, inductive and electromagnetic sensors, as will briefly be presented in the following.

4.7.2 Design of PD Couplers

4.7.2.1 Capacitive PD Couplers

To capture the PD signal from the terminals of the test object, the classical coupling device recommended in IEC 60270:2000 provides a HV coupling capacitor connected in series with a measuring impedance. The upper cut-off frequency of such a coupling device (Fig. 4.66a) is commonly limited below 10 MHz, which is equivalent to a rise time close to 30 ns. To increase the upper limit frequency, it is necessary to decrease the value of the coupling capacitor, because this is associated with a reduction in the internal inductance, which determines the achievable upper cut-off frequency (Fig. 4.66b). Consequently, the highest measuring frequency is achievable by means of capacitive sensors providing a simple metallic disc. The feasibility of this simple approach has initially been proven for the detection and

Fig. 4.66 Step voltage response measured for capacitive PD couplers.
a High-capacitive coupling capacitor designed according to IEC 60270:2000 (capacitance 2000 pF, rated voltage 24 kV).
b Low-capacitive coupling device (capacitance 50pF, rated voltage 12 kV).
c Disc-shaped C-sensors intended for PD probing (capacitance <5 pF)

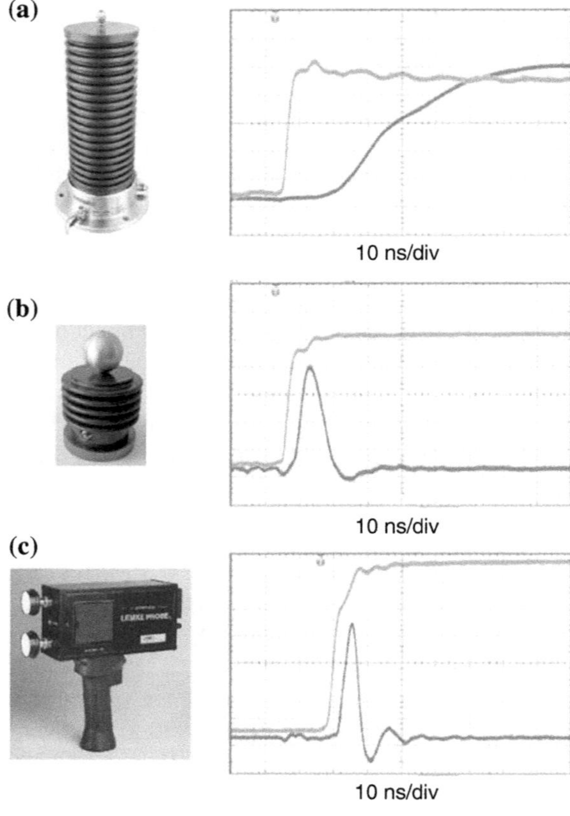

pinpointing of PD sources by means of a hand-held PD probe, as shown in Fig. 4.66c and described also in Sect. 10.4.4 (Fig. 10.43). Such kinds of capacitive sensors, often referred to as C-sensors, receive the electric field component of electromagnetic PD transients.

Figure 4.67 shows a sketch of a coaxial C-sensor designed for PD detection in power cable joints. Here a section of the outer semiconductive layer coating the cable insulation is removed to receive the electric field component, which is radiated from the inner cable conductor due to travelling waves excited by PD events. The achievable measuring sensitivity is mainly governed by the effective capacitance C_s between sensor electrode and inner cable conductor and lies in the pC range.

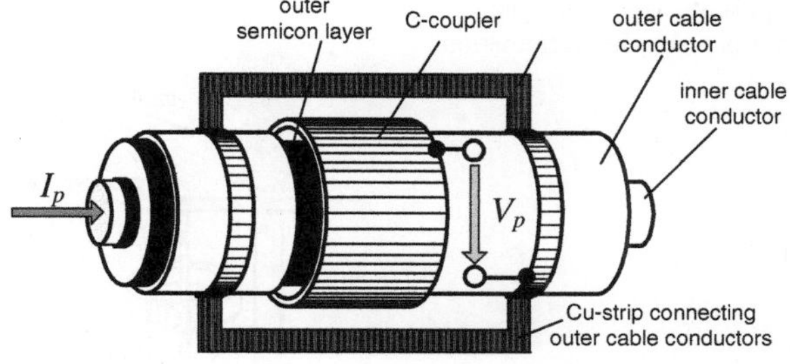

Fig. 4.67 Sketch of a capacitive PD coupler attached to a power cable

Example Consider a polyethylene-insulated power cable of dielectric permittivity of $\varepsilon_r = 2.2$. Assuming a ratio between outer and inner cable conductor of $r_a/r_i = e \approx 2.7$, one gets for a coaxial C-sensor of length $l_a = 100$ mm the following approximation:

$$C_s \approx 2 \cdot \pi \cdot \varepsilon_0 \cdot \varepsilon_r \cdot l_a \approx 12 \text{ pF}$$

Provided the received PD signal is transmitted via a measuring cable matched by its characteristic impedance Z_m, the peak voltage V_p appearing across Z_m and thus at the input of the peak detector can roughly be accounted for using the following approach:

$$V_p \approx C_s \cdot Z_c \cdot Z_m \cdot \frac{I_p}{t_r}.$$

Here are I_p and t_r the peak value and the rise time of the PD pulse current, respectively, and Z_c is the characteristic impedance of the power cable. Assuming, for instance, a cavity discharge creates a current pulse of rise time $t_r = 1$ ns and peak value $I_p = 1$ mA, one gets for the above-introduced circuit parameters ($C_s = 12$ pF, $Z_c = 30$ Ω, $Z_m = 50$ Ω) a detectable peak voltage of $V_p = 18$ mV, which is well measurable by means of digital oscilloscopes.

4.7.2.2   Inductive PD Couplers

The operation principle of inductive PD couplers, commonly referred to as L-sensors or or even "yoke coils", is comparable with that of high-frequency pulse transformers (HFCT) where the primary coil is formed by a conductor, which belongs to the test object, i.e. the turn number is $n = 1$. To capture the complete magnetic flux surrounding the primary conductor, the windings of the secondary coil are usually wounded around a high-permeable ferrite core. Under this condition, the transient PD current $i_p(t)$ through the primary conductor induces a transient voltage $v_p(t)$ in the secondary coil (Fig. 4.68). In this context the similarity to Rogowski coils should be underlined. The only difference is that the windings of the Rogowski coil are not wound around a permeable core (see Sect. 7.5.2).

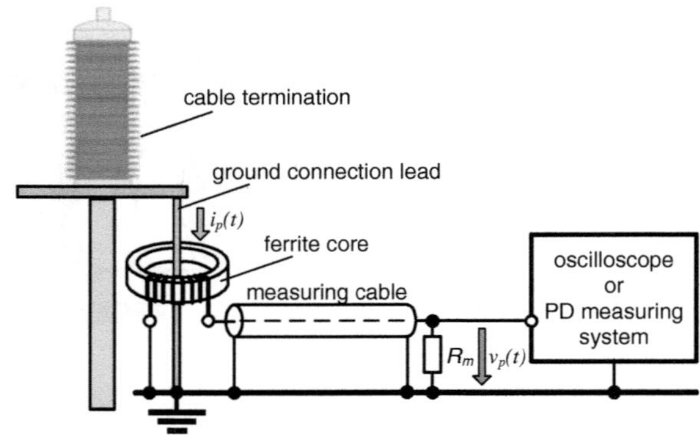

Fig. 4.68 L-sensor attached to a power cable termination

A simple PD coupling unit is shown in Fig. 4.68, where a L-sensor is attached around the ground connection lead of a power cable terminal. To characterize the dynamic behaviour, it is recommended to determine the step current response in the time domain according to Fig. 4.69. Comparing the oscilloscopic records shown in Figs. 4.69c and d it can be concluded that the transmitted pulse length is drastically reduced by decreasing the turn number of the secondary coil, i.e. from originally $n = 10$ down to $n = 1$, as has to be expected. From this follows also that by means of classical L-sensors a pulse length shorter than 1 μs is hardly achievable. Thus only PD signals in the RF range (3–30 MHz) but not in the VHF/UHF (>30 MHz) are transmitted.

4.7.2.3 Electromagnetic PD Couplers

The operation principle of electromagnetic (EM) PD couplers is in principle comparable to that of antennas operating in the near-field region. That means the output signal is determined by both vectors representing the electric field component $\vec{E}$ and the magnetic field component $\vec{H}$, as given by the Maxwell equations:

$$\text{rot}\,\vec{H} = \varepsilon \cdot \frac{\delta \vec{E}}{\delta t}, \quad \text{rot}\,\vec{E} = -\mu \cdot \frac{\delta \vec{H}}{\mathrm{d}t}. \tag{4.29}$$

Depending on the geometric configuration of the test object, various kinds of EM sensors are employed to detect PD signals in the VHF/UHF range, such as rod, disc or conical antennas. The latter are commonly used for PD diagnostics of gas-insulated substations (Pearson et al. 1991). A sketch of a conical UHF PD sensor installed in a flange of a GIS compartment is shown in Fig. 4.70. Generally such antennas are capable of receiving PD signals of frequency content up to about 1.5 GHz. The upper cut-off frequency is inversely proportional to the characteristic time constant, which follows from the inevitable stray capacitance between sensor electrode and ground

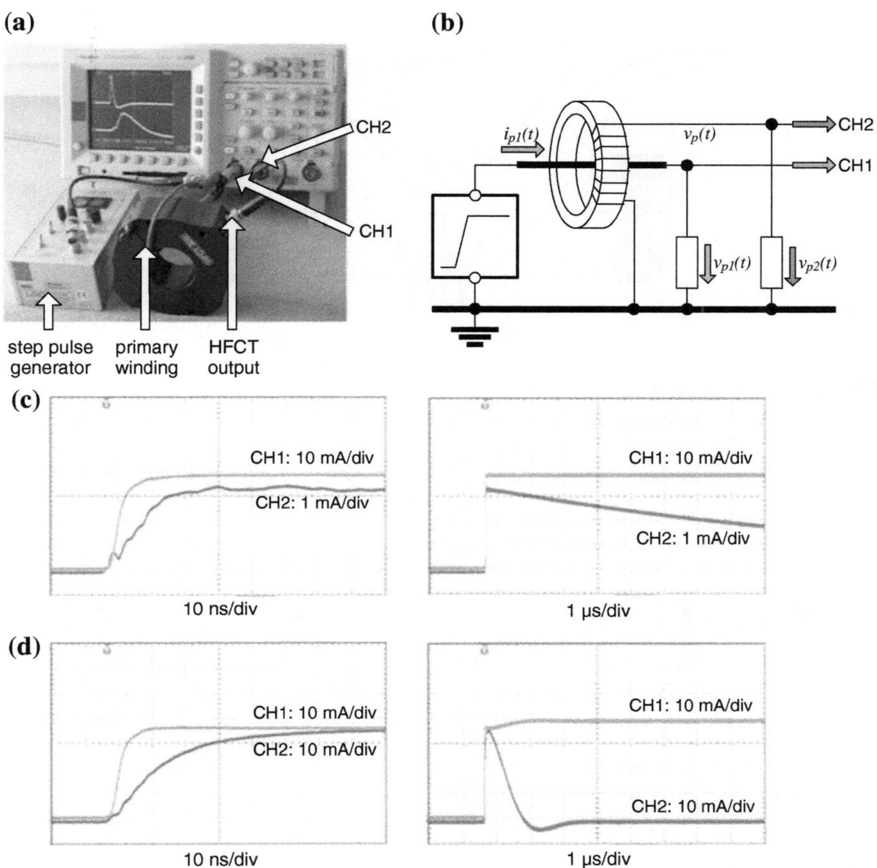

Fig. 4.69 Set-up for measuring the step current response of inductive PD sensors (**a, b**) and oscilloscopic records gained for pulse transformers having $n = 10$ windings (**c**) resp. $n = 1$ winding (**d**)

flange if multiplied by the characteristic impedance of the connected measuring cable (Meinke and Gundlach 1968; King 1983; Küpfmüller 1984).

To detect PD faults in power cable accessories, the feasibility of so-called directional coupler sensors (DCS), operating in the frequency range between 2 and 500 MHz, has also successfully been proven (Pommerenke et al. 1995). The design principle is comparable to that of capacitive sensors. The only difference is that the PD signal is captured from both sensor ends, which are commonly referred to as "ports". Installing a pair of DCS at both joint sides, the PD signal originating inside the joint can be discriminated from noise and even PD signals originating in the power cables connected to both sides. This is because a PD event inside the joint causes pulses at the ports "B" and "C" whose magnitudes are significantly higher than those occurring at the ports "A" and "D" (Fig. 4.71). Another benefit is that

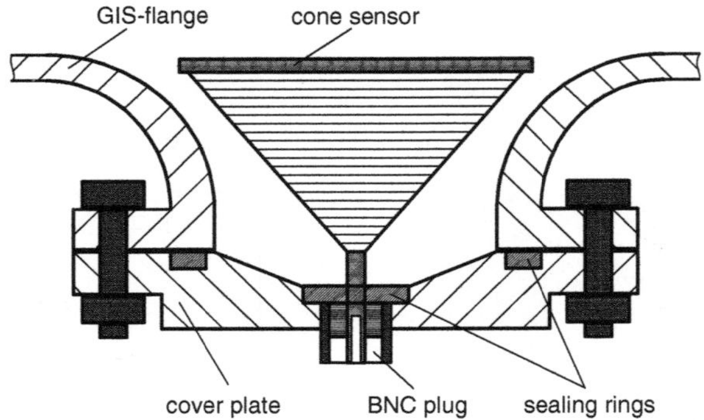

Fig. 4.70 Sketch of an UHF sensor installed in a GIS flange

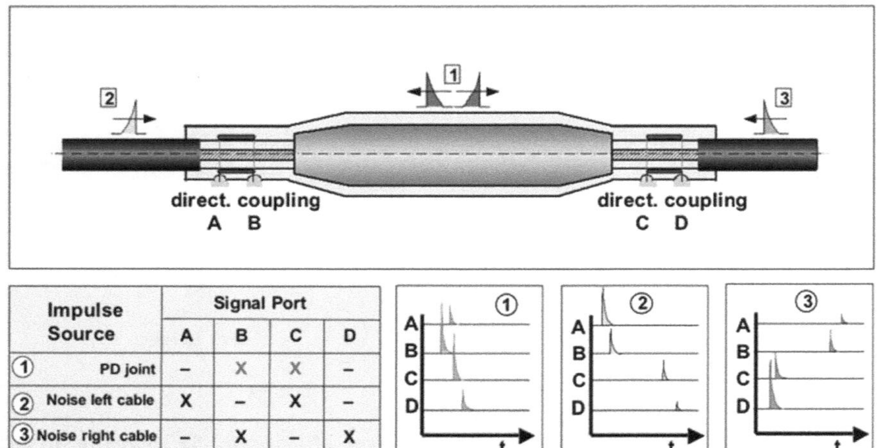

Impulse Source		Signal Port			
		A	B	C	D
①	PD joint	–	X	X	–
②	Noise left cable	X	–	X	–
③	Noise right cable	–	X	–	X

Fig. 4.71 Operation principle of a pair of directional coupler sensors (*DCS*) installed at both sides of a power cable joint

one single sensor can be used to calibrate the other one. Practical experience revealed that even under noisy on-site condition a measuring sensitivity in the pC range is achievable.

To capture the PD signal from grounding leads of the test object, high-frequency current transformers (HFCTs) can advantageously be used, as discussed previously. One obstacle of such kinds of PD couplers is, however, that the measuring frequency is commonly limited to the RF range (3–30 MHz). A promising alternative is the use of so-called pulse transformers, which are based on the concept of transmission line inverters, as illustrated in Fig. 4.72 (Lemke et al. 2003). Here, a coaxial cable of few cm in length is used, where the inner conductor is terminated at the output with its characteristic impedance, while the outer conductor is terminated

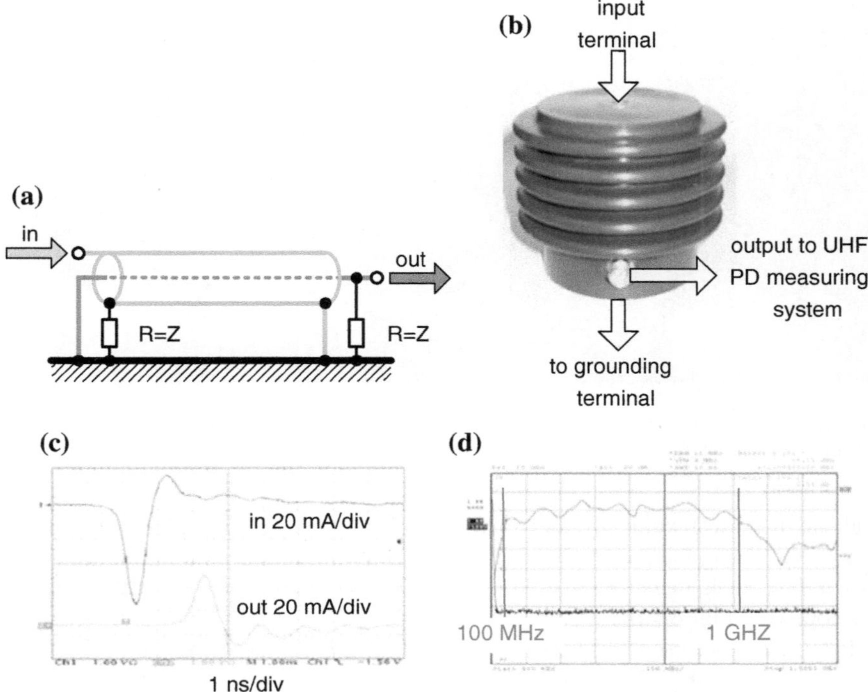

Fig. 4.72 UHF PD coupler based on the concept of transmission line inverters. **a** Equivalent circuit. **b** Technical design. **c** PD pulse response in the time domain recorded at time base of 1 ns/div. **d** Transfer function in the frequency domain

at the input. Moreover, the inner and outer conductor are grounded at the input and output, respectively. Under this condition both the magnetic and electric field components are transmitted at comparatively low attenuation up to the UHF-range (Lewis 1959), where the pulse polarity appears inverted. The feasibility of PD couplers based on this concept have successfully been proven in practice, especially for on-site PD monitoring of cable joints and terminations (Fig. 4.73).

4.7.3   Basic Principles of PD Detection in the VHF/UHF Range

As discussed previously in Sect. 4.7.1, the main benefit of the PD detection in the VHF/UHF range is the comparatively high signal-to-noise ratio. Using this technology, it can basically be distinguished between wide-band and narrow-band PD detection methods. Wide-band VHF/UHF measuring systems are equipped with a high-sensitive wide-band amplifier in combination with a very fast peak detector to

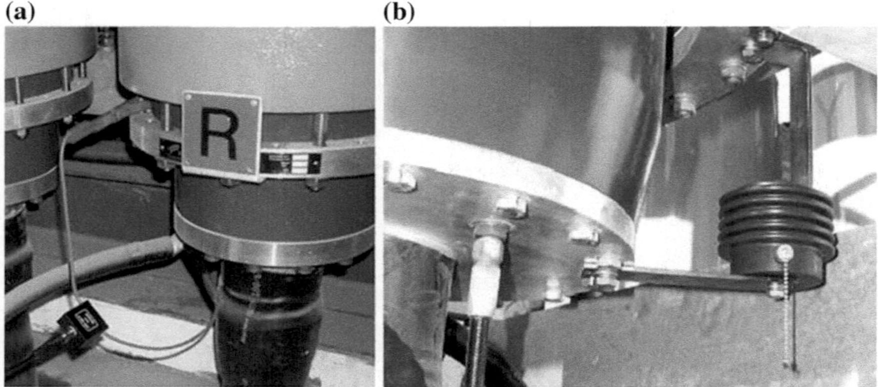

Fig. 4.73 UHF-PD coupler attached to a GIS-cable termination. **a** Flexible PD sensor used for periodical PD monitoring. **b** Fixed PD sensor used for continuous PD monitoring

evaluate the crest value of the amplified PD signal, as exemplarily shown in Fig. 4.74a. Using the narrow-band method, damped oscillations are excited, where also a fast peak detector is used to evaluate the maximum magnitude of the envelope, see Fig. 4.74b.

Peak detectors used for both the wide-band and narrow-band PD signal processing in the VHF/UHF range elongate the input signal, whose duration is usually in the ns range, up to the µs range. Under this condition, the further signal processing can be performed by means of classical PD measuring systems, so that phase-resolved PD patterns can conveniently be displayed. A survey on the basic principles commonly used for the PD detection in the VHF/UHF range is shown in Fig. 4.75.

In this context it should also be mentioned that for a wide-band PD detection in the VHF/UHF range also digital oscilloscopes can be applied, because these are nowadays commercially available for real-time measurements up to the GHz range.

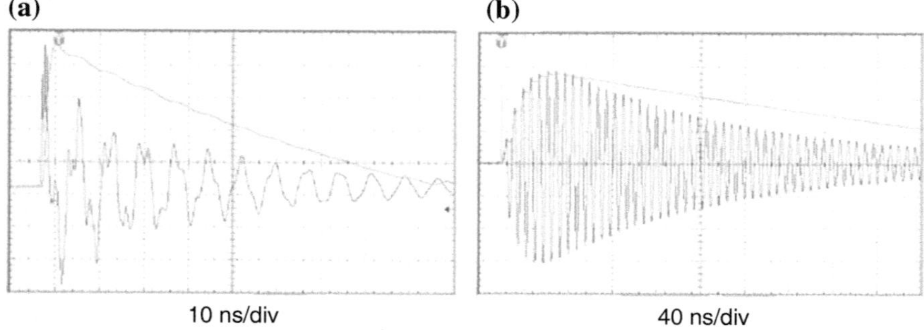

Fig. 4.74 PD pulse response of the investigated UHF amplifiers (violet traces) and the associated peak detector (green traces). **a** Wide-band measuring system. **b** Narrow-band measuring system

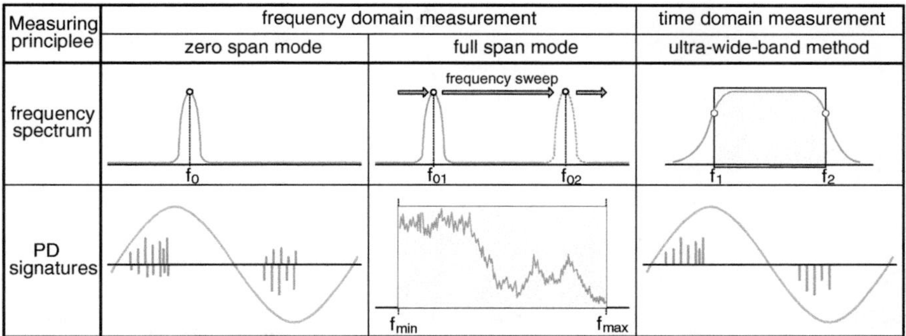

Fig. 4.75 Survey on UHF/VHF PD detection principles

For narrow-band PD detection classical spectrum analysers are applicable, where either the full-span mode or even at zero-span mode can be used, see Fig. 4.76.

Using the full-span mode, the frequency spectrum of the captured PD signal as well as the superimposed noise is recorded for the pre-selected start and stop frequencies. To discriminate disturbing noises from the signal caused by PD events, the background noise level is initially recorded just prior the actual PD test is performed, see Fig. 4.76e. The main obstacle of the full-span mode is that the classical phase-resolved PD pattern cannot be displayed. To overcome this crucial problem, the zero-span mode is often preferred, which is in principle comparable with that technique used for radio interference voltage (RIV) measurements (Sect. 4.3.3). That means the center frequency is adjusted such that the noise level becomes a minimum, which can conveniently be determined using the full-span mode.

4.7.4   Comparability and Reproducibility of UHF/VHF PD Detection Methods

As discussed previously, the main benefit of non-conventional UHF/VHF PD detection methods is the considerable enhancement of the signal-to-noise ratio if compared to the traditional apparent charge measurement in compliance with IEC 60270:2000. This offers the opportunity to perform sensitive PD diagnosis tests of HV apparatus under noisy on-site condition, which is commonly impossible by using the IEC method, which recommends a limitation of the upper limit frequency below 1 MHz. In this context it must be emphasized, however, that the magnitude of the PD pulses appearing at the output of PD instruments operating in the VHF/ UHF range is not correlated with the magnitudes of the apparent charge pulses. This is underlined by the measuring example shown in Fig. 4.77, which refers to a defective power cable termination.

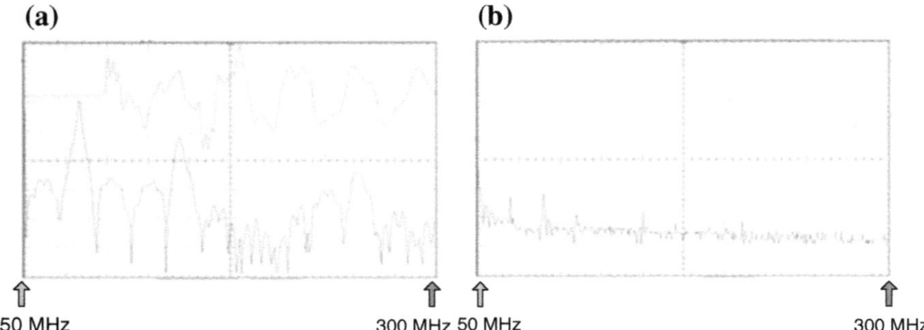

Fig. 4.76 Oscilloscopic screenshots gained by means of a spectrum analyser using the full-span mode (50–300 MHz). **a** Background noise level of the measuring surroundings. **b** Frequency response against a calibrating pulse

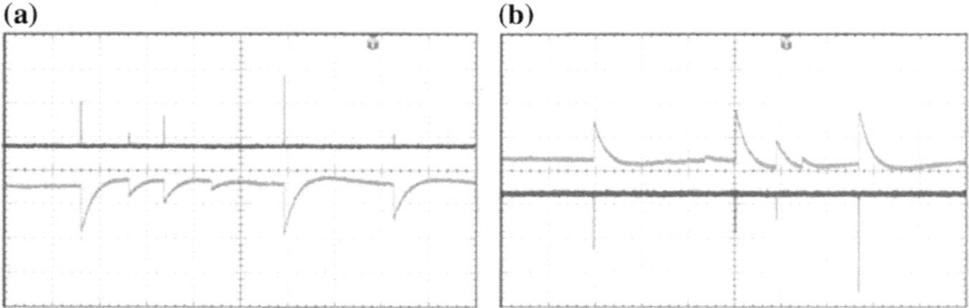

Fig. 4.77 Oscilloscopic screenshots of phase-resolved PD pulses captured from a defective power cable termination which were measured simultaneously by means of a VHF measuring system (pink trace) and a PD instrument designed according to IEC 60270 (green trace). **a** Positive half-cycle. **b** Negative half-cycle

As can be seen, the magnitudes of the PD pulses appearing at the output of the VHF measuring system (pink trace) are not proportional to those of the apparent charge pulses (green trace). This is underlined by the graphical presentation shown in Fig. 4.78, which results from 10 subsequent screenshots according to Fig. 4.77. Here the peak value of each current pulses appearing at the output of the UHV measuring system was plotted versus the magnitude of the associated apparent charge pulse.

Despite the drawback that the UHF/VHF PD detection method cannot be calibrated in terms of pC, there are also various benefits. So the signal-to-noise ratio is essentially enhanced if compared to the IEC method, as mentioned previously. This offers the opportunity to determine the PD inception voltage as well as the PD trend under noisy on-site conditions. Moreover, this technique can advantageously be used for the localization of potential PD defects in geometrical extended HV apparatus, such as GIS, using the time-of-flight measurement (Pearson 1991), as will be described in Sect. 10.4.1.

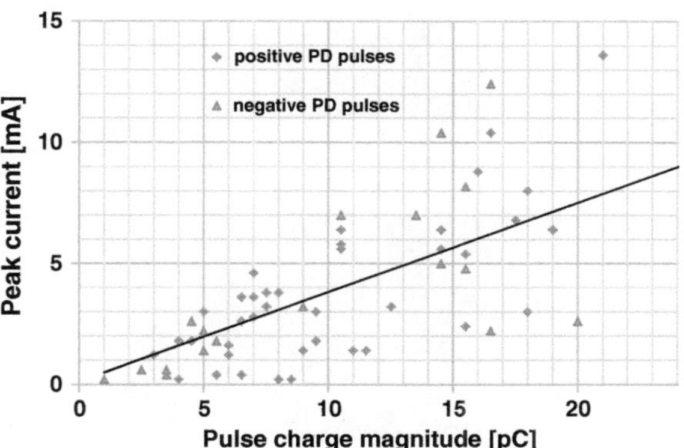

Fig. 4.78 Results of comparative PD studies showing the magnitudes of the PD current pulses evaluated by the VHF method versus the magnitudes of the apparent charge pulses

4.8   Acoustic PD Detection

PD events radiate not only electromagnetic waves but emit also acoustic pressure waves, where the acoustic signal covers a frequency range between some kHz and several hundreds of kHz. The main benefit of the detection of acoustic emitted (AE) waves is their immunity against electromagnetic interferences. To prevent an impact of other mechanical vibrations caused by pumps and fans as well as acoustic noises emanated from the iron core of transformers on account of magnetostriction and Barkhausen effect, commonly the ultrasonic frequency range, preferably between 40 kHz and few 100 kHz, is chosen to capture and acquire AE signals.

The ultrasonic PD detection has initially been employed to localize airborne noises due to corona discharges, for instance, to identify disturbing discharges at shielding electrodes of HV test facilities as well as harmful discharges igniting in broken cap-and-pin insulators. A photograph of a hand-held battery-powered ultrasonic PD detector designed for this purpose is shown in Fig. 4.79.

In the late 1950s, the ultrasonic PD detection technique was also employed to recognize and localize structure-borne noises emitted from PD defects in HV apparatus (Anderson 1956). Thereafter this technology became a widely established tool for preventive PD diagnostics of gas-insulated substations (Graybill 1974; Lundgaard et al. 1990; Albiez and Leijon 1991) and even for the localization of PD faults in large power transformers (Harrold 1975; Nieschwitz and Stein 1976; Howels and Norton 1978; Lundgard et al. 1989; Fuhr et al. 1993). Besides the magnitude, also the shape of the ultrasonic signal could be very informative to identify and localize potential PD defects due to the fact that the frequency content as well as the magnitude of the received acoustic signal appears considerably attenuated at increasing distance to the PD source.

Using only a single ultrasonic transducer, however, the localization procedure is extremely time-consuming, particularly in case of intermitting PD events. As an alternative, the so-called *triangulation* has nowadays become a common practice.

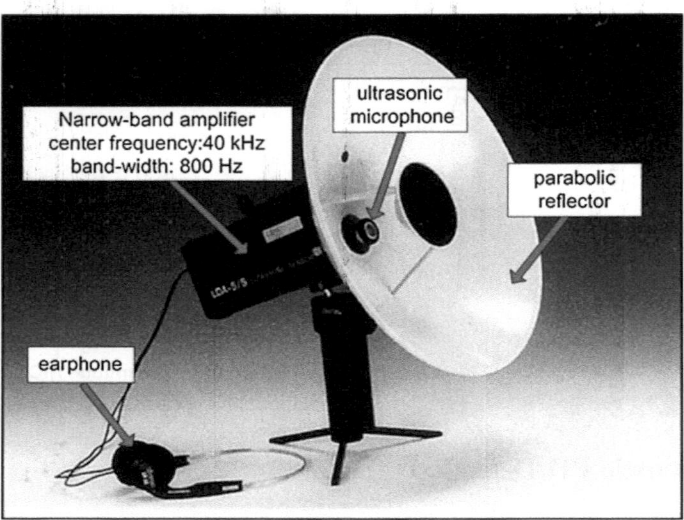

Fig. 4.79 Photograph of an ultrasonic PD detector. Courtesy of Doble Lemke

For this purpose three or even more transducers are used to perform *time-of-flight measurements*, as illustrated in Fig. 4.80. For a homogenous medium, the distances x_1, x_2 and x_3 between the PD source and the AE transducers are proportional to the measured time-of-flight denoted as t_1, t_2 and t_3, which is deduced from the oscilloscopic records. Thus, the trajectories shown in Fig. 4.80 are crossing that point where the PD source is located.

To minimize the localization uncertainty, it is a common practice to use the ultrasonic technique in conjunction with electrical PD measurements. This offers the opportunity to trigger the oscilloscope by an electrical signal at instant when the PD event ignites, see Fig. 4.81a. As the time lag of the captured electrical signal is below the μs range, this can be neglected if compared to the flight time of the acoustic signal, because this travels in oil only 1.2 mm per μs. Under noisy condition, the signal-to-noise ratio can considerably be enhanced by the use of the so-called averaging mode, as illustrated in Fig. 4.81b. The combined acoustic-electrical method confirms moreover that indeed a PD defect has been detected and not a disturbing acoustic noise.

Under laboratory condition, the electric signal required for triggering the oscilloscope is commonly captured from the test object via a coupling capacitor or even from the bushing tap, if available. Under noisy on-site condition, however, it is more beneficial to use the VHF/UHF technique to enhance the signal-to-noise ratio, as discussed in Sect. 4.7. In this context it must be emphasized, however, that the triangulation according to Fig. 4.80 provides only reasonable results for acoustic waves travelling in a continuum, where the velocity of acoustic pressure waves remains constant. For HV equipment of complex design, such as power transformers, the wave velocity is strongly affected by the very different

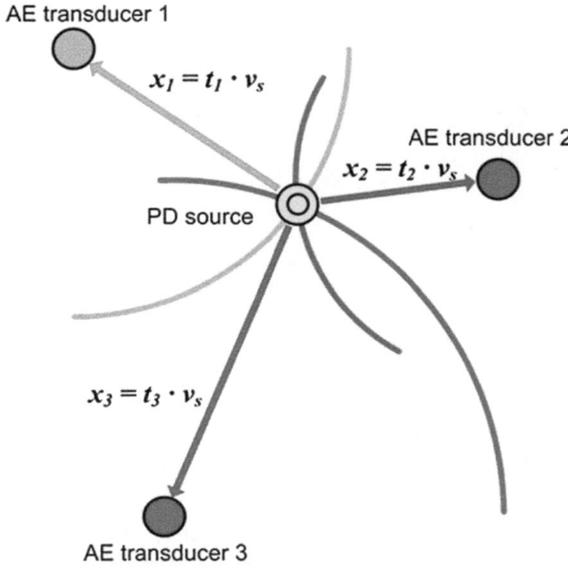

Fig. 4.80 Principle of triangulation used for localization the PD site

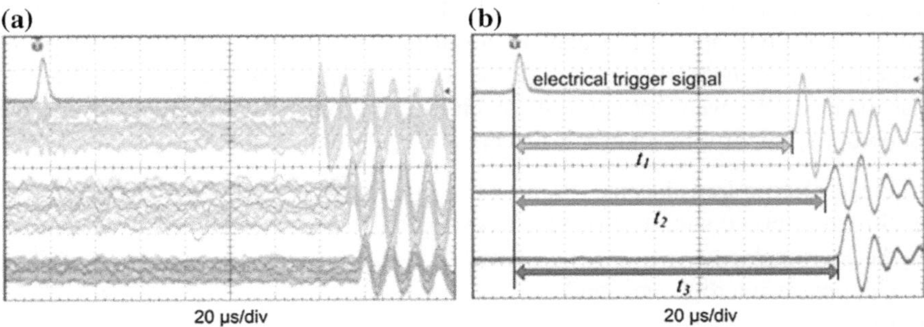

Fig. 4.81 Principle of time-of-flight measurement, where the acoustic signal is received by three ultrasonic transduces attached to the tank of a 110-kV instrument transformer. Here the oscilloscope was triggered by an electrical PD signal. **a** Single-pulse triggering. **b** Multi-pulse triggering (averaging)

construction materials, such as copper, steel, wood, pressboard and insulating oil. Thus, instead of the direct sound wave, propagating the shortest distance between PD source and ultrasonic transducer, two wave fronts of very different velocities have to be taken into consideration. These are commonly referred to as longitudinal (pressure) and transversal (shear) wave. Without going into further details, it should be mentioned that the shortest path is often not the fastest one, as illustrated in Fig. 4.82. This is due to the different velocities of the acoustic waves, which attain,

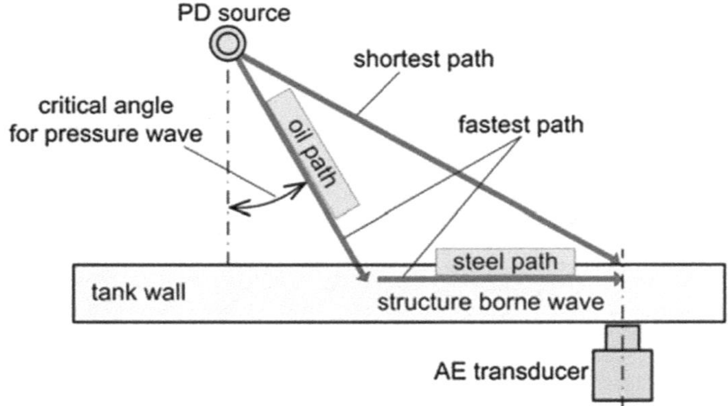

Fig. 4.82 Shortest and fastest path between a PD source in oil and an acoustic sensor placed on a power transformer tank

for instance, approx. 1.25 mm/µs in oil and 5.1 mm/µs in steel. To solve the very complex equations gained for the wave velocity in real HV equipment, nowadays advanced computer systems equipped with sophisticated software packages are available.

The facility required for the acoustic detection and location of PD sources in HV equipment comprises besides an array of AE transducers, a signal transmission unit (cabling or fiber optic link) and an acquisition system (digital oscilloscope or computer-based measuring system) to perform the signal processing as well as the visualization and even storage of the captured ultrasonic data. For this purpose the following types of ultrasonic transducers are commonly applied:

- Piezo-electric transducers,
- Structure-borne sound resonance transducers,
- Accelerometers,
- Condenser microphones, and
- Electro-optic transducers.

As the acoustic impedance of the transducers is very different from that of the metallic enclosure of the HV apparatus under investigation, the transducer surface is usually covered with hard epoxy resin to ensure an efficient signal transmission. This provides additionally the required insulation between transducer and metallic parts of the test object. Moreover, special attention should be paid to the coupling method due to the fact that the emitted acoustic wave might be reflected at the interface between transducer and the enclosure of the HV equipment. Thus, acoustic couplant gel or grease should be used to minimize the impact of reflections.

Generally, it seems beneficial to integrate a pre-amplifier in the ultrasonic transducer in order to enhance the signal-to-noise-ratio. As mentioned above, disturbing mechanical vibrations caused by pumps and fans as well as noises emanated

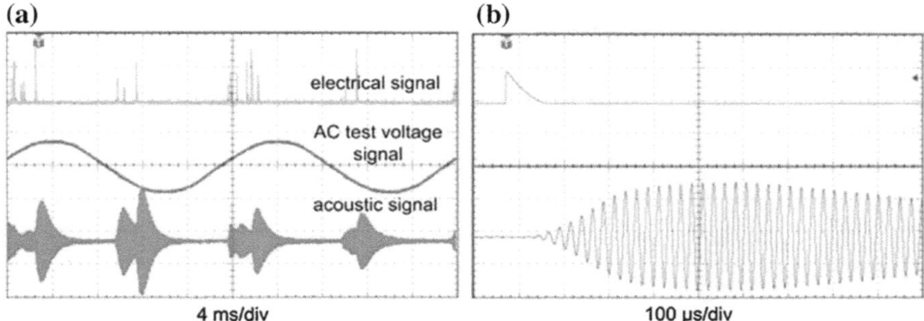

Fig. 4.83 PD pulse response of a narrow-band ultrasonic measuring system having a center frequency of 42 kHz and a bandwidth of 800 Hz. **a** PD pulse train leading to a superposition of the acoustic signal. **b** Response of the acoustic measuring system against a single PD pulse

from the iron core of transformer on account of magnetostriction and Barkhausen effect, can effectively be rejected, because such acoustic noises are not correlated to acoustic pressure waves emitted from real PD sources. Using narrow-band amplifiers operating at center frequencies around 40 kHz might also contribute to reasonable test results. Under this condition the pulse response is characterized by oscillations where the envelope covers often a time span longer than hundreds of µs. Thus, a superposition of subsequent acoustic signals might appear, as obvious from Fig. 4.83. Among others, this is the reason why the acoustic PD detection method is not capable of evaluating the PD magnitude quantitatively. Another drawback is the strong attenuation and dispersion of ultrasonic signals if travelling through various insulation structures because these feature a low-pass filter characteristics, where the signal is attenuated nearly proportional to the square of the characteristic frequency (Beyer 1987).

Chapter 5
Measurement of Dielectric Properties

Abstract The ageing of the insulation of HV apparatus is not only caused by the high electric field strength but also by thermal and mechanical stresses that evolve during normal operation condition. This may lead to chemical processes associated with a gradual deterioration of the integral insulation properties. Finally, weak spots and, in extreme case, an ultimate breakdown might occur, which causes not only an unexpected outage of HV equipment but also physical, environmental and financial damages. To ensure a reliable operation of HV assets encourages high standards of quality assurance tests after manufacturing as well as advanced tools for preventive diagnostics in service. As treated already in Chap. 4, PD measurements have become an indispensable tool to trace local dielectric imperfections since the 1960s, while the measurement of integral dielectric properties, such as capacitance and loss factor measurements, became of interest for insulation condition assessment of HV equipment already since the beginning of the last century when the first HV transmission systems above 100 kV were erected. In this context, it should be noted that the dielectric properties are often determined at test frequencies different from the service frequency (50/60 Hz). So valuable information on the insulation condition may also be gathered by measuring the dielectric response under DC voltage after this is switched on and even off, as will also be treated in the following.

5.1 Dielectric Response Measurements

Considering a capacitance, where the solid dielectric is arranged between plane parallel electrodes, which are suddenly subjected to a DC voltage ramp, the capacitance will rapidly be charged. Immediately thereafter, however, a comparatively low current is measurable, which is not only caused by the volume resistivity of the dielectric material but also by polarization phenomena. This is due to the Coulomb force (Coulomb 1785), which causes a displacement of the always present positive and negative charges neutralizing each other under zero field condition. That means a dipole moment is established after a certain time period. Vice versa, a depolarization occurs just after the DC voltage is switched off and the electrodes of

© Springer Nature Switzerland AG 2019
W. Hauschild and E. Lemke, *High-Voltage Test and Measuring Techniques*,
https://doi.org/10.1007/978-3-319-97460-6_5

the test sample are short-circuited. As the return of the charge carriers to its origin position occurs again after a certain time lag, a so-called *return voltage* (sometimes also referred to as "recovery voltage") would be measurable across the electrodes of the test sample, i.e. after a certain *relaxation* time, provided the electrodes of the capacitance under investigation are not short circuited. This phenomenon has originally been discovered by Maxwell in 1888 and explained more in detail by Wagner in 1914 based on the equivalent circuit shown in Fig. 5.1. This network is composed of the basic capacitor C_0 and the parallel resistor R_0 representing the DC resistance of the capacitance. Additionally, various R–C elements representing various relaxation time constants $\tau_1 = R_1\,C_1$, $\tau_2 = R_2\,C_2$, ... $\tau_n = R_n\,C_n$ have been introduced to characterize transition frequencies, which are associated with typical polarization phenomena, such as the trapping of charge carriers as well as interfacial and orientation polarization and even ion and electron polarization.

Nowadays, the return voltage measurement (RVM) is one of the most established diagnostic tools to assess the global insulation condition of HV equipment and their components (Boening 1938; Nemeth 1966, 1972; Reynolds 1985; Csepes et al. 1994; Lemke and Schmiegel 1995; Gubanski et al. 2002). The basic circuit used for the RVM method as well as typical voltage signals are illustrated in Fig. 5.2. Here the test sample is first excited by a constant DC voltage of magnitude V_e. For this purpose the switch S_1 is suddenly closed, while the switch S_2 remains still open. As a result of the continuous DC stress, the basic capacitance C_0 (Fig. 5.1) is rapidly charged. Different to this a certain time lag for charging the other capacitances C_1, C_2, C_3, ... C_n can be encountered This is due to the previously mentioned *relaxation time constants* τ_1, τ_2, ... τ_n. After a certain DC stressing time t_1, which lasts often several tens of minutes or even more, the switch S_1 is opened (Fig. 5.2). Immediately thereafter the test sample is short-circuited, closing the switch S_2 at instant t_1 for several minutes or even longer. Under this condition, the basic capacitance C_0 is completely discharged, while the capacitances C_1, C_2, C_3, ... C_n are only partially discharged due to the characteristic relaxation time constants τ_1, τ_2, τ_3, ... τ_n.

At instant t_2, when the switch S_2 is opened again, the residual charges stored in the capacitances C_2, C_3, ... C_n cause a re-charging of the basic capacitance C_0. Thus, a so-called return voltage $v_r(t)$ is measurable across the electrodes of the test

Fig. 5.1 Equivalent circuit of solid dielectrics

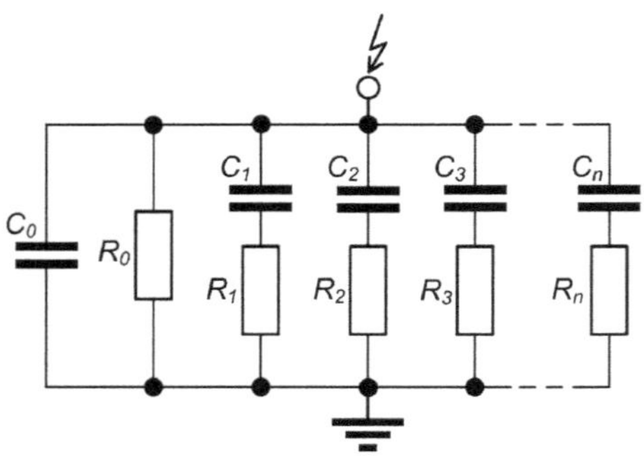

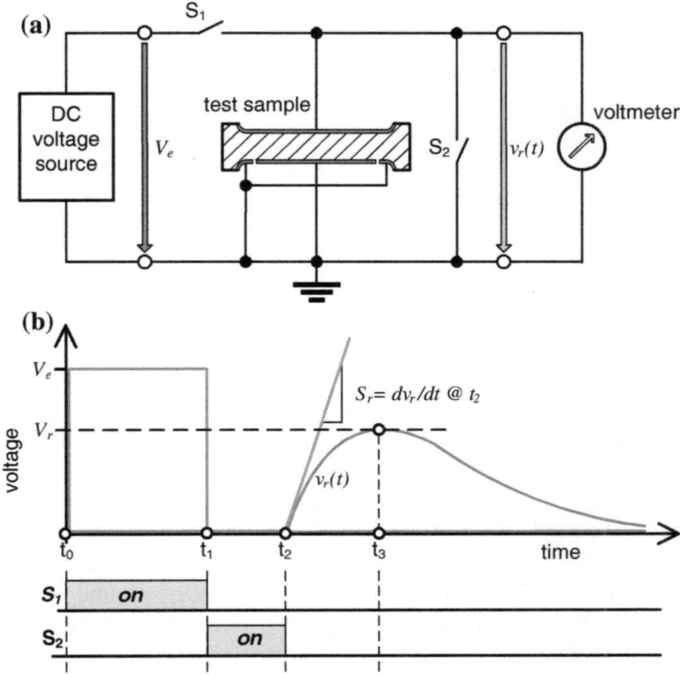

Fig. 5.2 Principle of return voltage measurement (*RVM*). **a** Basic circuit. **b** Characteristic signals

object. However, C_0 is also discharged due to the parallel connected resistor R_0 representing the volume resistance of the dielectric material under investigation. Consequently, after an initial voltage rise, which is characterized by the initial slope $dv_r(t)/dt$ and the maximum value V_r appearing at instant t_3, the return voltage decays again and approaches thus finally zero, see Fig. 5.2b.

Additional information on the insulation condition is gained by the establishment of so-called polarization spectra. For this purpose a series of recovery voltage measurements at constant DC test voltage is applied, while the DC stress period t_1 as well as the short-circuit duration (t_2-t_1) are stepwise increased. With respect to reproducible measurements the ratio $(t_2-t_1)/t_1$ is commonly kept constant and often chosen as 50%. Under this condition the initial slope $dv_r(t)/dt$ at instant t_2 and the peak value V_r of the return voltage versus the charging time t_1 is plotted for each test sequence. A practical measuring example is shown in Fig. 5.3, which was gained for oil-impregnated paper insulation with the moisture content as parameter. Generally, it can be stated that each maximum of the return voltage belongs to a dominant relaxation time constant, which is unambiguous correlated to a typical ageing phenomenon. This offers the opportunity to assess the condition of technical insulation, such as the moisture content and the depolymerization of oil-impregnated paper used for power transformers and power cables.

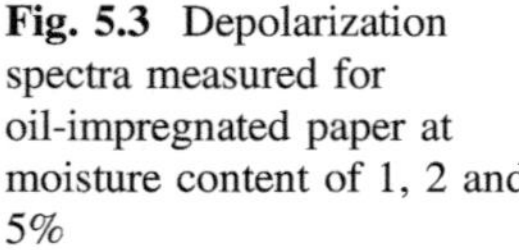

Fig. 5.3 Depolarization spectra measured for oil-impregnated paper at moisture content of 1, 2 and 5%

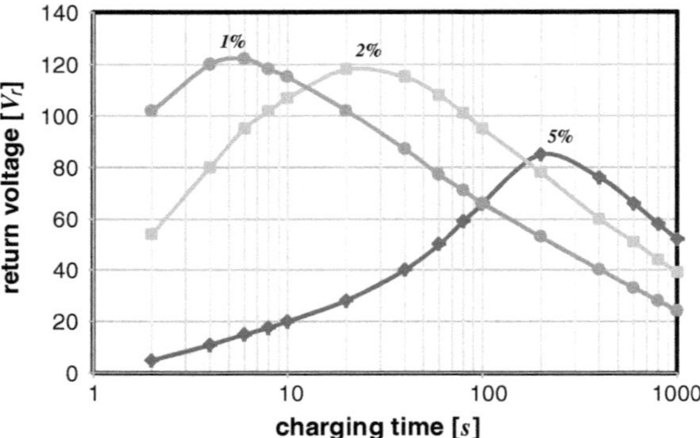

Another approach is the analysis of the *polarization* and *depolarization current* using the basic circuit sketched in Fig. 5.4a. In compliance with the above presented return voltage measurement the test sample is first subjected to a DC voltage step V_c where the switch S_1 remains closed up to the instant t_2. Under this condition, the main capacitance C_0 is completely charged, usually within less than one second.

To avoid an overload on account of the high magnitude of the charging current and thus a possible damage of the sensitive ammeter in the ground connection lead of the test sample, this instrument is initially short-circuited by the switch S_2, which is opened shortly delayed at instant t_1 in order to measure the time-dependent polarization current i_p (t) charging the partial capacitances C_1, C_2, C_3, ... C_n at various time constants, see Fig. 5.1. This current decays more or less exponentially down to a steady-state value I_{DC}, which is being governed by the DC resistance R_0.

The next step starts at instant t_2 when the switch S_2 is again short-circuited and the switch S_1 is opened to disconnected the test sample from the DC source. Just thereafter the switch S_3 is closed so that the main capacitance C_0 will be discharged almost completely. Few seconds thereafter, i.e. at instant t_3, the switch S_2 is opened, whereas S_3 remains closed. Under this condition, the time-dependent depolarization current $i_d(t)$ is measured. Even if the polarity of this current is opposite to that of the polarization current, the time function fits more or less the exponential function of the polarization current. The only difference is that a resistive current component occurring during the first phase is not detectable due to the short-circuited terminals of the test object.

In this context it should be noted that also other tools than the above are utilized to assess the insulation condition, such as the measurement of isothermal relaxation currents (Simmons et al. 1973; Beigert et al. 1991). In principle, the circuits adopted for this purpose provide more or less modifications of those presented above. As an example, the measurement of the so-called *recovery charge* shall briefly be presented in the following, which is based on an electronic integration of the depolarization current (Lemke and Schmiegel 1995). A schematic of the circuit developed for this procedure as well as typical voltage signals are depicted in Fig. 5.5.

As usual for both return voltage and polarization/depolarization measurements, the test object capacitance is first exposed to a DC voltage step V_e, usually in the

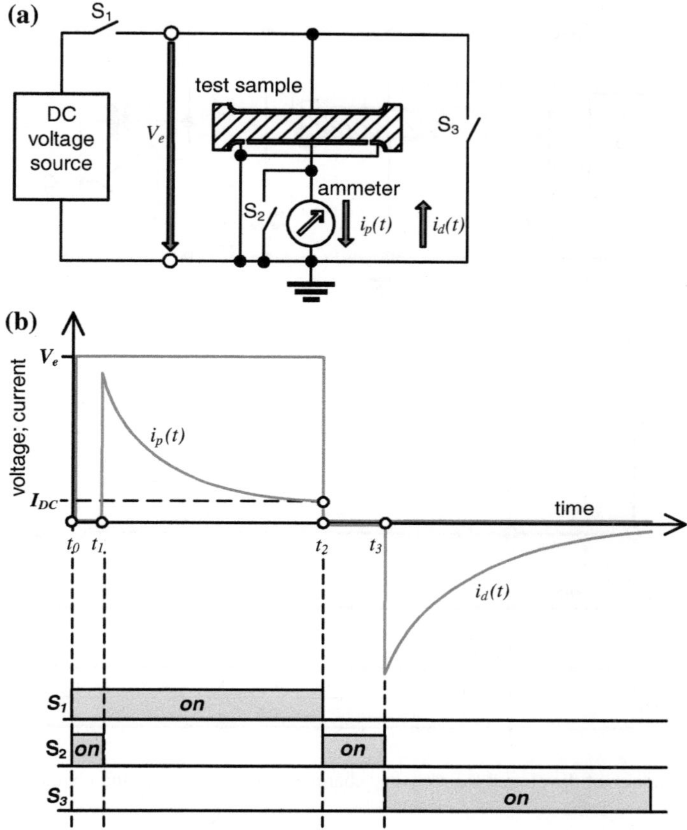

Fig. 5.4 Principle of polarization and depolarization measurement. **a** Basic circuit. **b** Characteristic signals

order of few kV. Based on practical experience, the duration of the pre-stressing time is preferably chosen between 2 and 10 min. During this time span the switch S_1 is closed, while all other switches remain still open. The second step starts at instant t_1 when the switch S_1 is opened and just thereafter the switch S_2 is closed. Provided, the condition $C_m \gg C_0$ is satisfied, the charge amount stored originally in the main capacitance C_0 of the test object will be transferred almost completely to the measuring capacitor C_m. Thus, the test object capacitance C_0 can simply be deduced from the ratio between the voltage V_m occurring across C_m within the time interval t_1–t_2 and the exciting voltage V_e occurring across C_0 during the time interval t_0–t_1:

$$C_0 \approx C_m \cdot \frac{V_m}{V_e}. \tag{5.1}$$

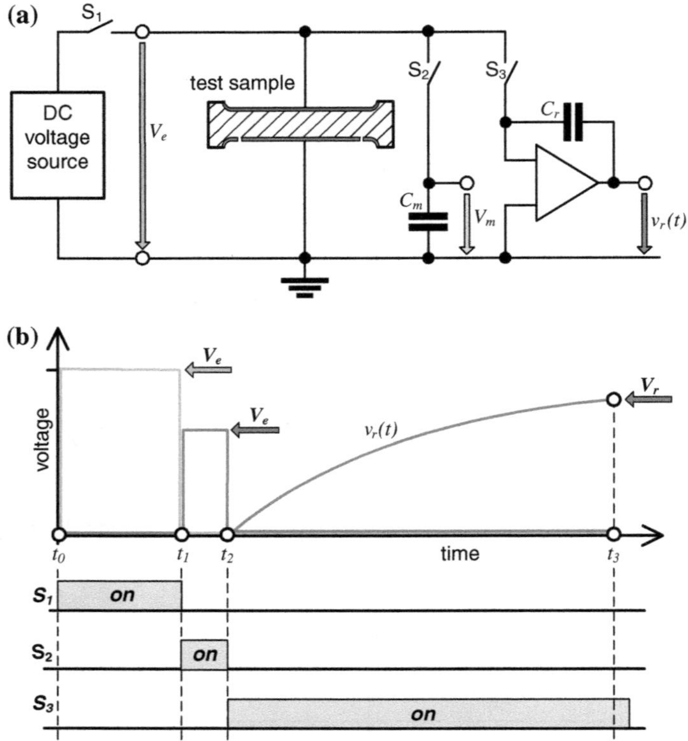

Fig. 5.5 Basic circuit developed for recovery charge measurements (**a**) and characteristic voltage signals (**b**)

The next step starts at instant t_2 when the switch S_2 is opened in order to disconnect the measuring capacitor C_m from the test object. Few seconds thereafter the switch S_3 is closed in order to transfer the residual charge, which was previously stored in the test object capacitance, to the integrating capacitor C_r. Fundamental studies revealed that both time intervals t_1-t_0 and t_2-t_1 should preferably be chosen close to 10 s. For a known capacitance C_r, which provides the decisive part of the active integrator, the recovery charge $q_r(t)$ stored in C_r can simply be deduced from the output voltage $v_r(t)$ using the following relation:

$$q_r(t) = C_r \cdot v_r(t). \tag{5.2}$$

The measurement is commonly finished at instant t_3, when steady-state conditions appear, i.e. the output voltage $v_r(t)$ of the electronic integrator remains nearly constant, which appears usually after approx. 10 min. An appropriate quantity used for assessing the insulation condition is the *polarization factor* F_p, which provides the ratio between the magnitudes the measured maximum recovery charge Q_r and the total charge Q_m stored during the DC stress period in the main capacitance C_0.

Combining the Eqs. (5.1) and (5.2), the *polarization factor* can be expressed as follows:

$$F_P = \frac{Q_r}{Q_m} = \frac{C_r}{C_m} \cdot \frac{V_r}{V_m}. \tag{5.3}$$

The ratio C_r/C_m provides in principle a scale factor of the measuring system so that the polarization factor F_P is proportional to the ratio between the voltage magnitudes V_r and V_m measurable across C_r and C_m, respectively. Based on practical experiences gained for XLPE-insulated power cables, the insulation condition can be assessed as "good" as long as the condition $F_p < 10^{-4}$ is satisfied.

Example Figure 5.6 shows typical voltage signals recorded for a service aged 20 kV XLPE cable. To get appropriate readings, the following settings of the polarization factor measuring instruments have been chosen:

Divider ratio	$R_2/R_1 = 1{:}400$
Measuring capacitance	$C_m = 100\ \mu\text{F}$
Integrating capacitance	$C_r = 1\ \mu\text{F}$

Using the Eqs. (5.1) and (5.3), the following values can be drawn from the record shown in Fig. 5.6:

Test voltage	$V_e = (5\ \text{V}){\cdot}400 = 2000\ \text{V}$
Main capacitance	$C_0 = (100\ \mu\text{F}){\cdot}(2.3\ \text{V})/2{,}000\ \text{V} = 145\ \text{nF}$
Polarization factor	$F_p = [(1\ \mu\text{F})/(100\ \mu\text{F})]{\cdot}[(0.64\ \text{V})/(2.3\ \text{V})] = 28 \times 10^{-4}$

As this value exceeds substantially the above mentioned limit $F_p < 10^{-4}$, a strong exposition to water trees was supposed, which could be confirmed based on microscopic investigations.

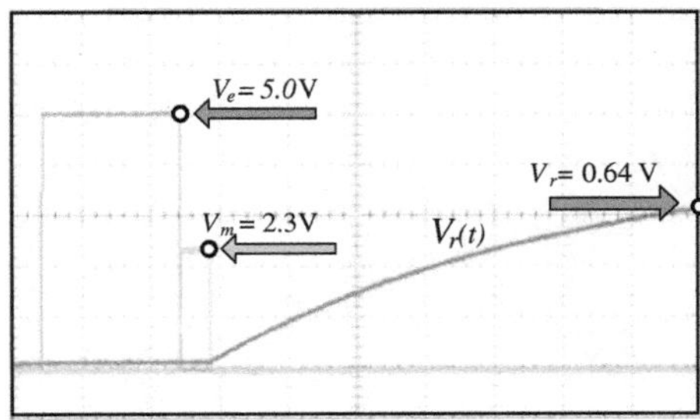

Fig. 5.6 Record of a recovery charge measurement performed on an aged MV XLPE power cable

5.2 Loss Factor and Capacitance Measurement

As the insulation of HVAC apparatus is mainly stressed by power frequency voltages of 50/60 Hz, the knowledge of dielectric properties exposed to such low-frequency alternating voltages is primarily of interest. Under this condition the equivalent circuit according to Fig. 5.1 can substantially be simplified, as shown in Fig. 5.7. Here the capacitance C_s of the test object can either be assumed as connected in series with the hypothetical resistance R_s (Fig. 5.7a), or even the test object capacitance C_p is bridged by a parallel resistance R_p (Fig. 5.7b). Here R_s or R_p reflect in principle the thermal losses dissipating in the dielectric material.

As the resistivity of technical insulations decreases at rising temperature, the current density may increase, particularly in regions where the convection of the dissipated power is limited. As a result, so-called hot spots may occur, which accelerate the insulation deterioration and lead finally to an ultimate insulation

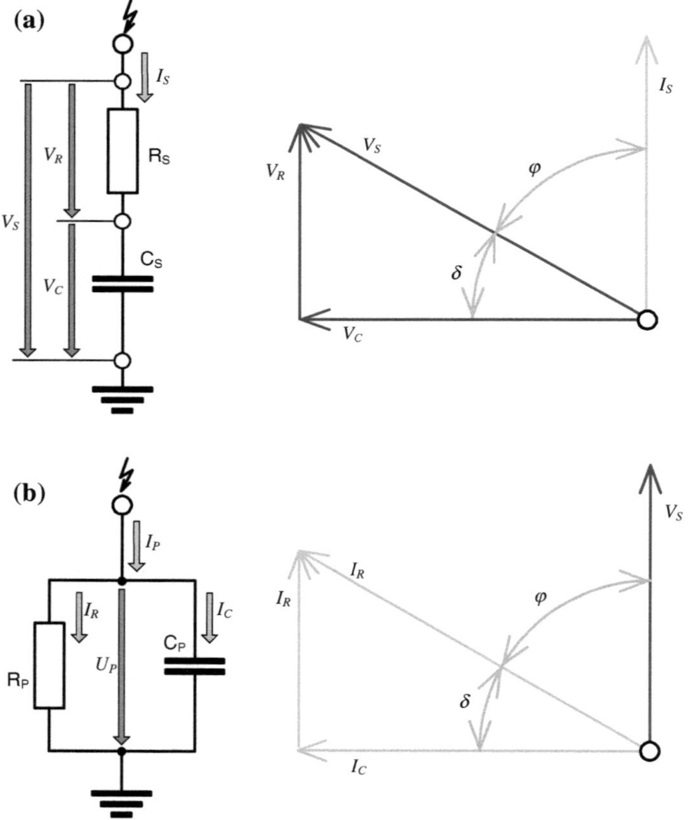

Fig. 5.7 Equivalent circuits and associated vector diagrams commonly used for the definition of the loss factor $\tan\delta$. **a** Series circuit. **b** Parallel circuit

puncture. Thus, the measurement of dielectric losses became an indispensable tool for quality assurance tests of HV apparatus since the beginning of the last century, when high alternating voltage was increasingly employed for long-distance power transmission links.

Considering the equivalent circuit shown in Fig. 5.7a, the current I_S flowing through both the capacitance C_S and the in series connected resistance R_S causes a phase shifting δ_S between the applied test voltage V_S and the voltage vectors V_C and V_R, respectively. From a practical point of view it seems convenient to evaluate the tangent δ, commonly referred to as *loss factor*, which follows from the ratio between the both voltage vectors V_R and V_C:

$$\tan \delta_S = \left/ \frac{V_R}{V_C} \right/ = \left/ \frac{I_S \cdot R_S}{I_S/j\omega \cdot C_S} \right/ = \omega \cdot C_S \cdot R_S. \tag{5.4}$$

With respect to Fig. 5.7b, where C_P and R_P are connected in parallel, the following relation applies:

$$\tan \delta_P = \left/ \frac{I_R}{I_C} \right/ = \left/ \frac{V_P/j\omega \cdot C_P}{V_P \cdot R_P} \right/ = \frac{1}{\omega \cdot C_P \cdot R_P}. \tag{5.5}$$

Multiplying Eq. (5.4) on one hand with the current I_S flowing through C_S and R_S and, on the other hand, Eq. (5.5) with the test voltage V_P dropping across C_P and R_P, it can readily be shown that the loss factor becomes equal for both circuits shown in Fig. 5.7 and can simply be expressed by the ratio between active (resistive) and reactive (capacitive) power:

$$\tan \delta_P = \left/ \frac{V_R \cdot I_S}{V_c \cdot I_S} \right/ = \left/ \frac{V_P \cdot I_R}{V_P \cdot I_C} \right/ = \frac{P_R}{P_C} \tag{5.6}$$

As known from the classical network theory, each series circuit can be converted into a parallel circuit and vice versa. Considering the equivalent circuits shown in Figure 5.7a and b, the following relations apply (Küpfmüller 1990):

$$C_P = C_S \cdot \frac{1}{1 + (\omega \cdot C_S \cdot R_S)^2} = C_s \cdot \frac{1}{1 + (\tan \delta_S)^2}. \tag{5.7}$$

$$R_P = R_S \cdot \left[1 + \frac{1}{(\omega \cdot C_S \cdot R_S)^2} \right] = R_S \cdot \left[1 + \frac{1}{(\tan \delta_S)^2} \right]. \tag{5.8}$$

$$C_S = C_P \cdot \left[1 + \frac{1}{(\omega \cdot C_P \cdot R_P)^2} \right] = C_P \cdot \left[1 + (\tan \delta_P)^2 \right]. \tag{5.9}$$

$$R_S = R_P \cdot \frac{1}{1 + (\omega \cdot C_P \cdot R_P)^2} = \frac{(\tan \delta_P)^2}{1 + (\tan \delta_P)^2}. \tag{5.10}$$

These relations enable the determination of the loss-factor based on the settings of the Schering bridge under balanced conditions, as will be treated in the following.

5.2.1 Schering Bridge

To measure the relative dielectric constant ε_r of insulating materials as well as the capacitance and loss factor under high voltage, the bridge circuit according to Fig. 5.8, which has been proposed by Schering in 1919, is commonly applied. This is in principle composed of two parallel connected voltage dividers, in the following referred to as measuring and a reference branch. As sketched in Fig 5.8, the HV arm of the measuring branch is formed by the test sample simulated by the series connection of C_S and R_S, see Fig. 5.7a, while the HV-arm of the reference branch is formed by the loss-free standard capacitor C_N.

As the voltage vectors V_S and V_3 appearing across the HV and LV arm of the measuring branch are proportional to their impedances, it can be written

$$\frac{V_S}{V_3} = \frac{(1/j\omega \cdot C_S) + R_S}{R_3} = \frac{1 + j\omega \cdot C_S \cdot R_S}{j\omega \cdot C_S \cdot R_3}. \tag{5.11}$$

Considering the reference branch, the ratio between the voltage vectors V_N and V_4 is given by

$$\frac{V_N}{V_4} = \frac{1}{j\omega \cdot C_N} \cdot \frac{R_4 + (1/j\omega \cdot C_4)}{R_4/j\omega \cdot C_4} = \frac{1 + j\omega \cdot C_4 \cdot R_4}{j\omega \cdot C_N \cdot R_4}. \tag{5.12}$$

To balance the bridge circuit, the tunable elements R_3 and C_4 are adjusted such that the reading of the indicating instrument Z_I shown Fig. 5.8 approaches zero. Obviously, this is accomplished when the voltage appearing across the LV arm of

Fig. 5.8 Circuit elements of a Schering bridge

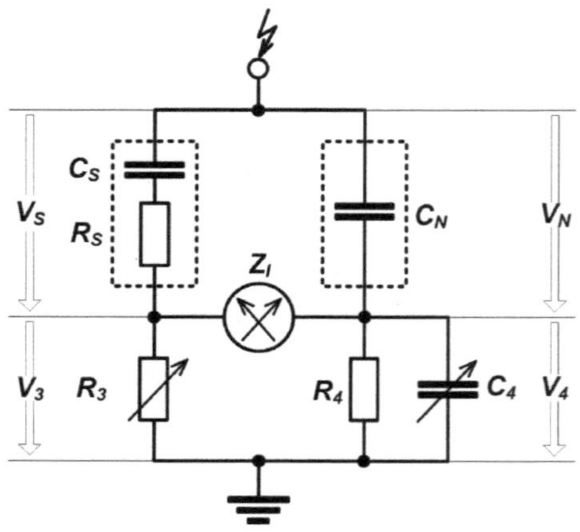

the measuring branch is equal in amplitude and phase to that appearing across the LV arm of the reference branch. Based on this the balance criterion can be expressed as follows:

$$\frac{V_S}{V_3} = \frac{V_N}{V_4}. \tag{5.13a}$$

$$\frac{1 + j\omega \cdot C_S \cdot R_S}{j\omega \cdot C_S \cdot R_3} = \frac{1 + j\omega \cdot C_4 \cdot R_4}{j\omega \cdot C_N \cdot R_4}, \tag{5.13b}$$

$$\frac{1}{j\omega \cdot C_S \cdot R_3} + \frac{R_S}{R_3} = \frac{1}{j\omega \cdot C_N \cdot R_4} + \frac{C_4}{C_N}. \tag{5.13c}$$

Comparing only those terms, which are divided by $j\omega$, one gets the following relation, which enables the determination of the equivalent series capacitance C_S from the settings of the Schering bridge under balanced conditions:

$$C_S = \frac{C_N \cdot R_4}{R_3}, \tag{5.14}$$

Comparing the remaining terms, one gets the following hypothetical series resistance R_S, which can be deduced from the settings of R_3, R_4, and C_4 under balanced conditions:

$$R_S = \frac{R_3 \cdot C_4}{C_N}, \tag{5.15}$$

Combining these Eqs., the loss factor can be calculated for using the following relation:

$$\tan \delta_S = \omega \cdot C_S \cdot R_S = \omega \cdot \frac{C_N \cdot R_4}{R_3} \cdot \frac{R_3 C_4}{C_N} = \omega \cdot C_4 \cdot R_4. \tag{5.16}$$

Commercially available Schering bridges are often equipped with a fixed LV resistor R_4 used for grounding the test object. Provided the measurement of C_S and $\tan \delta_S$ is performed under 50 Hz AC voltage, this resistor is adjusted exactly to $R_4 = 318.5$ Ohm. Inserting this in Eq. (5.12) one gets

$$\tan \delta_S = \left(314 s^{-1}\right) \cdot \left(318.5 V \cdot A^{-1}\right) \cdot C_4 = C_4 / (10\,\mu F).$$

Obviously, this simplifies the determination of the loss factor considerably. Assuming, for instance, that the balanced condition was accomplished for $C_4 = 0.027\,\mu F$, one gets

$$\tan \delta_S = (0.027\,,\mu F)/(10\,,\mu F) = 0.0027 = 2.7 \cdot 10^{-3}.$$

In this context it should be noted, that additional information on the insulation condition of HV equipment is obtained when besides the equivalent capacitance C_S according to Eq. (5.14) and the loss factor $\tan \delta_S$ according to Eq. (5.16) also the time-dependent trend of these quantities under service condition is determined.

In this context it must be emphasized that the above treatment refers to an equivalent circuit where the elements C_S and R_S are connected in series, see Fig. 5.7a. As already mentioned previously, each series circuit can be converted into the parallel circuit. Thus, to determine the values of the parallel connected elements shown in Fig. 5.7b, the Eqs. (5.9) and (5.10) have to be combined with Eqs. (5.14) and (5.15). From this follows for the parallel capacitance:

$$C_P = \frac{C_S}{1 + (\omega \cdot C_S \cdot R_S)^2} = C_N \cdot \frac{R_4}{R_3} \cdot \frac{1}{1 + (\omega \cdot C_4 \cdot R_4)^2} \tag{5.17a}$$

As for technical insulation the loss factor is usually considerably below 1, i.e. the condition $\tan\delta_S = \omega \cdot C_4 \cdot R_4 \ll 1$ applies, Eq. (5.17a) can be simplified as follows:

$$C_P \approx C_N \cdot \frac{R_4}{R_3} \tag{5.17b}$$

Based on Eq. (5.8) the parallel resistance R_P can be expressed as follows:

$$R_P = R_S \frac{1 + (\omega \cdot C_S \cdot R_S)^2}{(\omega \cdot C_S \cdot R_S)^2} = R_3 \cdot \frac{C_4}{C_N} \cdot \frac{1 + (\omega \cdot C_4 \cdot R_4)^2}{(\omega \cdot C_4 \cdot R_4)^2} \tag{5.18a}$$

Provided, the condition $\tan\delta_S = \omega \cdot C_4 \cdot R_4 \ll 1$ is satisfied, one gets the following simplification:

$$R_P \approx R_3 \cdot \frac{C_4}{C_N} \cdot \frac{1}{(\omega \cdot C_4 \cdot R_4)^2} = R_3 \cdot \frac{C_4}{C_N} \cdot \frac{1}{(\tan \delta_S)^2} \tag{5.18b}$$

Inserting the above expressions in Eq. (5.5), the loss factor can be expressed as follows:

$$\tan \delta_P = \frac{1}{\omega \cdot C_P \cdot R_P} = \frac{(\omega \cdot C_S \cdot R_S)^2}{\omega \cdot C_S \cdot R_S} = \omega \cdot C_S \cdot R_S = \omega \cdot C_4 \cdot R_4 = \tan \delta_S. \tag{5.19}$$

Obviously, this is equal to Eq. (5.16), as has to be expected.

For high-capacitive test objects, such as power cables, power transformers and rotating machines, it has to be taken into account that the capacitive current through the test object could heat up the measuring resistor R_3. To prevent a possible damage, this resistor must be shunted by a well-known parallel resistor, which has to be chosen considerably lower than R_3 in order to carry almost the entire load current flowing through the test object.

As the loss factor of technical insulation, such as polyethylene-insulated cables, is often below 10^{-4}, the impedances of the low voltage elements of the Schering bridge might be affected by stray capacitances. Therefore, these LV parts should be screened accordingly. Connecting the screening electrodes to the ground potential, however, acts like a bypass and causes thus a non-controlled phase shifting of the AC voltage dropping across the LV arms. To overcome this crucial problem, the LV parts are commonly screened using an *auxiliary branch* with adjustable potential (Wagner 1912). This can advantageously be done by shifting the potential of the screening electrode automatically using an impedance converter of extremely high input impedance and low output impedance, see Fig. 5.9.

Another challenge is to measure the loss factor of grounded HV apparatus, such as power transformers, rotating machines and power cables. This method, commonly referred to as *grounded specimen test* (GST), requires a disconnection of the HVAC test voltage supply from the earth potential (Poleck 1939). Under this condition, however, the test results may also strongly be affected by non-controlled stray capacitances between the HV test supply and the bridge circuit. To overcome this crucial problem, the Schering bridge has to be modified, as illustrated in Fig. 5.10. Just prior starting the actual loss factor measurement, a pre-balancing of the circuit is performed where the test specimen is disconnected from the HV terminal by opening the switch S_X and closing the switch S_4 in order to adjust the auxiliary elements R_5 and C_5 accordingly. After that S_4 is opened and S_X is closed to start the actual C-tanδ measurement under high voltage, where only the elements R_3 and C_4 are adjusted in order to balance the bridge circuit.

For GST measurements of three-phase arrangements, such as power transformers, various modifications of the bridge circuit have been proposed, such as GST-ground-guard, GST-ground-ground and GST-guard-guard. For more information on the basic configurations recommended for C-tanδ measurements of HV apparatus see the relevant standards IEC 60250: 1969 and IEC 60505: 2011.

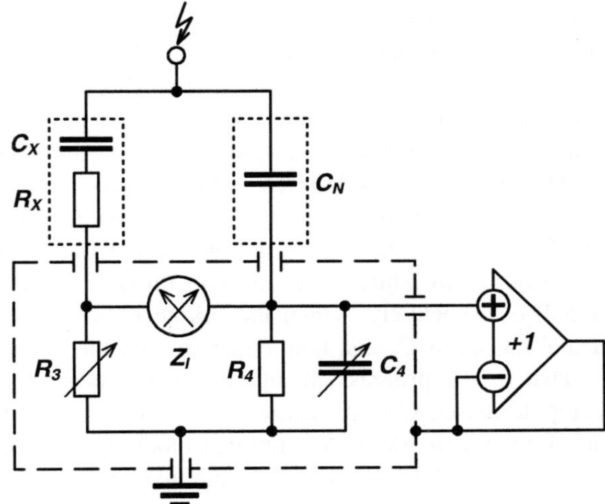

Fig. 5.9 Schering bridge with an auxiliary branch for automatic potential control of the screening electrodes

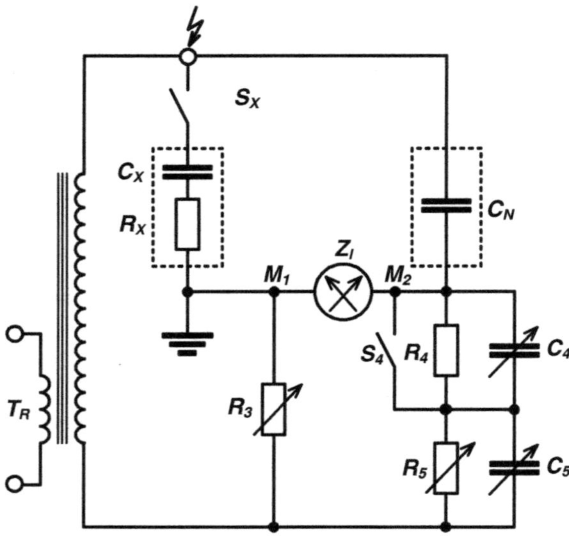

Fig. 5.10 Schering bridge modified for C-tanδ measurement of grounded specimen test (*GST*)

5.2.2   Automatic C-tanδ Bridges

Classical Schering bridges are capable of measuring tanδ variations down to 10^{-5} as well as capacitances below 1 pF. The measuring uncertainty is in the order of 1% for the loss factor and approx. 0.1% for capacitances. However, the main drawback is the time-consuming balancing procedure so that fast changing dielectric properties cannot be measured. To overcome this obstacle, fully automatic computer-based C-tanδ measuring bridges have been introduced in the 1970s when the first micro-computers were available (Seitz and Osvath 1979).

The basic concept employed is schematically illustrated in Fig. 5.11. Here the current I_X flowing through the test specimen is compensated by the current I_N flowing through the standard capacitor C_N. For this purpose a high precision differential current transformer is utilized, which is controlled by a micro-computer. The primary coils with windings W_1 and W_2 provide the low-voltage arms of the bridge and induce inverse magnetic fluxes in the magnetic core.

The residual flux is detected by the secondary coil with W_3 windings. The output signal controls the further data signal processing by means of a micro-computer. So the flux through the magnetic core of the differential transformer can fully be compensated by adjusting the currents through the auxiliary coils with windings W_4 and W_5, respectively. When the bridge is balanced, the computer calculates the actual values of C_X and tanδ depending on the applied AC test voltage level.

Due to the progress in digital signal processing (DSP), nowadays advanced computer-based C-tanδ measuring systems are available (Kaul et al. 1993; Strehl and Engelmann 2003). A simplified block diagram of such a measuring system is shown in Fig. 5.12. Basically, the circuit comprises two capacitive voltage dividers.

Fig. 5.11 Concept of an automatic C-tanδ bridge circuit

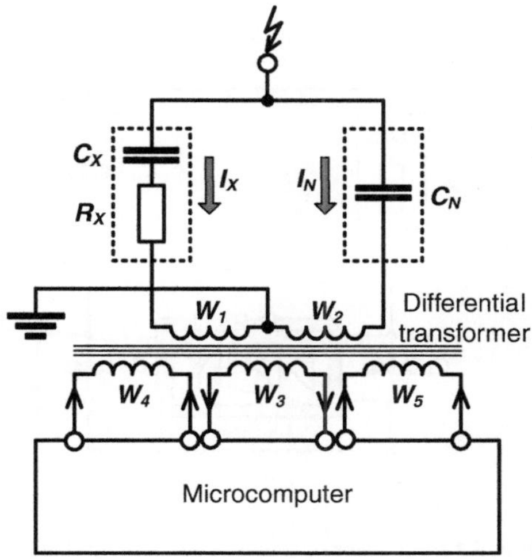

The measuring branch contains the test object symbolized by the series connection of $C_X + R_X$, which is grounded via the measuring capacitor C_M. The reference branch contains the loss-free HV standard capacitor C_N, which is grounded via the reference capacitor C_R.

Different to the automatic C-tanδ bridge, where the current I_X through the test object is compensated by the reference current I_N, an exact balancing is not required. This is because the loss factor is independent from the actual voltage magnitudes dropping across the LV arms comprising the capacitances C_M and C_R, so that the dielectric properties can also be determined by a direct measurement of the voltage vectors captured from C_R and C_M, see Fig. 5.13. Thus, it seems sufficient to adjust the voltage across C_M only to a peak value which approaches nearly the peak value of the voltage dropping across C_R where a deviation of several tens of percentages is acceptable.

The major components of a computerized DSP-based C-tanδ measuring system are shown in Fig. 5.14. The LV signals dropping across C_M and C_R are captured by the battery powered, potential-free operating sensors. Both sensors are equipped with a low noise differential amplifier of extremely high input impedance (>1 GΩ), a fast A/D converter (resolution 16 bit at 10 kHz sample rate), and an electro-optical interface. The digitized signals are transmitted via fiber optic links (FOL) to the computerized C-tanδ measuring system, which transmits also the control signals from the computer to both sensors. After a fast discrete Fourier transformation, which is performed by the use of a digital signal processor (DSP), the data are acquired, calculated, stored and displayed accordingly.

Real-time multitasking software enables a calculation of the dielectric quantities for each cycle of the applied HVAC test voltage. Using the averaging mode, a very high measuring accuracy is achieved. As the measuring principle is based on the determination of the phase angle of a non-balanced bridge in a frequency-independent but frequency-selective mode, the actual test frequency can be varied over a wide range, typically between 0.01 and 500 Hz.

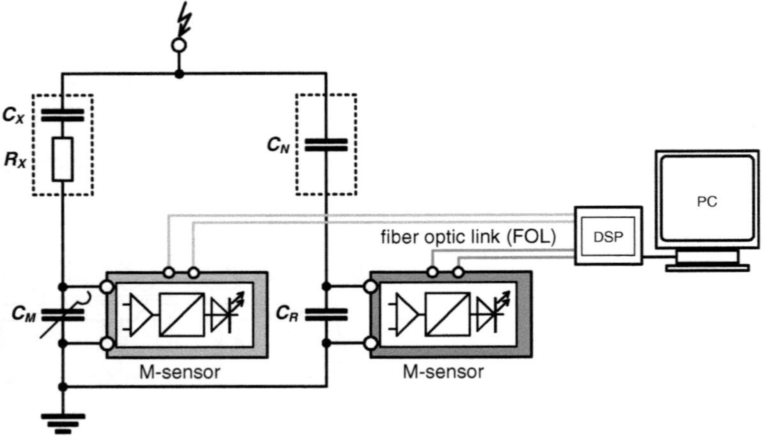

Fig. 5.12 Concept of a computerized DSP-based C-tanδ measuring system

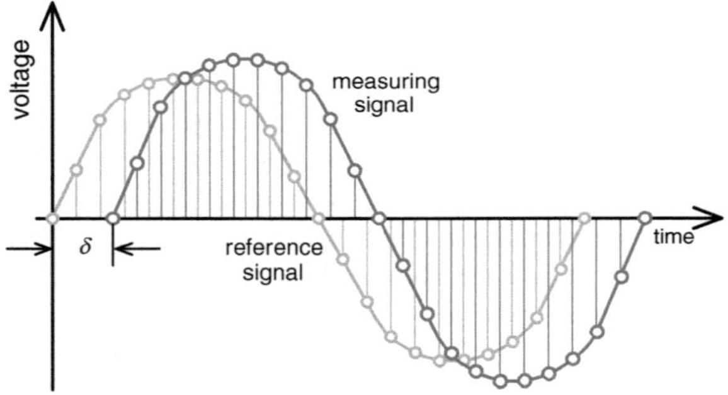

Fig. 5.13 Digitalization of the reference and measuring signal used by the *DSP-based* C-tanδ measuring system

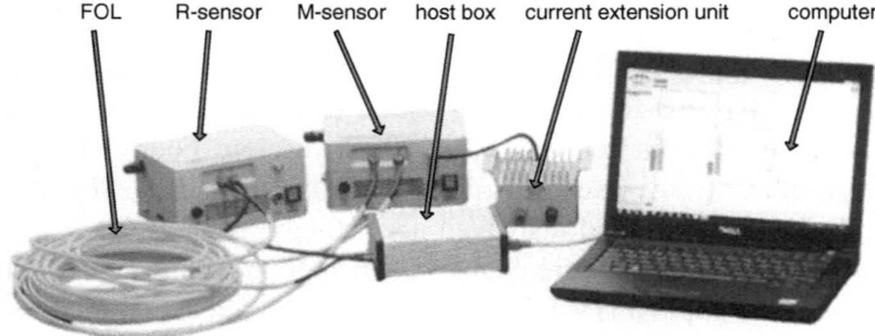

Fig. 5.14 Components of a DSP-based computerized C-tanδ measuring system (Courtesy Doble-Lemke GmbH)

All measured quantities can be displayed numerically in the real-time mode. Moreover, the data, such as the loss factor, the capacitance, the current through the test object, the test frequency and the test voltage level in terms of rms or even the peak value can be exported using an Excel-compatible data-format. Another benefit is that the dielectric quantities can be displayed on the PC screen depending on various parameters, such as the applied AC test voltage, as exemplarily shown in Fig. 5.15, and even versus the recording time, which is useful for trending purposes. Under power frequency (50/60 Hz) test voltages an "internal" measuring uncertainty below 10^{-5} for the loss factor is achievable. The "extended" measuring uncertainty, however, is mainly governed by the behaviour of the standard capacitor C_N (Schering and Vieweg 1928; Keller 1959).

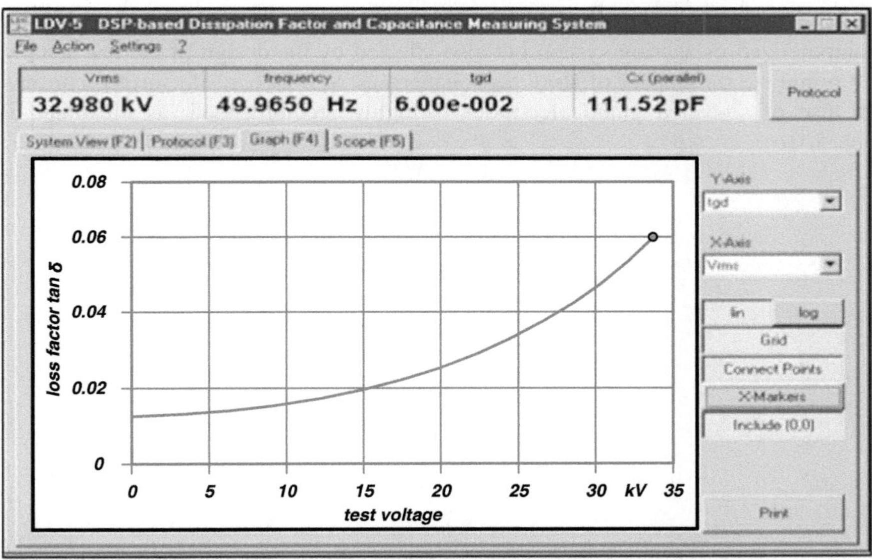

Fig. 5.15 Screenshot showing the loss factor versus the test voltage gained for an aged machine bar

Fig. 5.16 Electrode
configuration of a test sample
for C-tanδ measurement using
a guard electrode

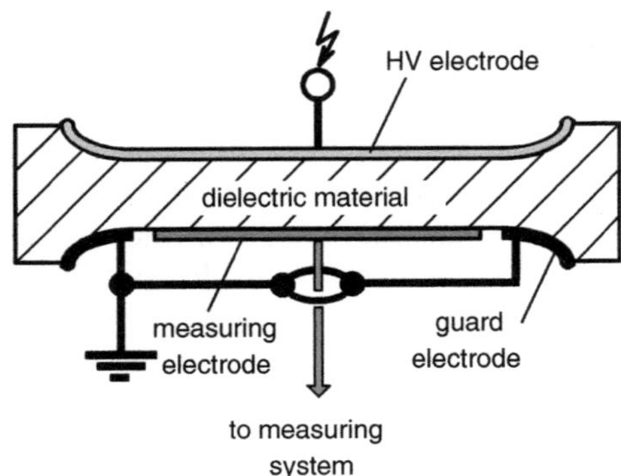

It has to be taken care that the measuring uncertainty is not only governed by the
computerized measuring system but also affected by the design of the test sample.
To minimize the impact of stray capacitances as well as parasite surface currents,
the use of so-called guard electrodes is highly recommended, as illustrated in
Fig. 5.16.

Chapter 6
Tests with High Direct Voltages

Abstract HVDC test voltages represent the stress of insulations in HVDC transmission systems. Today, HVDC transmission systems are long point-to-point connections for the transmission of high power. These links are realized by HVDC overhead lines and HVDC cables, especially submarine cables. In the near future, it is expected that the application of the HVDC technology will increase and also HVDC networks will be established (Shu 2010). Therefore, also HV tests as well as PD and dielectric measurements under direct voltage are becoming of increasing importance. Design and testing of HVDC insulation has to be performed under a difficult understanding of the acting electric fields. The reason are space and surface charges which are generated even below the PD inception voltages and strongly influenced by thermal effects. There are many publications on these phenomena, e.g. (Hering et al. 2017; Ghorbani et al. 2017; Christen 2014). This chapter starts with the different circuits for HVDC test voltage generation. Then, the requirements for HVDC test voltages according to IEC 60060-1:2010 and the consequences for the components of test systems are considered. The interactions between test generator and test object are investigated for capacitive load—e.g. of submarine cables—and for resistive load in case of wet, pollution and corona tests. A short description of test procedures with HVDC test voltages follows. Finally, it is described how direct voltages can be measured by suitable measuring systems of resistive dividers and suited measuring instruments, and how measurements—e.g. PD measurements—at DC voltage are performed.

6.1 Circuits for the Generation of HVDC Test Voltages

Today, HVDC test voltages are generated by rectification of HVAC voltages of transformers (see Sect. 3.1.1). Modern solid state rectifier elements, in the following *"diodes"* (silicium diodes), enable the generation of all necessary test voltages and currents, but the limitations in the reverse voltage to few 1000 V require the series connection of these diodes for HV rectifiers with reverse voltages up to several 100 kV's or even MV's. When the circuits for HVDC test voltage

© Springer Nature Switzerland AG 2019

W. Hauschild and E. Lemke, *High-Voltage Test and Measuring Techniques,*

https://doi.org/10.1007/978-3-319-97460-6_6

generation are considered, all mentioned rectifiers are HV rectifiers assembled from many diodes (see Sect. 6.2.2).

6.1.1   Half-Wave Rectification (One-Phase, One-Pulse Circuit)

When a HV rectifier, a load capacitor C (e.g. capacitive test object or smoothing capacitor) and a resistive load R (e.g. resistive voltage divider or resistive test object) are connected to the output of a simple HVAC transformer circuit (Fig. 6.1a), a HVDC voltage is generated. The rectifier opens for half-waves of one polarity and closes for the opposite polarity. It opens only as long as the HVAC voltage at the "stiff" transformer output is higher than the voltage of the charged capacitor. As soon as the capacitor carries any charge, its discharging starts and becomes significant when the rectifier closes after the voltage has reached its maximum V_{max} (Fig. 6.1b). The capacitor is discharged to a minimum voltage V_{min}, until the rectifier opens again for a short time ΔT (also expressed by a phase angel α). This means the output voltage is not constant, and it shows a so-called *"ripple" voltage* δV which is defined by

$$\partial V = \frac{V_{max} - V_{min}}{2}. \tag{6.1}$$

The discharging current $i_L(t)$ is connected with a charge Q which must be replaced within the short time ΔT by a current pulse $i(t)$ from the transformer

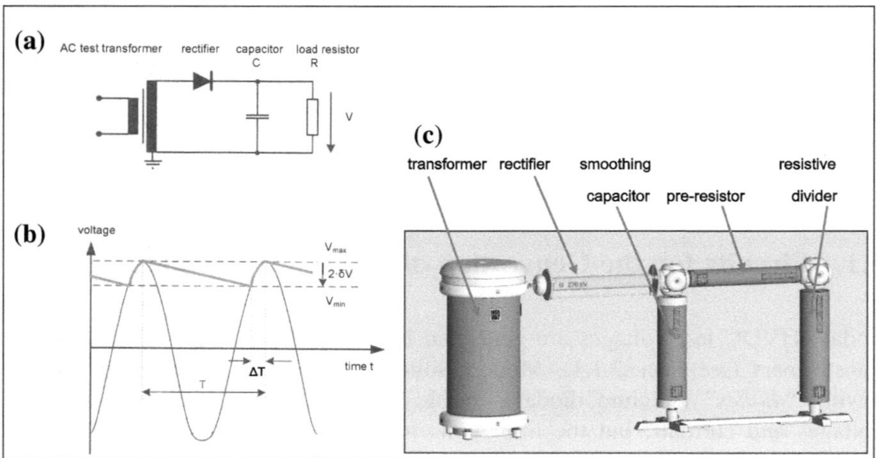

Fig. 6.1 One-phase, one-pulse half-wave rectification, **a** equivalent circuit diagram, **b** feeding AC voltage and DC voltage with ripple, **c** modular test system for HVDC 135 kV/10 mA and HVAC 100 kV/11 kVA

$$Q = \int_{T} i_L(t)\mathrm{d}t = I \cdot T = 2 \cdot \partial V \cdot C = \int_{\Delta T} i(t)\mathrm{d}t. \tag{6.2}$$

The term *"one-phase, one-pulse circuit"* reflects that the feeding voltage is a single-phase one and the DC charging is performed by one current pulse per period. With the mean direct current

$$I = \frac{V_{\max} + V_{\min}}{2 \cdot R}$$

and the relation between the duration of period T and frequency f of the feeding AC voltage, one gets an important relation for the ripple

$$\partial V = \frac{I \cdot T}{2 \cdot C} = \frac{I}{2 \cdot f \cdot C}. \tag{6.3}$$

The lower the ripple, the smoother is the HVDC test voltage. The ripple decreases with increasing load resistance (This means with decreasing load!), increasing load capacitance and increasing frequency of the charging AC voltage. But, in case of half-wave rectification, the ripple remains quite large. Equation 6.2 shows additionally that the feeding HVAC transformer circuit must be able to supply a sufficient current $i(t)$.

If the test object shows heavy pre discharges of relatively high pulse currents which may happen during wet and pollution tests, the voltage drops down as the transient energy demand cannot be supplied via the transformer, it must be taken from the smoothing capacitor C. To limit this *voltage drop* d_V, the smoothing capacitance should be large enough.

> **Note** According to IEC 60060-1:2010, the voltage drop is the "instantaneous reduction of the test voltage for a short duration of up to few seconds". Here, it will be used according to this definition. In literature, e.g. Kuffel et al. (2006) or Kind and Feser (1999), the term voltage drop is used for the continuous voltage reduction between the no-load case and the load case, especially of multi-stage cascades. The term *"voltage reduction"* will be used for this phenomenon in the following. The voltage reduction is caused due to the "forward voltage drop" and the internal resistance of the rectifiers.

HVDC sources with half-wave rectification are usually not powerful enough. They are used as HVDC attachments to HVAC test systems especially for demonstration circuits and for student's training (Fig. 6.1c). The half-wave rectification causes a non-symmetric load of the HVAC power supply. This can even lead to saturation effects in the transformer. If a transformer with symmetric output —this means grounded midpoint—of the winding is available (Fig. 6.2), the opposite polarity of the AC voltage contributes to the charging of the smoothing capacitor C and avoids saturation effects, since each of the rectifiers opens for one half-wave. Consequently, two charging current pulses appear, and this *one-phase, two-pulse circuit* halves the ripple. The two-way half-wave rectifier circuits are also basic stages for HVDC cascade generators.

Fig. 6.2 One-phase
two-pulse half-way
rectification

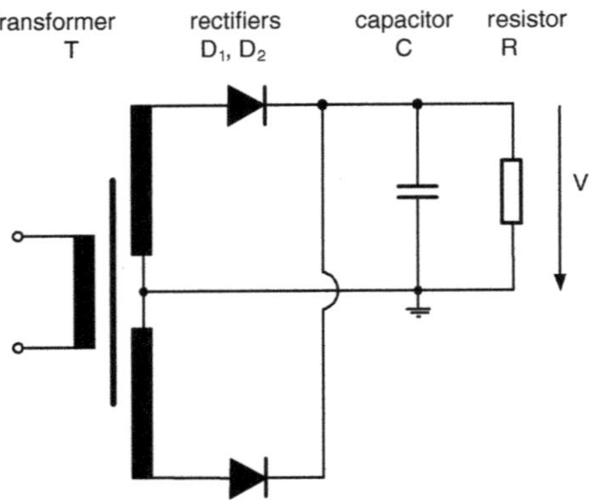

6.1.2   Doubler and Multiplier Circuits (Greinacher/ Cockcroft-Walton Cascades)

With the circuit of Fig. 6.3a, the output voltage can be doubled: The so-called doubler capacitor C_1 is charged to the voltage V_{C1}, and then, the voltage over the rectifier D_1 is oscillating around this value V_{C1}. Consequently, the smoothing capacitor is charged to a voltage which doubles the peak of the feeding AC voltage (Fig. 6.3b) if the losses in the circuit are neglected ($R \rightarrow \infty$). With losses—this means with a load resistor R—the output voltage is reduced below the theoretical no-load value. For the usual design of a HVDC attachment with doubler circuit connected to a HVAC generator (Fig. 6.3c) of a rated direct current of few 10 mA and a ripple factor $\delta V/V = \delta \leq 3\%$, this reduction might be in the order of 10%. Therefore, the smoothing capacitance should be selected large enough.

The *doubler circuit* shall be understood as the basic stage of a multiplier circuit first proposed by Greinacher (1920) for HVDC power supply for nuclear physics and improved by Cockcroft and Walton later on in 1932. The principle shall be discussed for a cascade of three stages (Fig. 6.4): The left column of the capacitors contains the doubler capacitors [sometimes also called "blocking capacitors" (Kind and Feser 1999)], and the right column contains the smoothing capacitors (Fig. 6.4a). When voltage reduction and load resistance are neglected, the AC voltage $V_{AC}(t)$ with a peak value of V_{max} is oscillating around a DC offset of $1 \cdot V_{max}$ at the first doubler capacitor and delivers a DC value of $V_1 = 2 \cdot V_{max}$ at the output of the first stage of the smoothing column (Fig. 6.4b). At the second stage, the DC offset is $V_{12} = 3 \cdot V_{max}$ and the voltage at the smoothing column is $V_2 = 4 \cdot V_{max}$. Consequently, the DC offset at input of the third stage is $V_{13} = 5 \cdot V_{max}$, and the output of the generator is a DC voltage of $V_3 = 6 \cdot V_{max}$.

The direct voltage per stage is twice the peak voltage V_{max} of the feeding alternating voltage. In stationary operation and rated voltage, the necessary *reverse voltage* of the rectifiers is $2 \cdot V_{max}$. Doubler capacitors (except that of the lowest stage) must be able to withstand a DC stress of $2 \cdot V_{max}$ plus the AC stress of the feeding voltage. Smoothing capacitors have to withstand a direct voltage of

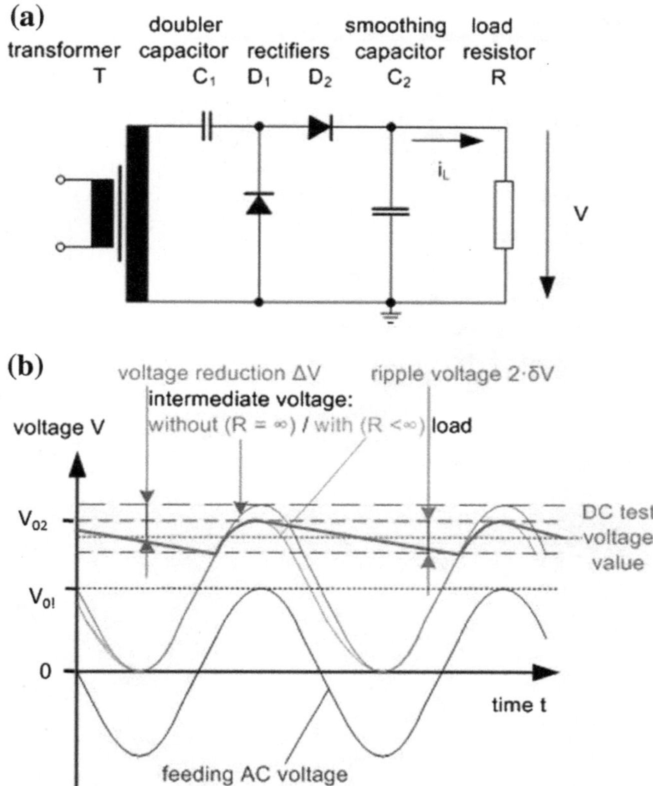

Fig. 6.3 One-phase, one-pulse doubler circuit with half-wave rectification, **a** equivalent circuit diagram, **b** AC feeding, oscillating intermediate and DC voltages ($R \rightarrow \infty$), **c** modular test system with doubler circuit 270 kV/10 mA

$2 \cdot V_{max}$. The lowest doubler capacitor is only stressed with half of that voltage but causes the largest contribution to the ripple. Therefore, it should have the double value of the capacitance, this fits to the voltage distribution and reduces the ripple. The 1500 kV cascade with a stage voltage of 500 kV—shown in Fig. 6.4c—is fed for its rated voltage with an AC voltage of 250 kV/$\sqrt{2}$ = 177 kV (rms).

The ripple depends for this *one-phase, one-pulse multiplier circuit* also from the number of stages. The continuous current I through the test object is supplied from the smoothing capacitors. Usually, all capacitors in the smoothing column have the same capacitance, which is necessary for a linear voltage distribution in case of transient stresses, e.g. at breakdown of the test object. The ripple caused by the discharging of each smoothing capacitor could be calculated with Eq. (6.3) if no charge is transferred through the rectifiers to the oscillating doubler capacitors. But, in reality, the charge transfer causes a higher ripple δV and a *voltage reduction* ΔV for cascades with n stages which can be estimated (Elstner et al. 1983) by:

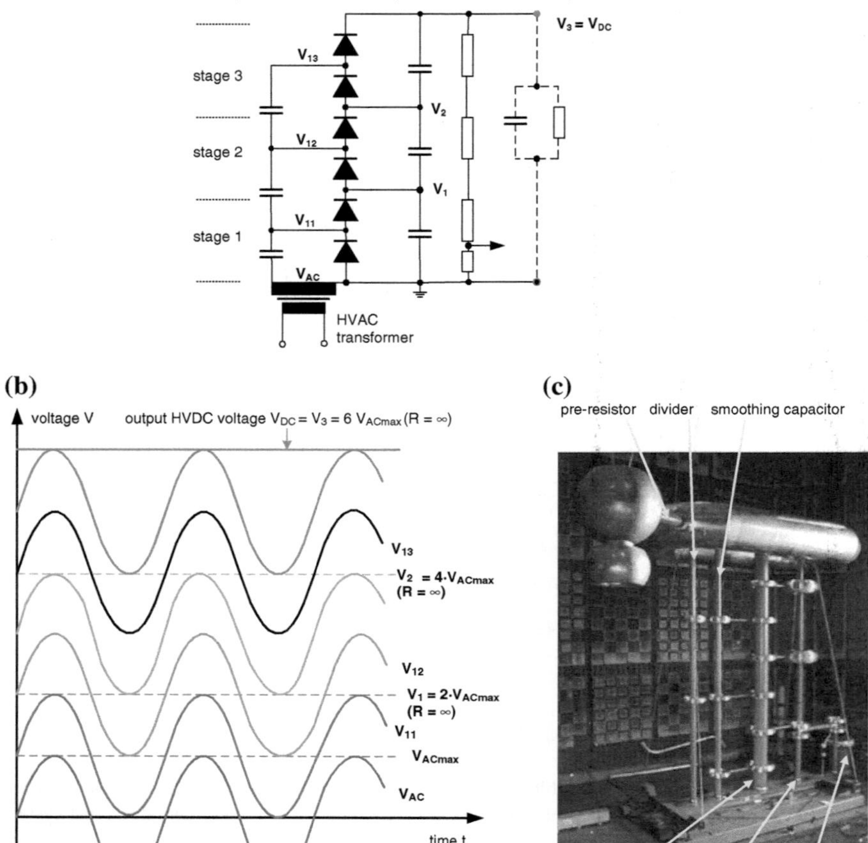

Fig. 6.4 Greinacher cascade (one-phase, one-pulse multiplier circuit), **a** equivalent circuit diagram, **b** potential at the three stages, **c** generator with 500 kV per stage for 1500 kV/30 mA

$$\partial V = \frac{I}{f \cdot C} \cdot \frac{n + n^2}{4}, \tag{6.4}$$

$$\Delta V = \frac{I}{f \cdot C} \cdot \frac{2 \cdot n^3 + n}{3}. \tag{6.5}$$

The relation between the real output voltage $(V_\Sigma - \Delta V)$ and the cumulative no-load charging voltage $V_\Sigma = nV_1$ of a HVDC multi-stage generator can be understood as an efficiency factor (as usual for impulse voltage generators)

$$\eta_{DC} = \frac{V_\Sigma - \Delta V}{V_\Sigma}.$$ (6.6)

For practical cases, the voltage reduction can be remarkably higher than expressed by Eq. (6.5) which considers only the parameters of the generator. Mainly, stray capacitances in the feeding circuit cause an additional voltage reduction (Spiegelberg 1984).

Greinacher cascades are the most applied generators in HVDC testing. The polarity can be reversed by turning the rectifiers inside the generator by hand or motor. They are well suited for capacitive test objects, but have limits in case of a resistive load. The ripple can be reduced and the efficiency factor improved when the smoothing capacitances C and the frequency f of the feeding voltage are increased. AC feeding voltages of higher frequencies are traditionally generated by motor-generator sets, but nowadays static frequency converters should be applied. Furthermore, the number of stages should be limited and the stage voltage increased as much as possible. For rated currents above 100 mA, more efficient circuits should be applied.

6.1.3 Multiplier Circuits for Higher Currents

Similar to the one-phase two-pulse circuit (Fig. 6.2), also doubler circuits can be designed. They are base stages of the so-called symmetric Greinacher cascade, a one-phase two-pulse multiplier circuit for currents of some 100 mA (Fig. 6.5). This

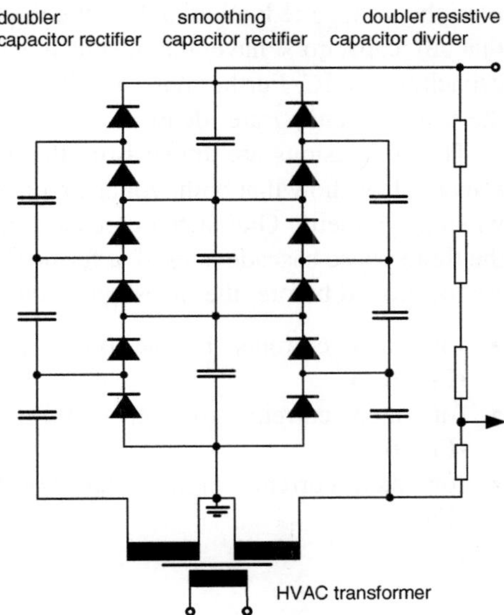

Fig. 6.5 Symmetric Greinacher cascade (one-phase, two-pulse multiplier circuit)

Fig. 6.6 Three-phase,
six-pulse multiplier circuit

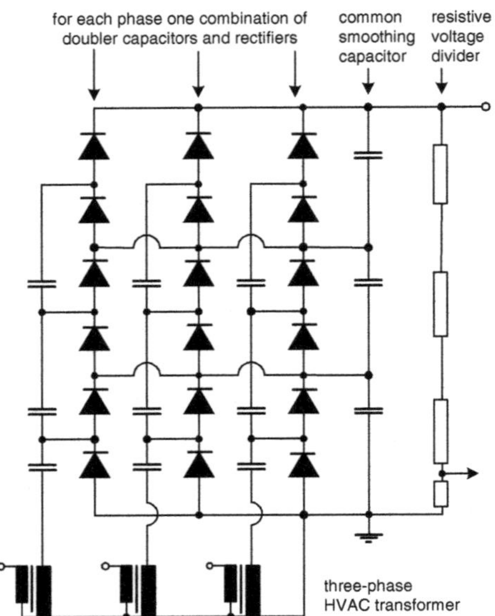

principle can also be applied with three phases. Figure 6.6 shows a three-phase, six-pulse multiplier circuit with six charging pulses within one period of the feeding three-phase voltage. Therefore, this circuit is suited for high test currents in the order up to few amperes.

Table 6.1 compares the ripple values and the voltage reductions for the conventional and the symmetric Greinacher cascades as well as for the three-phase, six-pulse multiplier circuit. They are related to the influence of the number of stages n on the voltage reduction and on the ripple. They are valid under the assumptions that all capacitors have the identical capacitance C, only the lowest doubling capacitor has $2C$. Furthermore, in all cases, the current I required by a load R and the test frequency f are identical.

The conclusions are drawn from the comparison of multiplier circuits of $n = 5$ stages. They show that both, voltage reduction and ripple, are about five times lower when a symmetric Greinacher cascade is used instead of a conventional one. When the three-phase cascade is used, a further improvement by a factor of three appears. As mentioned before, the investigated circuits should be applied as follows:

- for rated currents below 100 mA, the conventional Greinacher cascade (Fig. 6.4),
- for rated currents of some 100 mA, the symmetric Greinacher cascade (Fig. 6.5),
- for rated currents higher than 500 mA, the three-phase multiplier circuit (Fig. 6.6).

Table 6.1 Comparison of voltage reduction and ripple for multiplier circuits of n stages (The terms "x" and "y" are the factors according to lines 2 and 3 for the $n = 5$ stages.)

Type of circuit parameter	Greinacher cascade (one-phase, one-pulse multiplier circuit) (Fig. 6.4)	Symmetric Greinacher cascade (one phase, two pulse multiplier circuit) (Fig. 6.5)	Three-phase, six-pulse multiplier circuit (Fig. 6.6)
Voltage reduction ΔV	$\dfrac{I}{f \cdot C} \cdot \dfrac{2n^3 + n}{3}$	$\dfrac{I}{f \cdot C} \cdot \dfrac{2n^3 - 3n^2 + 4n}{12}$	$\dfrac{I}{f \cdot C} \cdot \dfrac{2n^3 - 3n^2 + 4n}{36}$
Ripple voltage δV	$\dfrac{I}{f \cdot C} \cdot \dfrac{(n^2 + n)}{4}$	$\dfrac{I}{f \cdot C} \cdot \dfrac{n}{4}$	$\dfrac{I}{f \cdot C} \cdot \dfrac{n}{12}$
Example $n = 5$: $\Delta V = \frac{I}{f \cdot C} \cdot x$	$x = 85$ (assumed to be 100%)	$x = 16.25$ (19%)	$x = 5.4$ (6.3%)
Example $n = 5$: $\delta = \frac{I}{f \cdot C} \cdot y$	$y = 7.5$ (assumed to be 100%)	$y = 1.25$ (16.7%)	$y = 0.42$ (5.6%)

The conclusions are drawn from the comparison of multiplier circuits of $n = 5$ stages. They show that both, voltage reduction and ripple, are about five times lower when a symmetric Greinacher cascade is used instead of a conventional one. When the three-phase cascade is used, a further improvement by a factor of three appears. As mentioned before, the investigated circuits should be applied as follows:

- for rated currents below 100 mA, the conventional Greinacher cascade (Fig. 6.4),
- for rated currents of some 100 mA, the symmetric Greinacher cascade (Fig. 6.5),
- for rated currents higher than 500 mA, the three-phase multiplier circuit (Fig. 6.6).

6.1.4   Multiplier Circuits with Cascaded Transformers (Delon Circuits)

When a suited transformer cascade supplies the AC feeding voltage into each stage of a HVDC multiplier circuit (sometimes called "Delon circuit"), the influence of the stages is compensated, and ripple and voltage reduction are reduced to the case $n = 1$ according to Table 6.1. Figure 6.7 shows the simplest circuit with $n = 2$ stages. All of them are based on a one-phase, two-pulse rectifier circuit (Fig. 6.2). The

Fig. 6.7 Multiplier circuit with cascaded transformers

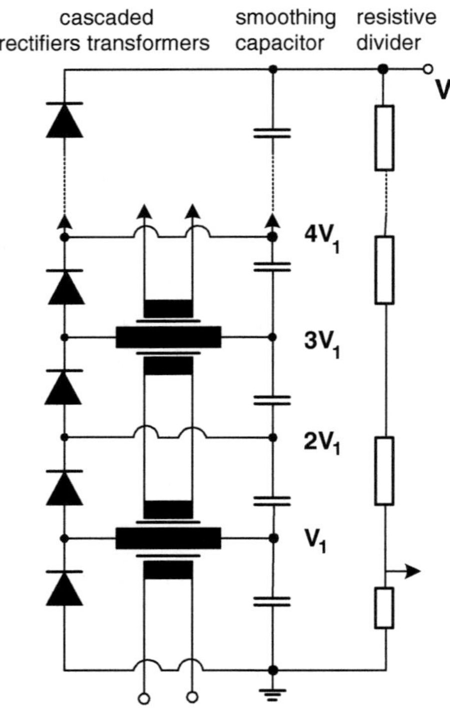

transformers of the cascade are not grounded and have to be isolated against the HVDC stress. The secondary winding of the lowest transformer (Fig. 6.7) is connected to the lowest smoothing capacitor and must be isolated in minimum for its DC potential V_1. The transformer winding (also on V_1) is connected to the primary winding of the next transformer which is on the DC potential $3 \cdot V_1$. Therefore, the second and all further transformers have to be isolated for a DC potential of $2 \cdot V_1$. Usually, all transformers are of the same design with a DC isolation of $2 \cdot V_1$ each. This DC insulation might be subdivided for the primary and the transfer (tertiary) winding.

The multiplier circuits with cascaded transformers enable the design of HVDC sources of medium rated currents with relatively low ripple and low voltage reduction. This can also be modified by feeding, e.g. a Greinacher cascade not only into the lowest stage, but also in one of the higher stages. Even the single charging into upper stages improves ripple and voltage reduction. As an example, Fig. 6.8 shows a cascade for 2000 kV with seven stages and feeding into the first and the fifth stage.

A second application of HVDC sources with cascaded transformers is that for modular DC test systems. One 400-kV oil-filled module may contain two 200-kV stages of a one-phase, one-pulse doubler circuit (Fig. 6.9), each equipped with its own stage of the transformer cascade (see Sect. 3.1.1.3), the elements of the rectifying circuit including smoothing capacitor. Also a voltage divider is arranged inside the module. Then, several modules can be arranged one above the next to a very space-saving HVDC test system. The polarity reversal is motor-driven.

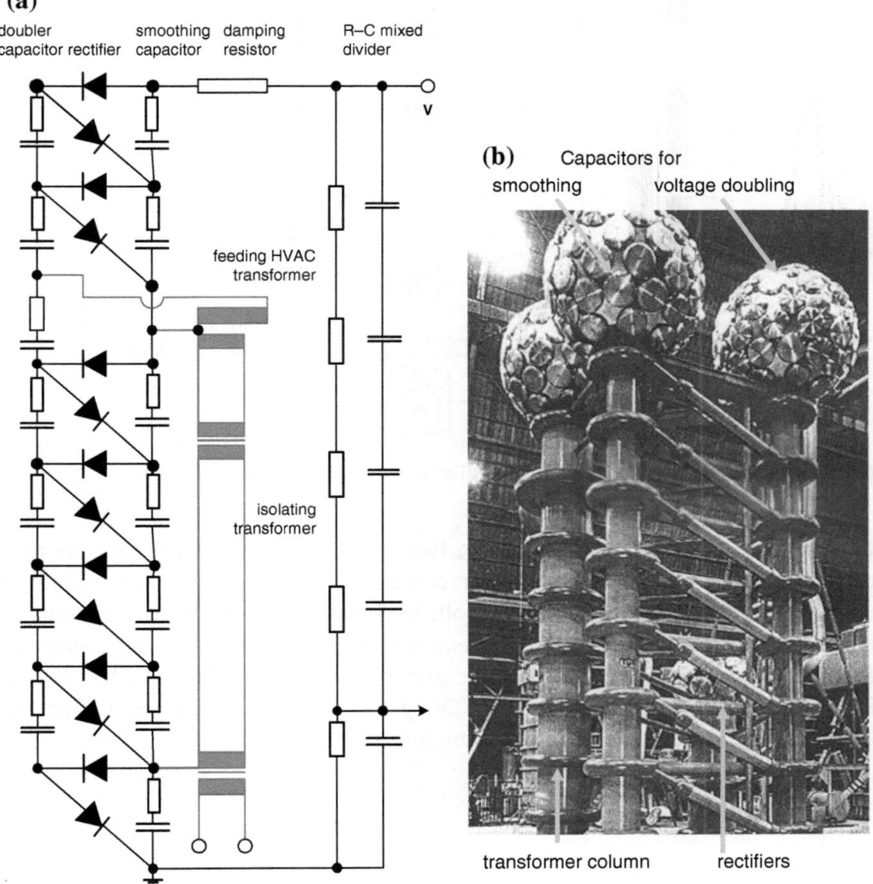

Fig. 6.8 Greinacher cascade with feeding into the sixth stage, **a** simplified circuit diagram, **b** generator for 2000 kV and for very fast polarity reversal up to 700 kV

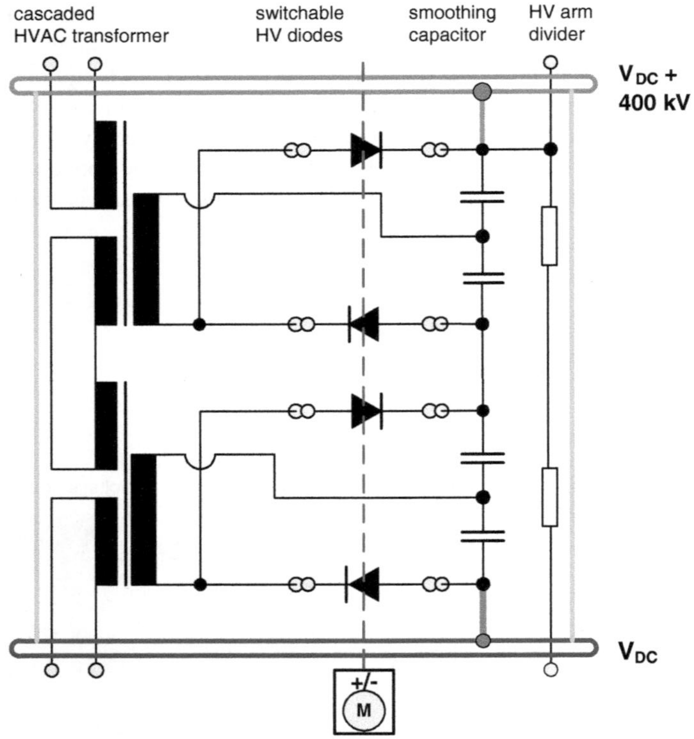

Fig. 6.9 Simplified circuit diagram of a HVDC module with two internal stages

Figure 6.10a shows such a system including blocking impedance and coupling capacitor for PD measurement. After connection of the power supply and the control unit, the system is ready for voltage testing. Such HVDC test systems for currents up to few 10 mA can easily be assembled and used for on-site testing, too. For generators of higher voltages and reasonable currents, modules can be switched in parallel. A stationary 2200 kV HVDC generator with five stages and parallel modules of the two lower stages is shown in Fig. 6.10b.

Fig. 6.10 Modular HVDC generators, **a** 800 kV/30 mA of two modules, **b** 2200 kV/10 mA of 7 modules in parallel-series connection. Both generators with external blocking impedance and PD coupling capacitor (size of modules is identical in both pictures!)

6.2   Requirements to HVDC Test Voltages

According to IEC 60060-1:2010 (Fig. 6.11), some necessary features of the design of HVDC test systems are discussed. There is a strong interaction between the test system and the test object (load). Therefore, the role of capacitive and resistive test objects is described for the selection of the generation circuit and its parameters.

Fig. 6.11 Definitions to direct test voltages

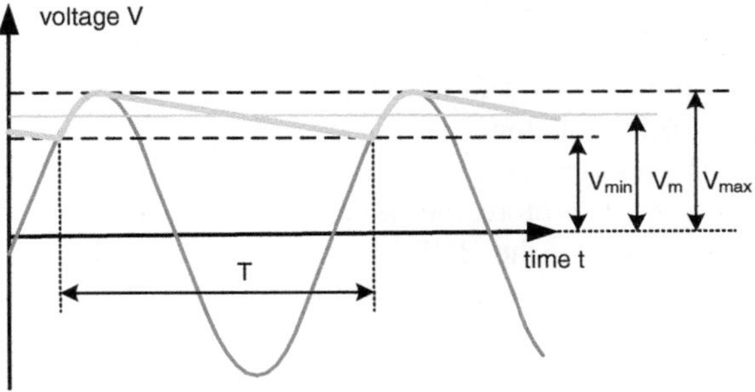

6.2.1   Requirements to HVDC Test Voltages

According to IEC 60060-1:2010, the *HVDC test voltage value* is not—as for the other test voltages—the peak value (V_{max}), but the *arithmetic mean value* during one period T of the charging process:

$$V_m = \frac{1}{T} \cdot \int_{t_1}^{t_1+T} v(t) \cdot dt. \tag{6.7}$$

The *tolerance* of the test voltage value is 1% for test durations up to 1 min, but for longer durations, 3% are acceptable. The definition of the mean value V_m results mainly from the voltage measurement with conventional voltmeters (moving-coil meters) which measure the mean value. One has to consider that the peak voltage V_{max} determines breakdown processes and must be applied for research work.

With the *ripple voltage* δV (Eq. 6.1), one gets the dimensionless *ripple factor* δ as a relation to the test voltage value

$$\delta = \frac{V_{\mathrm{max}} - V_{\mathrm{min}}}{2 \cdot V_m} \cdot 100 \le 3\%. \tag{6.8}$$

The requirement $\delta \le 3\%$ accepts that a remarkable difference appears between the peak voltage—which determines discharge phenomena in the insulation—and the test voltage value. A high ripple reduces also the inception voltage of partial discharges. Therefore, for both sides of an acceptance test, it is useful to have a ripple factor as low as possible.

Heavy pre-discharges, especially during wet and pollution tests, cause remarkable current pulses which may reduce the test voltage value V_m to a lower value $V_{m\,\mathrm{min}}$. The HVDC test system should be able to supply these transient discharge current pulses of up to few seconds without an instantaneous *voltage drop* higher than 10%:

$$d_v = \frac{V_m - V_{m\,\mathrm{min}}}{V_m} \le 10\%. \tag{6.9}$$

Unfortunately, IEC 60060-1:2010 does not specify the current value and its duration, e.g. for an assumed rectangular pulse or any indication of the required charge. For more details, see Sect. 6.2.3.2. Some publications consider $d_V \le 10\%$ as a too high value, (e.g. Hylten-Cavallius 1988; Köhler and Feser 1987). See also Sect. 6.2.3.2 below.

The reference value of the parameters δ and d_V is always the measured test voltage value V_m. Therefore, the *voltage reduction* ΔV (Eq. 6.5) is not a parameter of the test voltage, but of the HVDC test system. It characterizes the utilization factor of the used test system and has to be considered when a new test system is required.

6.2.2   General Requirements to Components of HVDC Test Systems

The circuit diagrams discussed above are simplified because they consider ideal elements and stationary conditions. Additionally, a HVDC generator has to withstand also transient stresses, e.g. in cases of a breakdown of the test object or a fast polarity reversal. If no countermeasures are taken, the stray inductances and capacitances would influence the distribution of the stressing voltages inside the generator. Furthermore, high breakdown currents have to be taken into consideration.

6.2.2.1   Protection Against Transient Stresses

The fast breakdown of the test object may excite oscillations of the HV test circuit consisting of the generator capacitances and/or the unavoidable stray capacitances and inductances. Also the diodes and the feeding transformer are not ideal elements and have certain impedances. All this forms a quite complicate equivalent network which should not be considered here, only the most important practical consequences of related calculations are summarized. The oscillations may cause non-linear voltage distributions in the generator and over-stresses of the components. The only way to avoid damages of components is a protection scheme of the generator. This starts with an *external damping resistor* between the generator and the voltage divider or the test object (Fig. 6.8a). Furthermore, there should be an internal damping resistor in all capacitances. Also the rectifiers should be equipped with grading capacitors for a linear voltage distribution at transient stresses and with internal damping resistors for both, over-current and voltage limitation. A rectifier consists of many diodes in series. Figure 6.12 shows an example of the *protection circuitry* of diodes for both cases, stationary and transient stress of the rectifier (Kind and Feser 1999). The protection scheme should be completed by *protection gaps* or surge *arresters* on especially endangered parts of the generator, e.g. the uppermost rectifiers or the output of the feeding HVAC transformer.

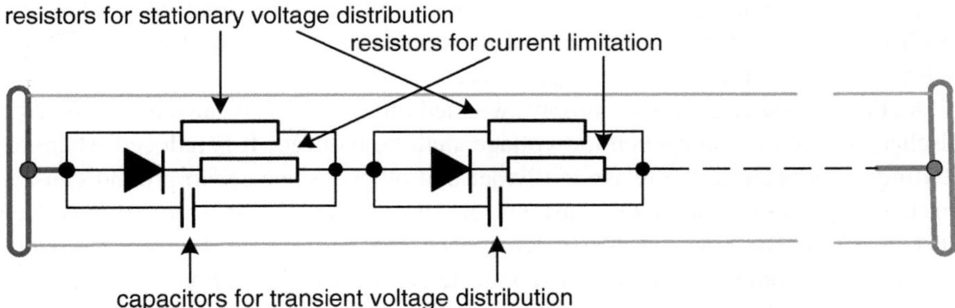

Fig. 6.12 Example of a protection circuitry of the diodes in a rectifier

Note The protection scheme operates only if a breakdown occurs at the test object connected via the external damping resistor. A breakdown between any point of the generator and any grounded or energized object, e.g. due to wrong arrangements in the HV test hall, may change the relation between the single components of the protection scheme in such a way that rectifiers and/or capacitors of the generator are endangered.

6.2.2.2 Polarity Reversal and Switch-off

A reversal of the voltage is a very hard stress for HVDC insulation, mainly because of the space and/or surface charges existing during the steady-state condition before and acting after the polarity reversal, too. (e.g. Okubo 2012; Wang et al. 2017; Tanaka et al. 2017; Azizian, A.et al.). Therefore, several standards require a polarity reversal in type tests. The used HVDC test system must have a motor-driven polarity reversal as, e.g. indicated in Fig. 6.9. The principle of the polarity reversal for a column of three rectifiers is shown in Fig. 6.13a: It is presumed that the rectifiers are connected to a column of smoothing capacitors and a capacitive test object (Frank et al. 1983; Schufft and Gotanda 1997). The reversal cycle starts at t_1 with switching-off the feeding AC voltage and a turn of the rectifiers. The capacitors are slowly discharged via the resistive part of the voltage divider (compare Fig. 6.8a). When the rectifiers approach the opposite electrodes at t_2, spark overs between their electrodes and the related opposite fixed electrodes occur. Within a very short time of some milliseconds, the capacitances are rapidly discharged to zero via the rectifier column. Now, the AC voltage is switched on again, and the capacitances are charged into the opposite polarity (t_3). The charging time depends on the time constant determined by the capacitances to be charged (generator plus test object), the impedances of the charging circuit and the available power of the HVAC feeding circuit. It may range from few seconds to few 10 s which is sufficient for most test applications.

Under certain conditions a much faster polarity reversal is required, e.g. within 200 ms. For that, the reversal time is defined as the interval between 90% of the outgoing voltage and the same 90% value of the opposite polarity (Fig. 6.13); whereas, the discharge phase is fast enough, the charging phase must be accelerated. This is realized by selecting a rated value for the charging much higher than necessary and interrupting the charging when 90% of the required voltage is reached (t_4). To avoid an overshoot of the opposite polarity, the feeding must be pulse-controlled to the exact voltage value.

A HVDC voltage cannot be simply switched off; the charged capacitance must be discharged via a resistor when the voltage shall be reduced. It is reduced when the feeding stops because the resistive divider and other resistances to ground cause a discharge process. The time constants are in a range of several seconds (see Sect. 6.2.3.1). The slow discharge can be accelerated by a *discharging bar*, consisting of a damping resistor, a hook (connected to the test field earth by a metal cord) and an isolating bar (Fig. 6.14). After the discharging, the same earthing bar is used for earthing by its grounding hook instead of its discharge contact (Fig. 6.14b).

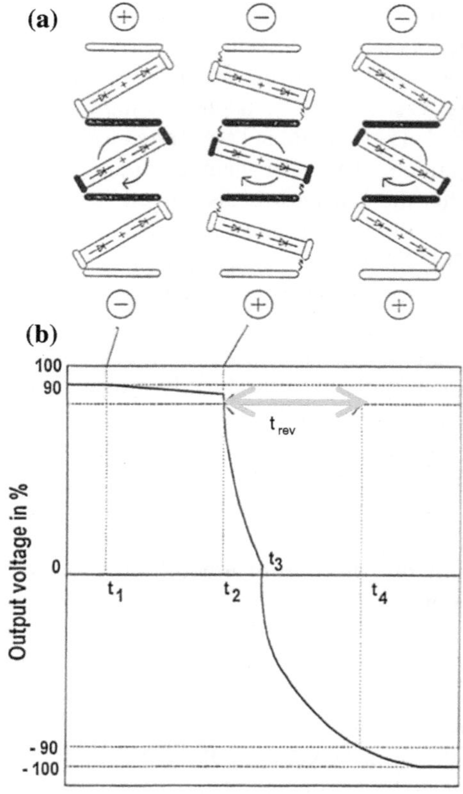

Fig. 6.13 Polarity reversal, **a** switching process of the rectifiers, **b** definition of the reversal time

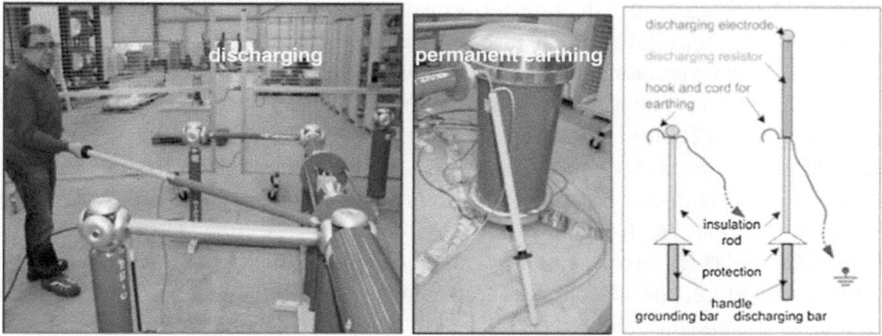

Fig. 6.14 Grounding of small HVDC generators

When the HVDC test system is designed for polarity reversals as described above, then the capacitors might be discharged via the turning rectifiers. After the feeding AC voltage is switched off, the grounding procedure starts with the discharge via the divider resistor down to about one-third of the rated voltage before the rectifiers are applied. They should not be applied for higher voltages to avoid an over-stress of the rectifiers. The permanent grounding of the capacitor columns shall only be used, after the generator is discharged. This can be done by earthing ropes or—for generators of rated currents up to several 10 mA by earthing switches (see Sect. 9.26 and Fig. 9.28). For discharging high capacitances see Sect. 6.2.3.1!

A new HVDC generator should be equipped with a well-established protection scheme and with a reliable discharge and grounding system. When the generator is not used, it shall be carefully grounded. For safety reasons, all capacitors of each stage of multi-stage generators must be directly grounded by a metal rope which can be moved through the generator by motor preferably. If protection scheme and grounding system of an older generator do not correspond to these requirements, an upgrade is urgently recommended.

6.2.2.3   Voltage Control, Selection of Smoothing Capacitances and Frequency

The output of the HVDC generator is controlled via the feeding AC voltage. This means that the control is one of the AC generation circuit. Traditionally, this is a regulating transformer and nowadays mainly a *thyristor controller* operating in a pulse-width mode. The wider the voltage pulse the higher is the output voltage. The shape of the AC voltage is not important for the output DC voltage because the HVDC generator can be understood as a filter which connects harmonics of the alternating voltage to ground. The DC output voltage is hardly influenced by conducted noise signals. The thyristor controller can be switched within less than a millisecond, what is necessary for the mentioned fast polarity reversals. When a very sensitive PD measurement shall be performed at direct voltage, the switching pulses of the thyristor controller could disturb. In such cases, the application of a regulating transformer can be recommended instead of or in addition to the thyristor controller.

With respect to voltage reduction and ripple, capacitance and frequency are interchangeable (Table 6.1). This means the system can be improved either by increasing the capacitances or by increasing the frequency of the feeding voltage. Of course, also both can be applied for improving a design. It must be taken into consideration that the frequency range of capacitors is limited. Capacitors for frequencies >300 Hz are remarkably more expensive than those for power frequency. Therefore, the selection of capacitances and frequencies of the feeding voltage should take the economic situation into account. There is no general rule for an optimum now. Each design and parameter combination has to be considered separately for a reasonable economic solution.

In the past, motor-generator sets have been applied for the generation of AC voltages of higher frequencies than 50/60 Hz for HVDC test systems. Today, static frequency converters are available (Figs. 3.26 and 3.35) and connect the selectable higher frequencies with the advantages of a thyristor controller. It can be assumed that in the next future, the power supply and control of HVDC test systems will be based on static frequency converters.

6.2.3   Interaction Between HVDC Test System and Test Object

HV testing requires the consideration of test circuits including test objects on the basis of the relevant standards. The standards of HVDC power systems are under a rapid development, a final survey cannot be given now. The IEC Technical Committee 115 on HVDC Power Transmission Systems is preparing the necessary basic standards. A survey on components of HVDC power transmission—including their HV testing—is e.g. given in the IEC Document 115/154/CD:2017. One should follow the related development of standards with the knowledge about the physical processes as tried to explain it in the following.

6.2.3.1   Capacitive Test Objects

Testing of HVDC cables: When a DC voltage is attached to an extruded cable, the externally applied voltage can be controlled, but the field strength distribution in the cable insulation depends additionally of material, time and temperature due to space charge formation (Maruyama et al. 2004; Pietsch 2012). This is well demonstrated by HVDC experiments with cable samples (Fig. 6.15):

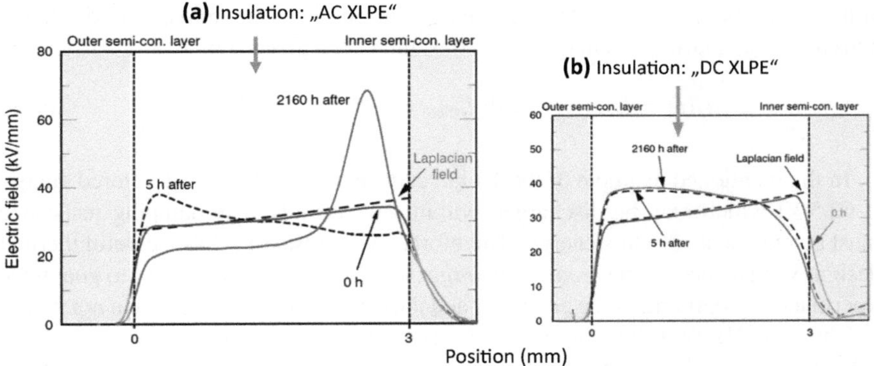

Fig. 6.15 Field strength distribution influence by space charges (explanations in the text), **a** sample made of "AC Polyethylene", **b** sample made of "DC Polyethylene"

(a) For *"AC XLPE"* as it is used in *HVAC* cables, one observes during and for a certain time after the HVDC charging of the cable sample a field strength distribution with the maximum at the inner electrode as expected for a coaxial system (Fig. 6.15a: Laplace field, blue). After 5 h due to space charges the maximum has shifted to the outer electrode (dotted, green), but when steady-state conditions are reached (e.g. after 2180 h, red) space charges of opposite polarity generate an extreme maximum of field strength near to the inner electrode. This would not be acceptable for real HVDC cables.

(b) Therefore *"DC XLPE"* has been developed using special additives to the polyethylene, which prevent a high space charge field near the inner electrode (conductor) (Fig. 6.15b). This guarantees a quite uniform and very stable steady-state of the insulation (field strength distribution after 5 h (dotted, green) and 2180 h (red) are practically identical). Also depending on the applied DC voltage the characteristic of the distribution is not changed. "DC XLPE" is well suited for HVDC cables.

The selection of the test duration has to consider the space charge behavior of the insulation and is quite difficult. This includes also different charging and discharging processes.

For charging currents of usually 3–10 mA, the charging may take up to several minutes. For example, the controlled charging of a 10 km long HVDC cable system, (about 2 μF) to a test voltage of 250 kV with a constant current of 5 mA would take nearly 2 min. Even during that short time, the generation of space charges cannot be excluded. After the test voltage value is reached, the shift of the field strength distribution to the steady-state (Fig. 6.15b) takes place.

At a test object of very low leakage current, the highest voltage value can remain for hours after switching-off the feeding voltage. To avoid serious safety problems, all capacitances must be discharged and grounded immediately (see Sects. 6.2.2.2 and 9.2.6.2). The necessary *discharge and grounding switch* must be equipped with a carefully designed damping resistor R_d which can be adapted to different capacitances of the test object. It transfers the discharge energy into heat. At a test voltage V_t, the time-dependent discharge current i.e., the maximum discharge current $I_{e\,max}$ and the discharge time constant τ_e are given by:

$$i_e(t) = I_{emax} \cdot e^{\frac{-t}{\tau_e}} \text{ with } I_{emax} = \frac{V_t}{R_d} \text{ and } \tau_e = R_d \cdot C_c. \tag{6.10}$$

In the mentioned example of the 10-km cable and $V_t = 250$ kV the stored energy is 62.5 kJ which may be discharged within few seconds. The damping resistance must be able to absorb this energy. Therefore, a wire resistor needs a careful thermal design. With respect to the *recovery voltage* (see Chap. 5), it is necessary to guarantee a permanent grounding of the generator and the capacitive test object when not in use.

The CIGRE Working Group B1.32 summarized the state of the art on the behavior of DC XLPE insulation and published the Technical Report 496 (2012) on the testing of extruded HVDC cables which acts as a certain standard. The report comprises the electrical tests for transmission cables using the phase-to-ground voltage V_0 under operational conditions as the reference voltage:

Prequalification tests of cable systems shall demonstrate the satisfactory long term performance of all components in an about 100 m long cable system. It is made only once during the development and includes a long duration voltage test at $V_t = 1.45\ V_0$ of different cycles with and without load current generated by a cable heating equipment (in total of 360 days). It is completed by polarity reversal $(2 \cdot 1.25 \cdot V_0)$ and composite voltage tests (superimposed voltage tests: DC/LI and DC/SI), Finally a detailed inspection shall be made.

Type tests are made before supplying cable systems on a general basis to demonstrate satisfactory performance characteristics. It is made at $V_t = 1.85 V_0$ for 30 days on cable loops typical for the cable system. The type test include load cycles, polarity reversal $(2 \cdot 1.45 \cdot V_0)$ test, DC/LI and a DC/SI composite voltage test followed by a HVDC test.

Routine tests (factory acceptance tests) are made on each manufactured component (cable or accessories) to verify that they meet the specific requirement. Each delivery length of cable shall be submitted to a negative DC voltage $V_t = 1.45 \cdot V_0$ for one hour. The CIGRE WG expresses that in addition to the DC test voltage an AC voltage test, which enables PD measurement, can be considered, provided the cable design allows AC application. Also for the cable accessories PD-monitored AC voltage tests might be useful.

On-site acceptance test (*after–installation tests*) shall demonstrate the integrity of the cable system as installed. The installed cable system shall be subjected to a negative HVDC test of $V_t = 1.45 \cdot V_0$, details must be agreed between supplier and user.

HVDC super-long cables: The charging and especially the discharging of the cable systems under test becomes really difficult if e.g. a large submarine cable (e.g. $V_m = 550$ kV, 200 km long, corresponding to 70 µF, e.g. see Fig. 1.3) is tested for commissioning ($V_t = 1.45 \cdot V_m = 800$ kV). The energy stored in the cable is about 22 MJ. The discharging of this example has been investigated by Felk et al. (2017):

The discharging due to the resistance of the cable insulation itself would take nearly 10 h. This means, the cable would be stressed much longer than during the test of e.g. 1 h. Furthermore no wire resistor has the necessary thermal capacity to overtake the energy. Therefore a HVDC *discharge device* based on an water filled vertical resistor is proposed. A water processing unit—as used for water end terminations for cable testing (see Fig. 3.44)—controls the value of the conductivity of the water. The processing unit increases the water conductivity by dosing salt into the water, and reduces it by a special resin bed for de-ionization. Because the water is heated due to its resistance, it is also cooled in the processing unit. The water circulates in the resistor, which consists of an inner tube (where the water goes up) and an outer tube (where the water goes down). The electric field between the two tubes as well as in vertical direction must be carefully designed. The thermal design should avoid dew on the outer surface of the resistor. The resistor can be connected of 400 kV modules with a water volume of about 70 litres each. For example, the 800 kV water resistor would consist of two modules and is always connected in parallel to the cable under test.

The water resistance is controlled so that it is very high during charging and testing the cable (which means with a very low influence on the HVDC voltage source) and low for discharging the cable. The thermal capacity of the water is so high that the energy of 22 MJ will increase the water temperature only by 40 K! For a new test the water must be cooled down and de-ionized.

Liquid-impregnated paper-insulated (LIP) HVAC cables: In some respect, HVDC testing of *HVAC LIP* cable systems is the classic example for direct voltage application to AC insulation, especially for testing on site (see Sect. 10.4.2). Therefore, the HVDC testing of HVDC cable systems does not cause new problems. Quite small HVDC test systems are able to charge the high capacitance of a LIP cable system. Also a certain relationship between the lifetime under AC operational stress and the results of suited HVDC tests has been found.

HVDC gas-insulated systems: For the connection of HVDC cables to HVDC power supplies, gas-insulated systems (GIS) are necessary to create safe disconnecter gaps, to measure the voltages and currents and to arrange arresters (Hering et al. 2017). The capacitance of such systems is not very high, but the insulation of the gas, preferably SF_6, in combination with solid spacers is quite sensitive. The reason is a change of the voltage distribution from a capacitive electrostatic field during the charging process, polarity reversals or over voltages to a resistive streaming field under steady-state DC conditions. Then the field is additionally influenced by the charge accumulation on the solid spacers as well as by thermal effects. Also the motion of particles may show the phenomenon of *"firefly"*, a hovering of sharp, PD-generating particles near to one of the electrodes. Partial discharges under HVDC conditions are very seldom, stochastic and difficult to measure (see Sect. 6.5). All these phenomena must be considered when test voltages are selected and test procedures are agreed (CIGRE JWG D.1/B.3.57, 2017). This CIGRE Joint Working Group recommends a very detailed electric *type test* consisting of

– a DC voltage withstand test,
– an AC voltage withstand test,
– PD measurement at DC and AC voltage,
– a polarity reversal DC test (see Sect. 6.2.2.2),
– a DC/LI composite voltage test (see Sect 8.2.3),
– a DC/SI composite voltage test (see Sect. 8.2.3),
– load condition test (withstand tests at rated current),
– an insulation system test on single components (withstand and PD $\leq$ 5 pC).

With respect to the high effort of the type test, the *routine tests* and the *on-site acceptance tests* shall be simple and time efficient. This cannot be reached with DC voltages. As an acceptable compromise, these tests shall be performed at AC voltages and completed by PD measurement.

Finally a *prototype installation test*, similar to the pre-qualification test of cable systems (see above) is being discussed (Neumann et al. (2017). This test shall demonstrate the long-time performance of the complete gas-insulated HVDC system (expected life time of 50 years). It is a long time test for 30 days containing load cycles and composite DC/LI and DC/SI voltage tests (see Sect. 8.2.3). For the load cycles a DC current corresponding to the rated current shall be injected. It requires a special current source for operating on HVDC potential (Neumann et al. (2017)).

6.2.3.2 Resistive Test Objects (Wet and Pollution Tests)

Wet and pollution tests require an active current due to a low surface resistance and/or to heavy predischarges. The limitation of the required current by the HVDC test system leads to a limitation of the test voltage (*voltage reduction* ΔV) in case of a permanent stress in stationary operation and to an instantaneous *voltage drop* (d_V) in case of transient stress. In both cases, the test cannot be performed correctly. Therefore, a lot of research work has been related to amplitude and shape of the required current (e.g. Reichel 1977; Rizk 1981; Matsumoto et al. 1983; Kawamura and Nagai 1984; Merkhalev and Vladimirsky 1985; Rizk and Nguyen 1987; Cigre TF 33.04.01, 2000). As a result, the IEC Technical Standard 61245:2015 gives hints for HVDC pollution testing, and the necessary specification of HVDC test systems (ripple factor $\leq 3\%$, voltage drop $\leq 10\%$, voltage overshoot $\leq 10\%$, voltage measurement for both, continuous and transient voltage components). The practice of HVDC pollution testing is described in several publications, e.g. Windmar et al. (2014).

Whereas the voltage reduction becomes only acceptable, when a HVDC test system of sufficient rated current is applied, the voltage drop d_V can be reduced to an acceptable value by a very large smoothing capacitor, possibly by an additional capacitor (Reichel 1977; Spiegelberg 1984) or by a feedback control with a higher feeding voltage (Köhler and Feser 1987).

According to the above references, the *leakage current impulse* (Fig. 6.17) increases with the surface conductivity from some 10 mA up to the order of 1–2 A, but its duration decreases from some 10 s with the increasing current amplitude down to the range of few 100 ms. The maximum charge of one-pulse might be in the order of 200–300 mC. The real pulse is replaced for calculations by triangular or rectangular current pulses (Fig. 6.16). The acceptable voltage drop at such an impulse is between 5% (Hylten-Cavallius 1988) and 8% (IEC 61245:1993; Köhler and Feser 1987). Merkhalev and Vladimirski (1985) estimated the relation of the measured flashover voltage (V_{FL}) of a polluted insulator when a current pulse with the charge q_p appears and an unregulated Greinacher generator with a limited smoothing capacitance C_{sl}, (charge Q_{sL}) has been used in comparison to a generator with values C_{s0} (of a charge Q_{s0}) which do not influence the flashover voltage (V_{F0}):

$$\frac{V_{FL}}{V_{F0}} = 1 + 0.5\left(1 - \exp\left(\frac{-q_p}{Q_{sL}}\right)\right). \qquad (6.11)$$

The deviation between the measured, too high flashover voltage V_{FL} (caused by an insufficient generator) and the correct value V_{F0} depends on the relation between

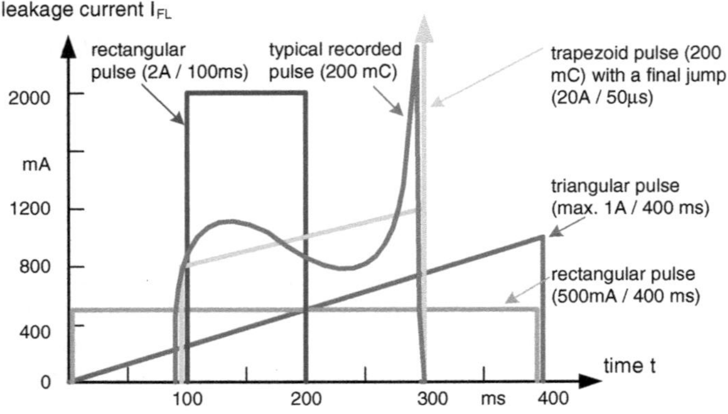

Fig. 6.16 Leakage current impulse (*red*) and its simplified replacements for calculations

the charge of the current pulse ($q_p \approx 200$ mC) and that of the smoothing capacitor Q_{sL}. If the charge stored in the smoothing capacitor exceeds the charge of the current pulse by a factor of 10 (this means $Q_{sL} \approx 10 \cdot q_p = 2000$ mC), then one gets the terms "$\exp(-q_p/Q_{sL}) \approx 0.9$" and "$V_{FL}/V_{F0} \approx 1.05$". The influence of the generator on the flashover voltage is below 5%. That means, for the above-considered generator of $V_r = 300$ kV, a smoothing capacitor of

$$C_s = \frac{Q_{sL}}{V_r} = \frac{2\,\text{As}}{0.3\,\text{MV}} \approx 6\,\mu\text{F}$$

would be necessary. This very high capacitance is too pessimistic (Hauschild et al. 1987; Mosch et al. 1988) when a powerful feeding and an optimized multiplier circuit (e.g. three-phase, six-pulse) are used. This shows that only the consideration of the required charge is too much simplified. A more detailed calculation shows that for $q_p = 200$ mC, a rectangular leakage current pulse of high peak and short duration causes a higher voltage drop than a lower current of longer duration (Fig. 6.17). This is related to the internal impedances of the test system. Consequently, the voltage drop—as also the above-mentioned voltage reduction— increases with the number of stages. The shape, maximum value and duration of leakage current pulses depend also from the height of the test voltage, the applied pollution test method and the parameters of the tested insulator.

For checking the suitability of a HVDC test system for pollution testing, a computer simulation of the behaviour of the whole test circuit under a leakage current pulse of, e.g. 200 mC, is recommended. Additionally, the assumed shape of that pulse must be selected (compare Fig. 6.16). This can be done using the results of earlier circuit simulations shown in Fig. 6.18 (Mosch et al. 1988): It compares the voltage drop caused by a trapezoid—recorded currents well describing—pulse of 200 mC including a very fast final jump—simulating the fast heating of the residual pollution layer and the immediate final leader discharge—with that of a

Fig. 6.17 Output voltage of a HVDC generator (Fig. 6.19) depending on amplitude and duration of a rectangular leakage current impulse

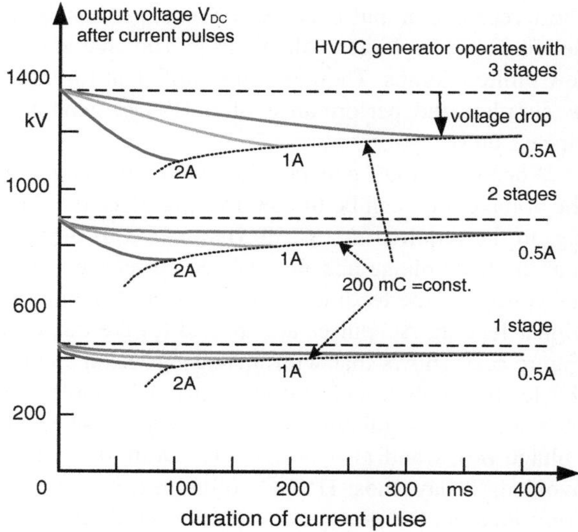

Fig. 6.18 Voltage drop for current pulses of rectangular and trapezoid shape

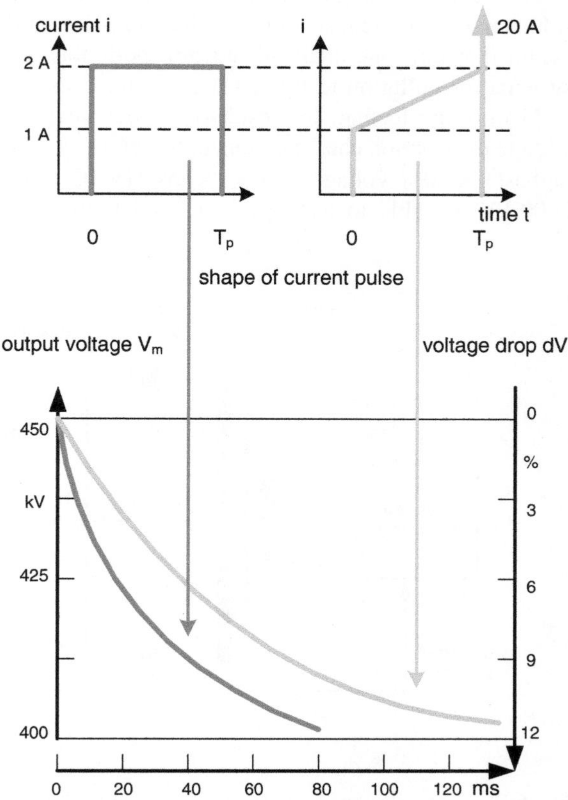

short rectangular pulse (2 A, 100 ms). It has been shown, a rectangular pulse delivers an even larger voltage drop. The charge demand of the final jump is only few millicoulombs. Therefore, it is sufficient for the estimation of the voltage drop, to simulate the performance of a HVDC generator assuming short rectangular current pulses.

When the HVDC generator is fed by a thyristor controller with feedback control, the voltage drop can be limited to a preselected value by a suited regulator interval, e.g. to the required $d_V \leq 5\%$ (Draft IEC 61245: 2013). Feedback control means that the test voltage and the leakage current are measured (Fig. 6.19 and used for the control of the feeding HVAC voltage. When a certain voltage drop is recorded, higher feeding AC pulses are applied for the duration of the leakage current pulse. Figure 6.20 shows that with increasing leakage current, the required frequency of the feeding pulses and, consequently, the ripple frequency of the output voltage increases. The regulating interval has to be selected in such a way that too high voltage drops and also overshoots (Draft IEC 61245:2013 requires $\leq 10\%$) are avoided. Today, most HVDC pollution tests are often performed with *feedback—controlled HVDC test systems* (Seifert et al. 2007; Jiang et al. 2010, 2011; Zhang et al. 2010a, b).

HVDC pollution tests must not be performed at complete insulator chains; it is sufficient to test one insulator of a chain because a linear voltage distribution can be assumed within one chain of uniform pollution. Therefore, powerful HVDC test systems for pollution tests are usually limited to test voltages up to 600 kV.

In opposite to that, for insulators under *artificial rain* conditions, no uniform voltage distribution can be assumed, therefore wet tests must be performed up to the highest DC test voltages. This means HVDC test systems for rated voltages of 2000 kV are able to test open air insulations for 1000 kV HVDC transmission

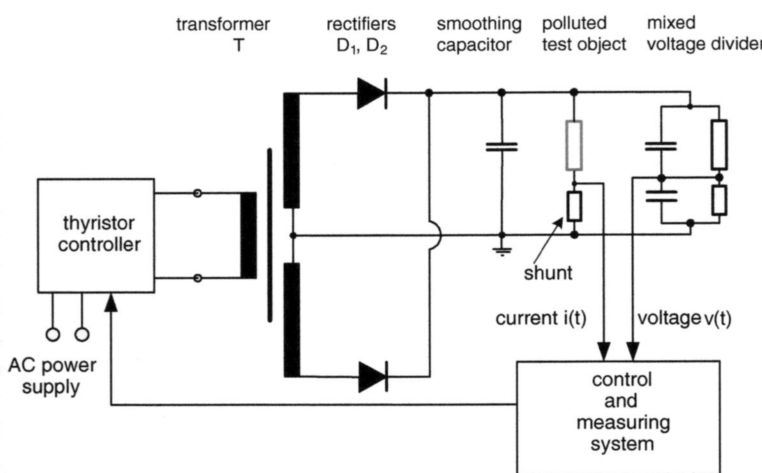

Fig. 6.19 Circuit diagram of a HVDC test system with feedback control

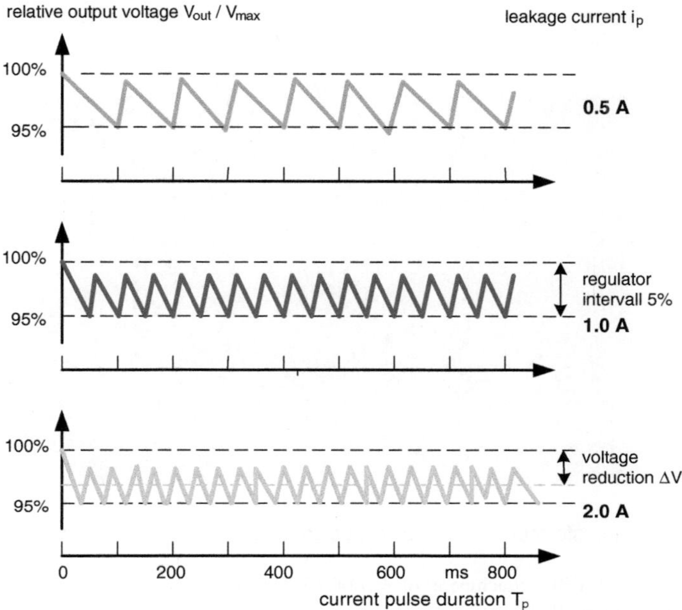

Fig. 6.20 Calculated output voltage of a feedback-controlled generator at different leakage currents, but identical pulse charge Q_p = 200 mC

systems. Current pulses due to heavy streamer discharges in transition to leader discharges are characterized by charges up to 10 mC. These charges can be supplied by generators without feedback control and rated currents of few 100 mA. If it shall also be used for wet testing of contaminated insulators higher rated currents and feedback control might be useful (Su et al. 2005).

6.2.3.3   Corona Cages and HVDC Test Lines

A remarkable part of the active losses of an air insulated HVDC transmission system is caused by partial discharges which are usually designated *"corona" discharges*. The design of the bundle conductors of a HVDC transmission line is usually verified by tests in a corona cage and/or on a test line. A corona *cage* is a coaxial electrode system with an outer electrode up to few metres diameter, realized by metal rods, and the bundle conductor to be investigated forms the inner electrode on HVDC potential. The outer electrode is grounded via an impedance for measurement of corona current pulses or the average corona current. A *HVDC test line* is a one-to-one model of a future HVDC overhead line. It shall demonstrate the performance of all components under operational conditions. Both, corona cages and test lines are outdoor arrangements and require outdoor HVDC test systems (Elstner et al. 1983; Spiegelberg 1984).

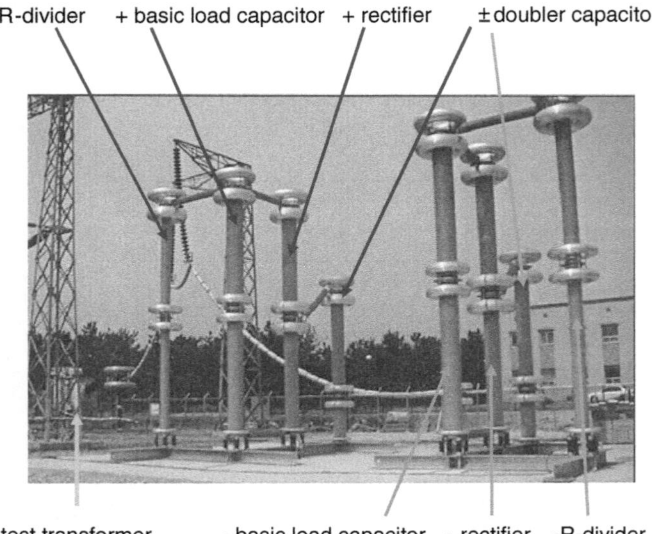

Fig. 6.21 Bipolar HVDC attachment of ±1000 kV/500 mA to a 600 kV/3.3 A HVAC test system. Courtesy of KEPRI Korea

Such test systems have to supply continuous currents of several 100 mA for avoiding a voltage reduction due to continuous corona discharges and short-term currents up to few amperes for avoiding voltage drops due to leakage current pulses. For a test line, both polarities must be supplied. This can be made by two separate HVDC test systems or by one system with bipolar output (Fig. 6.21). The shown system is a *bipolar HVDC attachment* to a 600 kV/3.3 A HVAC test system. It consists of two one-phase one-pulse doubler circuits, one for each polarity.

The test system itself must withstand its output voltage at all environmental conditions under which tests shall be performed. The details of those conditions (temperature range, atmospheric pressure range, humidity up to 100%, rain, natural contamination) must be carefully specified as well as a possible reduction of the rated voltage for certain test conditions. The external surface of all components shall be equipped with silicon rubber sheds (Fig. 6.21).

6.3 Procedures and Evaluation of HVDC Tests

Developed for all continuous voltages, the procedures and evaluations of HVAC tests (see Sect. 3.3 based on Sect. 2.4) can be applied for HVDC tests, too. Therefore, the described methods will not be repeated in this section, only few differences shall be mentioned.

The *progressive stress test* with continuously or step by step increasing direct voltage (Fig. 2.26) is used to determine a *cumulative frequency distribution* which can be approximated by a *theoretical distribution function*. The approximation can be performed according to the recommendations of Table 3.7. Also for *life-time tests* the remarks of Sect. 3.3.1 are applicable.

Compared with HVAC tests, it is more difficult to guarantee *independence* in HVDC tests. The reason is based on the phenomenon that partial discharges (and the trace of flashovers) at direct voltage cause surface and space charges of a long lifetime. Therefore, preceding stresses may influence the result of following stresses. It is absolutely necessary to check the independence of a test result before a further statistical evaluation will be made. The graphical check (Fig. 2.28) should be performed during the tests, and the test procedure should be modified if independence appears. Modifications are, e.g. the change of the rate of voltage rise, the careful cleaning of test objects after flashovers, the application of a new test object for each stress cycle or the application of a low alternating voltage between two stress cycles ("cleaning" by an alternating electric field). When a solid insulation is investigated, usually each test cycle requires a new sample.

In quality-acceptance testing, the procedure described in Sect. 2.4.6 and Fig. 2.39a is recommended also for direct voltages (IEC 60060-1:2010). The test should be performed at the polarity which delivers the lower breakdown voltages. If this is not clear testing at both polarities is necessary. Also PD-monitored withstand tests (Sect. 3.3.2) are applicable, but the randomness of partial discharges at direct voltage shall be considered (see Sect. 6.5). This may require longer durations on the different voltage levels (Fig. 2.39b). Other measurands than partial discharges, e.g. the leakage current or the insulation resistance, might be taken into consideration for diagnostic withstand testing.

6.4 HVDC Test Voltage Measurement

To measure high DC voltages up to some hundreds of kV originally *sphere gaps* have been used. As already pointed out in Sect. 2.3.5, based on experimentally determined breakdown curves of sphere gaps under clean laboratory conditions, a measuring uncertainty of about 3% is achievable for voltages ranging between 20 kV and about 2000 kV (Schumann 1923; Weicker 1927; Weicker and Hoercher 1938; IEC Publication 52:1960). However, the breakdown voltage of sphere gaps is not only affected by nearby earthed objects (Kuffel 1961) but also by the roughness of the electrodes as well as by dust and pollution deposited on the electrode surface, and as usual by the humidity and density of the ambient air. As charged particles are always attracted into direction of increasing electric field strength, which forces the deposition of dust particles on the electrode surface, the use of sphere gaps for measuring DC voltages above 200 kV is not recommended (see Table 2.7).

Based on experimental findings (Peschke 1968; Feser and Hughes 1988), DC voltages can also be measured at reasonable accuracy by means of *rod–rod gaps* in atmospheric air. Taking into account the correction factors for the humidity and density of air, a measuring uncertainty of about 2% is achievable for DC voltages ranging between approx. 20 and 1300 kV, as specified in the revised standard IEC 60052:2002 (see 2.3.5). However, the main drawback of air gaps used for measuring high voltages is the discontinuous measuring procedure, which is very time-consuming. Thus, since the 1920s spark gaps were increasingly replaced by *electrostatic voltmeters* (Starke and Schröder 1928), due to their capability to measure high voltages continuously. In the 1930s also measuring systems based on high-resistive converting devices have been introduced where the high voltage was indicated by either current or voltage meters, see Fig. 6.22 (Kuhlman and Mecklenburg 1935). Nowadays, the converting device provides commonly a resistive voltage divider, which is sometimes bridged by capacitances to record additional voltage changes superimposed on the stationary DC voltage, such as the ripple of DC voltages as well as the time-dependent voltage shape at polarity reversal.

An important design parameter of *resistive voltage dividers* used for DC measurements is the direct current flowing through the converting device. As dust and pollution deposited on the surface of the HV divider column may cause parasite leakage currents affecting the divider ratio, particularly at high air humidity, the direct current through the converting device should be chosen not lower than 0.5 mA (IEC 60060-2:2010). As already mentioned above, charged particles are always attracted into direction of increasing electric field strength, which forces thus

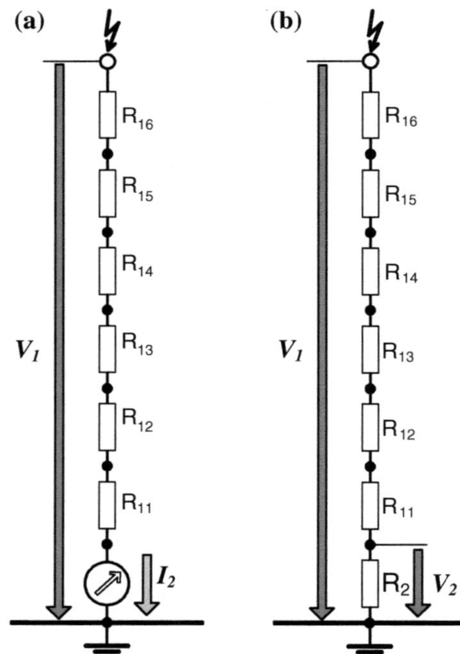

Fig. 6.22 Basic principles commonly used for indirect HVDC measurements, **a** ammeter in series with a resistive converting device, **b** voltage measurement across the LV arm of a resistive voltage divider

the deposition of dust particles on the surface of the divider column. However, reducing the resistance of the HV arm to minimize the impact of dust and pollution is limited not only by the additional load acceptable for the HVDC test supply, but also by the permitted operation temperature which increases drastically at decreasing resistance of the HV divider column.

Example Consider the HV column of a 100 kV voltage divider having a total resistance of 10 MΩ, which is composed of 100 low-voltage resistors in series, each rated 100 kΩ and 1 kV. Applying a DC voltage of 100 kV, the current through the resistive converting device would attain 10 mA, so that the power dissipating in the HV divider column approaches 1,000 W. This leads not only to an increase of the operation temperature, which might change the divider ratio, but could also damage the HV arm. This is the reason why the current through the HV arm should be kept as low as possible, but not lower than 0.5 mA to prevent an impact of dust and pollution on the divider ratio, as mentioned above. Under this condition the voltage drop across each LV resistor forming the HV arm attains 0.5 kV, which is equivalent to a specific value of 1 MΩ/kV.

This example underlines that a divider current in the order of about 0.5 mA seems to be a reasonable choice, because from a thermal point of view the current should be as low as possible. However, a current below 0.5 mA would lead to an impact of parasite leakage currents on the measuring uncertainty, as mentioned above.

To limit the radial and tangential field gradients along the resistive HV arm, the LV resistors connected in series are commonly wounded around an insulating cylinder like a helix, as obvious from Fig. 6.23. The photograph shows a 300 kV DC divider designed by Peier and Greatsch (1979). The HV arm is composed of 300 pieces of LV resistors, each of 2 MΩ, where the bottom 2 MΩ resistor provides the LV arm, so that the divider ratio amounts 1:300. To accomplish a potential distribution along the resistor helix comparable to the electrostatic field distribution caused by the top electrode alone, i.e. without resistive divider column, the pitch of the resistor helix was varied accordingly, as originally proposed by Goosens and Provoost (1946) to design impulse voltage dividers, see Sect. 7.4.2. The HV resistor column is arranged in an oil-filled PMMA cylinder to improve the convection of the power dissipating in the HV column, which attains only 150 W at 300 kV. That means, the maximum current through the HV resistor attains 0.5 mA, which complies with the requirements discussed above, i.e. on one hand this minimizes the impact of dust and pollution on the parasite leakage current along the HV divider column and, on the other hand, the power dissipating in the HV divider resistor and prevents hence a significant change in the resistance. Using wire-wounded LV resistors, which were artificially aged by a temperature treatment, a measuring uncertainty of about 3×10^{-5} can be accomplished.

In this context it should be mentioned that the above described resistive voltage divider, which is utilized by the German National Institute of Metrology (PTB Braunschweig) for high-precise reference measurements, is not applicable for DC voltage measurements in industrial test fields. This is because an unexpected breakdown of the test object is associated with fast changing over-voltages on account of the comparatively high self-inductance of the in series connected wire-wounded resistors. This causes a strong non-linear voltage distribution and

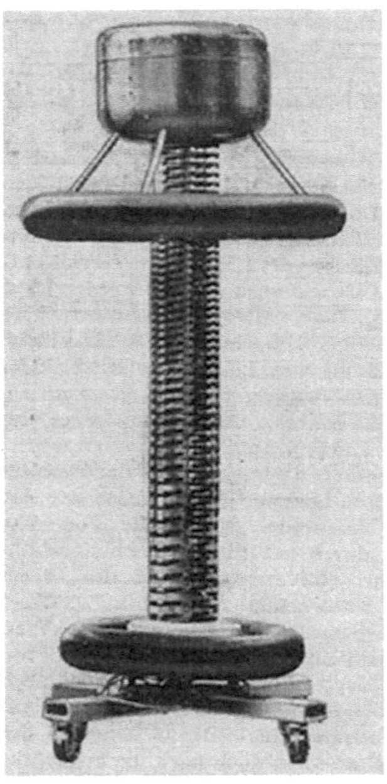

Fig. 6.23 Photograph of a high-precise DC voltage divider designed by Peier and Greatsch (rated voltage 300 kV, HV resistance 600 MΩ, total high 2.1 m)

would thus damage the HV arm. To enhance the energy capability, in principle metal-oxide film or carbon composition resistors could be used as alternative. However, the main obstacle is their comparatively high temperature coefficient, which would increase the measuring uncertainty, particularly under extended test duration causing a temperature rise in the HV arm. Even if this effect could be minimized by means of an artificial ageing, which is accomplished by a long-term temperature treatment, it has to be taken into account that such a conditioning is extremely time-consuming, which is thus applied only in very specific cases.

The best solution to avoid a possible damage of resistive dividers in case of unexpected breakdowns is the use of mixed voltage dividers, such as resistive–capacitive dividers or even capacitively graded dividers. An optimum performance is achieved by means of stacked capacitors connected to specific points of the resistive divider column, as obvious from the photograph shown in Fig. 6.24. As a rule of thumb, the capacitance of the HV column of such mixed voltage dividers should be chosen in the order of 200 pF. Moreover, shielding electrodes of comparatively large surface should be employed to prevent the occurrence of partial discharges, which could also affect the divider ratio.

To measure the output voltage of HVDC dividers, in principle classical *analogue instruments* indicating the arithmetic mean value can be employed. However, the better approach is the use of oscilloscopes or even digital recorders to measure DC voltages under real test condition, i.e. besides the static DC voltage also typical dynamic voltages, such as the *ripple* and the *voltage drop* as well as the parameters characterizing the *polarity reversal*. The requirements for approved DC voltage

Fig. 6.24 DC generator and a resistive–capacitive divider (2 MV, 1 mA) designed for the measurement of static and dynamic voltages

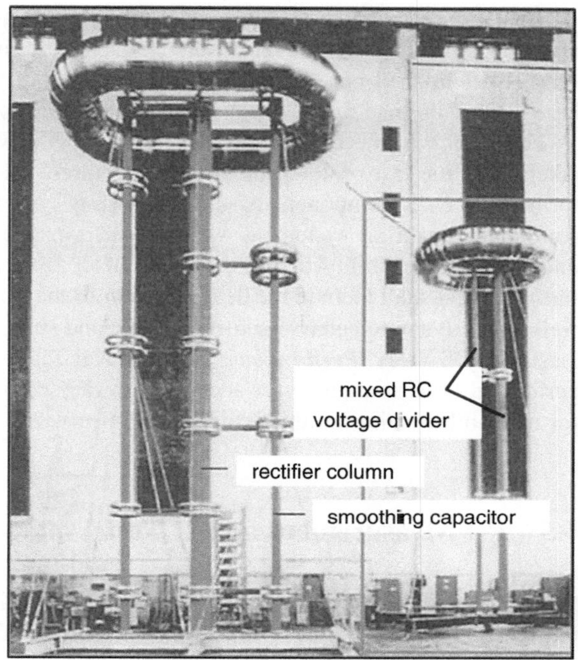

measuring systems are specified in IEC 60060-2:2010. So, the arithmetic mean value shall be measured with an expanded uncertainty $U_M \leq 3\%$, which corresponds to a coverage probability of 95%. To determine the dynamic behaviour of the measuring system, it is subjected to a sinusoidal voltage at the input. Changing the frequency between 0.5 and 7 times of the fundamental ripple frequency f_r, the difference of the measured output voltage magnitude shall be within 3 dB.

The above-given uncertainty limits shall not be exceeded in the presence of the maximum ripple specified in IEC 60060-1:2010. The ripple magnitude shall be measured with an expanded uncertainty $\leq 1\%$ of the arithmetic mean value of the DC test voltage or $\leq 10\%$ of the ripple magnitude, whichever is greater (IEC 60060-2: 2010). To measure the mean value of the DC voltage and the ripple magnitude, either separate measuring systems or the same converting device in connection with two separate measuring modes for both DC and AC voltages may be used.

The scale factor of the ripple measuring system shall be determined at the fundamental ripple frequency f_r with an expanded uncertainty $\leq 3\%$. The scale factor may also be determined as the product of the scale factors of the various components. Measuring the amplitude/frequency response of the ripple measuring system in a frequency range between 0.5 and 5 f_r; the amplitude shall not be lower than 85% of that value occurring at the fundamental ripple frequency f_r.

To measure rising and falling DC test voltages as well as the ripple and the voltage shape at polarity reversal, the characteristic time constant of the DC measuring system shall be ≤ 0.25 s. In case of pollution tests, the time constant shall be $\leq 1/3$ of the rise time typical for the appearing transients (voltage drop).

The results of the type and routine tests of HVDC measuring systems can be taken from the test protocol of the manufacturer, where routine test shall be performed on each component of the measuring system. *Performance tests* of the complete measuring system as well as performance checks must be performed under the responsibility of the user himself or by a calibration service. The performance test shall include the determination of the scale factor at the calibration as well as the dynamic behaviour for the ripple and shall be performed annually but at least every 5 years. *Performance checks* cover scale factor checks and should be performed at least annually or according to the stability of the measuring system; for more information on this issue see the Sects. 2.3.3 and 2.3.4.

6.5 PD Measurement at DC Test Voltages

The physics of gas discharges under direct voltage became especially of interest in the late 1930s when high DC voltage was increasingly used for physical, medical and military applications, for instance, to generate the operation voltage of X-ray equipment, cathode-ray tubes, electron-accelerators, image intensifiers, and radar facilities. At that time numerous technical papers and text-books have been published, mainly addressed to the fundamentals of discharges in various gases and even in vacuum (Trichel 1938; Loeb 1939; Raether 1939). An excellent survey on this subject can also be found in the textbook of Meek and Craggs (1978). In the 1960s, when the first long-distance HVDC transmission lines were put into operation, the measurement of partial discharges became also of interest, in particular to assess the insulation integrity of HVDC equipment after manufacturing, such as power cables and power capacitors (Rogers and Skipper 1960; Salvage 1962; Renne et al. 1963; Melville et al. 1965; Salvage and Sam 1967; Kutschinski 1968; Kind and Shihab 1969; Müller 1976; Densley 1979; Meek and Craggs 1978; Devins 1984).

The PD phenomena occurring in air gaps under DC voltage at both polarities are more or less comparable to those occurring under power frequency AC voltage. This applies in particular to so-called *Trichel pulses* (Trichel 1938) igniting under certain conditions at sharp negative electrodes in air, see Fig. 6.25. Due to their regular appearance in magnitude and repetition rate as well, Trichel pulses can advantageously be used for performance checks of PD measuring systems, for instance, to determine the actual pulse polarity, which appears often inverted when decoupled, transmitted and processed by a PD measuring system.

To assess the insulation condition of HV apparatus and their components, however, the mechanism of internal discharges, such as cavity, interfacial and surface discharges, is especially of interest. To explain the PD occurrence as well as the charge transfer due to cavity discharges under direct voltage, the classical *a-b-*

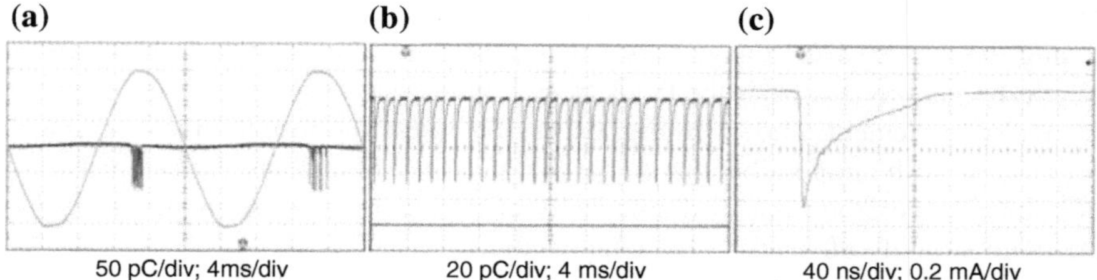

Fig. 6.25 Trichel pulses recorded for a point-to-plane air gap at test level close to the inception voltage, **a** under 50 Hz alternating voltage, **b** under direct voltage **c** shape of a single current pulse

c model illustrated in Fig. 4.13 (see Sect. 4.2) has been modified accordingly (Fromm 1995). So the three characteristic capacitances were bridged by high-ohmic resistances to simulate the voltage distribution between the electrodes under constant DC stress as well as to assess the time elapsing between consecutive discharges, which is commonly referred to as *recovery time*, and dominated by the time constant deduced from the intuitively assumed cavity capacitance and the resistivity of the dielectric material (Fromm 1995; Beyer 2002; Morshuis and Smit 2005).

As an alternative, the *dipole model* according to Pedersen (1986) can also be used to explain the PD charge transfer as well as the recovery time (Lemke 2016), as will briefly be discussed in the following. For this purpose a PD event in a gaseous inclusion shall be considered, which is embedded in the bulk dielectric between plane-parallel electrodes. Basically, it can be distinguished between the following three stages (Fig. 6.26):

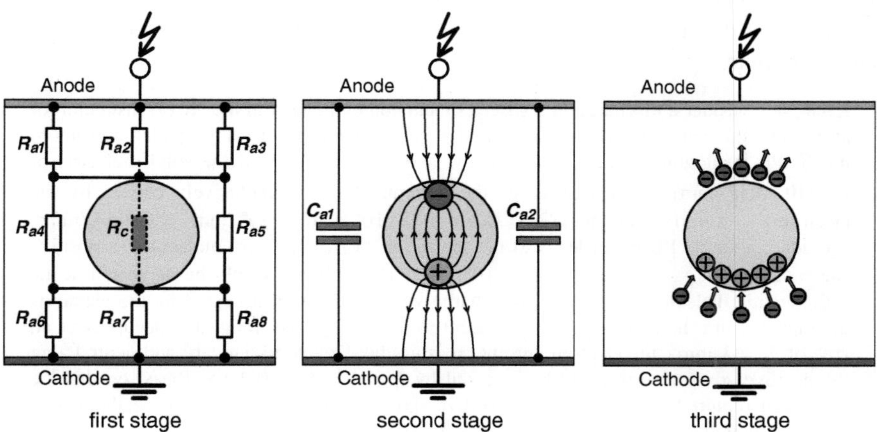

Fig. 6.26 Basic stages typical for cavity discharges under DC stress

Stage I: Initializing the ionization processes
Rising the applied DC voltage very slowly, the potential distribution between the electrodes, and in particular the field strength in the gaseous cavity, is governed by the resistive elements shown in Fig. 6.26a. To assess the inception voltage, however, seems to be impossible due to the fact that the volume and surface resistivity of solid dielectrics decreases dramatically at increasing field strength and is furthermore strongly affected by the temperature.

Stage II: Establishment of the dipole moment
Assuming that a self-sustaining discharge ignites at static inception voltage, the number of electrons released from neutral gas molecules equals always that number of positive ions, as discussed in Sect. 4.2. As a result, a dipole moment is established because all positive ions are deposited at the cathode-side cavity wall, while the electrons and negative ions (formed by the attachment of electrons to neutral molecules) are deposited at the anode-side cavity wall. As the dipole moment opposes the electrostatic field resulting from the applied DC test voltage, the field strength inside the gaseous inclusion is diminished, so that the self-sustaining discharge processes will be quenched suddenly.

Stage III: Dissipation of the dipole moment
Even if the resistivity of insulating materials is extremely high, it is supposed that the positive ions deposited at the cathode-side dielectric boundary are slowly neutralized by the attachment of electrons de-trapped from the solid dielectric, which is forced by a strong field enhancement adjacent to the positive space charge. Simultaneously, the electrons, previously deposited at the dielectric wall, will enter this boundary and propagate further through the solid dielectric into anode direction. As the drift velocity of the electrons and thus the associated *electron current* is extremely low, the *recovery time* required to establish the initial field conditions and thus to ignite the next discharge becomes extremely long. Practical experience revealed that the time elapsing between two consecutive PD events may approach several minutes.

> *Example* Consider a discharge in a gaseous inclusion embedded in the XLPE insulation of an extruded power cable. Assuming a field strength of $E_p = 20$ kV/mm, which appears in the XLPE insulation adjacent to the cavity, and a conductivity of the solid dielectric of $\kappa = 10^{-17} (\Omega \cdot \text{mm})^{-1}$, the density of the current, which is exclusively carried by the electrons, would attain $G_e = E_p \cdot \kappa = (2 \cdot 10^4 \times 10^{-17})$ A/mm^2 = 0.2 pA/mm^2. Provided, a single PD event leads to the ionization of $n_g = 10^8$ gas molecules, a positive space charge of $q_+ = e \cdot n_g = (1.6 \times 10^{-19} \text{ C}) \cdot 10^8 = 16$ pC would be deposited at the cathode-side dielectric boundary. This charge will slowly be neutralized by the electrons de-trapped from the solid dielectric adjacent to the cavity. Assuming this occurs over an area of $A_c = 1$ mm^2, the electron current crossing the dielectric boundary and neutralizing the positive space charge could be assessed by $I_e = G_e \cdot A_c = 0.2$ pA. Based on this the time span required to neutralize the positive space charge deposited at the cathode-side dielectric boundary gets $t_r \approx q_+ / I_e = (16 \text{ pC})/(0.2 \text{ pA}) = 80$ s.
>
> Considering now the negative space charge deposited at the anode-side dielectric boundaries, the current caused by the electrons when leaving the gaseous cavity can also be assessed as close to 0.2 pA, because this continues through the solid dielectric, i.e. from the

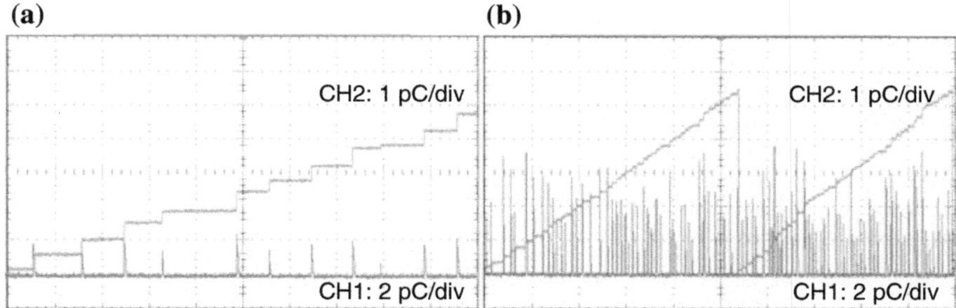

Fig. 6.27 Trains of charge pulses (CH1—pink trace) and accumulated pulse charge (CH2-green trace) recorded at direct voltage for discharges in a gaseous inclusion embedded in a PE cable sample, **a** recording time 20 s, test level slightly above PD inception voltage, **b** recording time 100 s, test level approx. 20% above PD inception voltage

anode-side dielectric boundary to the anode. Consequently, all electrons (including those attached to neutral gas molecules and thus forming negative ions) will leave the cavity within a time span of about 80 s. With other words: the initial space charge free field conditions will be accomplished after a relaxation time of about 80 s.

Performing PD tests under DC test voltage it has to be taken into account that only the following two PD quantities are measurable:

(i) The *pulse charge* q_i of a single PD event appearing at instant t_i.
(ii) The *recovery time* Δt_i, which is inversely proportional to the PD pulse repetition rate.

To measure these both PD quantities, in principle the basic measuring circuits including the coupling units (Fig. 4.23) and measuring systems as well as the calibration procedures, as specified in IEC 60270:2000 for PD tests under power frequency AC voltage, are applicable. In this context it should be remembered that the pulse repetition rate under constant DC stress is extremely low, as discussed above and obvious from Figs. 6.27 and 6.28. Thus, appropriately long testing times have to be chosen. However, under this condition individual charge pulses received from the output of conventional PD measuring systems can hardly be visualized by means of classical digital oscilloscopes due to their short duration, which is usually below 100 μs. Thus this signal must either considerably be stretched, usually up to the second range, or even displayed as "accumulated pulse charge", as illustrated in Figs. 6.27 and 6.28b. Another benefit of such a display mode is that the slope of the accumulated pulse charge is proportional to the mean PD current, which can be regarded as an additional valuable information (Lemke 1975). To prevent a saturation of the measuring device equipped with such a feature, a reset must be triggered just before an overload appears, as obvious from Fig. 6.27b.

Fig. 6.28 Graphs recommended in the Amendment to IEC 60270 to display PD tests results under constant DC voltage level, **a** consecutive charge pulses appearing during a 30 min test period, **b** associated accumulated pulse charge

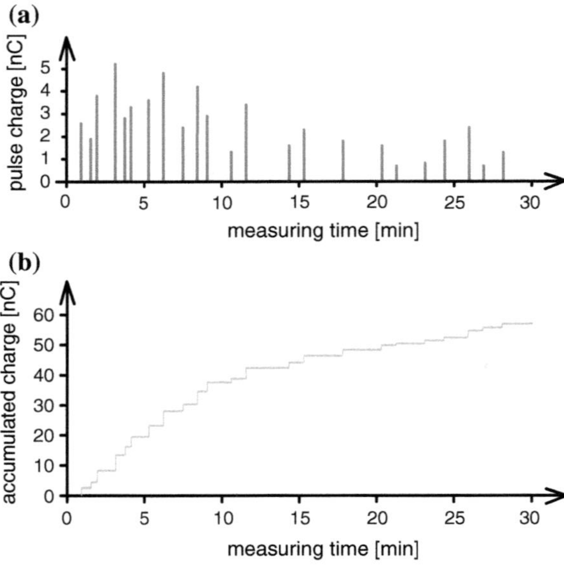

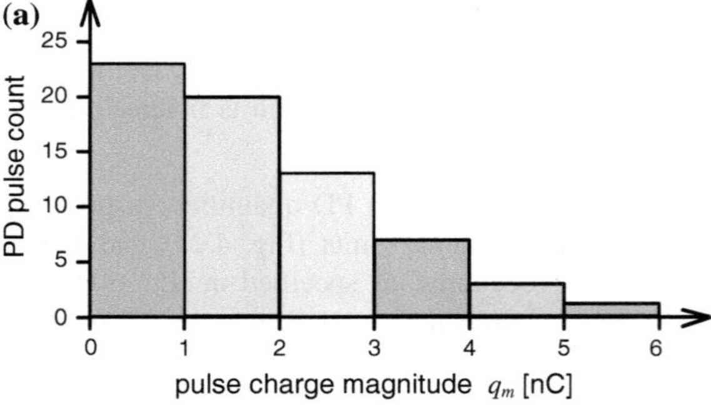

Fig. 6.29 Graphs recommended in the Amendment to IEC 60270:2000 for a statistical analysis of the PD data. **a** Count of PD pulses exceeding the pulse charges 0, 1, 2, 3, 4 and 5 nC. **b** Count of PD pulses, occurring within the pulse charge classes (0–1) nC, (1–2) nC, (2–3) nC, (3–4) nC and (4–5) nC

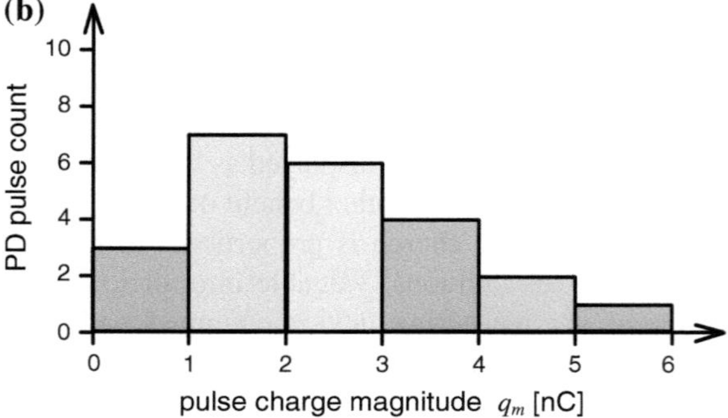

As the PD pulses are often scattering over an extremely wide range, which refers to both the magnitude and the repetition rate, a *statistical analysis* of the significant PD quantities is highly recommended. For this purpose, the PD measuring system should also be equipped with a unit, which enables to count the number of PD pulses exceeding preselected threshold levels (Fig. 6.29). In principle, such a pulse-high analyzer has already been employed earlier by Bartnikas and Levi in 1969. Practical experience revealed that PD tests of HVDC apparatus and their components should cover a minimum time period of 30 min. In this context the following definitions should be re-called, as specified in the Amendment to IEC 60270 (Ed. 3):2000, published in 2013:

- *Accumulated apparent charge qa*

  sum of the apparent charge q of all individual pulses exceeding a specified threshold level, and occurring during a specified time interval Δt_i.

- *PD pulse count m*

  total number of PD pulses, which exceed a specified threshold level within a specified time interval Δt_i.

Moreover it is recommended in the above mentioned Amendment to present the measured PD test results graphically, as exemplarily shown in the Figs. 6.28 and 6.29.

Chapter 7
Tests with High Lightning and Switching Impulse Voltages

Abstract *Lightning impulse (LI) over-voltages* and *switching impulse (SI) over-voltages* are caused by direct or indirect lightning strokes or even by switching operations in electric power systems, respectively. They cause transient stresses to the insulations, much higher than the stresses due to the operational voltages. Therefore, insulations must be designed to withstand LI and SI over-voltages, and the correct design has to be verified by withstand testing using *LI test voltages*, respectively, *SI test voltages*. This chapter deals with the generation of aperiodic and oscillating LI and SI impulse voltages and the requirements for their application in HV tests. Special attention is given to the interactions between the LI/SI generator and the test object. The deviations from the standardized impulse shape, e.g. by an over-shoot on the LI peak, are analysed, and the evaluation of recorded pulses according to IEC 60060-1:2010 and IEEE St. 4 (Draft 2013) is described. This is completed by the description of components and of the procedures for the correct measurement of LI/SI test voltages. Also the measurement of the test currents in LI voltage tests and the PD measurement at SI, LI and VFT test voltages are included.

7.1 Generation of Impulse Test Voltages

7.1.1 Classification of Impulse Test Voltages

A *lightning stroke* may cause—e.g. on a transmission line—a travelling wave of a current pulse with a peak value ranging from few kiloamperes up to about 200 kA (in very rare cases, even up to 300 kA). Investigations of Okabe and Takami on UHV transmission systems (Takami 2007; Okabe and Takami 2009, 2011) considered peak currents up to 300 kA and a front duration in the range between 0.1 and 5 µs for the calculation of LI over-voltages ("external over-voltage") based on the surge impedance of the overhead transmission line, the grounding resistance of a tower and the impedances of the involved components. Figure 7.1 shows the resulting over-voltages for a GIS and a power transformer. Whereas at the GIS, the shape and the peak of the over-voltage change with the front time of the current pulse, the over-voltage at the

© Springer Nature Switzerland AG 2019

W. Hauschild and E. Lemke, *High-Voltage Test and Measuring Techniques*,
https://doi.org/10.1007/978-3-319-97460-6_7

transformer is not influenced by the front time of the current impulse. In both cases, the front time of the over-voltage is in the order of 1 μs, but the over-voltage shows oscillations. IEC 60071-1:2006 and IEC 60060-1:2010 defines all impulse voltages with *LI front times* $T_1 < 20$ μs being LI voltages. Standard LI test voltages are aperiodic impulses. They are characterized by $T_1 = 1.2$ μs and an *LI timetohalf-value* of $T_2 = 50$ μs, abbreviation 1.2/50. In the case of Fig. 7.1 (Okabe and Takami 2011), the LI over-voltages are quite well represented by *standard LI test voltages* 1.2/50.

A switching process "on" or "off" in a power system causes an "internal over-voltage" due to the excitation of internal oscillation circuit(s) formed by the inductances and capacitances of the involved components of the system. The SI over-voltages are also oscillating with one or several frequencies which are remarkably lower than those of LI over-voltages (Fig. 7.2). All impulse voltages with front time $T_1 > 20$ μs are defined in IEC 60060-1:2010 being SI voltages. Shape and parameters of SI test voltages shall not only represent SI over-voltages, but should also be generated with the same test generator as LI test voltages (see Sect. 7.1.2). Their *SI time-to-peak* of about $T_p = 250$ μs should meet a minimum breakdown voltage of non-uniform air gaps with distances of about 5 m (Fig. 7.3, averaged

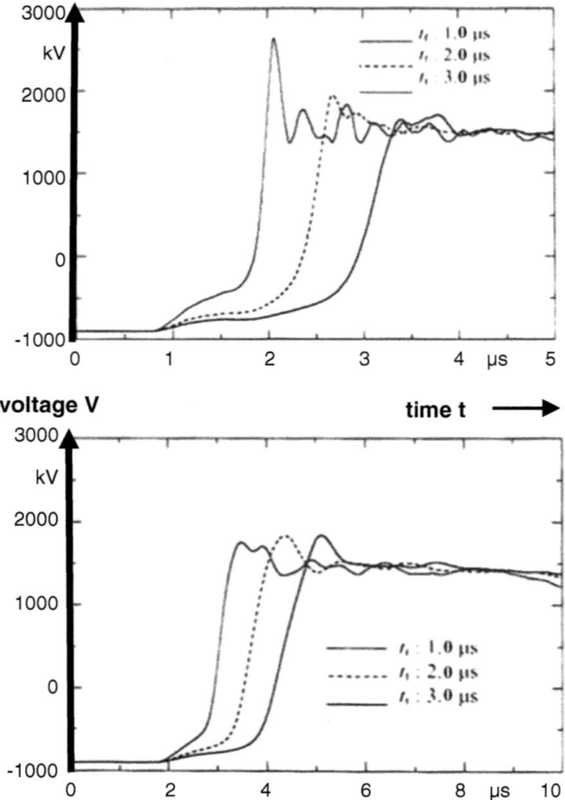

Fig. 7.1 LI over-voltage caused by 200 kA LI current pulse of different front times superimposed on the negative AC voltage peak at a GIS (*above*) and at a power transformer (*below*)

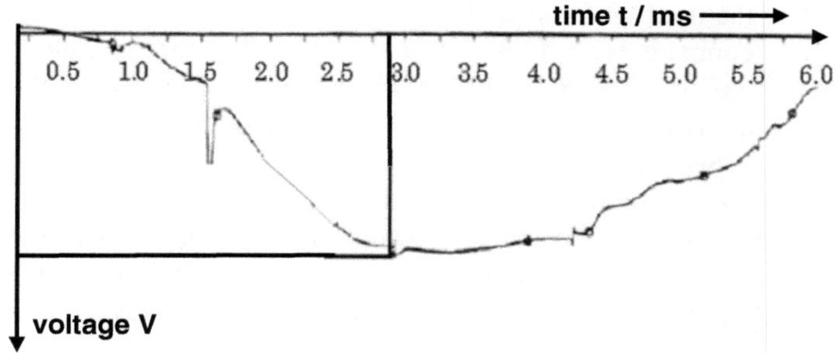

Fig. 7.2 Example for an SI over-voltage

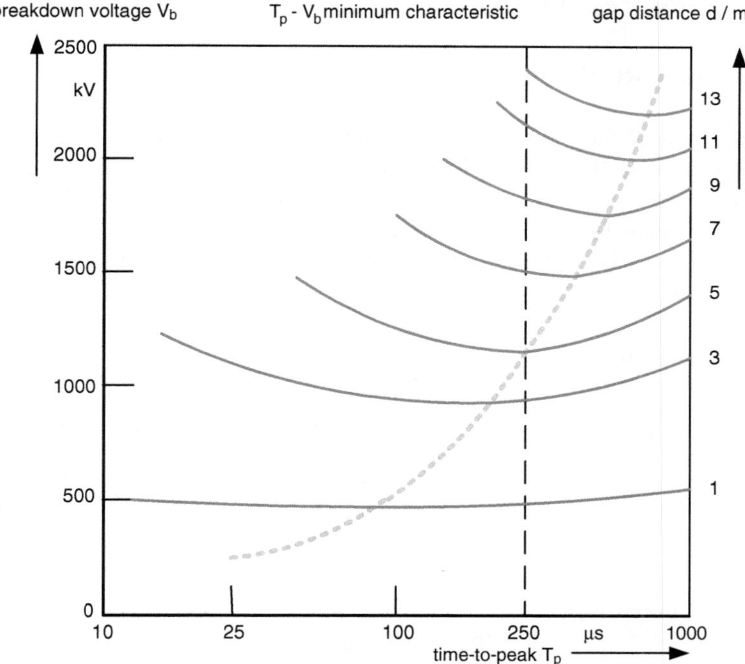

Fig. 7.3 SI breakdown voltage of long air gaps depending on the time-to-peak

characteristics according to Thione 1983). Therefore, *standard SI test voltages* are aperiodic impulses and characterized by T_p = 250 μs and a time to half-value of 2500 μs, abbreviation 250/2500. They shall represent all kinds of SI over-voltages.

For on-site testing, also *oscillating impulse voltages* (*OLI*; *OSI*) are applied (IEC 60060-3:2006). OLI and OSI test voltages can be generated with a generator of efficiency factors about twice of those for LI and SI voltages (see Sect. 7.1.3). Even if this is made for easier transportation and handling of the test system, it should be mentioned that the used OLI and OSI test voltages represent quite well the related over-voltages (compare with Figs. 7.1 and 7.2). Also the so-called "damped alternating voltage" (IEC 60060-3:2006; DAC) used for PD testing of cable systems in the field—is an OSI voltage. OSI voltages can also be generated by test transformers (see Sect. 7.1.4).

Last but not least, it should be mentioned that in case of disconnector switching in SF$_6$-insulated systems (GIS), over-voltages faster than LI over-voltages are generated. They are represented by *fast front test voltages* (*FFV*) and generated by switching the disconnector in the GIS under test (see Sect. 7.1.5).

7.1.2 Basic and Multiplier Circuits for Standard LI/SI Test Voltages

7.1.2.1 Basic RC Circuit

The operation of the basic circuit for impulse voltages shall be explained by its equivalent circuit (Fig. 7.4a): When an *impulse capacitor Ci* is charged via a charging resistor R_c up to the DC breakdown voltage V_0 of the switching gap SG, the impulse voltage V_i is generated by the connected elements (Fig. 7.4b): Then, the load capacitor C_1 is charged via the front resistor R_f which forms the front of the

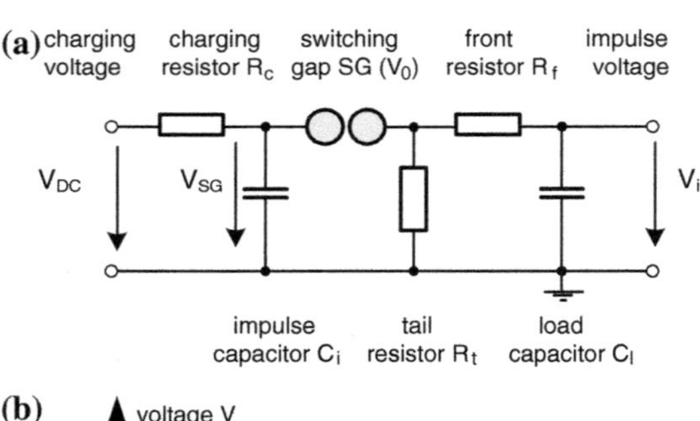

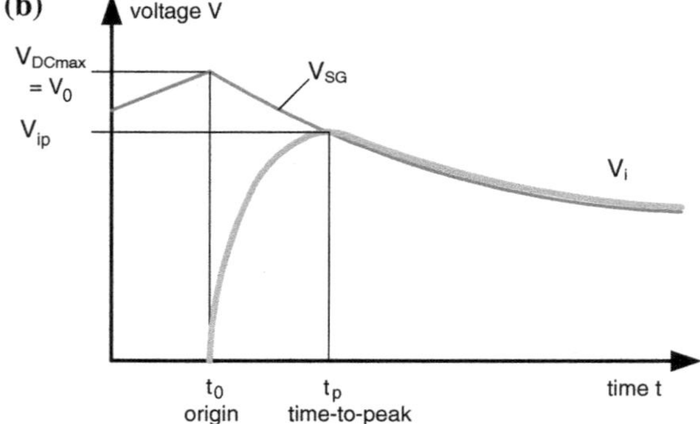

Fig. 7.4 Basic equivalent circuit for impulse voltage generation. **a** Equivalent circuit diagram. **b** Potential diagram

impulse voltage. At the same time, the impulse capacitor C_i is discharged via the tail resistor R_t and forms the tail of the impulse voltage. The superposition of both processes delivers a peak voltage V_{ip} which is lower than the breakdown voltage V_0 of the switching gap. The relation between the two voltages delivers the *efficiency factor* (also: *utilization factor*) of the "one-stage" (basic) impulse generator:

$$\eta = \frac{V_{ip}}{V_0} < 1; \quad \eta = \eta_s \cdot \eta_c. \tag{7.1}$$

The efficiency factor η can be understood as a product of the efficiency η_s depending on the impulse shape and the efficiency η_c depending on the circuit parameters (Hylten-Cavallius 1988). The shape efficiency η_s increases with the relation between tail and front time of the impulse to be generated. When an LI impulse voltage (1.2/50) shall be generated, the relation is, e.g. about 40, when an SI voltage (250/2500) is generated the relation is, e.g. 10 only. The circuit efficiency η_c depends mainly on the relation between impulse and load capacitor. The larger the impulse capacitance C_i in relation to the load capacitance C_1, the higher is the circuit efficiency η_c. The overall LI efficiency factor is relatively high ($\eta \approx 0.85...$ 0.95) and the SI efficiency factor remarkably lower, $\eta \approx 0.70...0.80$ only.

> **Note** In addition of the circuit of Fig. 7.4a, a second basic circuit which has the tail resistor not before, but after the front resistor, is sometimes discussed in textbooks. This circuit has a lower efficiency factor. Therefore, it is not used practically and not discussed here.

The front of the impulse voltage is mainly determined by the time constant τ_f and its tail by τ_t

$$\tau_f = R_f \cdot C_l; \quad \tau_t = R_t \cdot C_i. \tag{7.2}$$

Usually, an impulse voltage generator is equipped with an impulse capacitor C_i and a basic load capacitor C_1 of fixed values. The *front time* can be adjusted by a correctly selected front resistor R_f and the *time-to-half-value* by an appropriate tail resistor R_t. With the fixed values of C_i and C_1, the maximum SI peak voltage is only about 80% of the maximum LI peak voltage.

> **Note** The analytical calculation of the time parameters of impulse voltages and their efficiencies is given in older textbooks. Today, the analytical calculation is replaced by well-adaptable and commercially available software programs. This enables also the more detailed consideration of the characteristics of the test object and the stray capacitances in the test room and delivers more precise results.

The selection of the impulse capacitor C_i determines—together with the maximum charging voltage V_{0max} also the *impulse energy* of the generator:

$$W_i = \frac{1}{2} C_i \cdot V_{0max}^2. \tag{7.3}$$

Whereas the maximum charging voltage of a generator depends on the required test voltages, the impulse capacitance must be selected according to the expected total load (basic generator load plus test object load), to guaranty $C_i \gg C_1$.

7.1.2.2   Multiplier RC Circuit

The basic circuit (Fig. 7.4a) is usually applied for students training and demonstrations with voltages below 200 kV. For higher voltages, multiplier circuits proposed by E. Marx in 1923 are applied (Fig. 7.5a, without the red short-circuit bars).

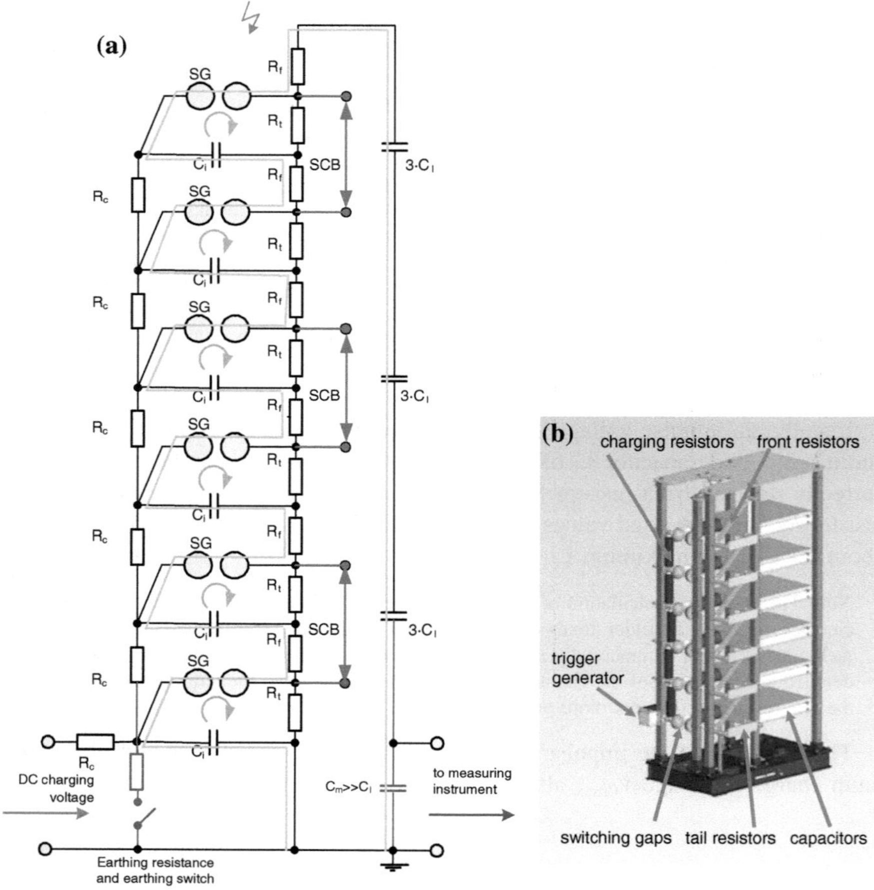

Fig. 7.5 Multi-stage impulse generator of n = 6 stages. **a** Multiplier circuit (explanations in the text). **b** Impulse generator

The impulse capacitors C_i of all n stages are charged via the charging resistors R_c which are connected in series to one column. When the charging resistors are dimensioned correctly, it does not play any role that the charging resistance of the highest stage is n-times larger than that of the lowest one, because the charging time is selected long enough that all impulse capacitors are equally charged. Today, a thyristor-controlled charging with a constant current up to a pre-selected voltage V_0 is used, at which the switching gap is triggered for breakdown. Now, the impulse capacitors start to discharge via the tail resistors on each stage (Fig. 7.5a: blue path). At the same time, the external load capacitance C_l (basic load of a capacitive divider plus stray capacitances of the generator to ground plus test object) is charged from the series connection of all impulse capacitors and front resistors (green path). An impulse voltage generator of n stages (Fig. 7.5b) charged with a DC voltage V_0 delivers with the efficiency factor η the output impulse voltage

$$V_{in} = n \cdot \eta \cdot V_0 \tag{7.4}$$

The term $V_{0n\ max} = n \cdot V_{0max}$ is called the *cumulative charging voltage* of the generator and usually used as the rated voltage of the impulse test system because $V_{0n\ max} > V_{in\ max}$ one has to be careful with the valuation of rated voltages for impulse test systems. It is always necessary to know the efficiency factor for all impulse voltage shapes of interest for the calculation of the related output voltages additionally.

For calculation of its circuit elements, a multi-stage generator (n stages, elements R_f; R_t, C_i; C_l) is usually transferred into an equivalent basic circuit with the elements

$$\begin{aligned}
R_f^* &= n \cdot R_f \\
R_t^* &= n \cdot R_t \\
C_i^* &= C_i/n \\
C_l^* &= C_l.
\end{aligned} \tag{7.5}$$

After the calculation of the circuit elements of the basic circuits, the above Eq. (7.5) are used for the determination of the multi-stage generator by re-transformation. The thermal design of the resistors—especially of the front resistors—determines the allowable impulse voltage repetition rate. The resistors are heated by the impulse current, which is flowing in case of the impulse voltage generation and should sufficiently cool down until the next impulse appears. A defined maximum temperature of the resistors must not be exceeded.

The controlled safe *triggering* characterizes a generator of high quality. Usually, only the lowest stage is equipped with a so-called "trigatron", a three-electrode arrangement (Fig. 7.6a). A small, battery-operated trigger device generates a voltage pulse of several kilovolts which causes a small trigger discharge at a pilot gap. This discharge triggers the breakdown of the main gap of the lowest stage. The trigger discharge delivers charge carriers and photons for the immediate, fast breakdown process, if the field strength in the main gap is high enough. This requires a certain minimum voltage, the so-called lower trigger limit (Fig. 7.6b). If the voltage at the trigger gap is too high, a breakdown is caused without triggering. This self-ignition delivers the upper trigger limit of the charging voltage. The *trigger range* between the two limits (Fig. 7.6b) should be as wide as possible. Usually, its width is between 5 and 20% of the withstand voltage of the non-triggered gap (upper curve). It depends on the design of the trigatron, the energy of the trigger discharge and the height of the DC voltage at the main gap.

The charging voltage and the trigger instant must be well controlled to guarantee safe triggering of the whole generator and to avoid "no-triggering" or self-ignition without triggering. As soon as the lowest switching gap breaks down, an over-voltage appears at the second stage, runs as a travelling wave through the generator (Pedersen 1967) and shall cause the breakdowns of all further gaps.

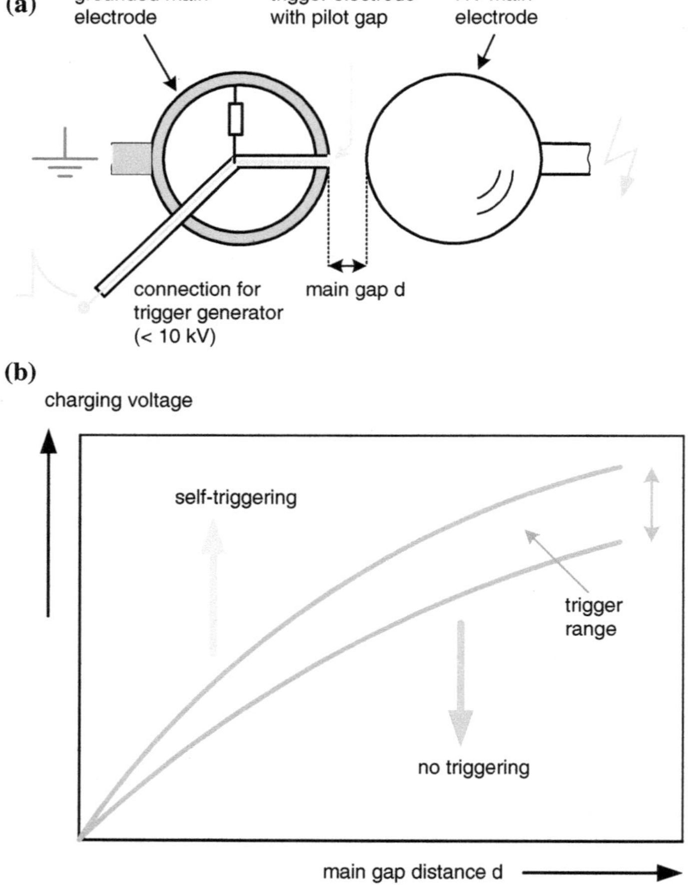

Fig. 7.6 Triggering of impulse voltage generators. **a** Trigatron spark gap. **b** Principle of the trigger range

The over-voltages must remain high enough to cause all necessary breakdowns. This depends on the impulse shape to be generated (e.g. damping front resistors) and on stray capacitances to ground which increase the over-voltages, whereas longitudinal stray capacitances reduce them (Rodewald 1969a, b). Based on such investigations, additional trigger measures (e.g. supporting gaps and ignition capacitors) have been introduced to maintain the height of over-voltages also for huge generators (Rodewald 1971; Feser 1973, 1974). Generators with symmetric charging (Sect. 7.1.2.4; Fig. 7.7) allow a save triggering without these additional measures (Schrader 1971).

The modular design of multi-stage generators is helpful for later extension to higher voltages by additional stages. It enables also the parallel connection of stages for higher impulse energy at lower voltages [Fig. 7.5a, red short-circuit bars (SCB)] as they are, e.g. required for testing the low-voltage winding of power transformers or medium-voltage capacitors. Also impulse test currents can be generated by impulse voltage generators with parallel stages.

7.1.2.3   Consideration of the Inductance in the Circuit

Till to this Subsection, all explanations have not considered the inductance in the impulse test circuit which cannot be avoided. The *inductance* is not a property of a conductor carrying a current, but of the magnetic field besides of it: It is a property of the magnetic flux generated by the electric current in a closed circuit and influenced by the geometries of the conductor(s) and of the considered circuit as well as of the positions of the circuit elements to each other (Rodewald 2017). For practical application the inductance is represented by a "frequency-dependent

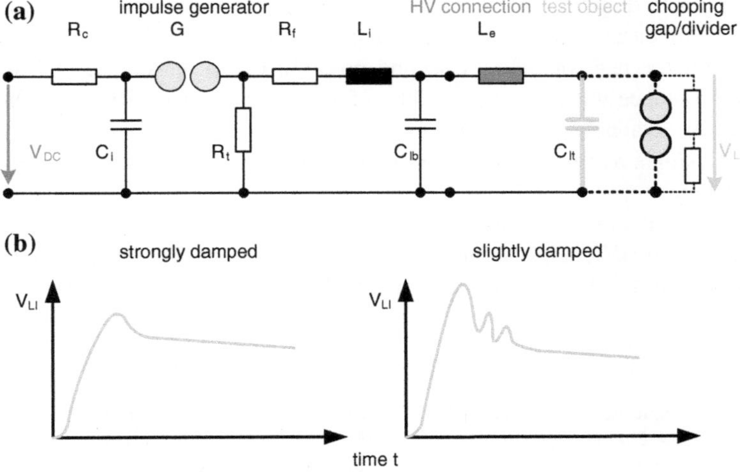

Fig. 7.7 LI voltage generation circuit considering the inductance **a** Equivalent circuit diagram with inductances. **b** Over-shoot superimposed on LI voltages (schematically)

resistor", usually simply called "inductance" (symbol L). In an equivalent circuit diagram, these inductances L are switched in series to elements carrying the current and generating the magnetic flux according to the conditions of the circuit (e.g. the HV lead from a generator to the test object and the ground return). The strongest influence on the inductance is caused by the magnetic field in the vicinity of the conductor. Therefore, e.g. the inductance of HV leads of larger length (l $\geq$ 1 m, distance to ground >1 m) can be estimated from their cross sections and lengths (Table 7.1).

Inductances form oscillating circuits with the capacitances and cause damped oscillations superimposed on the aperiodic pulses. The damping depends on the front resistor. This is several 100′ Ω for SI voltages and suppresses the oscillations completely. The more or less damped oscillations and the "*over-shoot*" (only less than one period of the oscillation) are found at LI voltages only, because the LI generator is equipped with front resistances of few 10′ Ω (Fig. 7.7). There are internal inductances of the generator and external of the test object and its connections.

Internal inductances L_i are those of the capacitors, the resistors and the connections between them. For estimations, the inductance for 1 m of the loop (e.g. green path in Fig. 7.5a) is about 1 μH. The reduction in the internal inductance of a generator requires its compact design with a loop as short as possible. A good generator should have an internal inductance of $L_i < 4$ μH per stage. Usually, the user cannot influence the internal inductance of the generator easily. When only a part of the stages is sufficient to generate the necessary voltage (so-called "part operation"), the loop should be short and should exclude the not-used part of the generator and of the basic load (voltage divider). For very old generators, one should check the inductance of the front resistors: The front resistors must be designed with low inductance, which can be reached by a bifilar winding. This means that two isolated, close together arranged wires are wound on a fibreglass tube in opposite directions. The magnetic fields of the wires have opposite directions and compensate each other to a remaining inductance which corresponds to the length of the resistor tube. A second possibility is a resistor band where the insulated resistance wires are woven into a fabric as a meander. Resistor bands are commercially available. The inductance of resistors can also be reduced when, instead of a single resistor, two or more parallel resistors are applied resulting in the same resistance.

External inductances L_e are those of the test object (even if this mainly a capacitance), the HV lead to the test object and the voltage divider as well as those of the earth return. HV lead and earth return shall be especially very short and can

Table 7.1 Inductances to be assumed for HV leads

Length of HV lead (m)	Single wire, $d = 2$ mm (μH/m)	Metal foil, $w = 10$ cm (μH/m)	Metal tube, $d = 10$ cm (μH/m)	Metal foil, $w = 50$ cm or 2 foils with spacer in between (μH/m)
1	1.37	0.70	0.59	0.40
10	1.83	1.26	0.96	0.84

often be influenced. With increasing LI test voltage, the distances between generator and test object become longer and oscillations and over-shoot cannot be controlled in testing UHV equipment (see Sect. 7.3). Up to a certain degree, also the inductance of the circuit can be reduced by an appropriate selection of the geometry of the HV lead. Table 7.1 gives some inductances depending on the shape and the length of the connection. Never a thin wire should be used for the HV lead or the ground return, because its inductance is higher than those of copper foil of a width $w \geq 10$ cm or metallic tube of a diameter $d \geq 10$ cm. A further reduction can be reached with a wider foil or two parallel foils and spacers with a distance d in between. Also quite useful is the application of the mentioned resistor bands as HV lead and external damping resistor to the test object. To maintain the impulse shape, the internal front resistor must be reduced, but the external resistor increases the damping efficiency including the efficiency factor.

Over-shoot compensations can be designed as low-pass filters of L/C/R combinations (serial or parallel compensation unit) arranged inside the generator or outside as separate components: Fig. 7.8 shows the principles of the two compensation units.

The series compensation unit (Fig. 7.8b; Wolf and Voigt 1997) prevents the penetration of higher-frequency contributions to the load capacitance which includes the test object. The series connection of compensation resistance R_c and compensation inductance L_c must be adjusted to that of the front resistor R_f and internal inductance L_i. Also the compensation capacitor C_c has to be related to the load capacitance C_l. The necessary adjustment covers a certain range of load cases,

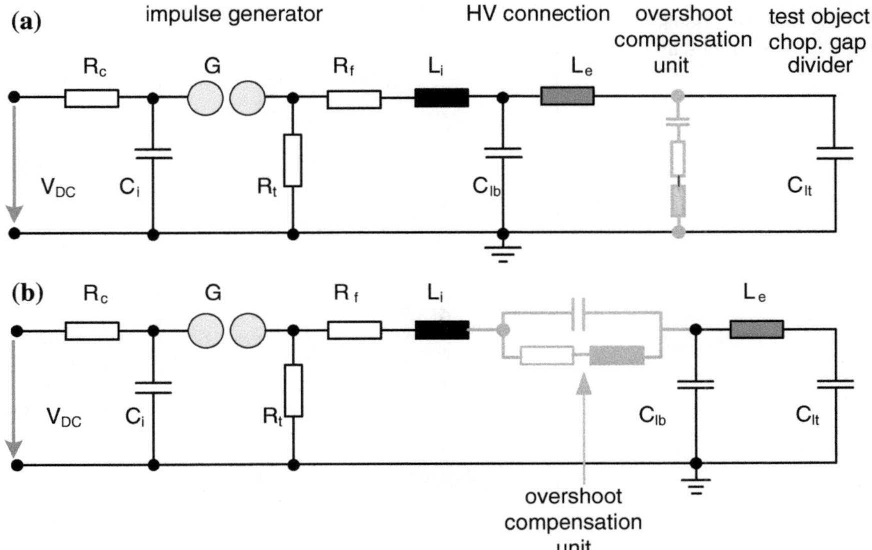

Fig. 7.8 Equivalent circuit diagrams of over-shoot compensation units. **a** Parallel compensation unit. **b** Series compensation unit

but if fine tuning is required, the compensation unit must be adapted. For larger impulse generators, the series compensation unit can be distributed to the different stages of the generator (with elements of the stage voltage, e.g. 200 kV) and without components of high-rated voltage (e.g. 3000 kV).

The *parallel compensation unit* (Fig. 7.8a, Schrader 2000; Hinow and Steiner 2009; Hinow 2011) is always a separate unit which might be combined with a chopping gap and a voltage divider to one compact unit (Fig. 7.9). Its adjustment has to consider the natural frequency range caused by internal and external inductances. The efficiency of the two principles is about the same. It seems that especially for LI testing of UHV equipment the handling of compensation units becomes too time-consuming and does not fit to the operation of an industrial test field. The problem can be easily solved by increased damping due to larger front resistors, but this requires larger tolerances for the front time of LI impulses (for details see Sect. 7.2.1)

7.1.2.4 Some Details of the Design of Impulse Voltage Test Systems

The *LI/SI voltage test system* (Fig. 7.10) includes the HV circuit consisting of the HV generator which is optionally completed by an over-shoot compensation unit, an HV *chopping gap* and a measuring system including an HV LI/SI divider (see Sect. 7.5). The test object is also a part of the HV circuit, but later considered under

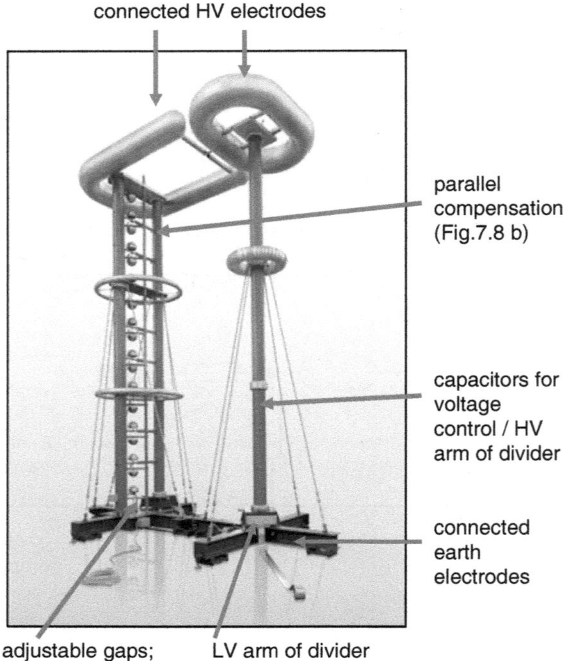

Fig. 7.9 Parallel compensation unit in combination with voltage divider and chopping gap

Sect. 7.3. Furthermore, it includes *the* control and measuring system, *the* switching cubicle with the thyristor controller and the DC voltage generator. In the following, some characteristics of the main elements are given. For the generator, see the above explanations.

Generator with symmetric charging: For larger impulse voltage generators, the charging voltage per stage is usually 200 kV. Because of the limited rated voltage of capacitors, usually two 100-kV capacitors are connected in series for 200 kV. To guarantee identical charging of both capacitors of a stage, potential resistors R_p must be arranged at the connection point of both capacitors (Figs. 7.11a and 7.12a). With a special circuit patented by Schrader (1971), a symmetric charging with ±100 kV is applicable (Figs. 7.11b and 7.12b). This requires a charging unit with symmetric output ±100 kV.

For each polarity, a separate column of charging resistors R_c, but no potential resistors R_p are required. There are some advantages of the symmetric charging for larger impulse test systems with two capacitors in series per stage: In the first line, a stage with a short HV loop of low inductance can be designed (Fig. 7.12d). The safe triggering of large generators with symmetric charging does not require the additional measures mentioned above. As there are no voltage-dependent circuit elements, the impulse shape (described by the time parameters) is independent from the peak voltage (Fig. 7.13). Also the parallel connection of stages for the generation of impulses with a higher energy is very simple.

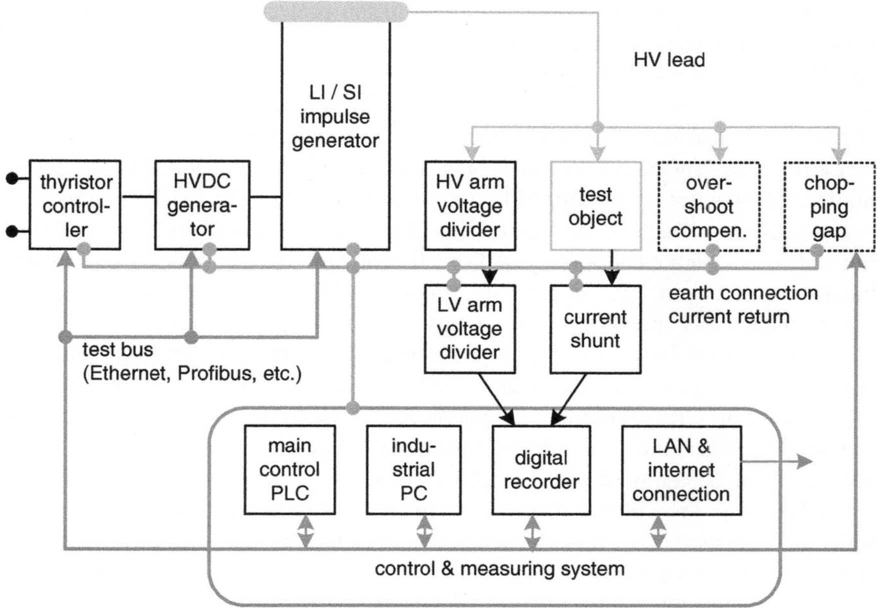

Fig. 7.10 Components of an LI/SI test voltage system

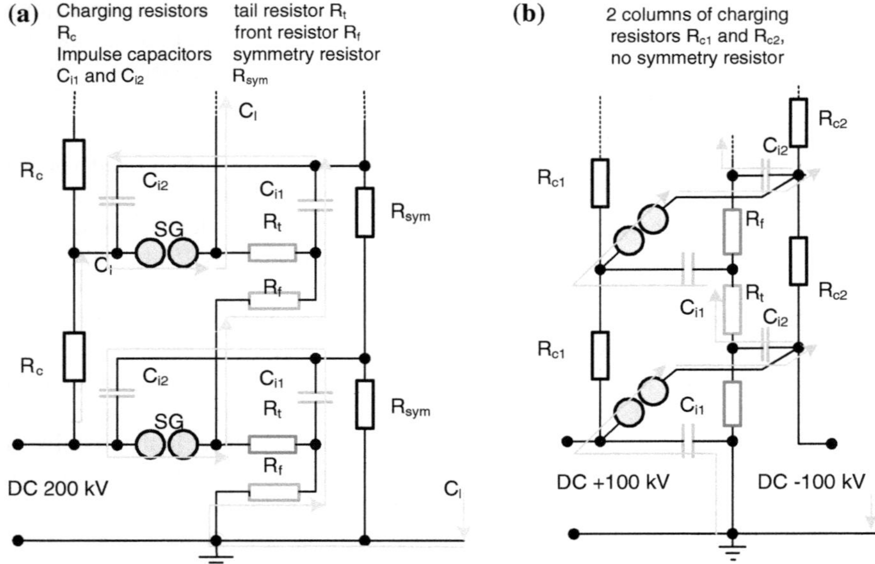

Fig. 7.11 Typical circuits for impulse voltage generators. **a** Unipolar charging (e.g. 200 kV). **b** Symmetric charging (e.g. ±100 kV)

Chopped lightning impulse (LIC) voltages and chopping gap: An external over-voltage in the power system is limited to the protection level by a lightning arrester. This means the over-voltage is chopped and collapses to this protection level. The duration of the voltage collapse is very short, its steepness very high. Such steepness causes very non-linear stresses in equipment with windings (power, distribution and instrument transformers, reactors, rotating machines). The mainly stressed insulation at the HV terminals of the equipment must be designed accordingly and verified by a test with chopped lightning impulse (LIC; Fig. 7.14a) voltages.

The LIC test voltage is generated as an LI test voltage described above and then chopped by a separate chopping gap. For LIC voltages up to about 600 kV, a usual sphere-to-sphere gap can be used; for higher voltages, *multiple chopping gaps* become technically mandatory (Fig. 7.14b, c). The voltage collapse of a multiple spark gap is much faster than that of a single large sphere gap. The chopping gap consists of in-series-connected sphere-to-sphere gaps, usually one gap for one stage of the generator. One sphere of each gap is fixed and arranged at a fixed insulating column. The other one—on suitable insulating support—is moveable by a motor drive and can be adjusted for the relevant voltage value. The parallel capacitor column controls the voltage distribution linearly (Rodewald 1972). This column might also be used as a damped capacitive voltage divider, which is usually a separate component (see Sect. 7.5). The instant of the chopping can be triggered as described above for the generator (Fig. 7.6). Also the combination with an over-shoot compensation unit is applied (Fig. 7.9).

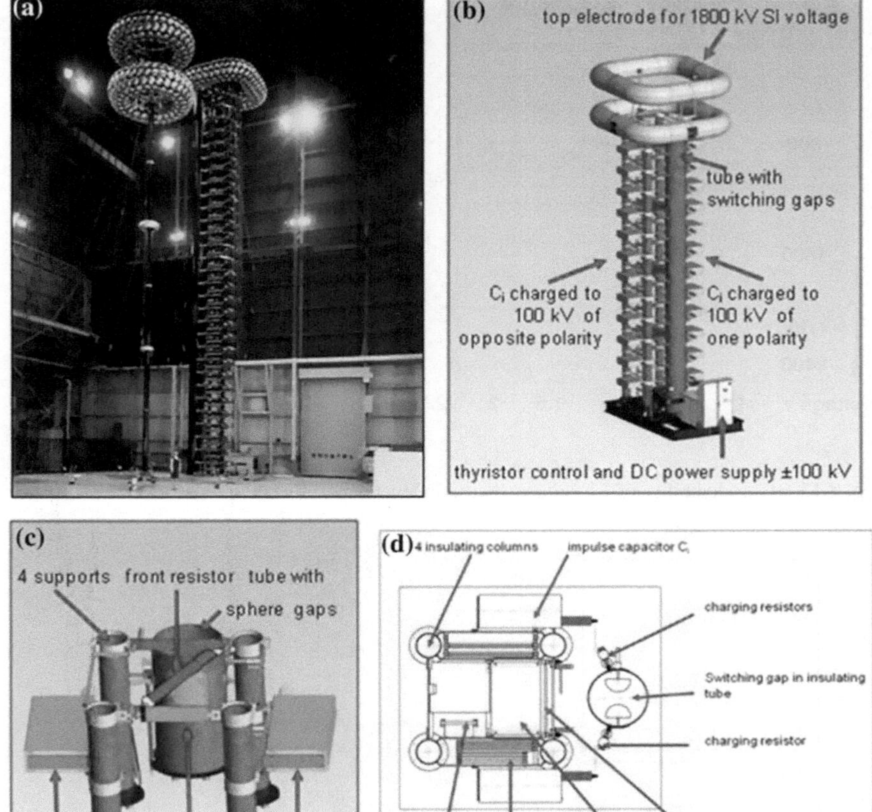

Fig. 7.12 Impulse voltage generators for unipolar and symmetric charging. **a** Generator with unipolar charging (Courtesy of Haefely, Basel). **b** Generator with symmetric charging. **c** One-stage of a multi-stage generator (inside view). **d** Cross section of a generator with symmetric charging

Electrodes for the HV components: An impulse voltage generator and the other HV components require a sufficient clearance D from grounded or energized objects in an HV test laboratory (Fig. 7.15a) to avoid breakdowns of the air gap between the HV circuit and the surroundings. The necessary clearance depends on the kind of pre-discharges which determine the breakdown process. The optimum design of the electrodes of the HV components enables not only the correct operation of an LI/SI test system but ensures also their minimum space in the test laboratory.

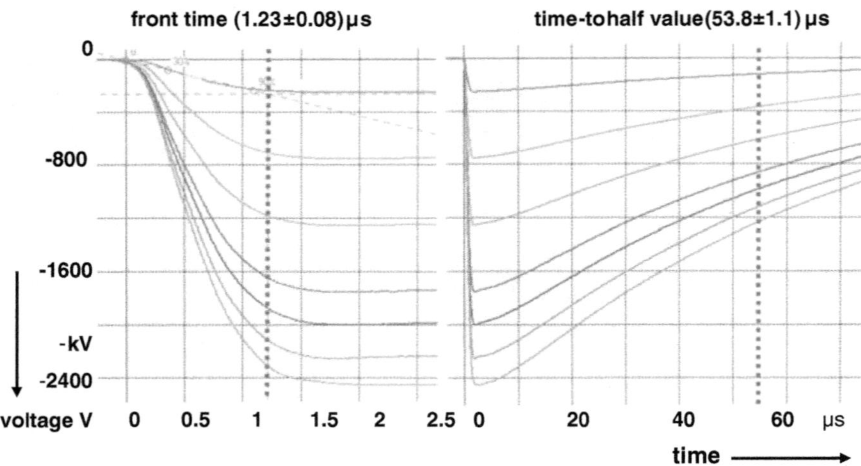

Fig. 7.13 Reproducibility of LI voltage shapes independent on peak voltage value

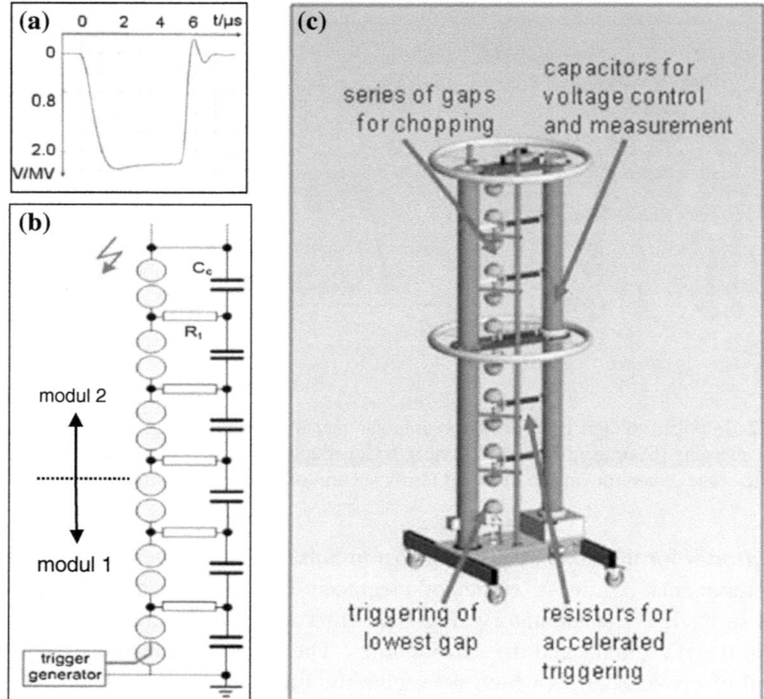

Fig. 7.14 Chopped lightning impulse generation. **a** Chopped lightning impulse (LIC) voltage. **b** Circuit of two modules of a multiple chopping gap. **c** Multiple chopping gap for 1200 kV (six single gaps)

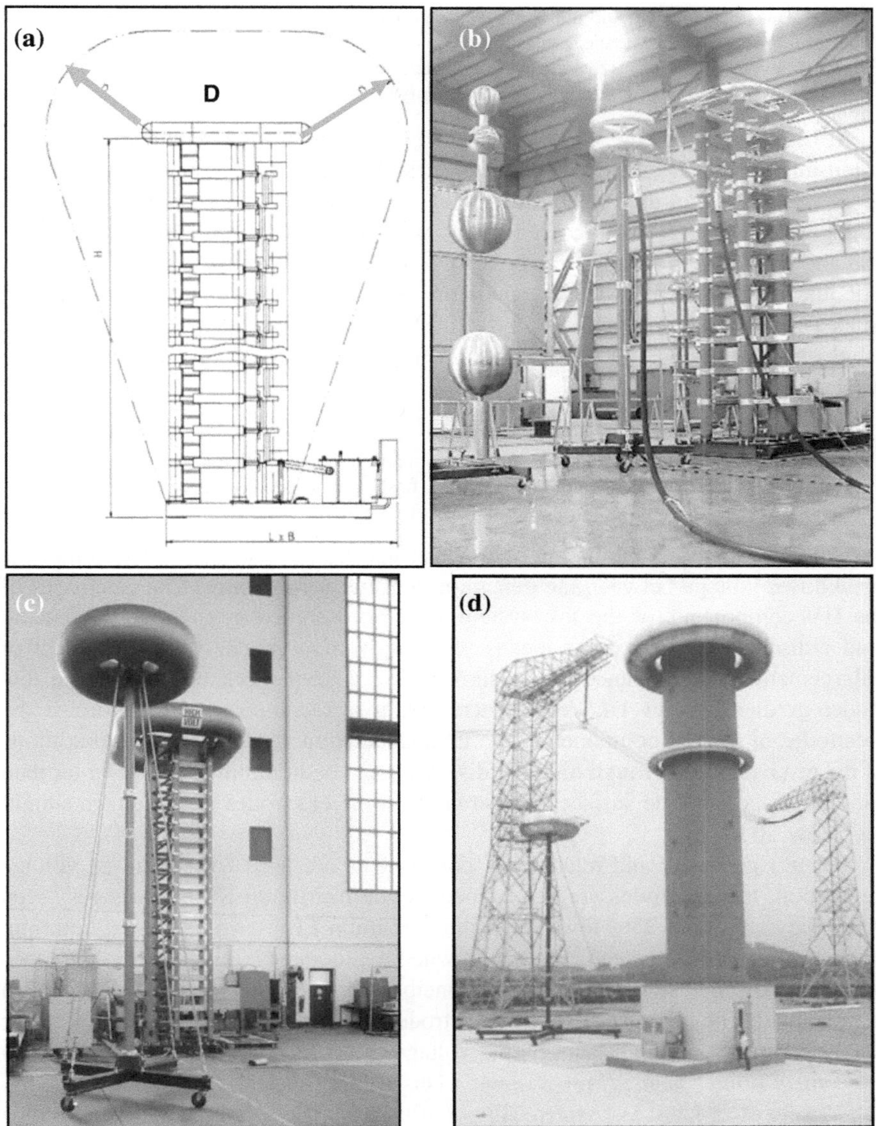

Fig. 7.15 Electrodes for the HV components of LI/SI voltage test systems. **a** Necessary clearance D around a LI/SI generator. **b** Test system only for LI voltage generation (2000 kV), **c** Test system 1800 kV for LI/SI voltage generation at limited clearance to the ceiling. **d** Outdoor test system 4200 kV for SI and LI voltage generation. Courtesy of KEPRI, Korea

Note This clearance should not be mixed up with the clearances for the test object according to Fig. 2.1. The clearances there consider that the voltage distribution at the test object is not influenced by the surroundings. Here, the operation of the generator shall not be disturbed by undesirable discharges or even breakdowns.

At LI test voltages, the *streamer discharge* determines the breakdown voltage of a non-uniform electric field in air. The electric field of an LI generator in an HV test room is such a non-uniform electric field. Consequently, the specific breakdown voltage is equal to the voltage demand of streamers of about 5 kV/cm for positive and about 10 kV/cm for negative polarity of the electrode with the larger curvature. This means that for a 3 MV LI generator, the minimum clearance should be 6 m (plus a certain safety margin of—say—20%). The curvature of the electrodes can be relatively small (Fig. 7.15b), because there is no need to avoid the streamer discharges.

Note The strong polarity effect is typical for streamer discharges in air. There is no remarkable polarity effect for internal insulation. If internal insulation, e.g. of a power transformer, shall be tested with LI voltage, a flashover of the air-part of the bushing is avoided when the test is performed at negative LI voltage.

At SI voltages, a combination of streamer and *leader discharges* determines the breakdown voltage between the generator and the surroundings. The electrodes of the HV components of the test system shall be designed in such a way that no leader discharge appears, this means with larger radii (Fig. 7.15c). The effect of enlargement of the distance to the surrounding is very week because of the low leader gradient (about 1 kV/cm). Therefore, it is recommended to optimize the electrodes of the HV components by a field calculation with the realistic conditions of the test room. As a rough hint, the distance must be in minimum 20% larger than for LI voltage, and the surface field strength of the electrodes at SI voltage should be below 20 kV/cm.

When a generator and related HV components are used for LI and SI voltage generation, the electrodes are determined by the maximum SI test voltages, even when they are about 25% lower than the maximum LI test voltages. An optimum utilization of a test area can be reached when a generator is moveable in the laboratory, e.g. by air cushions. The design principles for outdoor generators (Fig. 7.15d) are identical. They require also large electrodes for SI voltage generation, but under rainy conditions, the maximum output voltages must be remarkably reduced.

Control and measuring system: This—today usually computer-aided—sub-system of an LI/SI test system (Fig. 7.10; see also Sect. 2.2) of an LI/SI voltage test system enables the adjustment of the generator for the test voltage value and a certain test procedure (see Sect. 7.4), the measurement of LI/SI voltages and of related impulse currents (see Sects. 7.5 and 7.6). It is available for one, two or all three following modes:

1. *Manual operation* with measurement and evaluation of LI/SI parameters: The operator has to control the test system including adjustment of the voltages and duration of the breaks between impulses, and the evaluation and presentation of the test result (test record). The charging and triggering process must be

controlled. When control and measuring components are not connected to one system, this has to be done by the operator. This traditional mode is very seldom applied for industrial testing and research work, but applied for e.g. student's training.

2. *Computer-supported operation* and test result presentation: The test is performed manually, but the precise adjustment of test voltages—this means that of the distance of the switching gap as well as that of the charging DC voltage— and the test data presentation are overtaken by the system. Control and measuring components are connected. This mode is applied for larger and expensive test objects in industrial testing and for research work.

3. *Automatic testing* according to a pre-given test procedure by a computer control: The PC software for the test procedure is configured by the operator before, the HV test itself is performed, evaluated and presented automatically. Intervention of the operator is not necessary, but the test can be interrupted or terminated at any time by the operator. This mode is applicable for testing of very similar or even identical test objects in a larger scale or for statistical investigations in research work.

The control system delivers the commands for switching the breakers on and off, for adjusting the switching gaps of the generator for the pre-selected voltage and for the appropriate charging voltage adjusted by a thyristor controller. Based on the voltage measurement, the computer control checks that the voltage values are within the pre-given sequence and tolerances. Based on the evaluation of the voltage shape, breakdowns are recorded for the evaluation of the test. Also the evaluation of the related currents might indicate whether a test has been successful or has failed. The style of the *test record* depends fully from the intention of the user.

Switching cubicle *and DC rectifier unit*: A LI/SI test system has a relatively low power demand of some 10 kW. The switching cubicle contains the power switch and the operation switch, the instrument transformers for supply voltage and current measurement and protective equipment. The built-in thyristor controller enables a constant-charging current output of the connected rectifier unit. This DC rectifier unit is usually a doubler circuit (see Sect. 6.1.2 and Fig. 6.3) or for symmetric charging, a half-wave rectifier with symmetric output (modification of Fig. 6.2). Depending on the rated power and energy of the impulse generator, the charging voltage corresponds to the stage voltages (100–200 kV) and the charging currents are between few 10 mA and some 100 mA. The duration of the charging which determines the impulse voltage repetition rate depends on the total energy of the generator and is usually between 10 and 60 s. For special application, also faster charging processes and higher repetition rates can be realized.

7.1.3   Circuits for Oscillating Impulse Voltages

For factory testing with aperiodic LI voltages according to IEC 60060-1:2010 or IEEE St.4:1995, the inductances in the circuit are disturbing elements, but a defined inductance L_s in the circuit establishes an oscillating circuit of this series inductance and the load capacitances C_l. This circuit is excited by the triggered discharging of the impulse capacitors of the generator (Fig. 7.16a). The output voltage is a damped oscillation around the discharge curve of the impulse capacitances of the generator. The oscillating frequency is the natural frequency

$$f_0 = \frac{1}{2\pi \cdot \sqrt{L_s \cdot \frac{C_l \cdot C_i^*}{C_l + C_i^*}}}. \tag{7.6}$$

For a generator with n stages, one has to apply $C_i^* = C_i/n$ and $R_t^* = n \cdot R_t$ (Eq. 7.5). The total load $C_1 = C_{lb} + C_{lt}$ is the sum of the basic load and the test object load. The fixed series inductance L_s replaces the front (damping) resistors. According to IEC 60060-3:2006, impulse voltages with oscillations $f_0 > 15$ kHz are considered as "*oscillating lightning impulse (OLI) voltages*" (Fig. 7.16a), such with $f_0 < 15$ kHz as "*oscillating switching impulse (OSI) voltages*" (Fig. 7.17a). The damping is determined by the losses in the circuit, for pure capacitive test objects mainly by the tail resistors R_t of the aperiodic impulse. As these are higher for OSI than for OLI voltages, OSI voltages show not only a lower frequency, but also a larger damping (Fig. 7.17a).

Theoretically, the oscillating impulse voltage (OLI or OSI) can reach a peak value which is twice the peak value of the relevant aperiodic impulse (LI or SI) voltage. In practice, it reaches about 90% of that value. The efficiency factors are

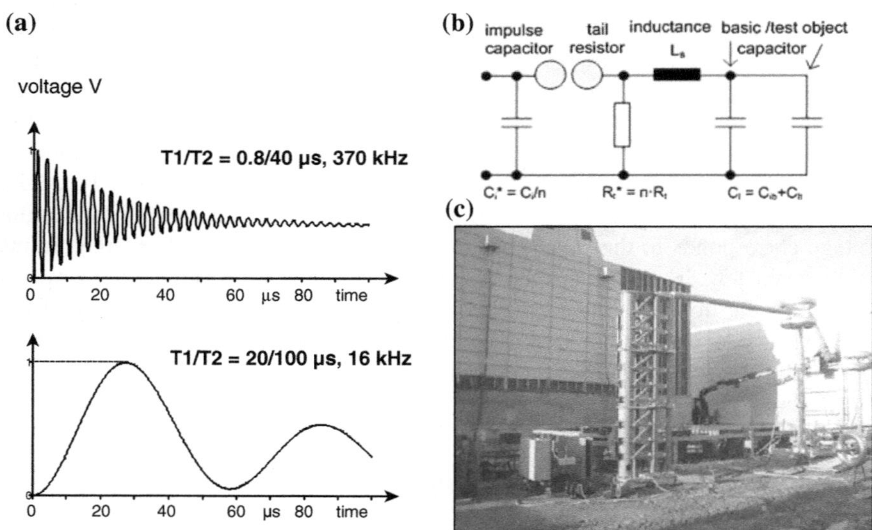

Fig. 7.16 Oscillating lightning impulse voltages (OLI). **a** OLI test voltages. **b** Equivalent circuit for oscillating impulse voltage generation. **c** 900 kV impulse test system for 850 kV LI and 1600 kV OLI voltages. Courtesy of Siemens Berlin

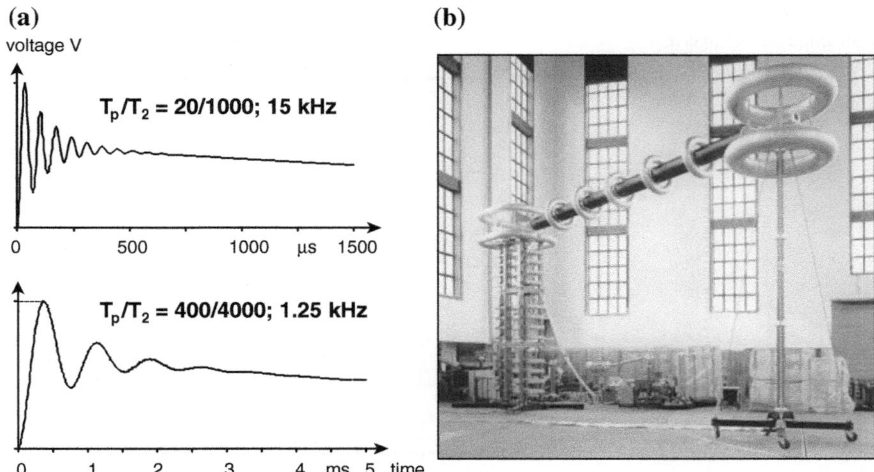

Fig. 7.17 Oscillating switching impulse voltages (OSI). **a** OSI test voltages. **b** 1200 kV impulse test for 900 kV SI and 1600 kV OSI voltages

$$\eta_{\mathrm{OLI}} = \frac{V_{\mathrm{OLI}}}{V_{0\Sigma}} \approx 1.7\ldots1.8 \quad \text{and} \quad \eta_{\mathrm{OSI}} = \frac{V_{\mathrm{OSI}}}{V_{0\Sigma}} \approx 1.3\ldots1.4. \tag{7.7}$$

As an example, Fig. 7.18 shows the remarkable influence of the test object (load) capacitance on the efficiency factor and the time-to-peak. The high-efficiency factors compared with those of the aperiodic impulse voltages are especially important when mobile impulse test systems are required for the testing in the field. Therefore, OLI and OSI voltages have been proposed for on-site testing (Kind 1974; Feser 1981), and meanwhile, they are standardized in IEC 60060-3:2006. For more details, see Sect. 10.2.1.

When the generator has to be designed for a maximum cumulative charging voltage $V_{0\Sigma\mathrm{max}}$, the basic load capacitance and also the series inductance must be able to withstand the maximum oscillating impulse voltage which is much higher than the $V_{0\Sigma\mathrm{max}}$ (Eq. 7.7). The insulation design of the basic load capacitance for OLI and OSI voltages is practically identical, whereas that of the series inductance is very different (compare Figs. 7.16c with 7.17b). For OLI voltages, a low inductance is required which can be made easily. Contrary to that the OSI generation requires a much higher inductance. Now, stray capacitances must be taken into consideration which would cause a non-linear voltage distribution along the coil. To avoid that, a longitudinal voltage control by toroid electrodes is necessary. The coil for OSI voltage is much longer, thicker and heavier than the one for OLI voltage. Furthermore, it has been found that the benefit of OSI testing is low; therefore, mainly OLI testing is applied (see Sect. 10.3.1).

It should be mentioned that also bipolar oscillating impulse voltages can be generated based on impulse voltage circuits (Schuler and Liptak 1980). They

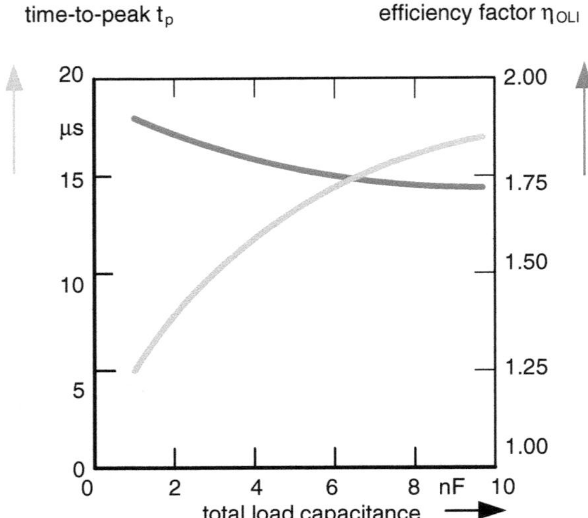

Fig. 7.18 OLI characteristic of an impulse voltage test system (250 kV/5 kJ)

arranged the inductance in parallel to the load capacitance and applied the bipolar OLI voltage for testing of rotating machines.

A special case of a bipolar OSI testing is applied for medium-voltage cables: The cable is charged with a DC voltage and then discharged via a suited switch (trigatron or semiconductor HV switch) in series with an inductance L_s and possibly a resistor R_d. The discharge causes a damped oscillation (Fig. 7.19). Under the term *"damped alternating voltage"* (*DAC*) (IEC 60060-3:2006), this bipolarly oscillating voltage is successfully used for *diagnostic PD measurements* on medium-voltage cable systems, but the whole test stress for the cable is a long DC ramp (duration in the order between 1 and 100 s) followed by the much shorter DAC voltage (duration in the order of few 100 ms). The duration of charging and the test frequency depends on the capacitance (length) of the cable, whereas the damping of the oscillation depends on the losses in the circuit. Therefore, the whole stress cannot be reproduced from cable test to cable tests. Occasionally, the voltage is used for withstand tests (Fig. 7.19c), but this cannot be recommended (for more details on DAC voltages, see Sect. 10.2.2.2).

7.1.4   OSI Test Voltage Generation by Transformers

When an *HV test transformer* is excited by controlled discharging a capacitor *bank* into its low-voltage (LV) side, this impulse causes an oscillation, which is transformed to the HV side according to the transformer ratio (Kind and Salge 1965; Mosch 1969). The schematic circuit diagram (Fig. 7.20a) is transferred with the transformer ratio into an equivalent circuit (Fig. 7.20b), which is used for calculations of the shape and frequency of the OSI voltage (Schrader et al. 1989).

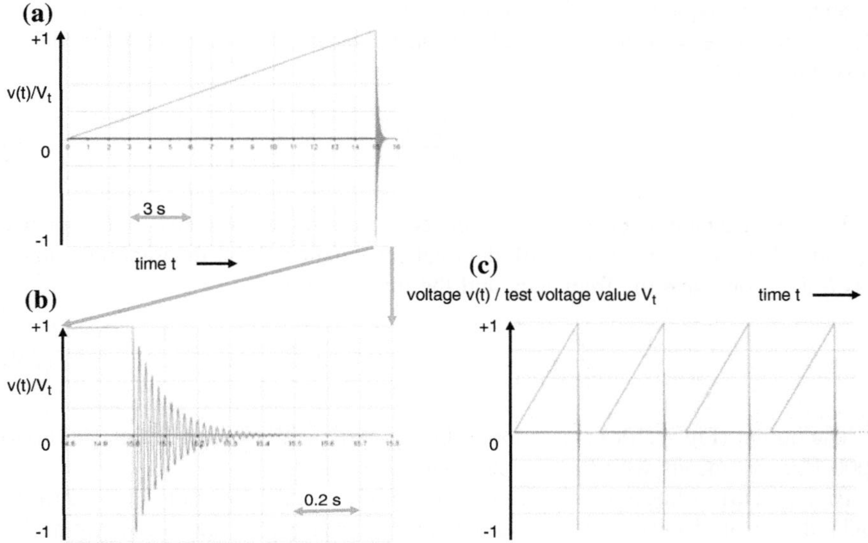

Fig. 7.19 Damped alternating (DAC) voltage. **a** A full DAC impulse (including the DC ramp for charging). **b** The short oscillating part of the DAC impulse. **c** A sequence of DAC impulses as occasionally used for withstand tests

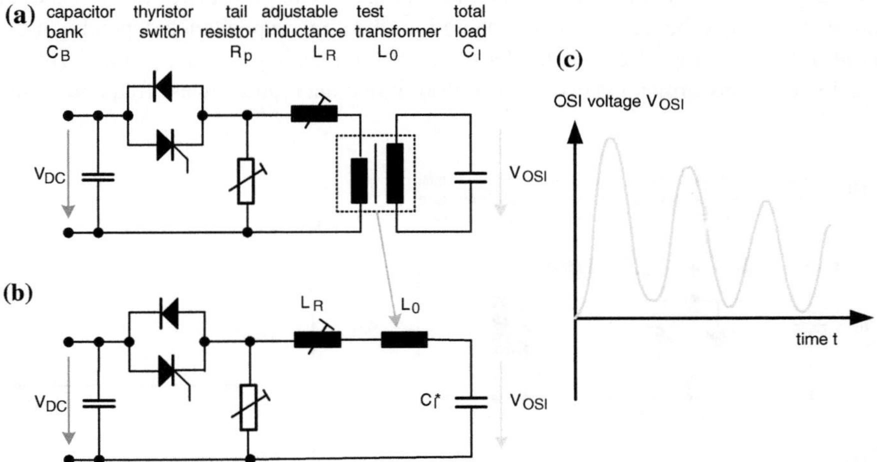

Fig. 7.20 Generation of unipolar OSI voltages by transformers (Schrader et al. 1989). **a** Schematic circuit diagram. **b** Equivalent circuit. **c** Unipolar OSI voltage

Shape and frequency are determined by the stray inductance of the transformer (L_o) and the series connection of the bank capacitance (C_B) with the transferred load capacitance (C_i^*)

$$C = \frac{C_B \cdot C_i^*}{C_B + C_i^*}.$$
(7.8)

The impulse parameters can be influenced by adjustable elements, an inductance L_R and a damping resistor R_p. With the total capacitance C and the total inductance $L = L_o + L_R$ one gets the frequency and the time-to-peak:

$$f = \frac{1}{2\pi\sqrt{L \cdot C}} \quad \text{and} \quad T_p \frac{1}{2f} = \pi\sqrt{L \cdot C}.$$
(7.9)

The output (Fig. 7.20c) is a unipolar OSI voltage with frequencies of 100 up to 1000 Hz; this means with time-to-peak $Tp > 500$ µs.

Bipolar OSI voltages can be generated with a modified circuit (Fig. 5.21a, b) (Schrader et al. 1989). The load capacitance is charged in the same way as for unipolar OSI voltages, but when the first peak is reached, a short-circuit switching by a thyristor causes the bipolar OSI voltage at the HV output (Fig. 7.21c). The capacitor bank is not any longer involved in the oscillation.

Even under optimum conditions, shorter time-to-peak than mentioned above cannot be generated because of the value of stray inductances of test transformers. In many cases, these times are remarkably longer. When a transformer cascade of three stages shall be used for OSI generation, the stray capacitance depends on the kind of feeding (Fig. 7.22) (Schrader et al. 1989). Feeding into the primary side of the lowest transformer means highest stray inductance and lowest frequency, say

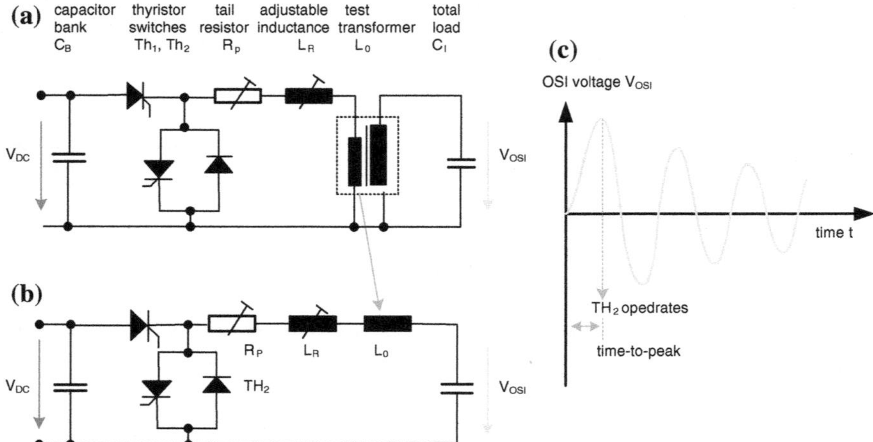

Fig. 7.21 Generation of bipolar OSI voltages by transformers (Schrader et al. 1989). **a** Schematic circuit diagram. **b** Equivalent circuit. **c** Bipolar OSI voltage

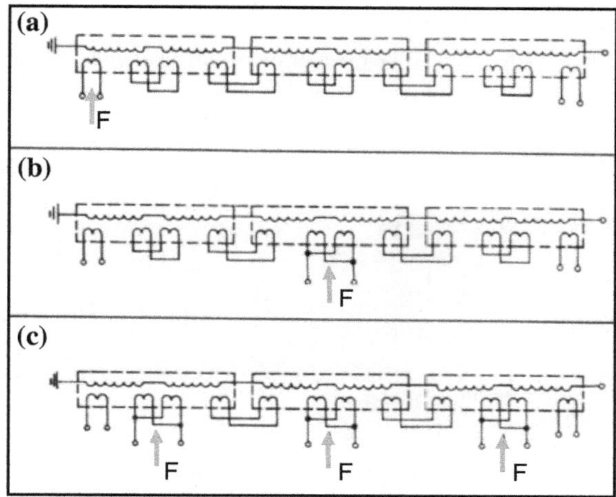

Fig. 7.22 Feeding modes of a three-stage cascade transformer for OSI voltage generation. **a** Single feeding into the primary winding of the lowest transformer. **b** Single feeding into the tertiary winding of the middle transformer. **c** Triple feeding into the tertiary winding of each transformer

100 Hz. When feeding is applied to the middle of the cascade (tertiary winding of the second transformer), the frequency increases by a factor of two (200 Hz). When feeding is realized into the tertiary windings of all three transformers, the stray inductance decreases to 1/40 compared with case a), and consequently, the frequency increases to more than 600 Hz. This principle has been applied to the mentioned 3-MV cascade transformer (Frank et al. 1991), (Figs. 3.15 and 7.23). On each stage, there is a capacitor bank with a rectifier unit. The DC voltage is generated on the stages from a low-frequency AC voltage supplied via the windings of the transformer in a special mode. The three capacitor banks are discharged each into the tertiary winding of one transformer at the same time. This enables the generation of OSI voltages up to 4.2 MV. The leader discharges in air generated by this extremely high OSI voltage are very similar to natural lightning (Hauschild et al. 1991).

7.1.5   Circuits for Very Fast Front (VFF) Impulse Voltages and Solid-State Generators

VFF over-voltages are generated by switching GIS disconnectors and consecutive reflections in the GIS busbars, by steep LI voltage breakdowns of the insulation of overhead lines or by the operation of a lightning arrester. Similar voltages are expected in case of a nuclear explosion (EXO-EMP). They might be characterized by an oscillating impulse with a first front of some 10 ns up to few 100 ns and superimposed contributions of higher frequencies (Feser 1997). Figure 7.24 shows a typical example of a VFF voltage (CIGRE WG 33.03 1998). The development of UHV AC and especially UHV DC transmission systems strengthens the interest in VFF testing (Szewczyk et al. 2016; Rodrigues Filho et al. 2016).

Fig. 7.23 3-MV transformer cascade with OSI attachment and triple feeding mode

Fig. 7.24 Time characteristic of a VFF voltage

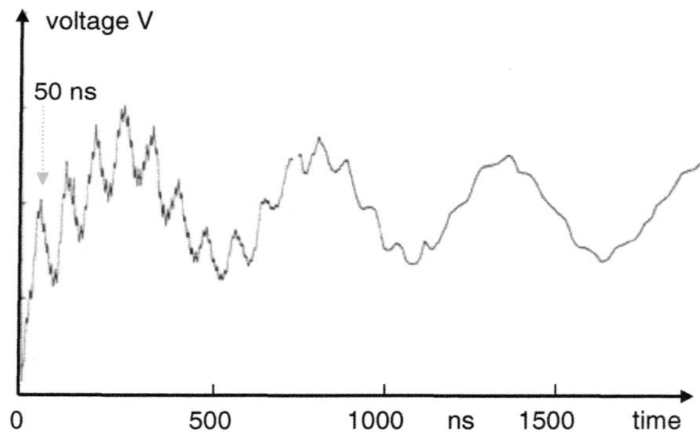

VFF voltage testing of GIS is performed by defined switching of the disconnectors (IEC 61259:1994). There is no horizontal standard related to the requirements of VFF test voltages. For research and development of components which might be stressed by VFF over-voltages in service, VFF impulse voltages are usually generated by a Marx impulse voltage generator with a connected steeping circuit (Kind and Feser 1999) (Fig. 7.25), consisting mainly of a capacitor and a fast sphere gap with compressed-gas insulation and high breakdown field strength. This gap is connected in series with the test object and enables front times in the order of few 10 ns. An impulse voltage generator without steeping circuit, operating without front resistors, can generate impulse voltages with front times down to about 100 ns.

The latest development of UHV equipment has directed the attention to the behaviour of compressed-gas insulation under VFF stress (Ueta et al. 2011; Wada et al. 2011). For VFF voltage generation, a metal enclosed, compressed-gas-insulated

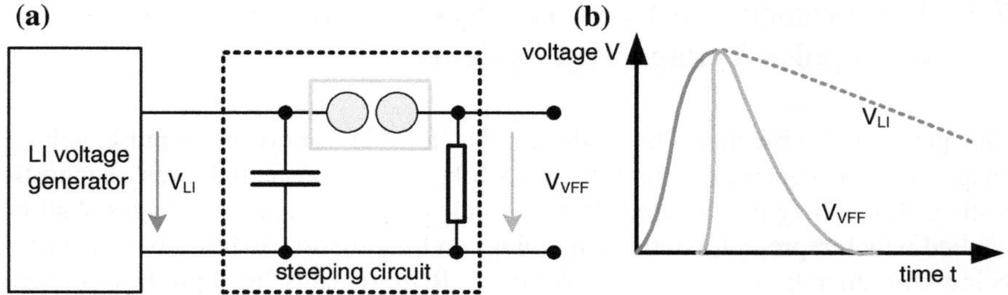

Fig. 7.25 Generation of VFF test voltages. **a** Equivalent circuit. **b** Potential diagram

steeping circuit is applied (Fig. 7.26). In the field compartment before the series gap, an impedance (resistor or inductor) is arranged to generate different superimposed oscillations. When its position is changed related to the gap different superimposed oscillations appear (few MHz to 20 MHz). The test object is inside of the same metal enclosure.

Solid state generators: For low output voltages and special applications, the progress of power electronics enables the development of impulse generators based on solid state elements instead of switching gaps [e.g. Shi et al. (2015), Elserougi et al. (2015), Kluge et al. (2015)]. They can also be recommended for HV testing e.g. of low voltage equipment (IEC 61180). This trend should deserve attention, even if it is not yet applicable for testing of high- and medium-voltage equipment.

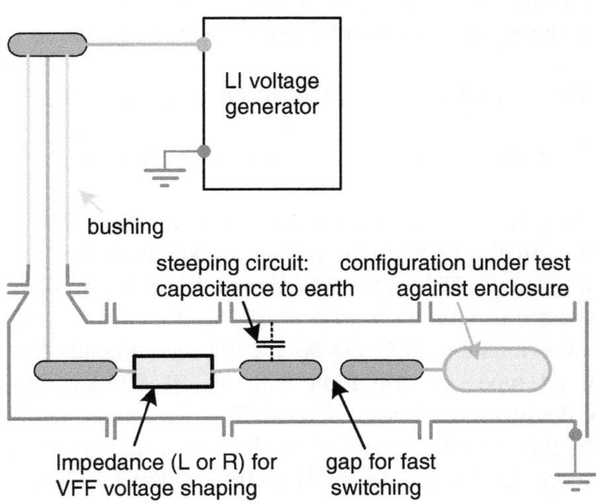

Fig. 7.26 Principle circuit for VFF investigation of gas insulation

7.2　Requirements to LI/SI Test Systems and Selection of Impulse Voltage Test Systems

The preceding subsections have shown that a wide variety of impulse voltage shapes can be generated. For research work, development and even diagnostic testing, this variety can be used. But for quality testing, impulse voltages shall be applied which represent external (lightning) and internal (switching) over-voltages being reproducible within certain tolerances. Requirements for these test voltages are given in standards like IEC 60060-1:2010 or IEEE Std. 4 (Draft 2013) and will be explained in the following.

7.2.1　LI Test Voltage and the Phenomenon of Over-Shoot

7.2.1.1　Requirements of IEC 60060-1 and IEEE Std. 4 to Standard LI Voltages 1.2/50

For a smooth LI impulse voltage, the direct parameter evaluation can be made according to the parameter definitions as described below. But when a LI test voltage shows oscillations or an overshoot, the parameters of a LI test voltage shall be evaluated using the so-called "*k-factors*" derived from "*test voltage functions*" (Fig. 7.27).

The k-factors used for equipment up to $V_m \leq 800$ kV (Eq. 7.10a; Fig. 7.27a) respectively for testing UHV equipment (Eq. 7.10b; Fig. 7.27b) are empirically determined from the comparison of LI voltages with and without oscillations:

$$k(f) = 1/(1 + 2.2 \cdot f^2/MHz) \textit{for testing } V_m \leq 800 \; kV \qquad (7.10a)$$

$$k(f) = 1/(1 + 7.5 \cdot f^2/MHz) \textit{for UHV testing} \qquad (7.10b)$$

The k-factors express that an overshoot of long duration (low frequency) has a stronger influence on the breakdown voltage of an insulation than one of short duration (high frequency). This is the well-known breakdown voltage—breakdown time characteristic of insulations (Kind et al. 2016). As a first step of introduction of this new type of evaluation, IEC 60060-1:2010 recommends the application of Eq. 7.10a, b to all types of insulation with $V_m \leq 800$ kV, for the necessary improvement of the method see Sect. 7.2.1.2. The determination of the test voltage curve $V_t(t)$ from the recorded curve $V_r(t)$ shall be made in the following steps (Fig. 7.28, for more details, see Annex B of IEC 60060-1:2010):

Fig. 7.27 Test voltage functions **a** According to IEC 60060-1:2010 with empirical data and limit value of IEC 60060-1:1989. **b** Comparison with proposal of CIGRE WG D1.36 (2017) for UHV LI tests

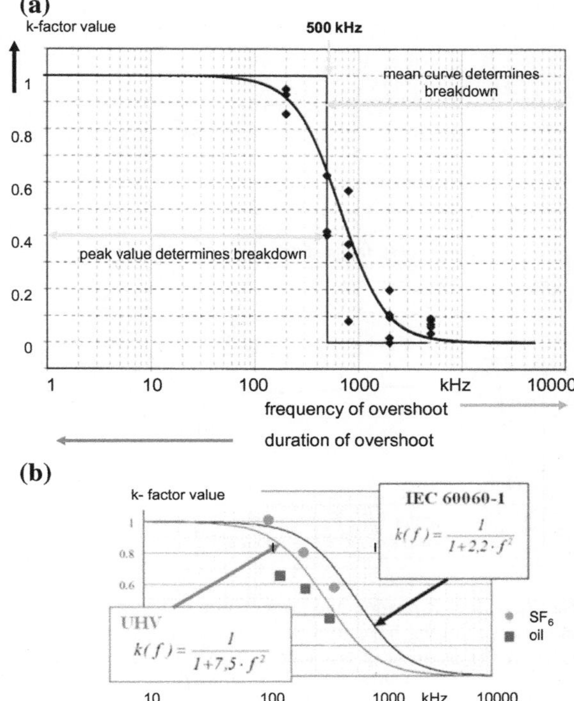

1. Determine the *base curve* $V_b(t)$ as an estimate of the exponential function with the parameters V_0, τ_1 and τ_2:

$$V_b(t) = V_0 \cdot \left(e^{(t/\tau_1)} - e^{-(t/\tau_2)} \right). \tag{7.11}$$

The base curve (Eq. 7.11) represents the recorded curve without over-shoot and shall be characterized by its peak value V_B, whereas the full recorded curve is characterized by its extreme value V_E (Fig. 7.28a).

2. Find the *residual curve* as the difference between the recorded curve and the base curve (Fig. 7.28a):

$$V_R(t) = V_r(t) - V_b(t). \tag{7.12}$$

3. Use a digital filter with a transfer function (amplitude–frequency response) equal to the test voltage function $H(f) = k(f)$ (7.10a, b, as described in detail in IEC 60060-1:2010, Annexes B and C) and use it for filtering the frequency spectrum $V_R(F)$ of the residual curve. The result is the filtered residual curve in the frequency domain

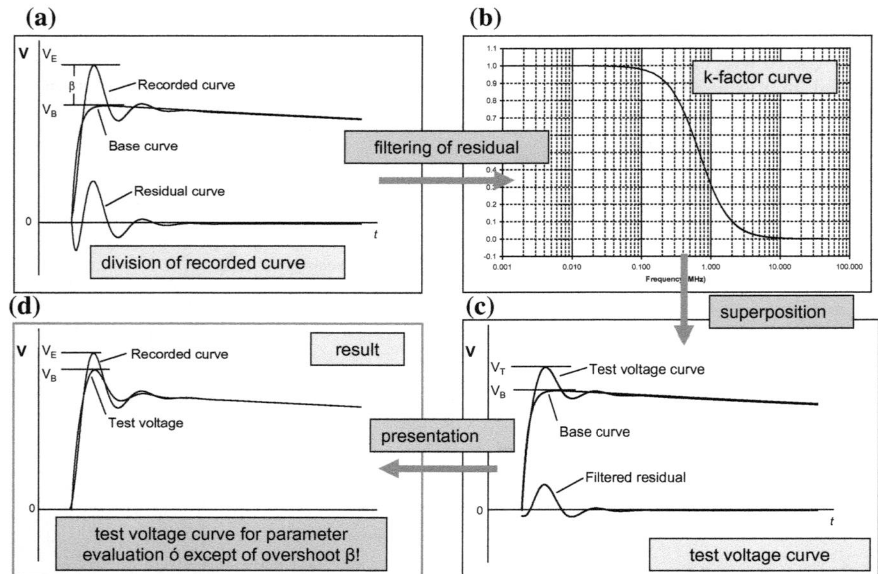

Fig. 7.28 Determination and presentation of the test voltage curve. **a** Recorded curve, base curve and residual curve. **b** Test voltage function. **c** Base curve, filtered residual curve and test voltage curve. **d** Presentation of test voltage curve and recorded curve

$$V_{RF}(f) = k(f) \cdot V_R(f), \tag{7.13}$$

and—after re-transformation to the time domain—the *filtered residual curve* $V_{RF}(t)$ (Fig. 7.28b).

4. Superimpose the filtered residual curve on the base curve to get the *test voltage curve* (Fig. 7.28c):

$$V_t(t) = V_b(t) + V_{RF}(t). \tag{7.14}$$

5. In a presentation of the result, both—the recorded curve and the test voltage curve—shall be shown (Fig. 7.28d).

Note It should be mentioned that the handling of the zero-level problem is not considered here. For the zero level and the details of the implementing the evaluation software, see IEC 60060-1:2010, Annexes B and C and IEC 61083-2:2013, see for the filter curve also Lewin et al. (2008).

It should be mentioned that in addition to the computer-aided evaluation, also a manual calculation of the test voltage value V_T is described in IEC 60060-1:2010 (Annex B.4) as well as by Berlijn et al. (2007). This procedure considers not the whole frequency spectrum of the residual curve, rather its corresponding value

$k(f_{os})$ at the single main frequency f_{os} of the over-shoot which is simply multiplied with the maximum of the residual voltage $V_{Rmax}(t)$. The result is superimposed on the estimated base curve to get the test voltage value V_T. In contrast, the filtering (Eq. 7.13) works also when the over-shoot is the result of a mixture of frequencies and when noise signals of higher frequencies are superimposed on the recorded curve. The manual evaluation cannot be recommended.

In case of a smooth recorded curve (Eq. 7.11) with $V_E = V_B$, one gets $V_R(t) = V_{RF}(t) = 0$ because of $k(f) = 1$. IEC 60060-1:2010 requires that all parameters of the LI test voltage are evaluated from the test voltage curve. When an LI test voltage fulfils the following requirements, it is a standard LI test voltage 1.2/50:

The *test voltage value VT* is the maximum value of the test voltage curve (Figs. 7.28c and 7.29). In an LI voltage test, the required test voltage value must be adjusted with a tolerance of $\pm 3\%$.

The *front time T_1* is a virtual parameter defined as 1.67 times the interval between the instants when the impulse voltage is 30 and 90% of the test voltage value (A and B in Fig. 7.29). The front time $T_1 = 1.2$ µs has a tolerance of $\pm 30\%$, this means the real front time has to be within (0.84–1.56) µs.

> **Note** For test objects of high capacitance as cables or capacitors, the upper tolerance limit might be significantly enlarged to 5 µs or even more. Also for UHV equipment, an upper tolerance limit in the order of 2.5 µs is under discussion.

The *time-to-half-value T_2* is a virtual parameter as the time interval between the virtual origin which is the intersection between the time axis and the straight line drawn through the points A and B in Fig. 7.29, and the instant when the voltage crosses the half of the test voltage value (Fig. 7.29): It is required $T_2 = 50$ µs with a tolerance of $\pm 20\%$, this means the real value has to be within (40–60) µs.

The *relative over-shoot magnitude β* is the difference between the extreme value of the recorded curve and the maximum of the base curve related to the extreme value (IEC 60060-1:2010):

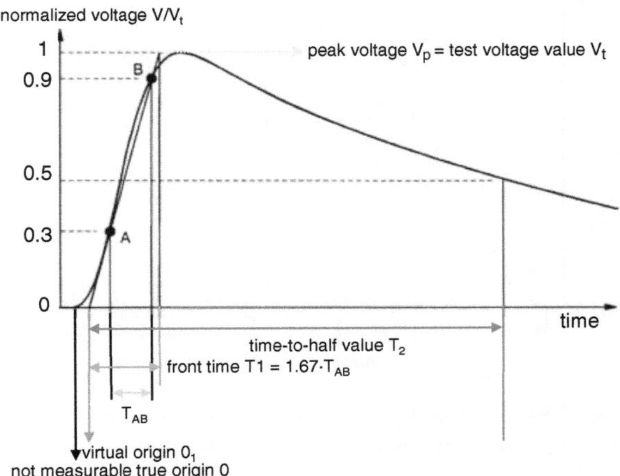

Fig. 7.29 Parameter definition for full LI test voltages

$$\beta = \frac{V_E - V_B}{V_E} \leq 10\% \tag{7.15}$$

The latest edition of IEEE Std. 4 (2013) recommends an over-shoot up to 5%, but allows an increase to 10% for reasons "to allow waveforms accepted by the historical" smooth curve of the "over-shoot method" (IEEE Std. 4—1995 and IEC 60060-1:1989).

Quality tests require in addition to full LI test voltages also *chopped LI test voltages* (LIC) which represent the stress of the insulation after a protecting device (e.g. an arrester or a protection gap) has operated. An LIC voltage is also caused by any breakdown in the HV circuit, but in the following, only controlled breakdowns with a *chopping gap* will be considered (see Sect. 7.1.2.4).

The LI voltage can be chopped in the front (Fig. 7.30a) or on the tail (Fig. 7.30b). The instant of chopping is defined as the intersection of the line through the points C ($0.7\ V_{\mathrm{CH}}$) and D ($0.1\ V_{\mathrm{CH}}$) with the voltage level immediately before the collapse. The time to chopping T_C is the interval between the virtual origin O_1 and the instant of chopping. The duration of the voltage collapse T_{CO} is defined as 1.67 times the time interval between the points C and D. The virtual steepness S_C of the chopping is calculated with the voltage V_{CH} at the instant of chopping and the duration of the voltage collapse T_{CO}:

$$S_C = \frac{V_{\mathrm{CH}}}{T_{\mathrm{CO}}}. \tag{7.16}$$

Whereas IEC 60060-1:2010 specifies only a tail-chopped LIC voltage of $T_C = 2$–5 µs (Fig. 7.30b), the IEEE Std. 4 specifies also a standard front-chopped LIC voltage of $T_C = 0.5$–1.0 µs (Fig. 7.30a).

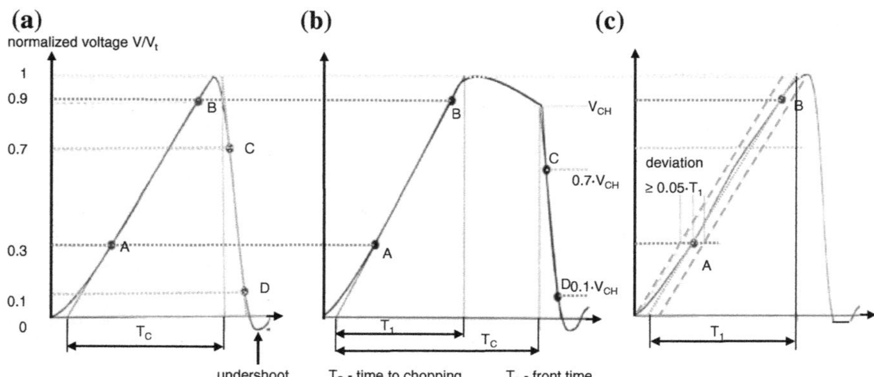

Fig. 7.30 Chopped lightning impulse voltages. **a** Front-chopped LIC voltage. **b** Tail-chopped LIC voltage. **c** Linearly rising front-chopped impulse

Note According to the IEC opinion, a front-chopped LIC voltage is not required for testing objects with windings, because the tail-chopped LIC voltage causes a higher steepness and consequently a more non-uniform voltage distribution in the test object. The traditional IEEE opinion considers more the representation of high over-voltages limited by protection devices. It seems that in a future version of IEEE Std. 4, the IEC practice will be applied, too.

A linearly rising front-chopped LIC voltage is a voltage rising with an approximately constant steepness until it is chopped at a voltage V_E. The linearly rising front-chopped LIC voltage is mainly applied in the test practice according to the IEEE standards. It is defined by the extreme value V_E, the front time T_1 and the *steepness*

$$S_F = \frac{V_E}{T_1} \tag{7.17}$$

The voltage increase is considered to be approximately linear from 30% up to the instant of chopping. The tolerance of the steepness is characterized by a band of $\pm 0.05 \cdot T_1$ from the mean line through AB (Fig. 7.30c). The parameters of linearly rising LIC voltages are not specified in the horizontal standards, but in the relevant apparatus standards.

Voltage value: The evaluation of tail-chopped LIC voltages can be made with a method adapted to the k-factor calculation. For this, two records are needed, one of the tail-chopped LIC voltage from the performed test and one full reference LI voltage on lower voltage without changing the set-up of the HV test circuit (except of the switching gaps and the charging voltage of the generator) and the measuring system. The reference LI curve is used for the determination of the base curve. The recorded LIC curve is treated with that base curve similar as described above. For more details, see IEC 60060-1:2010 (Annex B.5). The problem of over-shoot does not appear for front-chopped impulses, they can be evaluated as shown in Fig. 7.30a.

7.2.1.2   Situation and Future of the Treatment of Over-Shoot

The evaluation method according to the IEC and IEEE standards as described above is an important first step into the direction of a physically correct evaluation of LI test voltages with overshoot, but it is also a compromise between new ideas and traditional thinking. Therefore, it shall be tried to explain in the following the possible directions of the further improvement of the k-factor method. Let us consider the new evaluation method in comparison with that of IEC 60-1:1989 (Fig. 7.27).

As considered in Sect. 7.1.2.3, the inductance and capacitances in the circuit may cause oscillations which are damped by the resistive losses in the circuit. The oscillations have remarkable influence on the breakdown behaviour when they appear in the region of the peak and increase the peak value of the LI test voltage. In case of a strong damping, the oscillation is reduced to a single half-wave, which is called "over-shoot". In the following, the term "over-shoot" shall also include oscillations of lower damping.

The previous version of IEC 60-1:1989 tolerated oscillations and over-shoot up to 5% of the smooth peak. If their frequency "is not less than 0.5 MHz or the duration of the over-shoot not more than 1 μs, a mean curve should be drawn …for the purpose of measurement". The test voltage value was the peak of the recorded curve for over-shoot frequencies $f < 0.5$ MHz; for $f > 0.5$ MHz, it is the maximum of the drawn mean curve. There was no rule how to estimate the mean curve. This abrupt change of the evaluation at 0.5 MHz (Fig. 7.27) is physically wrong, causes an error of up to 5% at that frequency, a certain arbitrariness for the operator or for provider of evaluation software. Therefore, a change had been urgent. On the mid of the 1990s, the CIGRE-Working Group 33.03 and a related European research project started experiments on the influence of the over-shoot (e.g. Garnacho et al. 1997, 2002; Berlijn 2000; Simon 2004). A combined voltage (see Sect. 8.1.1) of a smooth LI voltage and an oscillating short impulse were applied to insulation samples of air, SF_6, oil-impregnated paper and polyethylene. The test voltage values of test series were usually limited up to 200 kV. The experiments delivered 50% LI breakdown voltages for the breakdown of the impulse with over-shoot (extreme value V_E), for the smooth standard impulse (peak value V_{LI}) and enabled the determination of a well-defined base curve (Eq. 7.11; maximum V_B). For each sample, the results at different over-shoot frequencies have been combined as the frequency-depending test voltage factor (test voltage function)

$$k(f) = \frac{V_{LI}(f) - V_B(f)}{V_E(f) - V_B(f)}. \tag{7.18}$$

A clear decrease in the test voltage factor with increasing frequency has been found (Fig. 7.27, measuring points), but the dispersion of the results was so large, that no clear influence of the different types of insulations has been identified. Therefore, a common k-factor curve has been evaluated (Fig. 7.27 and Eq. 7.10a, b), which is overtaken into the standards (see Sect. 7.1.2.1).

The results of the LI parameter evaluation according to the valid IEC 60060-1:2010 differ from those according to IEC 60-1:1989. Even if the new evaluation delivers physically better results, the differences may have certain consequences for design and testing of equipment. The results of the evaluation of numerous LI voltages (e.g. of IEC 61083-2:2013) according to the old and the new procedure (Table 7.2) show the consequences. If there is an over-shoot with $f < 0.5$ MHz, an up to 3% higher LI test voltage would be necessary now. The front time would become shorter and the time to half-value longer. For $f > 0.5$ MHz, an up to 6% lower LI test voltage can be applied now, the front time increases and the time to half-value decreases. These results show the tendency, but they include not only the differences in the procedures, but also the uncertainties caused by the software. Further comparisons of the old and the new method are published by Pfeffer and Tenbohlen (2009).

Table 7.2 Differences of LI parameter evaluation according to IEC 60060-1:2010 and IEC 60-1:1989

Parameter	Over-shoot frequency $f < 0.5$ MHz	Over-shoot frequency $f > 0.5$ MHz
Test voltage value	0…−3%	+2%…+6%
$(V_{2010} - V_{1989})/V_{1989}$		
Front time	0…−6%	0…+15%
$(T_{1\ 2010} - T_{1\ 1989})/T_{1\ 1989}$		
Time to half-value	0…+5%	−4%…7%
$(T_{2\ 2010} - T_{2\ 1989})/T_{2\ 1989}$		
Over-shoot	Independent on the frequency	
$(\beta_{2010} - \beta_{1989})/\beta_{1989}$	−10%…+40%	

An unexpected result was found for the values of the over-shoot which became usually higher according to IEC 60060-1:2010 than according to the old version. The reason is the missing rule for the old "*mean curves*" which got higher maxima by the old "user-friendly" software than the well-defined "*base curves*" (Eq. 7.11) by the new software. Also the definition of the over-shoot (Eq. 7.15) is physically *not* correct. It does not consider the duration of the over-shoot (Hinow et al. 2010), because the extreme value V_E of the recorded curve is the reference value. An over-shoot definition which considers the duration could follow a German proposal to TC 42 when the over-shoot magnitude would be defined from the test voltage curve—as the other parameters of LI voltage, too (Hinow et al. 2010):

$$\beta^* = \frac{V_T - V_B}{V_T} \tag{7.19}$$

The comparison of the two definitions (Fig. 7.31) shows that β^* delivers always lower values than β. For an over-shoot frequency of $f < 0.5$ MHz, β^* is typically higher than for the case $f > 0.5$ MHz. With respect to the duration of the over-shoot, this is a plausible characteristic.

A further point is the dependence on the frequency. The frequency is estimated from the duration of the over-shoot (Garnacho et al. 1997), but the breakdown process is influenced by the available time according to the well-known breakdown voltage—breakdown time characteristic (see Sect. 7.3). This characteristic can be described by a statistical time-lag followed by the formative time-lag. The latter can be described by the *formative voltage–time area* (Kind 1957; Kind et al. 2016). which might be also applied to the over-shoot treatment (Hauschild and Steiner 2009; Garnacho 2010; Ueta et al. 2011c). Then, the over-shoot will be characterized by the voltage–time area above a certain voltage value V_X (Fig. 7.32). This might be a certain percentage of the extreme value (Fig. 7.32a, b) or the test voltage value V_T (Fig. 7.32c). Then, for example, the relative over-shoot magnitude would be calculated by:

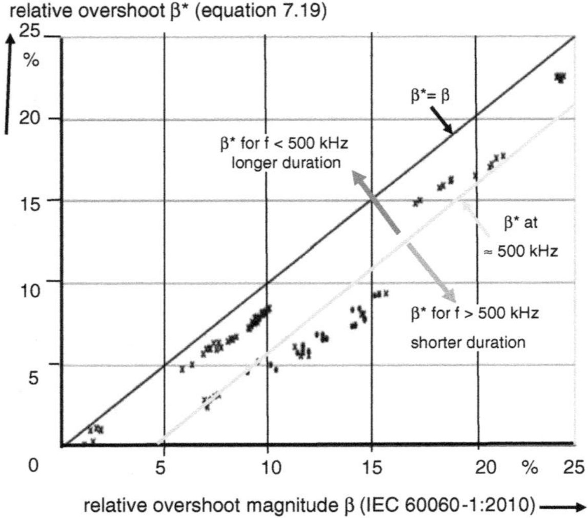

Fig. 7.31 Comparison of the over-shoot magnitudes of β (IEC definition Eq. 7.15) and β^* (Eq. 7.19)

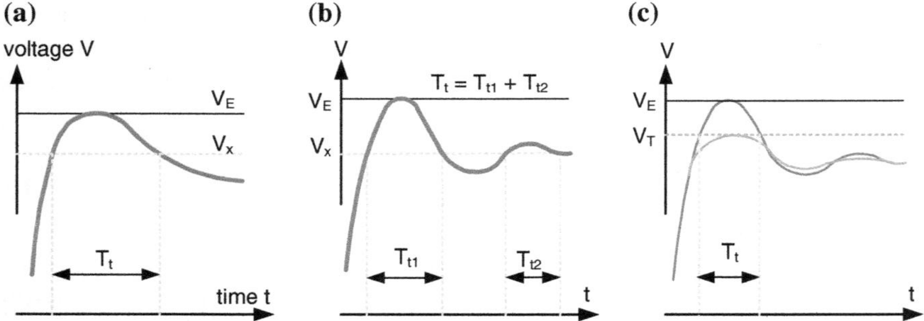

Fig. 7.32 Definition of over-shoot in conjunction with the formative time model. **a** Example for an aperiodic over-shoot using a percentage of extreme value V_E. **b** Example for oscillating over-shoot using $0.9 \cdot V_E$. **c** Example for oscillating over-shoot using the test voltage value V_T

$$\beta^\circ = \frac{1}{V_T \cdot T_t} \cdot \int\limits^{T} (V_t(t) - V_T)\mathrm{d}t, \tag{7.20}$$

with the over-shoot duration: $T_t = \sum_{i=1}^{n} t_i(V_i \geq V_T)$.

A limitation of β° would mean a limitation of the formative voltage–time area which considers the really acting stress combination of voltage and time.

When an *over-shoot* definition related to the duration of the over-shoot is applied, it seems to be appropriate to apply also the test voltage function depending on the duration. The duration of an aperiodic over-shoot is related to the frequency by $T_t \approx 0.5/f$, and one can derive from Eq. (7.10), the new *test voltage function*

$$k^*(T_t) = \frac{4T_t^2/\mu s^2}{4T_t^2/\mu s^2 + 2.2}.$$

(7.21)

This function (Fig. 7.33) would not only improve the physical understanding. It has a linear scale with a direct relation to the time parameters of the LI voltage. The case $T_t = 0$ means no over-shoot. The determination of the duration is even simpler than that of the frequency. It would also consider the case that the oscillations are only slightly damped (Fig. 7.32b), and the second peak contributes to the formative voltage–time area. Instead of using the test voltage function of IEC 60060-1:2010 (Eq. 7.10a, b), a different definition of the relative over-shoot magnitude can be used—e.g. as that of Eq. 7.20.

The *test voltage function* for higher voltages and different insulation samples is under investigation till now, e.g. by Garnacho (2010), Garnacho et al. (2014), Hinow with TU Cottbus (2011), Ueta et al. (2010, 2011b), Diaz and Segovia (2016) The present function (Eq. 7.10a, b) is based on experiments of small samples and voltages mainly $V_T < 200$ kV. Therefore, the breakdown is quite fast and the assumption to have a certain average characteristic seems to be not correct. It must be shifted (Fig. 7.34)—to the left (lower frequencies) for large insulation and/or relatively slow breakdown processes as for long air gaps or larger transformer insulation used for UHV transmission (Tsuboi et al. 2011, 2013; Ueta et al. 2012b) and only a bit to the right (higher frequencies) for small and compact insulation of very fast breakdown processes (SF_6; solids, vacuum, etc.), but till now, IEC TC42 has not yet drawn the final conclusion of that international research work.

It can be assumed that different characteristics will be found for different electric fields, different insulation materials and possibly even for different over-shoot magnitudes. Also the evaluation algorithms and the remarkable uncertainty of the determination of the test voltage function (Okabe et al. 2015) must be taken into consideration. Possibly for a future standard IEC 60060-1, all available

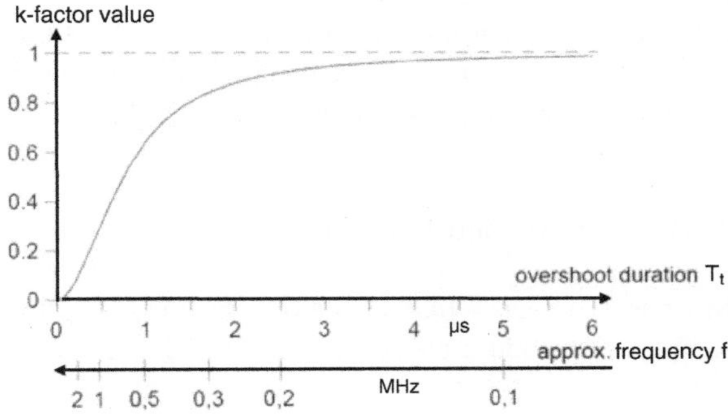

Fig. 7.33 Test voltage function depending on over-shoot duration (Eq. 7.21)

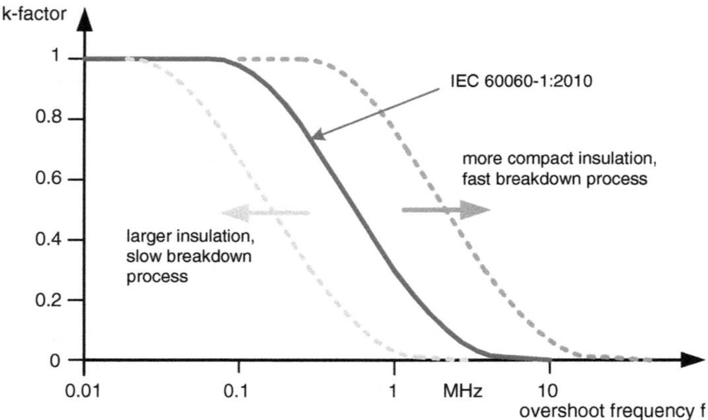

Fig. 7.34 Expected modifications of the test voltage function

experimental results should be used to find *one common test voltage function* applicable within acceptable tolerances for parameter evaluations of all insulations.

The *fitting method* for extracting the base curve as described above (Eq. 7.11) is also subject of ongoing investigation, (e.g. Satish and Gururaj 2001; Kuan and Chen 2006; Ueta et al. 2011a, b, c, d, 2012a, b; Garnacho et al. 2013; Pattanadesh and Yutthagowith 2015). The introduction of a new evaluation method for the base curve could cause differences in the LI parameter evaluation. Contrary to the test voltage function, there seems to be no urgent need for the introduction of a new method for the determination of the mean curve. First, it must be shown that the present method is insufficient for practical cases.

Related to testing of *UHV equipment*, the CIGRE Working Group D1.36 (2017) has summarized the present knowledge: It has been shown that for UHV test objects the requirements $T_1 \leq 1.56$ µs and ß $\leq 10\%$ cannot be met, when the capacitance of the test objects reaches the order of 10 nF (see also the following Sect. 7.2.1.3). It is recommended to increase the upper tolerance limit to $T_1 = 2.5$ µs and the acceptable overshoot to ß $\leq 15\%$ when testing UHV equipment with $V_m > 800$ kV. Furthermore the test voltage function for UHV LI testing is changed (Fig. 7.27b) to the above mentioned Eq. (7.10b) and an improved procedure for the base curve estimation is proposed. Hopefully IEC TC42 will provide on that basis the next edition of IEC 60060-1.

Last but not least, there are discussions to reduce the over-shoot in LI testing by the over-shoot compensation (see Sect. 7.1.2.3).

7.2.1.3   Interaction Between HVLI Test System and Test Object

Most test objects—like insulators, bushings, GIS, power transformers or cable samples—provide a capacitive load for the test system. In few cases, test objects have an inductive characteristic, like the low-voltage winding of power transformers. Resistive test objects do not play any role, because outdoor insulations are not tested under wet or polluted conditions with LI voltages. The influence of the test object shall be explained by the equivalent single-stage circuit (Fig. 7.4).

Capacitive test objects: The load capacitance $C_1 = C_b + C_i$ (usually consisting of the basic load C_b of the generator plus the test object capacitance C_{to}) determines the front time constant τ_f (together with the front resistor R_f, Eq. 7.2), the time constant for the tail τ_t (together with the impulse capacitor C_i and the tail resistor R_t) and the circuit efficiency factor η_c (together with the impulse capacitor C_i of the generator, Eq. 7.1):

$$\tau_f = R_f\left(\frac{C_i \cdot (C_b + C_{to})}{C_b + C_{to} + C_i}\right), \tag{7.22a}$$

$$\tau_t \approx R_t(C_i + C_b + C_{to}), \tag{7.22b}$$

$$\eta_c \approx \frac{C_i}{C_i + C_b + C_{to}}. \tag{7.23}$$

The *front time* T_1 (characterized by τ_f, Eq. 7.22a) depends strongly from the test object capacitance because usually $C_i \gg C_b + C_{to}$ and consequently $\tau_f \approx R_f \cdot (C_b + C_{to})$ with often $C_{to} > C_b$. In many cases, the front time is nearly doubled when the test object capacitance is doubled. Considering that the width of the front time tolerance (0.84–1.56 µs) is less than doubling its lower limit (Fig. 7.35), the front resistors R_f of a generator must be adapted accordingly. For a fixed value of τ_f—respectively, of the front time T_1—the necessary value of R_f for the upper, respectively, the lower tolerance limit of the front time can be calculated (Fig. 7.35a). A horizontal line through the resulting band corresponding to a certain range of load capacitance gives the range of load, for which one resistance R_{f1} can be applied to deliver standard LI voltages. For higher load capacitance, a lower resistance R_{f2} has to be applied and so on. If the front resistance becomes too low, an unacceptable over-shoot is generated which limits the generation of standard LI impulses. For the example of Fig. 7.35a, this is reached when the front resistance approaches about 100 Ω (This means e.g. 10 Ω per stage for a 10-stages generator).

The simplified consideration of Fig. 7.35 neglects the important influence of the circuit inductance which causes the overshoot (accepted by standards up to ß = 10%). The overshoot is damped by the front resistor which must exceed a minimum value. According to CIGRE WG D1.36 (2017), this minimum acceptable front resistor (in Ω) R^*_{fmin} for $T_1 \leq 1.56$ µs can roughly be estimated from the total capacitive load (in µF) ($C_1 = C_b + C_{to}$) and the total circuit inductance (in µH) ($L_s = L_{generator} + L_{HV\ lead} + L_{test\ object} + L_{current\ return}$) by

$$R^*_{fmin} = 1.45\sqrt{L_s/C_l} \tag{7.24}$$

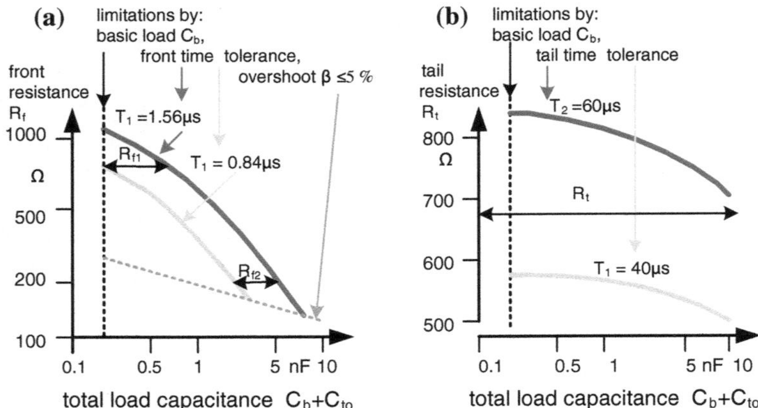

Fig. 7.35 Influence of the load capacitance on selection of front and tail resistors of a generator 2000 kV/400 kJ of ten stages. **a** Principle of the selection of front resistance. **b** Situation of the tail resistance (Kind and Feser 1999)

With the characteristic UHV load capacitances (up to more than 10 µF) it might be difficult to meet the limit of the overshoot $\beta \leq 10\%$. Therefore the test standards for UHV equipment (now under preparation) shall allow a higher LI overshoot (up to $\beta = 15\%$) and/or longer LI front time (higher front resistors) for sufficient attenuation of the overshoot.

For the adaptation of the front resistance, each LI generator has several sets of resistors. Usually, these sets are designed in such a way that resistors can also be switched in parallel or even in series to vary the available values of front resistance by combinations for a wide range of capacitive load. For each single-resistance value, a certain load range is covered. Usually, a LI test system is equipped with three sets of front resistors enabling seven different resistance values by different parallel connections. For efficient handling of a larger LI test generator, the not-used resistors should be stored on the stages which should be easily reached by internal fixed ladders (Fig. 7.12d). Each impulse test system should be equipped with an instruction which resistors should be applied for which test object capacitance to cover a certain range of load capacitances (Fig. 7.36).

The *time to half-value* T_2 (characterized by τ_t, Eq. 7.22b) is not strongly influenced by the test object capacitance because it is dominated by the impulse capacitance C_i (Fig. 7.35b). One tail resistance is sufficient to cover the whole tolerance band of the time to half-value between 40 and 60 µs.

The circuit *efficiency factor* can easily be estimated by Eq. 7.23. The generator with its basic load reaches usually a total efficiency factor $\eta \approx 0.95$ for $C_i \gg C_{to}$. It decreases slightly with increasing test object capacitance and values $C_{to} > 0.2\ C_i$ cannot be recommended.

An *inductive test object* combined with a capacitance is, e.g. found when the low-voltage side of a power transformer is tested (Fig. 7.37a). The inductance L_{to} forms an oscillating circuit especially with the impulse capacitance C_i and causes an

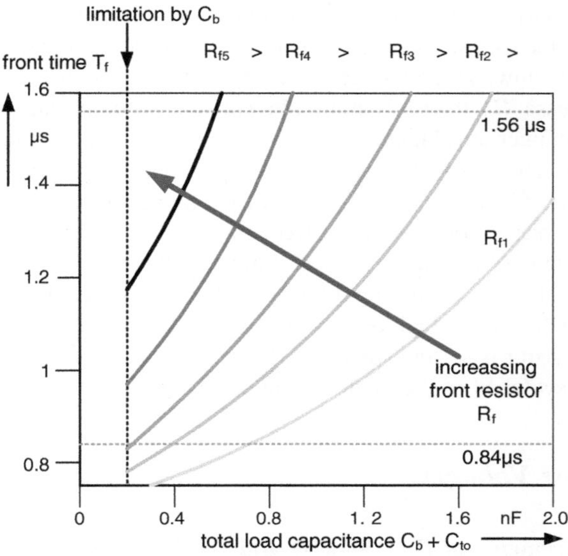

Fig. 7.36 Selection of front resistance depending on test object capacitance

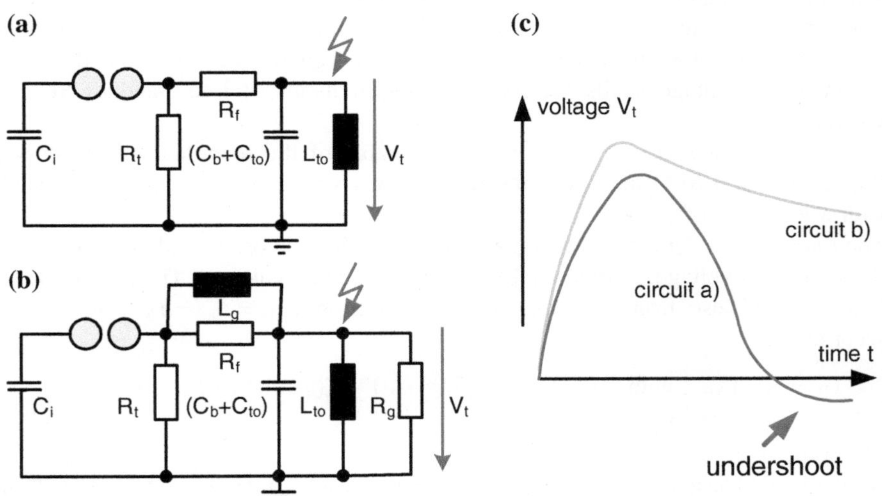

Fig. 7.37 Testing of objects with inductances. **a** Equivalent circuit. **b** Equivalent circuit with "Glanninger" attachment. **c** Comparison of the impulse shape

impulse tail which is characterized by an oscillation instead by an exponential function. The oscillation shortens the time to half-value T_2, often outside the tolerance ($T_2 < 40$ µs), causes an undershoot to the opposite polarity as well as a reduction of the efficiency factor (Fig. 7.37c). The influence on T_2 is especially serious and increases with decreasing impulse capacitance C_i of the generator (Fig. 7.37c). As the LI test voltages of the low-voltage side of the transformer under test are relatively low, several stages of a multi-stage generator are connected in parallel to increase C_i. If this is not sufficient, the problem can be solved by a so-called "Glanninger attachment": An inductance L_g is switched in parallel to the front resistor R_f and a resistor R_g in parallel to the test object inductance L_{to} (Fig. 7.37b). The Glanninger inductance $L_g < L_{to}$ is correctly selected when it bridges the front resistor only for lower frequency (tail of the impulse) and when the time to half-value is sufficiently extended, but the voltage is divided between $L_g//R_f$ and the test object ($L_{to}//R_g$) causing a further reduction of the efficiency factor. The "Glanninger" inductance shall be in the order of $L_g = (0.01–0.1)\ L_{to}$, and the "Glanninger" resistor is selected as $R_g \approx R_f \cdot L_{to}/L_g$. The Glanninger attachment is a separate unit switched to the parallel connected stages of the generator.

7.2.2   SI Test Voltages

7.2.2.1   Requirements of IEC 60060-1 and IEEE Std. 4

Over-shoot is not a problem for SI test voltages because a front resistor of quite high resistance is necessary to meet the front time parameter. The recorded curve has also a sharp beginning, the so-called "true origin", and does not require a virtual origin (Fig. 7.38). Therefore, the parameters are evaluated from the recorded curve directly. When an SI voltage fulfils the following requirements, it is a standard SI test voltage:

The *test voltage value Vt* is the maximum value of the recorded voltage curve (Fig. 5.39). In a SI voltage test, the required test voltage value must be adjusted with a tolerance of $\pm 3\%$.

The *time-to-peak Tp* is the time interval between the true origin and the maximum value of an SI voltage. It replaces the front time T_1 at LI voltages. The time-to-peak is determined also from the time of the intersection $T_{AB} = t_{90} - t_{30}$ (Fig. 7.39a) according to

$$T_P = K \cdot T_{AB} \quad \text{with} \quad K = 2.42 - 3.08 \cdot 10^{-3}\, T_{AB} + 1.51 \cdot T_2 \cdot 10^{-4}$$
$$\text{with} \quad T_p, T_{AB} \quad \text{and } T_2 \text{ in } \mu s \tag{7.25}$$

A standard SI test voltage requires $T_p = 250\ \mu s$ with a tolerance of $\pm 20\%$. This means the real front time has to be within (200–300) μs.

> **Note** When SI testing was introduced in the late 1960s, the evaluation of the time-to-peak has been made according to all appearances. This principle has not been suitable for computer evaluation, and therefore, the evaluation based on the intersection has been introduced with IEC 60060-1:2010.
>
> The definition front time—which would be in the order of $T_1 \approx 170\ \mu s$—has not been introduced because the equivalent time-to-peak of $T_p = 250\ \mu s$ is used for decades, and for traditional reasons it should not be changed.

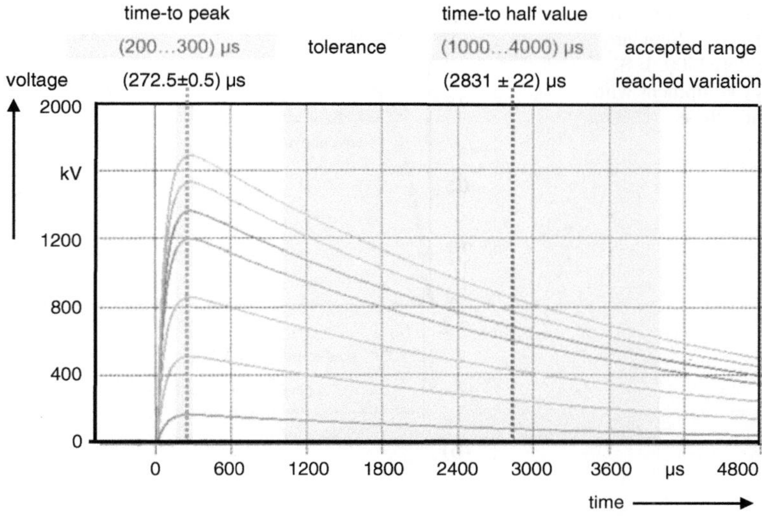

Fig. 7.38 Recorded curves of SI test voltages 250/2500 and their reproducibility for different peak values

As shown by Sato et al. (2015) Eq. (7.25) is only valid for standard SI voltages, its application to non-standard SI voltages (e.g. 20/4000 µs as accepted for SI voltage tests on site, see Sect. 10.2.1.3) might lead to failures up to 25%. Sato et al. give an formula better adapted to the full range of SI voltages.

The *time to half-value* T_2 is a virtual parameter as the time interval between the true origin, and the instant when the voltage crosses the half of the test voltage value (Fig. 7.39a): It is required $T_2 = 2500$ µs with a tolerance of $\pm60\%$, this means the real value has to be within (1000–4000 µs).

> **Note** The tremendous tolerance of T_2 is related to tests on equipment with saturation phenomena, e.g. caused by the iron cores of transformers and reactors. There the SI voltages cause a saturation of the magnetic core which leads to the immediate collapse of the voltage (Fig. 5.39b). For such tests additional parameters, the time above 90% and the time to zero have been introduced (see below). The wide tolerance interval is acceptable because the breakdown process in air—development of leader discharges (see Fig. 7.3)—is determined by the steepness of the front of the SI voltage and very few by development of its tail.

The *time above* 90% T_{90} is the interval during which the SI voltage exceeds 90% of its maximum value (Fig. 7.39b).

The *time to* zero T_Z is the interval between the true origin and the instant when the SI voltage has its passage through zero (Fig. 7.39b).

The parameters shall not been mixed up, because a standard SI test voltage 250/2500 has to be characterized by the set $(V_t; T_p; T_2)$ and for testing equipment with saturation effects the different SI set $(V_t; T_p; T_{90}; T_Z)$ may be used.

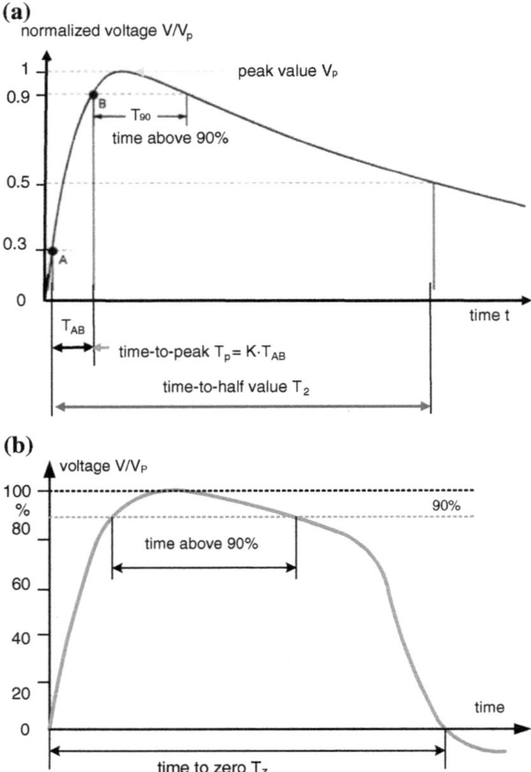

Fig. 7.39 Definitions of SI test voltages. **a** Standard impulse 250/2500. **b** SI testing of equipment with saturation effects

7.2.2.2 Interaction Between HVSI Test System and Test Object

Consideration of polarity effects: Both, LI and SI voltage tests on internal insulation shall be performed at negative polarity to avoid flashovers at bushings or terminations in air. Whereas non-uniform air insulations have much lower breakdown voltages at positive polarity, there is approximately no, sometimes even a slight opposite polarity effect for internal insulation. Also the design of the control electrodes and of the insulation structures of the HV components of an LI/SI test system is determined by the maximum positive SI voltage to be generated. Positive specific SI breakdown voltages of non-uniform fields in air can go down to the order of 1 kV/cm, whereas at LI voltage one can assume a value of 5 kV/cm.

Capacitive test objects: It has been shown in Sect. 7.1.2 that the efficiency factor can be understood as the product of the shape efficiency and the circuit efficiency (Eq. 7.1). The shape efficiency factor η_s of SI voltage is much lower than of LI voltage. The circuit efficiency factor η_c can be estimated according to Eq. 7.24 as for LI voltages. For a load capacitance $C_1 = C_b + C_{to} = 0.2 \cdot C_i$, the circuit efficiency becomes $\eta c = 0.83$. With a shape efficiency of $\eta_s = 0.75$ one gets a total efficiency factor of only $\eta = 0.62$ (Eq. 7.3). The time-to-peak is less sensitive to changes of the load capacitance than the front time of LI voltage. The tolerance of the time to half-value is so large that changes of the test object do not play any role for the selection of the tail resistors. When an LI/SI test system is ordered the maximum expected test object capacitance must taken into consideration for the selection of the impulse energy per stage (Eq. 7.3).

Resistive test objects: Wet and pollution tests of external insulation are also performed at SI test voltages, especially for EHV and UHV equipment. The influence of the resistance of a polluted test object to the total front resistance of an SI generator (up to some 10 k′ Ω) remains negligible. But during testing heavy discharges with currents of some Amperes may cause remarkable voltage drops. This is not only a problem for wet and pollution tests, but also when long air gaps with leader discharges are investigated. As described above for AC and DC voltage, the voltage drop can be calculated when the height and duration of the current pulse are known or assumed. Impulse currents of peak values above 10 A and duration of some 10 μs have been observed (Les Renardieres Group 1977). There is not yet any standard on the acceptable voltage drop at a pre-given current impulse. In any case for the mentioned tests, an impulse generator of high impulse energy should be applied.

Inductive test objects with saturation effects: For testing power transformers, the IEC 60076-3:2013 requires $T_p > 100$ μs; $T_{90} > 200$ μs and $T_Z > 1000$ μs. For the prolongation of the time to zero, a pre-magnetization of the core with SI voltage of opposite polarity and lower maximum value ($V_e > 0.7 \cdot V_t$) can be performed if necessary. Also after a SI voltage test, the core should be demagnetized by applying lower SI voltages of opposite polarity.

7.3 Procedures and Evaluation of LI/SI Voltage Tests

Breakdown and standardized withstand voltage tests as well as the statistical background are already described in Sect. 2.4. The following explanations are related to special LI/SI test procedures and refer to that section.

7.3.1 Breakdown Voltage Tests for Research and Development

The multiple-level method (MLM) is the most important and mainly used test method for the determination of the performance function of insulation samples (see Sect. 2.4.3 and Fig. 2.31). The performance function describes the relationship between stressing LI/SI voltages and breakdown probabilities completely. The MLM test can be easily performed on self-restoring insulation in air. The *independence* might be checked by independence tests. A procedure for that test is explained in Table 2.8. Usually the test results for air gaps are independent when the break between two LI/SI voltages is not shorter than several seconds.

For air or SF_6 insulation with surfaces to solid insulation (e.g. insulators), before the further evaluation of the test, a check of independence is more important than for gases alone. If dependence is indicated, the test procedure should be modified, e.g. by longer breaks between two impulses following each other. Also the short application of a lower impulse voltage of opposite polarity or of a low AC voltage during the break may help to get independent results.

For liquid impregnated and solid insulation the MLM application requires for each LI/SI voltage stress a new test sample. This is a remarkable effort and the limited reproducibility of the test samples causes an additional contribution to the dispersion of the measured performance function. For such test objects it can be checked to apply the *progressive stress method* (PSM; Fig. 2.26c). where one sample is applied to a series of impulses until breakdown. A single test sample delivers more information in a PSM test than in a MLM test.

The statistical evaluation should be made with a powerful software package based on the maximum-likelihood method (Speck et al. 2009). This delivers (Fig. 7.40) estimations of the breakdown probabilities including confidence regions for the used seven voltage levels, point estimations of the performance function (red line in the middle), its confidence limits (blue limits) and also confidence limits of the quantiles (pink lines). Figure 7.40 shows the evaluation based on the normal (Gauss) distribution. It can be repeated for a different distribution function to find the optimum adaptation. All possible conclusions can be drawn from a diagram like Fig. 7.40.

When instead of the whole performance function, the estimation of a certain quantile (e.g. V_{50} or V_{10}) is sufficient, the *up-and-down method* can be applied (Sect. 2.4.4 and Figs. 2.38 and 2.39). Also for that test procedure a maximum-likelihood evaluation is possible and can be recommended.

7.3.2 LI/SI Quality Acceptance Tests

A passed quality acceptance LI/SI test verifies the insulation coordination (Sect. 1.2) by performing the test according to a standardized procedure, see Sect. 2.4.6; Fig. 2.43 and IEC 60060-1:2010.

For external, self-restoring insulation (mainly of atmospheric air), two procedures are acceptable (see also Sect. 2.4.6):

A1. The LI/SI test voltage is lower than the 10%—breakdown voltage: $V_t < V_{10}$.
A2. The LI/SI test voltage is 15-times applied and the number of breakdowns is
   $k \leq 2$.

For internal, non-self-restoring insulation (all solid or liquid impregnated insulation) no breakdown may occur during the quality acceptance test; therefore, the following procedure is applied:

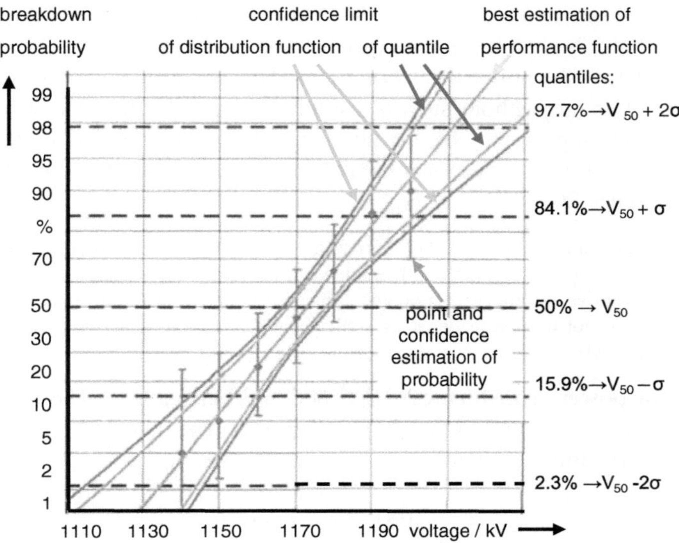

Fig. 7.40 Performance function of an air gap evaluated by the maximum-likelihood method based on a normal (Gauss) distribution

B. The LI/SI test voltage is three times applied and no breakdown may occur ($k = 0$).

For insulation as that of a GIS consisting of a self-restoring (SF_6 gas) and a non-self-restoring (solid insulators) part, it has to be shown that the breakdown had happened in the self-restoring part. This can be indicated by a certain number of withstands at the LI/SI test voltage after the breakdown. A further indication can be a PD measurement which shows no increased PD level. The valid procedure is specified by the relevant apparatus committee of IEC or IEEE. In the following some examples for LI/SI acceptance test procedures are given:

Testing of external insulation (*IEC 60071-1*): The procedure A1 ($V_t < V_{10}$) requires the estimation of V_{10} from a performance function measured down to this low breakdown probability or for known V_{50} and standard deviation σ from $V_{10} = V_{50}$–1.7 σ or from an up-and-down test with seven impulses per voltage level (see Sect. 2.4.4). The procedure A2 ($n = 15/k = 2$) is applicable for all other cases when the above necessary parameters are not available and the effort for their determination is assumed to be too high. It is necessary to mention that the LI voltage tests are applied to dry insulation (see Sect. 2.1.2), whereas the SI voltage tests are applied to wet insulation (see Sect. 2.1.3).

Testing of gas-insulated substation (GIS) (IEC 62271-203:2003): The GIS insulation is characterized by a self-restoring part, the SF_6 gas, and a non-self-restoring part, the surface and the epoxy resin of the spacer insulators. The LI/SI voltage tests are combined with numerous dielectric, thermal, power and mechanical measurements and tests. The standard waveforms are applied and the procedure A2 (15/2) shall be applied. It has to be guaranteed that no breakdown or flashover occurs in the non-self-restoring part. This is considered as verified, if the last five impulses are without such a disruptive discharge. If the first breakdown appears after the impulse no. 10, the number of total impulses has to be extended accordingly. In the worst case of a breakdown at no. 15, the total number of test impulses becomes 20.

Note It is important to state that the "horizontal" IEC 60071-1:1993 requires only three impulse voltages and in case of one breakdown nine additional voltage impulses during which no disruptive discharge is tolerated. This procedure is of higher uncertainty, and therefore, it had not been considered as sufficient when the "vertical" GIS standard was established in 2003.

Testing of power transformers (IEC 60076-3/FDIS:2013): The used impulse generator shall exceed a minimum impulse energy. The standard recommends an energy $W_{i\ min}$ (in Joules) according to the empiric formula:

$$W_{i\,min} > \frac{100 \cdot 2\pi f \cdot T_2^2\left(V_t^2\right) \cdot S_r}{Z \cdot V_m^2 \cdot \eta^2}, \tag{7.26}$$

with the parameters of the transformer under test:

V_m rated phase-to-phase voltage in Volts;
f rated frequency in Hertz;
Z short-circuit impedance in % related to the test terminals;
S_r three-phase power rating in Volt-Amperes;

and with the test parameters:

V_t LI test voltage value in Volts;
T_2 time to half-value of he LI voltage in μs;
η efficiency factor in per unit.

The LI voltage test is a routine test for power transformers $V_m > 72.5$ kV, for lower rated voltages a type test. For transformers $V_m > 170$ kV, the test includes also chopped LI voltages (LIC). An SI voltage test is a routine test for transformers $V_m > 170$ kV and for lower rated voltage a special test. The impulse shall be of standard shape 1.2/50 within the usual tolerances. The evaluation according to the k-factor method is allowed, but alternatively IEC 60076-3:2013 defines some strange differences to IEC 60060-1:2010:

As long as the over-shoot does not exceed 5% ($\beta \leq 5\%$), the extreme value may be taken as the test voltage value. If the over-shoot exceeds 5%, the front time might be extended up to $T_1 = 2.5$ μs and a test with chopped lightning impulses must be performed. Remains $\beta \leq 5\%$ also now, the test voltage value is the extreme value. Only for rare cases, when the over-shoot $\beta > 5\%$ cannot be avoided, the evaluation according to IEC 60060-1:2010 shall be applied for transformers of rated voltage ≤ 800 kV. For UHV transformers, even longer front times can be agreed between the parties of an acceptance test. Also the lower tolerance limit of the time to half-value can be reduced up to $T_2 = 20$ µs by agreement. The test shall be performed in the following sequence:

- one full LI reference voltage of $(0.5\text{–}0.6) \cdot V_t$,
- one full LI test voltage V_t,
- two chopped (LIC) test voltages of $1.1\ V_t$,
- two full LI test voltages V_t.

If an SI voltage test is performed, it follows after the LI/LIC test and before the tests with AC voltages (see Sect. 3.2.5). It consists of

- one SI reference voltage $(0.5\text{–}0.7)\ V_t$ and
- three SI test voltages V_t.

Both tests are successful if no internal breakdown collapses the voltage. For the LI test, additionally, the normalized voltage shapes of the reference LI voltage and of the LI test voltage, as well as the normalized shapes of the measured impulse currents (see Sect. 7.5) at the two voltage levels should be identical.

LI/SI testing of cables (IEC 62067:2006): The *routine test* of cables does not include an LI/SI test voltage. The AC withstand test and a sensitive PD measurement are considered to be sufficient for verifying a correct production, but in defined intervals, more detailed *cable sample tests* (which repeat parts of the type test) are performed to confirm the correct production.

A *cable type test* includes a long set of single tests including LI/SI tests which are performed at warm cable samples after a heating cycle voltage test, which includes 20 single cycles of one day each, and a PD measurement. First, the SI test is performed with 10 positive and 10 negative SI voltages. If no breakdown occurs, the sample has passed and can be stressed with LI voltages according to the same procedure. The PD test is repeated after the LI/SI test.

Additionally, there are *pre-qualification tests* on complete cable systems of about 100 m length with joints and terminations. The total test duration is about one year with in minimum 180 heating cycles. The LI test voltage shall be applied to the whole test assembly or to samples with a length of in minimum 30 m after each heating cycle. The test temperature shall be in an interval between maximum conductor temperature and 5 K above. The test consists of 10 positive and 10 negative LI voltage applications again. It is passed if no breakdown occurs.

As the capacitances of the cable samples are between 0.15 and 0.3 nF/m, the test object capacitance for 30 m samples may reach 9 nF and for 100 m up to 30 nF. These are quite high capacitances, and therefore, the standard allows front times $T_1 = 1\ldots5$ µs, T_2 is within the standard values $T_2 = 40\ldots60$ µs. For SI test voltages, the usual tolerances shall be applied.

7.4 Measurement of LI and SI Test Voltages

To measure the peak value of high impulse voltages, originally sphere gaps have been employed, as already presented in Sect. 2.3.5. However, this direct measurement method is nowadays only recommended for performance checks including linearity tests. Occasionally, also field probes are applicable, as described in Sect. 2.3.6, particularly for measuring very fast transient voltages characterized by time parameters down to the nanosecond range (Feser and Pfaff 1984). This Section is addressed to indirect measuring methods using a HV converting device, which is connected to the LV measuring device via a transmission system, such as a BNC cable or even a fiber optic link. As the design of the converting device is the most challenging task, the following treatment focuses mainly on this issue. More details on this subject can also be found in the textbook of Schon published in 2013.

7.4.1 Dynamic Behaviour of Voltage Dividers

Lightning impulse voltages, particularly if chopped in the front, cover a frequency spectrum up to 10 MHz and even above. To prove whether the scale factor remains constant within such a wide frequency band, the dynamic behaviour of the entire HV measuring system must be known. As this is mainly governed by the voltage divider providing the converting device, only this component will be investigated in the following. Basically, the transfer function can be determined in either the frequency or the time domain. The set-up commonly used for the second method is sketched in Fig. 7.41.

Usually, a step voltage of some 100 V in magnitude and a rise time in the nanosecond range is applied. To accomplish such a short rise time commonly a mercury-wetted relay is used, which is often referred to as Reed-Relay. If switching at repetition rate around 100 Hz, disturbing background noises can effectively be rejected by the so-called averaging mode. Due to the proximity effect, it has to be taken care that the voltage divider is arranged in agreement with real HV test conditions where neither the HV connection lead nor the measuring cable should be replaced after the performance test has been carried out.

Among others, the dynamic behaviour is mainly affected by the stray capacitances between divider column and earth as well as other grounded structures. For a better understanding, consider the equivalent circuit of a resistive voltage divider, as shown in Fig. 7.42a. Here, the HV arm is subdivided in n equal elements where the partial resistors as well as the associated earth capacitances are assumed as linearly distributed along the entire HV divider column:

$$R_{11} = R_{12} = \cdots = R_{1n} = R_1/n \quad C_{e1} = C_{e2} = \cdots = C_{en} = C_e/n$$

Under this condition the potential distribution along the HV divider column can be approximated by a hyperbolical function, which is equivalent to that of a long

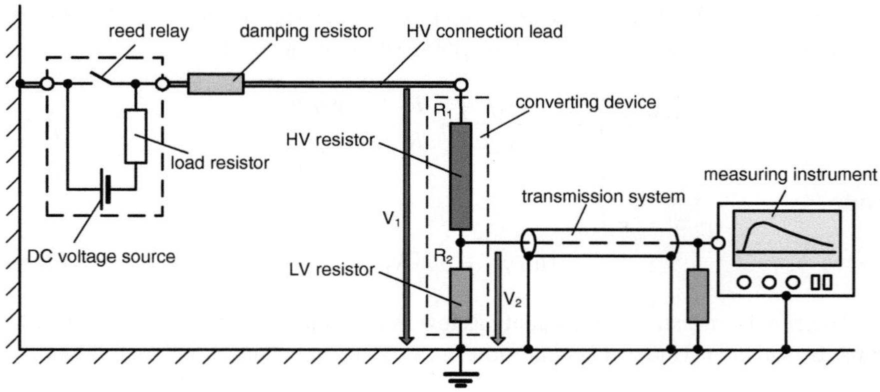

Fig. 7.41 Set-up for measuring the step voltage response of LI voltage dividers

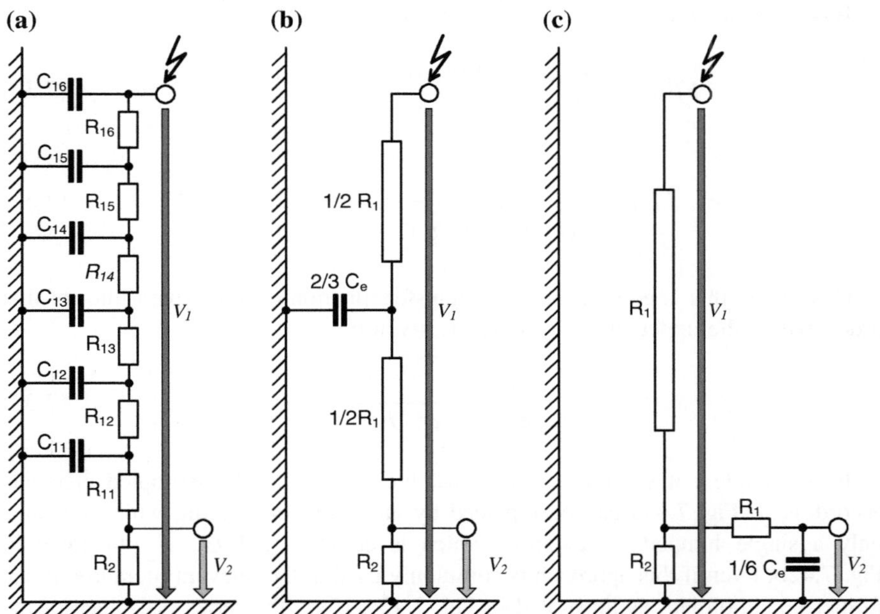

Fig. 7.42 Equivalent circuits of unshielded resistive voltage dividers. **a** Distributed capacitances (transmission line). **b** Lumped capacitance connected to the HV arm. **c** Lumped capacitance integrated in the LV arm (low-pass filter) with $R_s = R_1$ and $C_p = C_e/6$

transmission line (Raske 1937; Elsner 1939; Asner 1960). As the resistance R_1 providing the HV arm is much greater than the resistance R_2 providing the LV arm, the time-dependent voltage $v_2(t)$ appearing across R_2 can be deduced from the high voltage $v_1(t)$ applied to the top electrode using the classical Laplace transformation.

For the here considered network, which is composed of distributed elements according to Fig. 7.42a, the following approximation applies:

$$F(j\omega) = \frac{V_2(j\omega)}{V_1(j\omega)} \approx \frac{R_2}{R_1} \cdot \frac{\sin h(\gamma)}{\sin h(n \cdot \gamma)}, \tag{7.27}$$

with

$$(n \cdot \gamma)^2 = j\omega \cdot R_1 \cdot C_e. \tag{7.28}$$

To keep the measuring uncertainty as low as possible, the inequality $(n \cdot \gamma)^2 \ll 1$ must be satisfied. Under this condition Eq. (7.27) can be simplified as follows:

$$F(j\omega) \approx \frac{R_2}{R_1} \cdot \frac{1}{1 + (n \cdot \gamma)^2/6} = \frac{R_2}{R_1} \cdot \frac{1}{1 + j\omega \cdot R_1 \cdot C_e/6} \approx \frac{R_2}{R_1} \cdot \frac{1}{1 + j\omega \cdot \tau_f} \tag{7.29}$$

Based in this, the amplitude-frequency spectrum can be expressed as follows:

$$F_r(\omega) = \left| \frac{F(j\omega)}{F(0)} \right| = \left| \frac{F(j\omega)}{R_2/R_1} \right| = \frac{1}{\sqrt{1 + (\omega \cdot \tau_f)^2}}. \tag{7.30}$$

$$F_r(f) = \frac{1}{\sqrt{1 + (2\pi f \cdot \tau_f)^2}} = \frac{1}{\sqrt{1 + (f/f_2)^2}}. \tag{7.31}$$

Obviously, this is equivalent to the transfer function of a low-pass filter of first order, where the upper limit frequency is given by

$$f_2 = \frac{1}{2\pi \cdot \tau_f} = \frac{1}{2\pi \cdot R_1 \cdot C_e/6}. \tag{7.32}$$

In this context it should be noted that the network with distributed elements according to Fig. 7.42a can be replaced by an equivalent circuit, which contains only a single lumped capacitance, which is equal to 2/3 C_e, as illustrated in Fig. 7.42b. Even if this approach is commonly used in the relevant literature, it can further be simplified, as illustrated in Fig. 7.42c. Here, the series resistance of the low-pass filter is equal to the resistance R_1 of the HV arm, while the parallel capacitance amounts 1/6 of the total earth capacitance C_e of the HV divider column, which can roughly be estimated using the so-called antenna formula, as will be discussed below more in detail.

The *amplitude-frequency response* of a low-pass filter of first order given by Eq. (7.31) is plotted in Fig. 7.43a. For comparison purpose, the *unit-step response* in the time domain is drawn in Fig. 7.43b, which reads

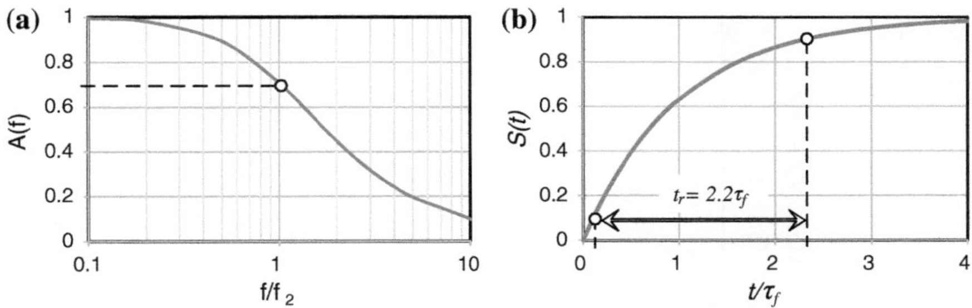

Fig. 7.43 Dynamic behavior of a low-pass filter of first order. **a** Amplitude–frequency response. **b** Unit-step response

$$g(t) = 1 - \exp\left(-t/\tau_f\right). \tag{7.33}$$

Example Consider the HV arm of an LI divider of resistance $R_1 = 10$ kΩ and stray capacitance of $C_e = 30$ pF, one gets the following values providing an equivalent low-pass filter of first order (Fig. 7.42c): $R_s = R_1 = 10$ kΩ and $C_p = C_e/6 = 5$ pF. Based on this the following values of the circuit elements are obtained:

• Response time constant	$\tau_f = R_1 \cdot C_e/6 = 50$ ns
• Rise time	$t_r = 2.2 \cdot \tau_f = 110$ ns
• Upper limit frequency	$f_2 = 1/(2 \cdot \pi \cdot \tau_f) = 3.2$ MHz

To estimate the deviation from the real values of a front-chopped LI voltage, a linearly rising voltage ramp according to Fig. 7.44 (red curve) shall be considered. Under this condition, the recorded voltage (blue curve) follows nearly the applied voltage but delayed by the time constant $\tau_f = 50$ ns. Thus, at chopping time T_c, the following peak value has to be expected:

$$V_p = V_c\left(1 - \tau_f/T_c\right).$$

Assuming the LI voltage is chopped at instant $T_c = 500$ ns, one gets the following relative deviation of the output voltage from the true value of the applied LI test voltage:

$$\frac{V_c - V_p}{V_c} = \frac{\Delta V}{V_c} = \frac{\tau_f}{T_c} = \frac{50\,\text{ns}}{500\,\text{ns}} = 0.1 = 10\%.$$

In this context it must be emphasized that under realistic condition instead of an exponential step response often an oscillating step response is encountered, as exemplarily shown in Fig. 7.45. This is mainly due to the interaction between the inherent inductance of the divider column with the earth capacitance. Thus, for comparison purposes, the so-called area time constant τ_a has been introduced. This approach is based on the fact that the area between the applied unit-step voltage and

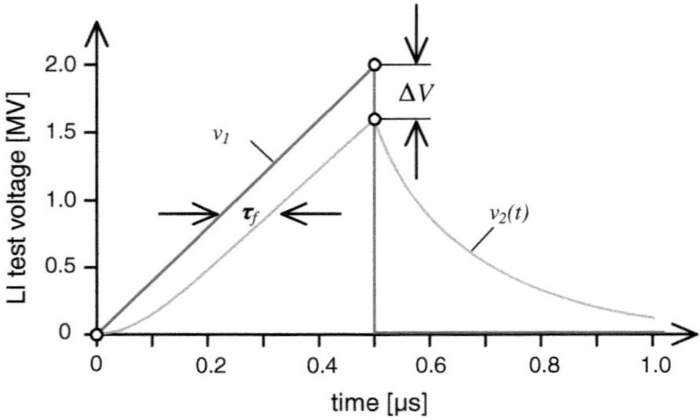

Fig. 7.44 Voltage ramp $v_1(t)$ chopped at time $T_c = 500$ ns and output signal $v_2(t)$ after passing a low-pass filter of first order characterized by the time constant τ_f

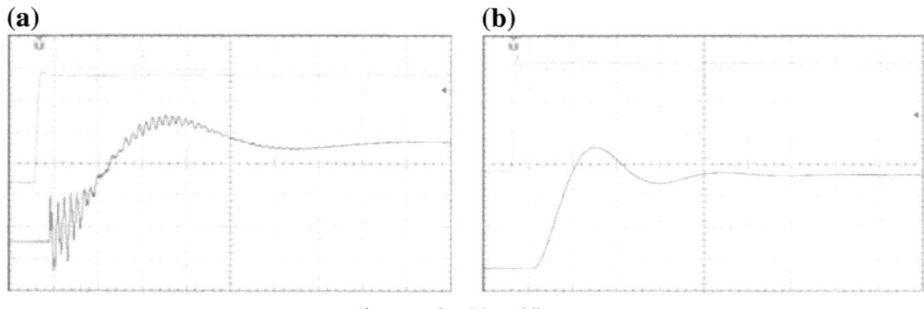

time scale: 20 ns/div

Fig. 7.45 Experimentally determined response (*red curves*) of a capacitive (**a**) and a resistive (**b**) divider subjected to a step voltage (*blue trace*)

the related response function $g(t)$ of a low-pass filter according to Eq. (7.33) is direct proportional to the characteristic time constant τ_f of the low-pass filter:

$$\tau_a = \int_0^\infty \{1 - [g(t)]\}\mathrm{d}t = \tau_f. \tag{7.34}$$

The decisive *unit-step response parameters* recommended in IEC 60060-2, are illustrated in Fig. 7.46b and defined as follows:

• Experimental response time	$T_N = T_\alpha - T_\beta + T_\gamma - T_\delta + T_\varepsilon$
• Partial response time	T_α
• Residual response time	$T_R = T_\beta - T_\gamma + T_\delta - T_\varepsilon \ldots$

(continued)

(continued)

• Over-shoot	β_{rs}
• Settling time	t_s *(definition see text below)*
• Origin of step response	O_1 *(definition see text below)*

The unit-step response is usually recorded for a nominal epoch, which is defined as the range of values between the minimum (t_{min}) and the maximum (t_{max}) of the relevant time parameters of the impulse voltage for which the measuring system is to be approved, such as the front time T_1 for full and tail-chopped LI voltages, the time to chopping T_c for front-chopped LI voltages, and the time to peak T_p for SI voltages. Moreover, the so-called reference level epoch has to be considered, which is defined as the time interval in which the reference level of the step response is determined with its lower limit being equal to 0.5 times of the lower limit of the nominal epoch (0.5 t_{min}) and its upper limit being equal to 2 times the upper limit of the nominal epoch (2 t_{max}). The above listed parameters are determined for the time span, which elapses between the origin O_1 of the recorded unit step response and the settling time t_s, with O_1—the instant at which the recorded response curve starts a monotonic rise above the amplitude of the noise at the zero level of the unit step response, and t_s—the shortest time for which the relative contribution of the residual response time becomes and remains less than 2% of t_s, that means $/T_N - T_R/ \leq 0.02\ t_s$, which must be satisfied for $t > t_s$.

In the past, it was a common practice to estimate the measuring uncertainty of voltage dividers by the so-called convolution method, which is based on the characteristic unit-step response parameters. Results of international round robin tests revealed, however, that this method is not capable of determining the measuring uncertainty of technical HV impulse dividers within the limits specified in the relevant IEC standards. Among others (e.g. assumption of a low-pass filter) this is due to the fact that the origin of the recorded curve, which is denoted in IEC 60060-2010 as O_1, appears more or less delayed if compared with that instant $t = 0$ when the voltage step is applied, see Fig. 7.45. This signal delay is mainly

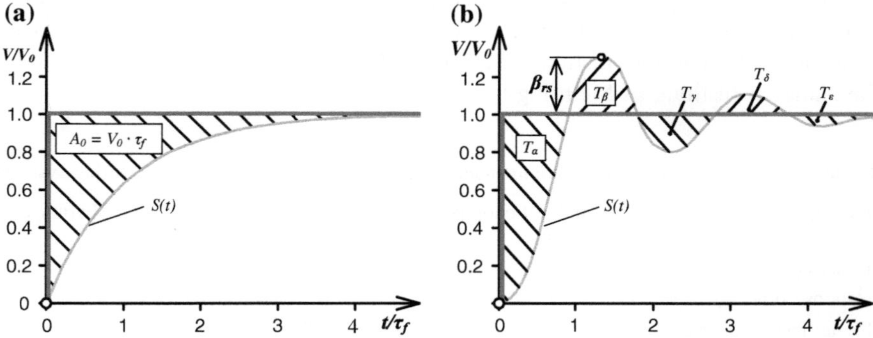

Fig. 7.46 Unit-step response parameters of voltage dividers. **a** RC circuit (low-pass filter of first order). **b** RLC circuit

governed by the travelling wave velocity, which is in the order of 30 cm/ns for HV connection leads and the divider column as well, and almost 20 cm/ns for measuring cables. Moreover, disturbing oscillations could be excited (Fig. 7.45), for instance, due to reflections at the ends of the HV connection leads, or even caused by a poor design of the ground return leads. Consequently, the above mentioned origin O_1 of the recorded response function cannot exactly be determined and leads thus inevitably to an erroneous determination of the characteristic response time parameters.

Despite the problems presented above, the determination of the unit-step response parameters has become a widely established procedure to optimize the dynamic behaviour of voltage dividers. Moreover, this method is recommended as a "finger print" for performance checks (see Sect. 2.3.2). To calibrate the *scale factor* as well as to prove the adequate dynamic behaviour, the standard IEC 60060-2:2010 recommends the comparison with *reference measurement systems* (*RMS*, see Sect. 2.3.3), which shall be capable of determining the expanded measuring uncertainty within the following limits:

$\leq 1\%$ for the peak values of full LI and SI voltages as well as tail-chopped LI voltages
$\leq 3\%$ for the peak values of front-chopped LI voltages
$\leq 5\%$ for the time parameters of LI and SI voltages, in its range of use.

To prove the appropriate performance of a HV measuring system, the calibration shall be based on comparison against such a RMS. Its calibration shall be traceable to the standards of a National Metrological Institute. For this purpose, LI voltages of various front times shall be applied covering the nominal epoch range. Alternatively, the *scale factor* of the reference measuring system shall be established for one impulse voltage shape by means of a higher-class reference measuring system at the relevant test voltage level. Additionally, the measured *unit-step response parameters* of a RMS shall satisfy the recommendations summarized in Table 7.3.

Due to the fact that the measuring uncertainty of a HV measuring system is mainly governed by the voltage divider itself, the above recommendations can in principle also be adopted for reference voltage dividers alone either to be approved by a National Metrology Institute or even by a Calibration Laboratory accredited by the National Institute of Metrology.

Table 7.3 Response parameters recommended for LI and SI voltage reference measuring systems

Parameter	Full and tail-chopped LI voltages (ns)	Front-chopped LI voltages (ns)	SI voltages (µs)
Experimental response time T_N	≤ 15	≤ 10	
Partial response time T_α	≤ 30	≤ 20	
Settling time t_s	≤ 200	≤ 150	≤ 10

7.4.2   Design of Voltage Dividers

To measure high impulse voltages, in principle resistive, capacitive or even mixed voltage dividers are applicable, see Fig. 2.10. As the design of LI voltage dividers is the most challenging task, the following treatment is addressed mainly to this subject. Generally, it has to be taken into account that LI dividers should always be arranged "after" the test object, as illustrated in Fig. 7.47, i.e. never "between" impulse generator and test object. Using the latter arrangement the transient current through the test object would cause an additional inductive voltage drop across the HV connection lead. Consequently, a peak voltage higher than that applied to the test object would be measured.

7.4.2.1   Resistive Voltage Dividers

To measure high LI voltages, in particular when chopped in the front, resistive voltage dividers should preferably be employed. Such dividers are in use since the beginning of the last century (Binder 1914; Peek 1915; Grünewald 1921; Marx 1926; Bellaschi 1933; Burawoy 1936; Finkelmann 1936; Hagenguth 1937; Raske 1937; Elsner 1939). As discussed previously, the dynamic behaviour and thus the measuring uncertainty of HV dividers are mainly governed by the stray capacitance C_e between the HV divider column and earth. This can roughly be assessed using the so-called antenna formula (Küpfmüller 1990), where the divider column is replaced by a vertical metallic cylinder of high h and diameter d. Under the generally satisfied condition $d \ll h$ one gets the following approximation:

$$C_e \approx \frac{2\pi \cdot \varepsilon_0 \cdot h}{ln(h/d)} \approx 56(\text{pF/m}) \cdot \frac{h}{ln(h/d)}. \tag{7.35}$$

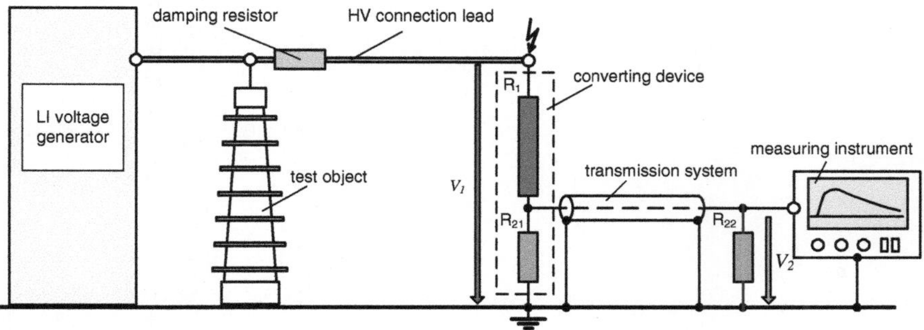

Fig. 7.47 Arrangement of a measuring system for LI test voltages

Example Consider a HV divider column of $h = 4$ m in high and $d = 0.1$ m in diameter. Inserting these values in Eq. (7.35), the earth capacitance attains $C_e \approx 58$ pF. Provided, the resistance of the HV arm amounts $R_1 = 10$ kΩ, one gets the following characteristic time constant of the equivalent circuit according Fig. 7.42: $\tau_f = R_1 \cdot C_e/6 \approx 100$ ns. Measuring a LIC voltage chopped at instant $T_c = 0.5$ μs, the measured peak value would be about 20% lower than the true value applied to the test object, see Fig. 7.44. Consequently, the here investigated voltage divider is not capable of measuring front-chopped LI voltages in compliance with in IEC 60060-2.

A photograph of a resistive divider designed for measuring LI voltages up to 2.2 MV is shown in Fig. 7.48. Despite of the comparatively high resistance being $R_1 = 10$ kΩ, an experimental response time as low as 15 ns was achieved, which is considerably lower than that gained for the above investigated divider of comparable geometric dimensions and equal resistance of the HV arm. Among others, this was accomplished by the use of a "shielding" electrode, as originally proposed by Davis in 1928 and Bellaschi in 1933, because this reduces the impact of the earth capacitance on the response time constant τ_f. So the partial currents I_{e1}, I_{e2} ... and I_{e6} between divider column and earth are more or less compensated by the partial currents I_{h1}, I_{h2} ... and I_{h6} between divider column and HV electrode, as schematically shown in Fig. 7.49.

Additionally, the field grading along the HV divider column was optimized, on one hand by the mentioned larger top electrode (Fig. 7.50) and, on the other hand, by modifications of the windings. For the latter purpose the pitch of the helix

Fig. 7.48 Photograph of a shielded resistive 2.2-MV LI voltage divider (Courtesy of TU Dresden)

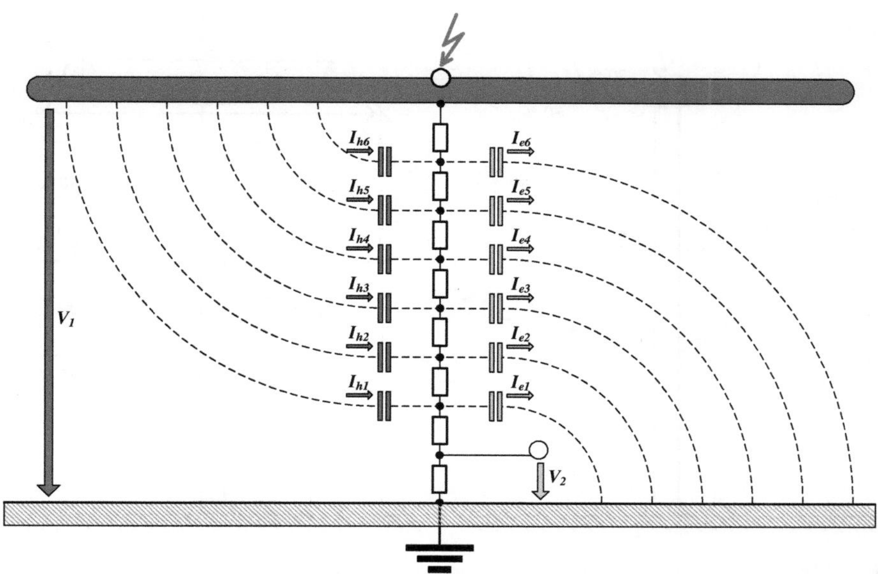

Fig. 7.49 Principle of the compensation of the partial currents between divider column and earth by the currents between HV electrode and divider column

formed by the resistive wire wounded around an insulating core was varied accordingly, as originally proposed by Goosens and Provoost in 1946. Optimum conditions are accomplished when the field distribution along the divider column equals the field distribution between the top and earth electrodes alone, which would occur in absence of the divider column. Moreover, the inductance of the resistive HV arm was minimized by means of two bifilar windings wounded in opposite direction around a cylindrical core (Spiegelberg 1966). Another option earlier employed to minimize the inductance of the HV divider column is the use of meander-like resistors (Mahdjuri-Sabet 1977) known as *Schniewind-band*.

Based on practical experiences it can be stated that shielded resistive dividers having a HV resistance in the order of 10 kΩ are the only one applicable for measuring front-chopped LI voltages in compliance with IEC 60060-2:2010 up to about 2 MV. For measuring higher LI voltages, the tall dividers must be equipped with excessively large and thus expensive shielding electrodes, which would reduce the mechanical stability. A carefully adapted shielding and potential grading is also limited under practical conditions due to the fact that the clearance to grounded and even energized structures must be chosen as large as possible to avoid any disturbances of the optimized field grading.

To improve the dynamic behavior, in principle the resistance of the HV arm could also be reduced significantly below 10 kΩ. However, this would substantially decrease the tail time and even the output voltage of the LI generator. Moreover, it has to be taken into account that the power dissipation in the HV divider column

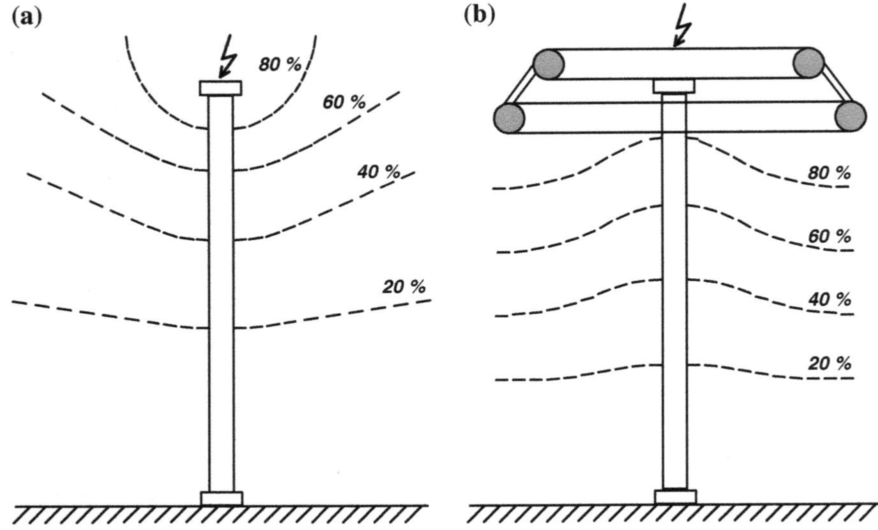

Fig. 7.50 Distribution of equipotential lines along the HV column of resistive LI voltage dividers. **a** Without field grading. **b** With field grading using a large shielding electrode

increases with the square of the applied voltage, which leads to a serious temperature rise. Based on practical experience it can be stated that an experimental response time below 15 ns can only be achieved when the resistance of the HV arm is reduced to few kΩ. Such voltage dividers, however, are only applicable for measuring LI voltages below 500 kV.

7.4.2.2 Damped Capacitive Dividers

To overcome the above discussed obstacles of pure resistive voltage dividers, the use of capacitive dividers seems to be a reasonable alternative because neither the output voltage of the LI generator nor the time parameters are adversely affected by a capacitive load. Moreover, the divider column will never be heated up, even if voltages up to several MV are applied. The main disadvantage of pure capacitive dividers is, however, that heavy oscillations might be excited, as exemplarily shown in Fig. 7.51a.

This is due to the fact that capacitive dividers act in the high-frequency range like an oscillating circuit, because under this condition the divider column composed of stacked capacitors can be simulated by a metallic cylinder and thus by an inductance, which interacts with the earth capacitances. To prevent such oscillations Zaengl (1964) and Spiegelberg (1964) proposed an interconnection of the stacked capacitors via serial resistors. The main benefit of this approach is not only that disturbing oscillations can drastically be attenuated, as exemplarily shown in Fig. 7.51b, but that the experimental response time constant can be reduced

(a)   **(b)**

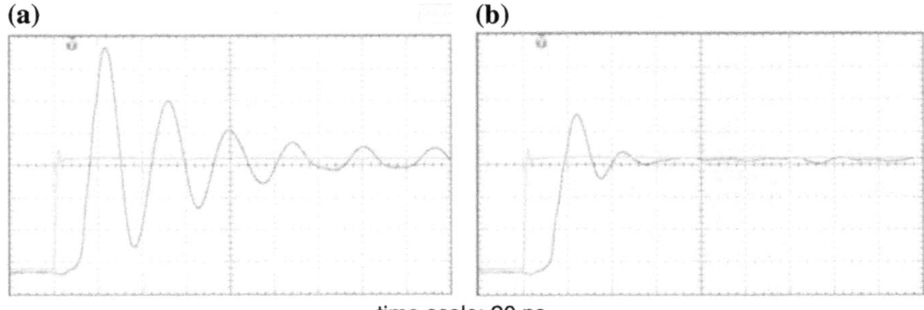

time scale: 20 ns

Fig. 7.51 Step-pulse response of a capacitive divider without and with a damping resistor

drastically, which is in contrast to the transfer characteristics of classical RC low-pass filters where the experimental response time constant increases proportional to the series resistance.

For a better understanding, the equivalent circuit shown in Fig. 7.52b shall be analysed in the higher frequency range. Under this condition the series capacitances act as short-circuited, so that only the following circuit elements must be considered:

$$R_d = R_{11} + R_{12} + \cdots + R_{1n} = n \cdot R_{11},$$
$$L_d = L_{11} + L_{12} + \cdots + L_{1n} = n \cdot L_{11},$$
$$C_p = C_e/6.$$

To simplify the following treatment, the divider column shall be replaced by a metallic cylinder of high h and diameter d. Based on the classical antenna formula (Küpfmüller 1990) the equivalent inductance can roughly be assessed by

$$L_d \approx \frac{\mu_0 \cdot h \cdot ln(h/d)}{2\pi} \approx 0.2(\mu H/m) \cdot \frac{h}{ln(h/d)}. \tag{7.36}$$

Combining this with Eq. (7.35), the traveling wave impedance of the here considered rod antenna can be assessed as follows:

$$Z_a \approx \sqrt{\frac{L_d}{C_p}} \approx \sqrt{\frac{\mu_0}{\varepsilon_0}} \cdot \frac{1}{2\pi} \cdot ln\left(\frac{h}{d}\right) = \frac{Z_0}{2\pi} \cdot ln\left(\frac{h}{d}\right). \tag{7.37}$$

Inserting the characteristic wave impedance $Z_0 = \sqrt{\mu_0/\varepsilon_0} \approx 377\,\Omega$, it can also be written:

$$Z_a \approx \frac{377\,\Omega}{2\pi} \cdot ln\left(\frac{h}{d}\right) \approx (60\,\Omega) \cdot ln\left(\frac{h}{d}\right). \tag{7.38}$$

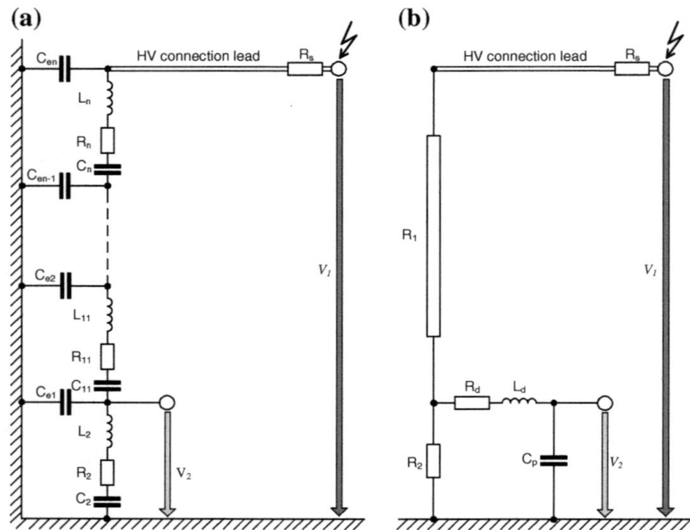

Fig. 7.52 Equivalent circuit of a damped capacitive divider. **a** Damped C-divider with distributed elements. **b** Ideal R-divider connected to a RLC network

For further simplification it shall be assumed in compliance with the classical network theory that a transition from an oscillating response to a monotonic one is accomplished when the series resistance is chosen as

$$R_d > 2\sqrt{\frac{L_d}{C_p}}.$$

(7.39)

Replacing C_p by $C_e/6$ according to Fig. 7.42 and combining Eq. (7.39) with Eq. (7.38) one gets the following approach:

$$R_d > 2\sqrt{\frac{6 \cdot L_d}{C_e}} \approx \frac{5 \cdot Z_0}{2\pi} \cdot ln\left(\frac{h}{d}\right) \approx 0.8 \cdot Z_0 \cdot ln\left(\frac{h}{d}\right) \approx (300 \ \Omega) \cdot ln\left(\frac{h}{d}\right).$$

(7.40)

Even if this approach is the result of strong simplifications, it has successfully been proven as a fundamental design criterion for damped capacitive voltage dividers. In this context it has to be taken into account, however, that in addition to the "internal" damping resistor following from Eq. (7.40) also an "external" HV resistor of about 300 Ω has to be connected between test object and top electrode of the divider to prevent the occurrence of travelling waves.

Example Consider a 2 MV divider which is composed of five stacked capacitors, where the total divider column shall be $h = 5$ m in high and $d = 0.25$ m in diameter, i.e. $h/d = 20$. Based on Eq. (7.40) one gets $R_d > (300 \ \Omega) \cdot \ln (20) \approx 900 \ \Omega$. Thus the five stacked capacitors should be connected by four series resistors, each of $(900 /4) \ \Omega \approx 225 \ \Omega$. Using an additional external resistor of 300 Ω, the total series resistance of the HV divider arm attains $(900 + 300)\ \Omega = 1{,}200\ \Omega$. This has to be taken into account to choose the resistance to be integrated in the LV arm.

In this context it should be emphasized that the design of the LV arm of damped capacitive dividers is a challenge because the ratio between the inductances of the HV and LV arm must comply with the divider ratio, which equals the resistance ratio and is inversely proportional to the capacitance ratio:

$$L_2/L_1 = R_2/R_1 = C_1/C_2.$$

Example Applying Eq. (7.36) the effective inductance of the HV arm of the above considered 2 MV divider column of $h = 5$ m in high and $d = 0.25$ m in diameter gets $L_1 = 0.2$ (μH/m) $\cdot$ 5 m $\cdot$ ln (20) $\approx 3\ \mu$H. Assuming, for instance, a divider ratio $R_2/R_1 = C_1/C_2 = 1/1000$, the inductance L_2 of the LV arm must be chosen as low as 3 nH.

The most effective way to minimize the inductance of the LV arm is to design this like a disc, i.e. to connect a great number of elements in parallel, as obvious from Fig. 7.53. Her each parallel connected element comprises a series connection of a capacitor with a resistor.

Comparing the different divider types, it can be stated that damped capacitive dividers provide an excellent dynamic behaviour. Another benefit is that the power dissipating in the resistors inserted between the stacked capacitors is quite low due to the short duration of the transient current through the HV divider arm. Moreover, the effective capacitance of the HV arm provides a basic load for the LI generator. As the charging current is inversely proportional to the frequency, damped capacitive dividers are also well suited for measuring SI and AC voltages as well as composite voltages. If equipped with a high-ohmic resistor, which is connected in parallel to the HV arm, such an "universal" divider, as shown in Fig. 2.10, is also capable of measuring DC voltages including superimposed voltage ripples covering a frequency spectrum up to several kHz. Due to the wide field of application, the here presented damped capacitive divider is often also referred to as *"multi-purpose divider"*.

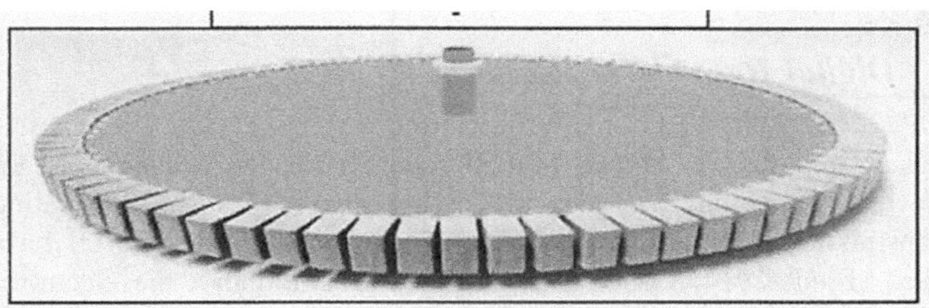

Fig. 7.53 LV arm of a damped capacitive divider comprising 60 parallel elements each composed of a resistor connected in series with a capacitor

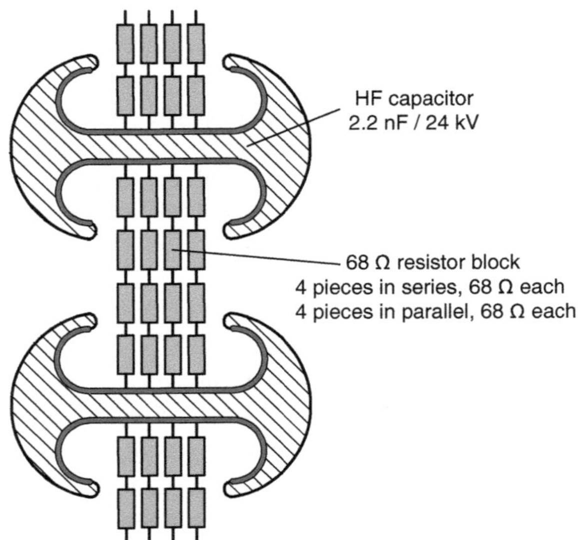

Fig. 7.54 Section of a damped capacitive divider, rated voltage 200 kV

Due to the benefits mentioned above, most reference measuring systems (RMS) are nowadays equipped with damped capacitive dividers. In the following the general design principle of a 200 kV LIC divider shall briefly be described. The HV arm is composed of 10 ceramic capacitors in series each rated 25 kV/2 nF. The individual capacitors are connected in series via 9 resistor blocks, as illustrated in Fig. 7.54. Step pulse response measurements revealed that an optimum dynamic behaviour was achieved when the resistance of each resistor block is chosen to about 70 Ω. This was realized by means of low-inductive metal-oxide resistors rated 68 Ω. Four of them are connected in series and four of them in parallel, so that the resulting resistance of each resistor block attains also 68 Ω. Using the classical step-voltage response method the following time parameters according to Table 7.3 were determined:

Experimental response time: $T_N \approx 2$ ns
Partial response time: $T_\alpha \approx 5$ ns
Settling time: $t_s \approx 120$ ns.

7.4.3 Digital Recorders

To measure fast transient signals in high-voltage technology, such as LI test voltages as well as travelling waves excited by lightning surges and propagating along HV transmission lines, originally *cathode-ray oscilloscopes (CRO)* have been employed (Binder1914; Gabor 1927; Krug 1927). Initially, the electromagnetic capability (EMC) was not a problem because the first available CRO were especially designed for HV measurements and thus not equipped with sensitive amplifiers, which have later been employed for the vertical and horizontal deflection of the electron beam. In the 1940s, peak voltmeters have been employed, using the first available vacuum tubes to rectify the fast LI signal. Digital recorders, originally referred to as "digitizers", entered in HV measuring technique in the early 1980s (Malewski et al. 1982). After the initial EMC problems have successfully been solved (Strauss 1983 and 2003; Steiner 2011) digital recorders are nowadays almost exclusively used for recording and processing LI and SI test voltages (Fig. 7.55). Due to the recent achievements in digital signal processing (DSP), computerized digital recorders are also increasingly employed not only for LI and SI test voltage measurements but also for AC and DC voltage measurements as well as for measuring composite and combined test voltages.

A photograph of a stand-alone device operating in connection with an industrial PC is shown in Fig. 7.56. As obvious from the simplified block diagram sketched in Fig. 7.57, the main units of a digital recorder are the attenuator at the input followed by a low-noise amplifier, an analogue–digital converter (ADC), a memory unit and an industrial computer (IPC). A micro-controller serves for the adjustment of the input sensitivity as well as for controlling the various units performing the digital signal processing, acquisition and data storage. The industrial computer runs with a specific software package, which enables the acquisition of the stored raw data and the visualization of the time-dependent input signal. Simultaneously, the relevant impulse parameters are indicated, such as the peak value of the measured LI/SI test voltage, the front and tail time and occasionally the chopping time.

The measuring uncertainty may be affected by both the hardware and the software. Due to the comparatively high quantization rate of nowadays available digital recorders, which attains usually 12 or 14 bits and even more, as well as the high sampling rate being in the order of 100 MS/s and above, the contribution of the hardware to the measuring uncertainty is much lower than that of voltage dividers. As the quantization error is given by 50% of the *least significant bit* (LSB), this is approx. 0.012% for a slow-rising signal at quantization rate of 14 bit. However, the deviation from the true value increases significantly when front-chopped LI voltage according to Fig. 7.58 are measured at a sample rate of 100 MS/s. This is because the voltage difference ΔV_s between each sample is inversely proportional to the sampling rate f_s. For an assumed chopping time of, for instance, $T_c = 0.5$ μs and a crest voltage V_c, the voltage difference between each sample can be approximated by $V_s = V_c/(T_c \cdot f_s) = V_c/50 = 0.02 \cdot V_c$. Under this condition the maximum deviation of the measured value from the true value, which is given by 0.5 LSB, attains 1% of the crest value V_c.

Performing a digital signal processing, generally the classical *Shannon theorem* has to be taken into account (Shannon 1949; Stanley 1975; Robinson and Silvia 1978). Based on this it can be stated that the sampling rate should exceed twice the maximum frequency content of the signal to be measured, where the analogue bandwidth must substantially exceed this maximum frequency content. With reference to IEC 61083-1:2001 the analogue bandwidth should be not lower than 6 times the sampling rate.

To validate the measuring uncertainty, various calibration procedures are recommended in IEC 61083-1:2001, which refer to both the hardware and the

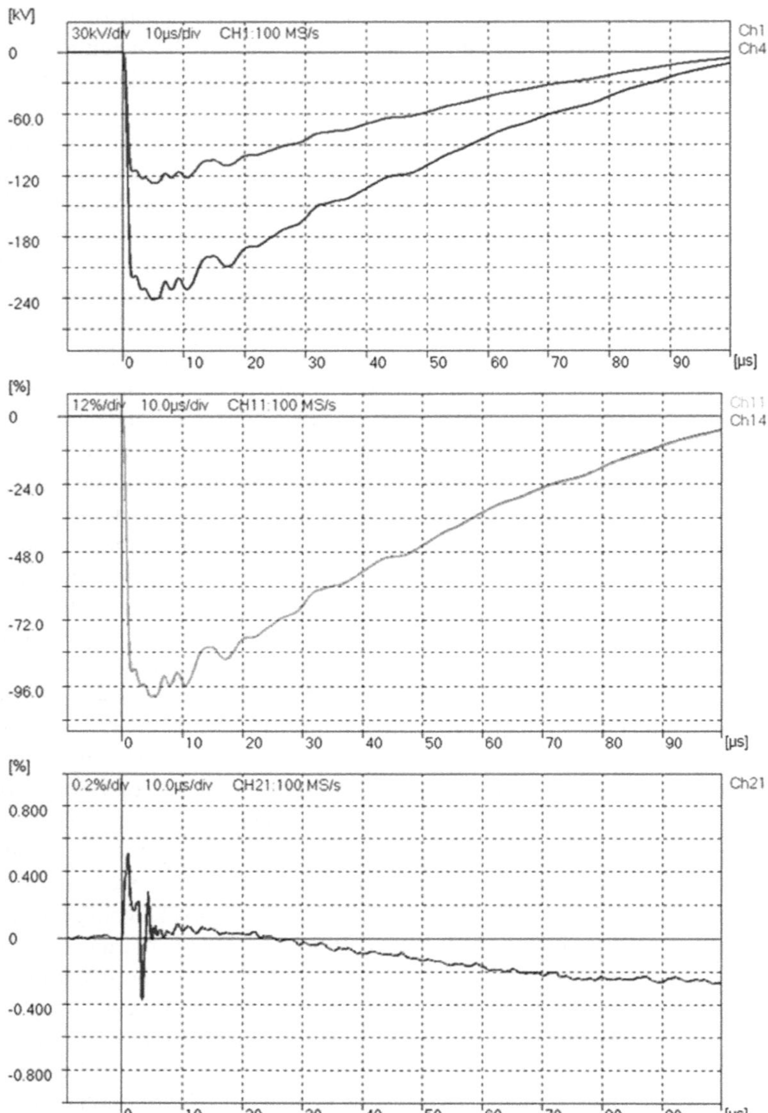

Fig. 7.55 Recording and processing of LI test voltages (From the top: records of 50 and 100% test voltages; comparison of the normalized voltages; difference of the normalized voltages)

software. To validate the uncertainty caused by the hardware, the *differential non-linearity (DNL)* as well as the *integral non-linearity (INL)* have to be determined using an "ideal" ADC as reference. The transfer function of such a reference ADC is characterized by a stepwise-increasing output code $k = 1, 2 \ldots 2^N$, where the input voltage is increased stepwise by $w(r) = V_{fsd}/2^N$. That means, for an output

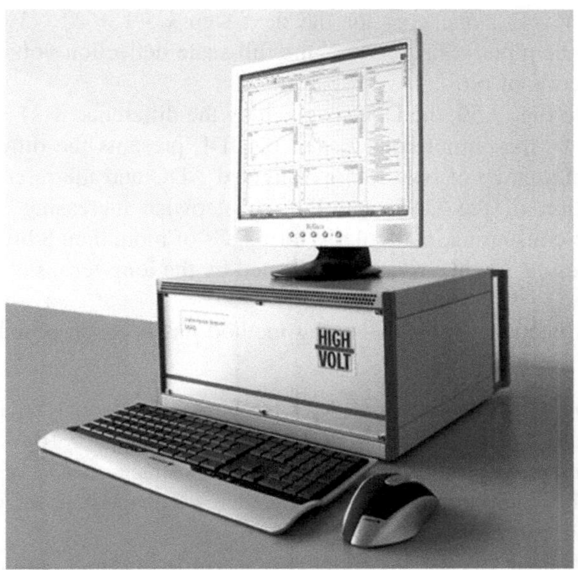

Fig. 7.56 Photograph of a stand-alone digital recorder

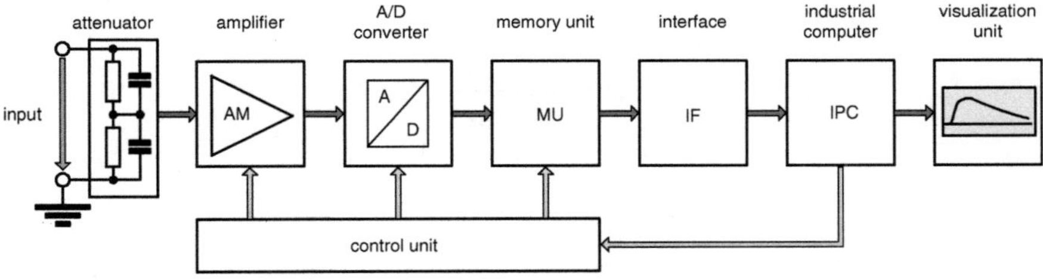

Fig. 7.57 Simplified block diagram of a digital recorder for HV measurements

Fig. 7.58 Quantization error
of a front-chopped LI test
voltage due to the analogue–
digital conversion

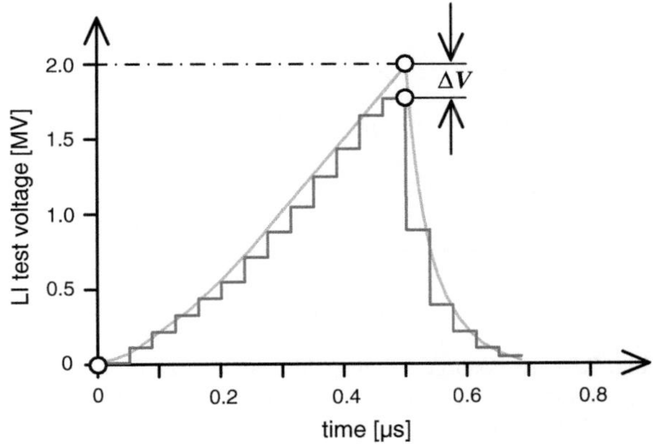

code $k = 2^N \cdot V_{in}/V_{fsd}$ one gets for the next step $k+1 = 2^N \cdot (V_{in}+w(r))/V_{fsd}$. Here are V_{in}—the input voltage, V_{fsd}—the full-scale deflection voltage and N—the resolution in terms of bit.

According to Fig. 7.59, the DNL is given by the difference $w(k) - w(r)$ for each possible value of the output code k, and the IDL presents the difference $s(k)$ between the input voltage of both the investigated ADC and the reference ADC. As the determination of the DNL and IDL at stepwise increasing DC voltage is extremely time-consuming, particularly for ADCs of more than 8 bit resolution, the experimental results might strongly be affected by the long-term stability of the DC voltage source.

Thus, as an alternative, the following method has been proposed by Steiner in 2011, see Fig. 7.60:

1. Apply a precise sine wave of low frequency (e.g. 50 Hz) and adjust the amplitude close to the full-scale deflection (e.g. between -10 V and $+10$ V).
2. Record in total ten periods of the applied AC voltage.
3. Compare the count of occurrences of each code (code rate) with the ideal number of occurrences.
4. Divide each code rate by the number of recorded periods, i.e. $n = 10$ for the here-considered case.
5. Calculate the differential non-linearity (DNL) for each code:

$$d(k) = \frac{h(k)_{\text{nonlin}}}{h(k)_{\text{lin}}} - 1.$$

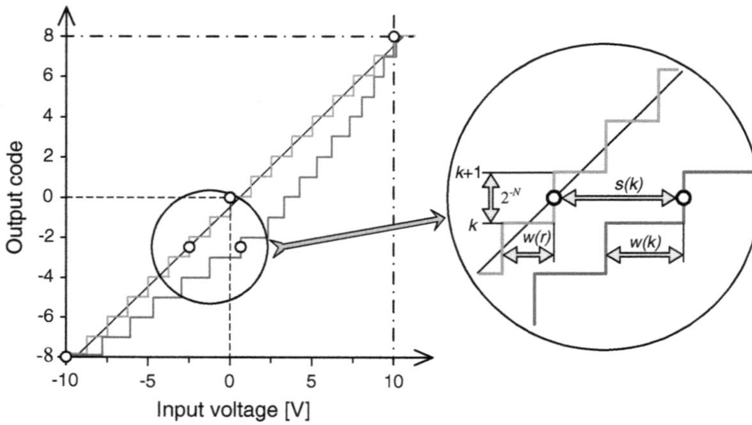

Fig. 7.59 Estimation of the differential non-linearity given by $w(k) - w(r)$ and integral non-linearity $s(k)$ given by the deviation of the actual input voltage from the reference voltage of an "ideal" ADC

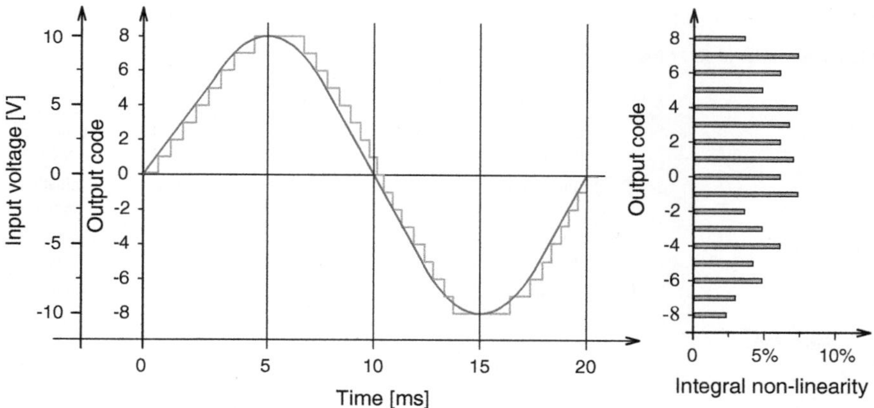

Fig. 7.60 Comparison of a sin-shape input voltage with the output code of a non-linear 4-bit ADC

6. Calculate the integral non-linearity (INL) for each code:

$$s(k) = \sum_{i=0}^{k} d(i).$$

The test is passed if at full-scale deflection the requirements $d(k) = 10.81\%$ and $s(k) = 10.51\%$ are satisfied.

The software package is tested according to IEC 61083-2:2013. It recommends new rules for the calibration including the calculation of the k-factor relevant for LI test voltages of substantial over-shoot. To verify the performance of digital recorders intended for LI voltage measurements, in total 52 reference curves are recommended to be created by a computerized test data generator (TDG). These artificial curve shapes are available from either a compact disc (CD) or even an USB stick. Basically, the reference impulse wave-shapes applied should be representative for the test voltage shapes specified in IEC 60060-1:2010 and IEC 60060-3:2006, such as:

- Full lightning impulse voltage (LI)
- Front-chopped lightning impulse voltage (LIC)
- Tail-chopped lightning impulse voltage (LIC)
- Oscillating lightning impulse voltage (OLI)
- Switching impulse voltage (SI)
- Oscillating switching impulse voltages (OSI).

After digital signal processing of the data received from the TDG, the significant impulse parameters are determined by means of the implemented software. The procedures shall follow the evaluation principles given in IEC 60060-1:2010. These refer to the measurement of the peak value and the over-shoot as well as to the

Table 7.4 Test of software with the test data generator (TDG) according to IEC 61083-2:2013

Reference number of IEC 61083-2	Parameter	IEC reference value	IEC acceptance limits	Example: evaluated values	Remarks
LI – A4[a]	Test voltage value V_t	−856.01 kV	−(855.15... 856.87) kV	−856.4 kV	Evaluation accepted
	Front time T_1	0.841 µs	(0.824... 0.858) µs	0.851 µs	Evaluation accepted
	Time to half-value T_2	47. 80 µs	(47.32... 48.28) µs	47.88 µs	Evaluation accepted
	Relative over-shoot β	7.9%	(6.9...8.9) %	7.2%	Evaluation accepted
LI – M7	Test voltage value V_t	1272.3 kV	(1271.0... 1273.5) kV	1272 kV	Evaluation accepted
	Front time T_1	1.482 µs	(1.452... 1.512) µs	1.390 µs	Evaluation rejected[a]
	Time to half-value T_2	50.03 µs	(49.53... 50.53) µs	5 0.10 µs	Evaluation accepted
	Relative over-shoot β	11.2%	(10.2... 12.2) %	9.9%	Evaluation rejected[a]

[a]The tested software must be improved with respect to the evaluation of reference LI—M7
Compared with IEC 61083-2, the polarity of the reference impulses is opposite

characteristic time parameters, such as the front and tail time and the chopping time for LIC, as well. The software is assumed as properly working when the results obtained are within a tolerance band specified in IEC 61083-2:2013, where any manipulation of the test data cannot be accepted.

Example Table 7.4 shows the comparison between the evaluated parameters of two reference LI wave shapes (wave shapes no. LI-A4 and LI-M7 of the test data generator (TDG)) and those required according to IEC 61083-2:2013. Whereas the evaluation of no. LI-A4 delivers results within the tolerance band required by the standard, not all parameters of no. LI-M7 are correctly determined. This means the software has to be improved before it can be applied for practical tests.

Generally, it can be stated that the measurement of the peak value of full LI test voltages at measuring uncertainty of 2% is accomplished when the amplitude resolution amounts 10 bits and the sampling rate attains 100 MS/s, where the analogue bandwidth should not be lower than 100 MHz. Moreover, the integral non-linearity should be below 0.5%, and the internal background noise level should not exceed 0.4% of the full-scale deflection. For more information in this respect see Hällström (2002), Hällström et al. (2003), Wakimoto et al. (2007) and Schon (2013).

7.5   Measurement of High Currents in LI Voltage Tests

Performing LI voltage tests, high *impulse currents* may occur not only as consequence of a breakdown but also when the test object passes the withstand test. This is due to the capacitive load current, which is direct proportional to the comparative high steepness of the applied LI voltage. Moreover it should be mentioned that the magnitude and shape of current pulses associated with LI voltage tests enables the identification of potential insulation failures. For this reason, *impulse current measurements* are mandatory for LI withstand voltage tests of power transformers. Performing such tests, just prior the specified (100%) LI test voltage level first a reference voltage level is applied, which is being 50–70% of the specified test level, see Fig. 7.61. To compare the shapes of both the LI voltage and the associated current pulse, the records are normalized to their respective extreme values.

The LI withstand test is passed when no significant differences are recognized between the normalized reference values and the normalized test values (IEC 60076-3:2013).

Circuits used for LI current measurement are composed of the following components (Fig. 7.62):

- *Converting device*, usually a shunt or even a fast current transformer to convert the transient current into a convenient measurable low voltage,
- Transmission system, usually a BNC measuring cable or a fiber optic link,
- Measuring instrument, usually a digital recorder.

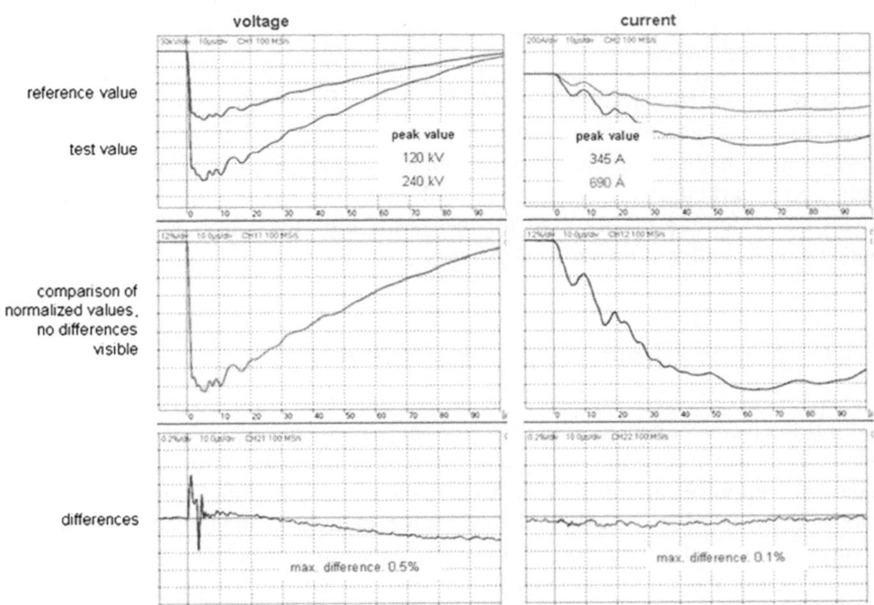

Fig. 7.61 Voltage and current recorded during a LI withstand voltage test of a distribution transformer

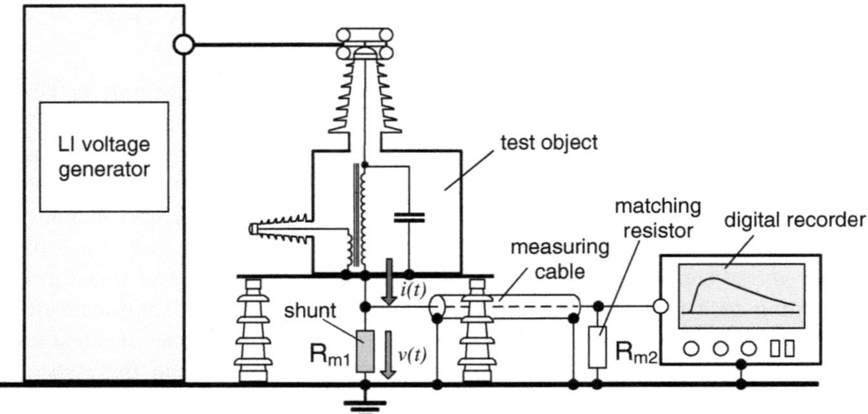

Fig. 7.62 Basic components required for a LI current measuring system

The requirements for digital recorders comply in principle with those specified in IEC 61083-1:2006 for LI voltage measurements (see Sect. 7.4.3) and shall thus not be considered here. This Section is thus only addressed to the design principles of converting devices. For more information on this issue see also the textbooks of Schon (2010, 2013).

Note The classic application of high current measurements is addressed to performance tests of protecting devices, such as *lightning arresters*, where peak values up to some hundreds of kA have to be measured. Measuring such high currents, however, is not the subject of this book. For more details in this respect, see IEC 62475:2012 and the series of the IEC 60099 standards.

To measure high-current pulses associated with LI voltage tests various types of converting devices are in use, such as resistive shunts, Rogowski coils, current transformers, hall-sensors, and magneto-optic sensors. In the following, however, only some particularities of resistive and inductive converting devices will be reviewed, which are mostly used for measuring high transient currents associated with LI withstand and breakdown tests.

7.5.1 Resistive Converting Device (Shunt)

Even if the physical background of current measurements are based on Ohm's law, which is hence easily understandable, it has to be taken into account that for measuring impulse currents in the kA range the resistance of the converting must be chosen extremely low, usually in the $m\Omega$ range, in order to attenuate the measurable voltage down to the Volt range. However, under this condition, the voltage collapsing across the shunt can substantially be affected by parasite inductances, as obvious from Fig. 7.63. Here, the left oscilloscopic record is obtained for a resistive shunt grounded via thin wire, whereas the right record is obtained for a shunt grounded via a Cu-foil. As the inductance of the thin wire is much greater than that of the Cu-foil, the left record shows a signal enhancement in the front region, which is the consequence of an inductive voltage component $v_L(t)$ superimposed on the resistive voltage signal $v_R(t)$, as qualitatively illustrated in Fig. 7.64.

Generally, the measurable voltage signal $v_m(t)$ following from the time-dependent current $i_m(t)$ can be expressed by

$$v_m(t) = v_R(t) + v_L(t) = R_m \cdot i_m(t) + L_m \cdot \frac{\mathrm{d}i_m(t)}{\mathrm{d}t}. \tag{7.41}$$

To prevent erroneous measurements, the inductive voltage component $v_L(t)$ and thus the parasite inductance L_m of the measuring circuit must be kept as low as possible. For a better understanding consider Fig. 7.65, where the inductive measuring loop is indicated by the shaded area. For the here given geometric parameters

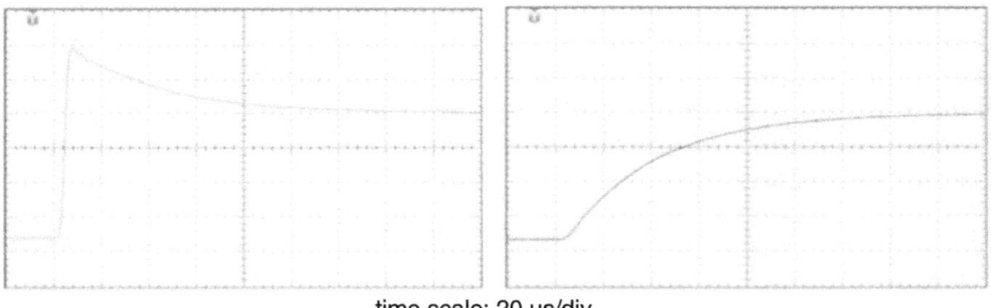

Fig. 7.63 Response of a current measuring system against an exponential rising current where the transducer was grounded via either a thin wire (left) or a low-inductive Cu foil (right)

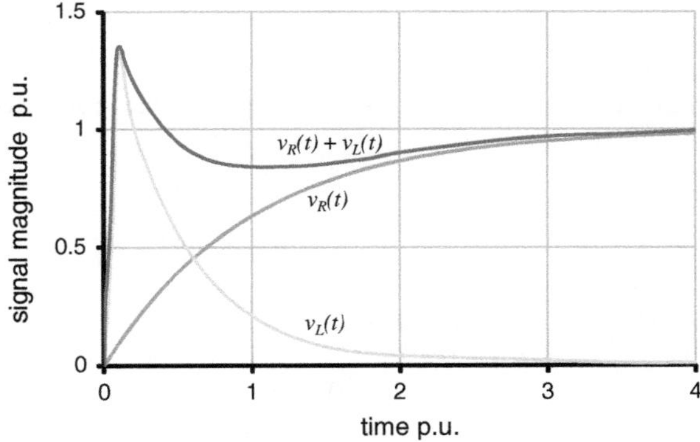

Fig. 7.64 Partial voltages $v_R(t)$ and $v_L(t)$ as well as the measurable voltage $v_m(t) = v_R(t) + v_L(t)$ appearing across a resistive shunt in case of an exponential rising current pulse

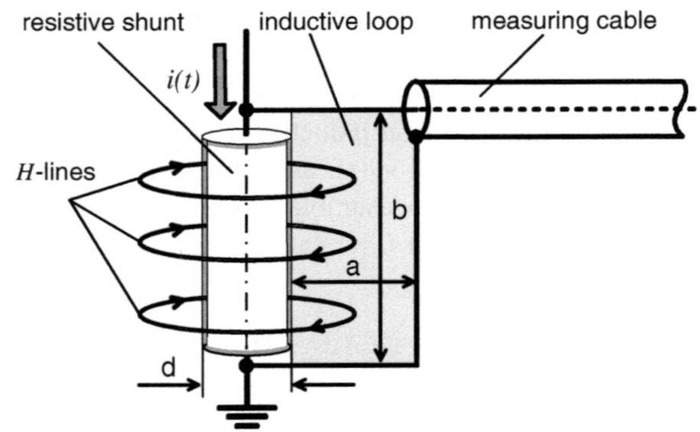

Fig. 7.65 Equivalent circuit used to estimate the inductive voltage component $v_L(t)$ induced in the measuring loop between resistive shunt and BNC measuring cable

a and b the parasite inductance L_c can simply be assessed using the following approach (Küpfmüller 1990):

$$L_c = \frac{\mu_0 \cdot b}{2\pi} \cdot ln\left(\frac{2a+d}{d}\right) \tag{7.42}$$

with $\mu_0 = 0.4\pi$ (nH/mm)—the permeability of air. Additionally, the self-inductance of the resistive transducer has also to be taken into account. Replacing this by a cylindrical conductor of diameter d and length b, the following approach can be adopted (Küpfmüller 1990):

$$L_s \approx \frac{\mu_0 \cdot b}{2\pi} \cdot ln\left(\frac{b}{2d}\right) \tag{7.43}$$

Combining Eqs. (7.42) and (7.43) and inserting $\mu_0 = 0.4\pi$ (nH/mm) one gets the following resulting inductance of the measuring circuit:

$$L_m = L_c + L_s \approx \frac{\mu_0 \cdot b}{2\pi} \cdot ln\left(\frac{a \cdot b}{d^2}\right) = \frac{0.4\pi \cdot b \cdot (1\,\text{nH})}{2\pi} \cdot ln\left(\frac{a \cdot b}{d^2}\right) \tag{7.44}$$

Example Consider a resistive shunt characterized by the following parameters:

Shunt resistance	$R_s = 10$ mΩ
Shunt diameter	$d = 10$ mm
Measuring loop	$a = b = 50$ mm

Inserting these values in Eq. (7.44) one gets

$$L_m \approx (0.2\ \text{nH/mm}) \cdot (50\,\text{mm}) \cdot ln\left(\frac{110 \cdot 50\,\text{mm}^2}{200\,\text{mm}^2}\right) = (10\,\text{nH}) \cdot ln(27.5) \approx 33\,\text{nH}.$$

Provided, an exponential rising current pulse of time constant $\tau_c = 2$ μs attains a crest value of 1 kA at instant $t_p \approx 3 \cdot \tau_c = 6$ μs, the voltage appearing across an "ideal" shunt of 10 mΩ could be described by the following time function:

$$v_r(t) = [(1\,\text{kA}) \cdot (10\,\text{m}\Omega)] \cdot [1 - exp(-t/\tau_c)] = (10\,\text{V}) \cdot [1 - exp(-t/\tau_c)]$$

Due to the above estimated inductance of $L_m = 33$ nH an inductive voltage component would be induced (Fig. 7.64). As the steepness attains a maximum value of 1 kA/2 μs = 0.5 kA/μs at the very beginning of the current pulse, the superimposed inductive voltage component would approach a crest value of $V_c = (33\,\text{nH}) \cdot (0.5\,\text{kA/μs}) = 16.5$ V. Obviously, this is almost 70% greater than the maximum value, which attains a peak value of $V_p = (1\,\text{kA}) \cdot (10\,\text{m}\Omega) = 10$ V across a pure resistive shunt.

To minimize the measuring error, the parasite inductance and thus the shaded area shown in Fig. 7.65 must be kept as low as possible. In practice this is realized by means of coaxial or even disc-like designed shunts, see Fig. 7.66.

7.5.2   *Inductive Converting Device (Rogowski Coil)*

Another way to eliminate the disturbing inductive voltage component is the use of pure inductive converting devices consisting of either a single turn (Fig. 7.67a) or even of numerous turns (Fig. 7.67b), as proposed by Rogowski in 1913. With reference to the geometric parameters obvious from Fig. 7.67, the output voltage of a Rogowski coil without load can roughly be assessed using the following approach (Küpfmüller 1990):

$$v_l(t) = M \cdot \frac{di}{dt} = \left[\frac{\mu_0 \cdot n \cdot b}{8} \cdot \ln\left(\frac{2a+b}{2a-b}\right)\right] \cdot \frac{di}{dt}, \qquad (7.45)$$

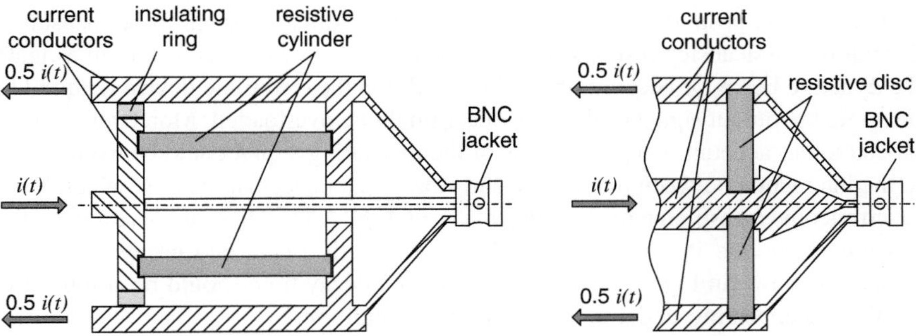

Fig. 7.66 Design principles of low-inductive shunts

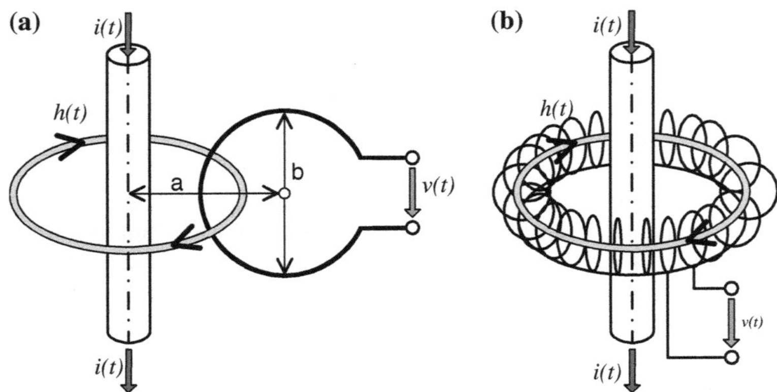

Fig. 7.67 Principle of impulse current measurement using inductive converting devices. **a** Single-turn current transducer. **b** Multiple-turn current transducer (Rogowski coil)

with n—the turn number, a—the distance between the axis of the conductor carrying the time-dependent current $i(t)$ to be measured and the axis of the perpendicular arranged Rogowski coil, and b—the core diameter.

As the output voltage $v_i(t)$ is proportional to the derivate of the time-dependent current, an integration must be performed to get a signal, which is direct proportional to the time-dependent current (Fig. 7.68).

In principle the dynamic behaviour of Rogowski coils is comparable to that of a high-pass filter, i.e. a stationary direct current is not measurable. Consequently, the measuring error increases at decreasing frequency content and thus at increasing length of the current pulse (Fig. 7.68). To reduce the lower-limit frequency, the coil inductance must be increased accordingly, which can effectively be achieved by winding the turns around a high-permeable core. Based on practical experience it can be stated that Rogowski coils are capable of measuring current pulses up to 100 kA/μs, where the pulse duration is limited below some tens of μs. Using classical current transformers, pulse lengths up to the ms range can be captured at reasonable measuring uncertainty, while the steepness is limited to approx 1 kA/μs.

As already discussed above, the output voltage of inductive converting devices without load is proportional to the derivate of the primary current to be measured. Therefore classical step response measurements are not recommended to investigate the dynamic behavior because under this condition the electronic device required to integrate the output signal of the transducer might be overloaded. Moreover it has to be taken into account that oscillations of the measuring signal could be excited due to the interaction between inductance and stray capacitances of the coil, which leads to erroneous measurements, as illustrated in Fig. 7.69. Thus, to investigate the dynamic behavior, it is recommended to use linearly rising and decaying current ramps, as shown in Fig. 7.70, where the rise and decay time should be comparable with those appearing under actual measuring conditions.

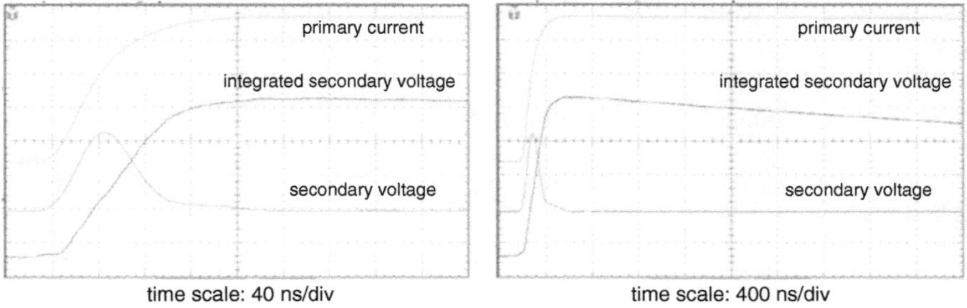

Fig. 7.68 Transfer characteristics of a Rogowski coil

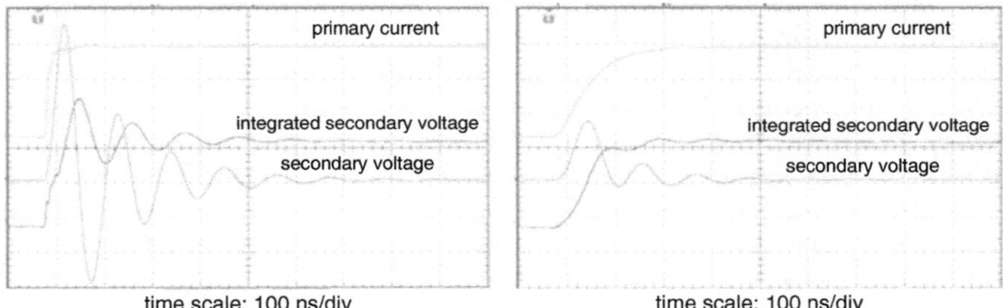

Fig. 7.69 Dynamic behavior of a current transformer at fast rising (left) and a slowly rising (right) primary current

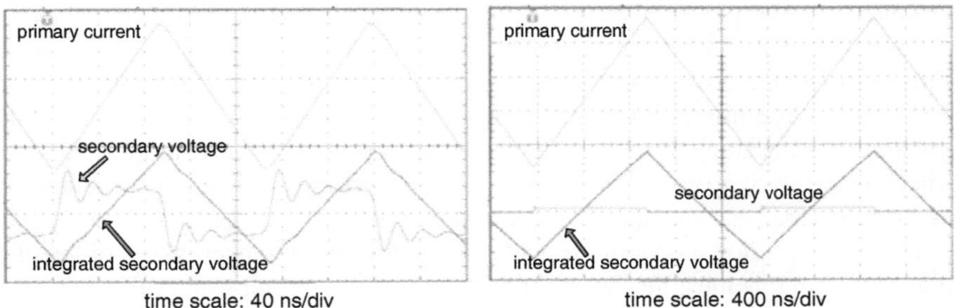

Fig. 7.70 Dynamic behavior of a current transformer subjected to linearly rising and decaying current ramps at repetition rates of 5.4 MHz (left) and 0.54 MHz (right)

7.6 PD Measurement at Impulse Voltages

From a physical point of view, PD measurements of HV equipment should be carried out under test voltages, which are representative for service conditions. This is the reason why the relevant standard IEC 60270:2000 is mainly addressed to the measurement of partial discharges under power frequency AC voltages, see Sect. 4.3. As known, under service condition the insulation of HV equipment is not only stressed by the continuously applied operation voltage but also by different kinds of over-voltages, such as switching impulse (SI) and lightning impulse (LI) voltages as well as very fast front (VFF) voltages. In the following specific aspects of PD measurements under such kinds of test voltages will briefly be highlighted.

7.6.1 SI Test Voltages

In the late 1980s, when paper-insulated lead-covered cables (PILC) were increasingly substituted by cross-linked polyethylene-insulated (XLPE) cables, the measurement of partial discharges became an inevitable tool for quality assurance tests of XLPE cables performed not only in laboratory after manufacturing but also in the field after installation. The reason for that was the fact that the high-polymeric insulation is very sensitive to partial discharges, because a PD activity in the pC range causes already irreversible degradation processes. However, energizing the comparatively high cable capacitance by the use of classical AC test transformers needs a high power and is thus very expensive, in particular if PD tests are performed under on-site condition. Therefore, to reduce the test expenditure, the feasibility of alternative test voltages have been extensively investigated (Dorison and Aucort 1984; Auclair et al. 1988; Lefèvre et al. 1989). Among others the use of switching impulse (SI) voltages has been found as a promising alternative due to the fact that the PD signatures of internal discharges are quite well comparable to those encountered under power frequency AC voltage (Lemke et al. 1987).

The major units of a single-stage SI generator originally used for on-site PD tests of medium voltage cables are shown Fig. 7.71. The most critical component is the spark gap because this emits heavy electromagnetic interferences and disturbs thus sensitive PD measurements, which has thus been replaced by a low-noise solid state switch. Basically it has to be taken into account that the entire charge required to energize the cable capacitance up to the desired test level must previously be stored in the surge capacitor shown in Fig. 7.71. Thus, for testing power cables of lengths up to some km this must be chosen in the µF range to accomplish a sufficient high utilization factor.

Example Fundamental PD studies under SI voltages revealed that potential PD defects in extruded power cables can be recognized at reasonable probability when the peak value of the applied SI voltage attains twice the phase-to-earth voltage, where the latter is expressed in terms of rms. Testing, for instance, a 12/20 kV XLPE cable of 1 km in length, a cable capacitance of nearly 0.2 µF must be charged up to a SI crest value of 24 kV. Using a single-stage SI generator according to Fig. 7.71, which is equipped with a 1 µF surge capacitor and a front resistor of 5.6 kΩ, the utilization factor attains 0.8, as obvious from the oscilloscopic screenshot shown in Fig. 7.71b. Consequently, the 1 µF surge capacitor must be charged up to a crest voltage of 31 kV to accomplish the desired SI crest voltage of 24 kV. Testing a cable of 5 km in length, a cable capacitance of approx. 1 µF must be

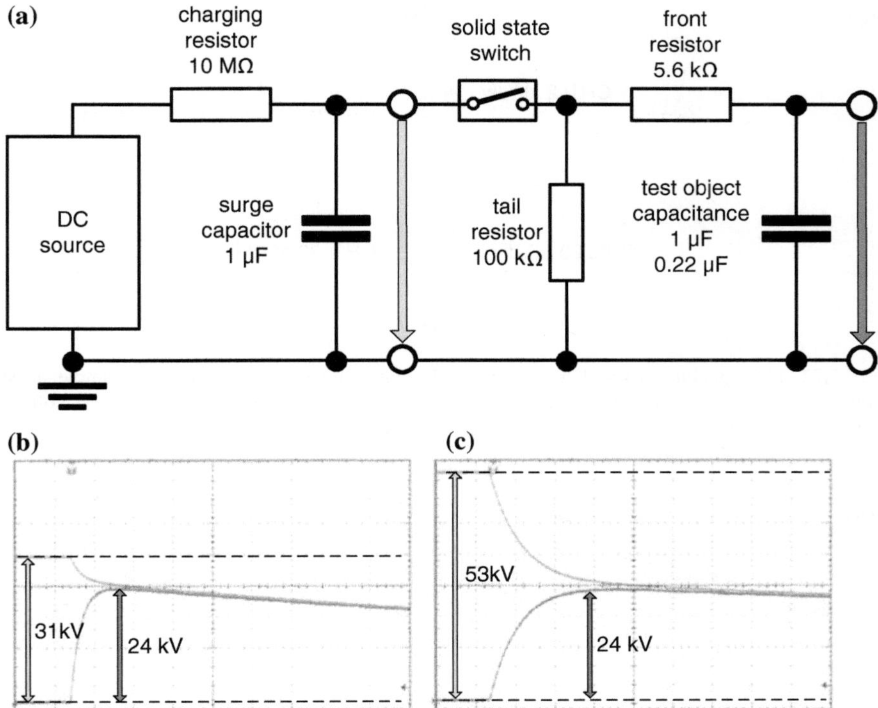

Fig. 7.71 SI voltage circuit for PD measurement. **a** Simplified block diagram of a single-stage SI generator. **b** Record of characteristic voltages with a 1 µF surge capacitor and a 0.22 µF cable capacitance. **c** Record gained for 1-µF cable capacitance

charged. Under this condition, the utilization factor approaches 0.45, that means to accomplish the desired SI test voltage of 24 kV the 1 µF surge capacitor of the SI generator must be charged up to a crest voltage of 53 kV, see Fig. 7.71c.

Increasing the cable lengths leads not only to a reduction of the utilization factor, as discussed previously, but also to an increase of the time-to-crest of the SI voltage, which follows also from the oscilloscopic records displayed in Fig. 7.71b, c. Experimental studies revealed, however, that the characteristic PD signatures of typical PD defects in extruded power cables and their accessories are not significantly affected by the time to crest if increased from few hundreds up to some thousands of µs, as exemplarily shown in Fig. 7.72. Thus it is not necessary to change the front resistor of the SI generator when power cables lengths ranging between some hundreds and some thousands of meters are PD tested. In this context it should also be mentioned, that additional information on the PD activity is available when besides the main PD quantity "apparent charge" additionally the "accumulated pulse charge" is measured. Different to the apparent charge, this quantity is only marginally scattering and seems to be correlated to the applied SI test voltage, as obvious from Fig. 7.73b–f.

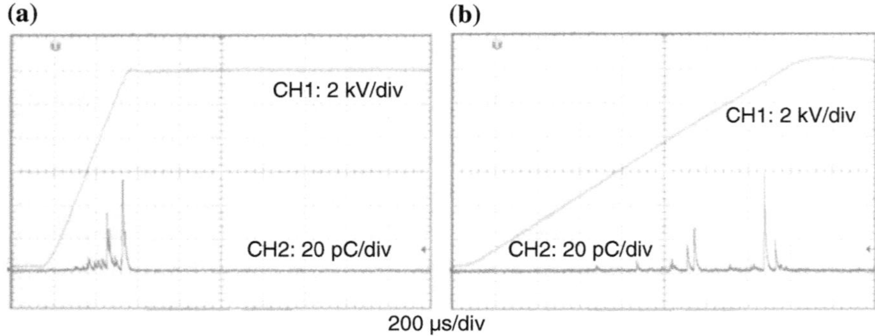

Fig. 7.72 PD signatures caused by an artificial cavity implemented in a cable sample at SI voltages of different steepness and almost identical peak values. **a** Rise time 400 µs, test voltage level 12 kV. **b** Rise time 1,600 µs, test voltage level 12.4 kV

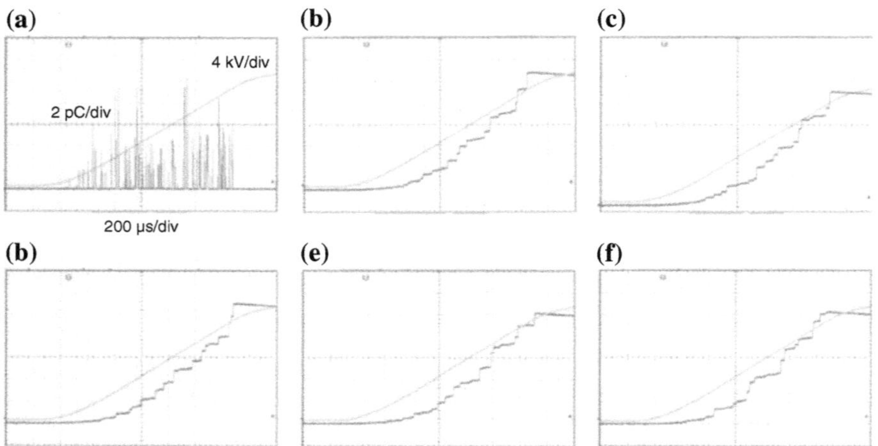

Fig. 7.73 Reproducibility of accumulated PD charges. **a** PD pulse train during the front time of an SI voltage. **b–f** Accumulated pulse charges recorded at identical front time and test voltage level (time scale: 200 µs/div, voltage scale: 4 kV/div, front time 1.6 ms, test voltage level 12.4 kV)

7.6.2 DAC Test Voltages

As discussed above, a main drawback of SI voltages used for PD tests of power cables is the strong reduction of the utilization factor at increasing cable length (Fig. 7.71). To overcome this crucial problem it seems to be very promising to use the cable capacitance itself as a surge capacitance because under this condition the utilization factor approaches 100%. This benefit has originally been utilized for after-laying tests of extra-high-voltage (EHV) XLPE cables (Dorison and Aucort 1984; Auclair et al. 1988; Lefèvre et al. 1989). Performing withstand tests, the cable capacitance was slowly charged by means of a DC test facility up to the desired test level and just thereafter discharged via an inductor in series with a spark gap (Fig. 7.74a). Under this condition the cable insulation is stressed by an oscillating switching impulse (OSI) voltage (see Sect. 7.3.1), later named "damped alternating (DAC) voltage" (IEC 60060-3:2006). The oscillations appearing in the kHz range are attenuated due to the dielectric losses dissipating in the cable insulation and additionally by the winding resistance as well as the magnetization losses in the iron core of the inductor. The reason behind the oscillating discharge of the cable capacitance was to avoid the accumulation of unipolar space charges on account of the pre-stressing DC voltage and thus to prevent local field enhancements in in the polymeric insulation, which could trigger an unexpected breakdown.

Initially attempts were made to combine such withstand tests with PD measurements. However, these failed due to strong electromagnetic interferences radiated from the spark gap shown in Fig. 7.74a. Another obstacle was that the characteristic PD patterns as well as the apparent charge measured under DAC voltage were not comparable to PD test results gained under power frequency AC voltage. Without going into further details it can be stated that this is due to an

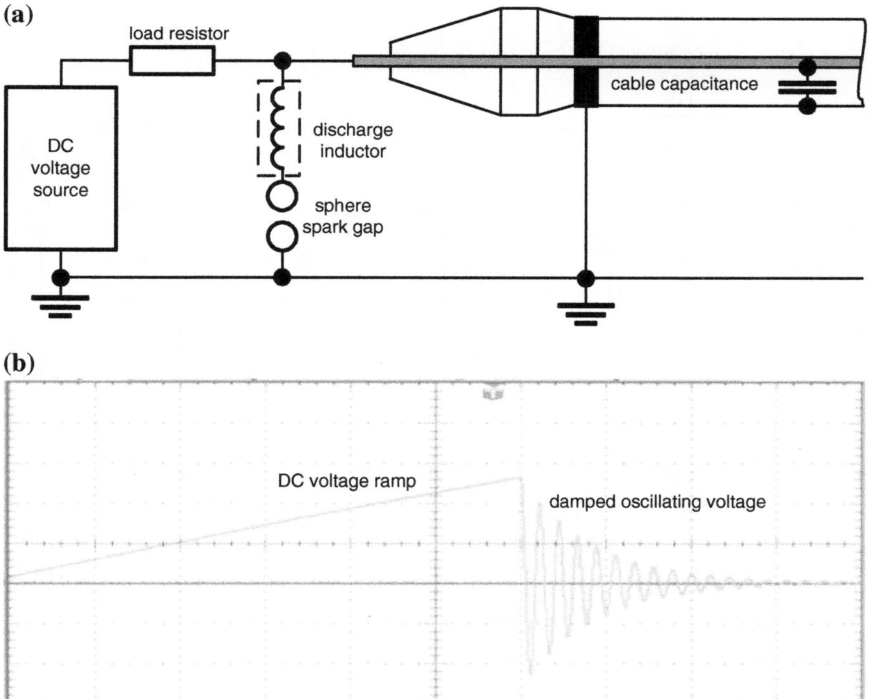

Fig. 7.74 Block diagram of a DAC test generator (**a**) and characteristic oscilloscopic record (**b**) showing the continuous rising DC stress and the damped AC voltage occurring just after the sphere spark gap has been triggered

enhancement of the PD inception voltage significantly above the static inception voltage as consequence of the statistical time-lag i.e. the time span which elapses between the instant when the static inception voltage is achieved and the instant at which an initiatory electron is available to initialize a self-sustaining discharge (Grey Morgan 1965). Experimental studies revealed that the impact of the time-lag on the PD inception voltage can substantially be reduced when the natural frequency of the applied DAC voltage is decreased from the originally used kHz range down to the power frequency (Lemke and Schmiegel 1995a, b). For this purpose the DAC test set was equipped with a 1 µF surge capacitor and a 4 H discharge inductor, as illustrated in Fig. 7.75. Under this condition the test frequency attained 80 Hz, where the PD signatures of typical imperfections in extruded power cables became quite well comparable with those gained under power frequency (50 Hz) AC voltage, as exemplarily shown in Fig. 7.76. The here displayed phase-resolved PD pulses, which refer to a defective MV cable termination, were recorded in a first test series under 50 Hz AC voltage adjusting the test level slightly above the inception voltage. To display the characteristic PD pattern like a waterfall diagram, the charge pulses occurring in only 18 AC voltage cycles were recorded in Fig. 7.76a. In a second test series, the cable termination was initially pre-stressed by a negative DC voltage ramp followed by a DAC voltage, where only those PD

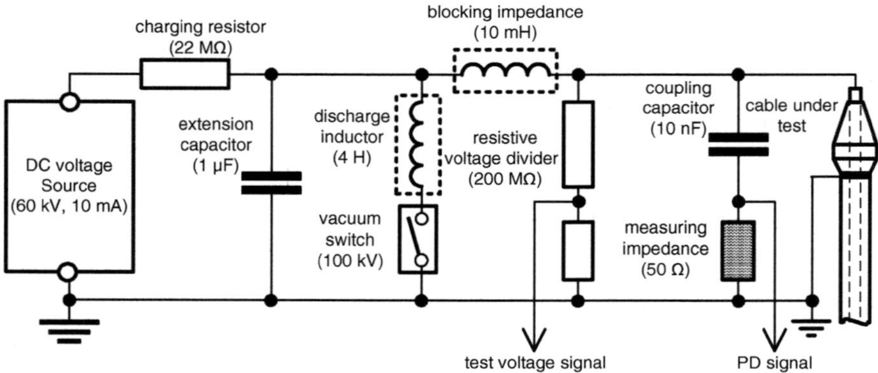

Fig. 7.75 Block diagram of a DAC test facility used for fundamental PD studies

(a)   **(b)**   **(c)**

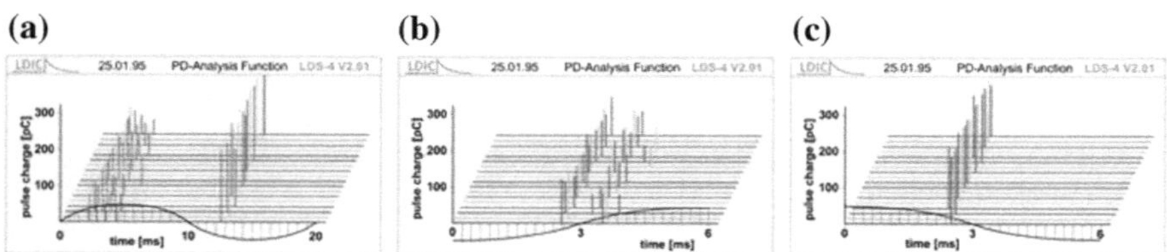

Fig. 7.76 Phase-resolved PD patterns of void discharges in a cable sample. **a** Continuous 50-Hz AC voltage. **b** 80-Hz DAC voltage stress displayed for the first positive voltage sweep. **c** 80-Hz DAC voltage stress displayed for the first negative voltage sweep

pulses occurring during at the first positive voltage sweep were displayed. This procedure has been repeated 18 times to create the waterfall diagram shown in Fig. 7.76b. In a third test series a positive DC voltage ramp (see Fig. 7.74b) was applied and the PD pulses occurring during the first negative DAC voltage sweep were displayed, see Fig. 7.76c.

The comparability of the characteristic PD patterns found for damped and continuous AC voltage was also confirmed by the PC screenshots shown in Fig. 7.77, which were also gained for a defective MV cable termination. Here the PD test was first performed under damped AC voltage of 80 Hz natural frequency, where the test level was again adjusted slightly above the inception voltage. The PD pulses displayed in Fig. 7.77a are the result of 180 DAC voltage applications. In a second test series, the cable termination was subjected to a continuous AC voltage of 50 Hz, where the test level was adjusted also slightly above the inception voltage. For comparison purpose only PD pulses occurring within 180 positive half-cycles were displayed in Fig. 7.76b. This was realized by means of a windowing unit, which was triggered once per 10 s, so that the entire test period covered 1800 s (90,000 AC cycles). Despite the large scattering of the PD pulse magnitudes (Fig. 7.77b) it can be stated that the PD test results gained under both damped and continuous AC voltage are reasonably comparable with each other.

As already discussed above, applying DAC voltages for PD diagnostic tests of XLPE cables, the high-polymeric insulation will be pre-stressed on account of the slowly rising DC voltage ramp (Fig. 7.74b). Depending on the cable length and thus the capacitance being charged, the pre-stressing period may vary between the second and minute range. Experimental studies revealed, however, that the PD signatures of PD defects representative for extruded medium-voltage power cables are not significantly affected by the duration of the DC pre-stress, as exemplarily shown in Fig. 7.78.

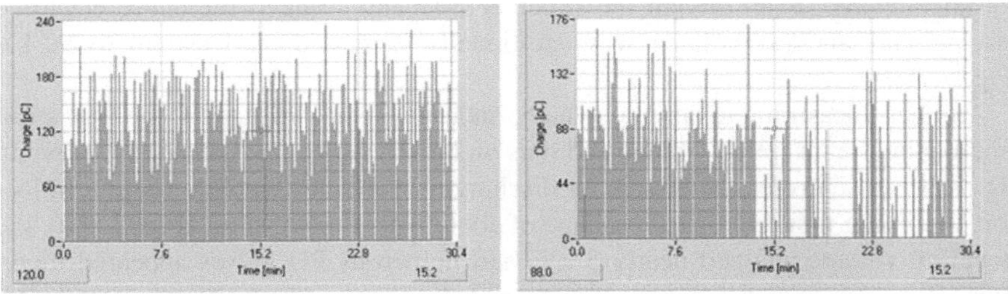

Fig. 7.77 Pulse charge magnitudes measured for a MV cable termination. **a** Damped AC voltage of 80 Hz. **b** Continuous AC voltage of 50 Hz

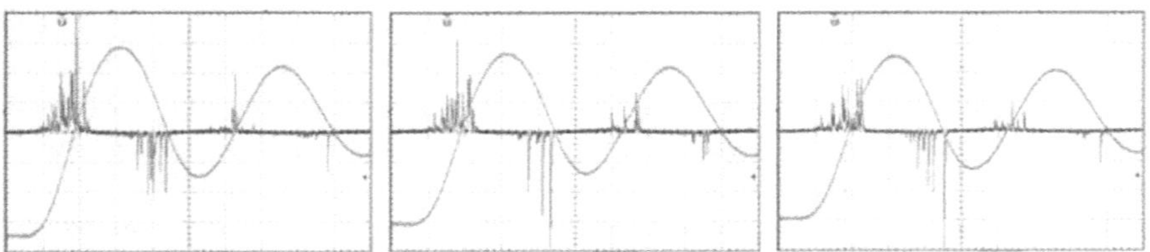

Fig. 7.78 PD signatures of a defective 20-kV cable termination at DAC test voltage using DC voltage ramps of 10, 50, and 250 µs rise time (from left to right)

7.6.3 Short Impulse Voltages (LI and VFF Test Voltages)

As known from the physics of gas discharges, the inception field strength of very fast changing test voltages, such as lightning impulse (LI) voltages and very fast transient (VFT) voltages may substantially be increased on account of the so-called statistical time-lag. This is the time span elapsing between the instant at which the static inception field strength is achieved and that one at which initiatory electrons liberated by natural processes are available to ionize gas molecules (Grey Morgan 1965). As a typical example compare the ignition of streamer discharges under both comparative slowly rising SI voltage and very fast rising impulse voltage according to Fig. 7.79, which refers to a rod-plane gap in ambient air of 20 cm in electrode spacing (Lemke 1967). Applying a negative impulse voltage of approx. 150 µs time-to-crest to the plane electrode, the first streamer discharge appeared at the positive cone electrode, where the inception voltage scattered around 10 kV. At rising voltage further streamer discharges of increasing length and pulse charge were recognized (Fig. 7.79a). However, applying a fast rising impulse voltage approaching an initial test level of 50 kV within less than one µs, only a single streamer discharge igniting at 50 kV was observed, i.e. the inception voltage was about 5-times greater than that measured under SI voltage. Despite the huge pulse charge of 180 nC created under the fast rising voltage, an ultimate breakdown was not encountered, even though the streamer filaments bridged the entire electrode spacing and the test voltage was additionally increased, at least up to 92 kV (Fig. 7.79b).

Another measuring example, which underlines the impact of the statistical time-lag on the inception voltage and thus on the discharge mechanism, is shown in Fig. 7.80, which refers to interfacial discharges in a defective MV cable termination. Applying first an impulse voltage of about 8 µs time-to-crest (Fig. 7.80a), the inception voltage attained nearly 3 kV, and numerous PD pulses appeared thereafter. However, increasing now the steepness of the applied impulse voltage by decreasing the time-to-crest stepwise, at least down to about 3 µs (Fig. 7.80b), the inception voltage increased considerably, and only a single PD pulse of comparatively high charge was recognized, i.e. the previously observed PD pulse train disappeared completely.

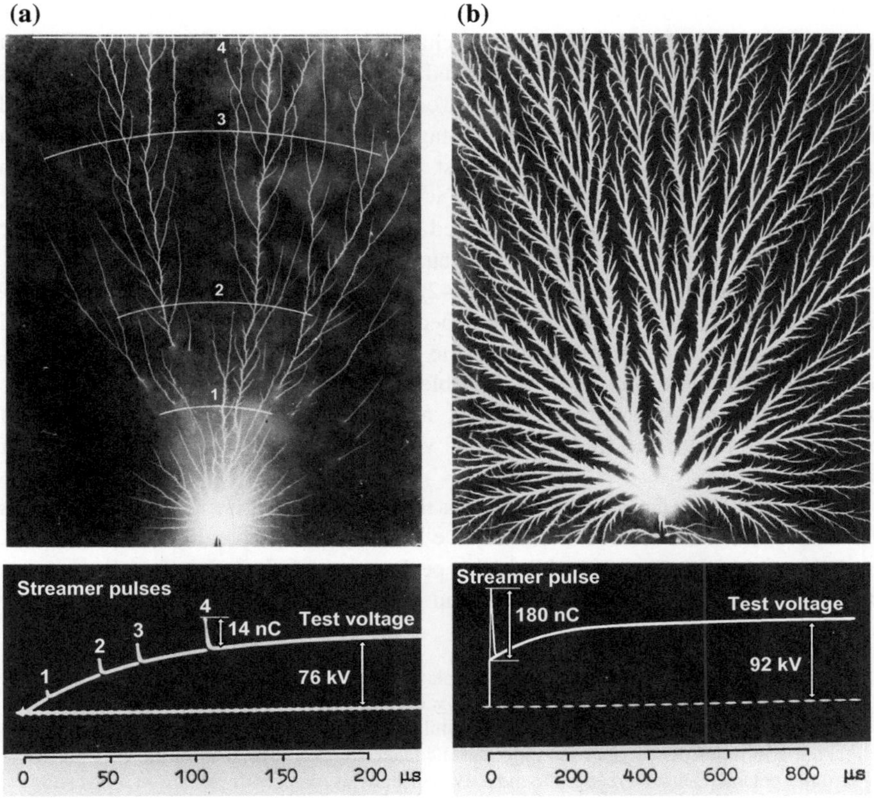

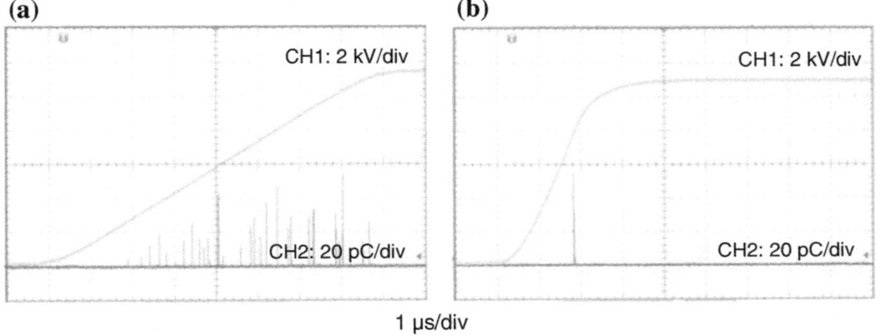

Fig. 7.79 Lichtenberg figures (fotograms) of streamer filaments developing in a 20 cm cone-plane air gap under SI voltage (**a**) and steep front voltage (**b**) and associated oscilloscopic records (Lemke 1967), explanation in the text

Fig. 7.80 Impact of the front time of SI voltages on the PD signature of a cavity discharge. **a** 8 μs front time, 11.6 kV test voltage. **b** 3 μs front time, 11 kV test voltage

The above considered impact of the statistical time-lag on the inception voltage becomes especially dominant for discharges under very fast front transient (VFT) voltages, which might be excited by power electronic converters. This advanced technology was originally introduced to adjust the speed of industrial drives using square wave operating voltages. As the operation principle of such power converters is based on the fast chopping of inductive currents, high over-voltages may be excited associated with a severe PD activity as, for instance, in rotating machines, which is associated with an accelerated degradation of the exposed insulation (Stone et al. 1992; Kaufhold et al. 1996; IEC 61934:2006; IEC 60034-18-41:2014; IEC 60034-18-42:2008; IEC 60034-25:2007; IEC 61934:2006). Nowadays this problem became also an issue for other industrial networks and even smart grids due to the increasing use of power electronic systems, for instance, to convert the variable frequency of wind farm turbines into power frequency AC voltage, as well as to convert the low DC voltage of photo-voltaic plants into high AC or DC voltage required for long-distance power transmission links.

As known from the classical network theory, chopping an inductive current is associated with a transfer of the "inductive" energy into the parallel connected load. In case of a pure capacitive load, the peak value of the excited voltage surge becomes often considerably higher than the operation voltage, as exemplarily shown in Fig. 7.81.

Example The oscilloscopic screenshots displayed in Fig. 7.81 refer to a stator of a low voltage motor having a coil inductance of $L_c = 0.17$ H. One coil termination was connected to a constant direct voltage of $V_1 = 210$ V, while the other coil termination was periodically (about 150-times per second, see Fig. 7.81a) connected for a time span of $T_d = 2.2$ ms to ground potential. Just thereafter the inductive current was chopped by means of a fast solid state switch realized by an insulated gate bipolar transistor (IGTB). At time period when the stator coil is connected to ground potential, the inductive current rises almost linearly and attains finally a crest value of $I_c = 2.7$ A, which follows also from Faraday's induction law:

(a) (b)

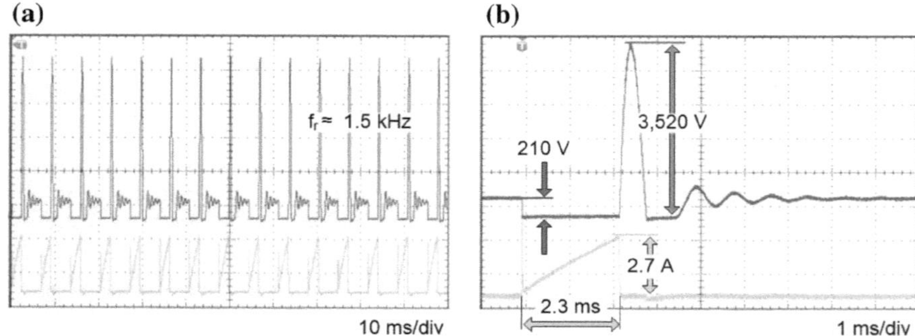

Fig. 7.81 Commutating voltage surges (pink trace) excited by repetitive chopping of inductive currents (blue trace) recorded at time scales of 10 ms/div (**a**) and 1 ms/div (**b**), explanation in the text

$$V_1 = L_c \cdot \mathrm{d}i/\mathrm{d}t \approx L_c \cdot I_c/T_d,$$

$$I_C = V_1 \cdot T_d/L_c = (210\,\mathrm{V}) \cdot (2.2\,\mathrm{ms})/0.17\,\mathrm{H} \approx 2.75\,\mathrm{A}$$

At instant $T_d = 2.3$ ms when the inductive current through the stator coil is chopped, the entire inductive energy is transferred to the capacitive load, which was chosen as $C_p = 0.1$ μF. Under this condition the peak value V_2 of the voltage appearing across the parallel connected elements L_c and C_p can be calculated for using the following known relation:

$$\frac{1}{2} \cdot L_c \cdot I_c^2 = \frac{1}{2} \cdot C_p \cdot V_2^2;$$

$$V_2 = I_c \cdot \sqrt{L_c/C_p} \approx (2.7\,\mathrm{A}) \cdot \sqrt{(0.17\,\mathrm{H})/(0.1\,\mathrm{\mu F})} \approx 3.52\,\mathrm{kV}.$$

This exceeds the applied direct voltage by more than one order and is in well agreement with the measured peak voltage, as indicated in Fig. 7.81b.

Very fast transient voltages excited by power electronics, sometimes also referred to as commutating surges, are characterized by rise and decay times in the μs range as well as repetition frequencies up to some tens of kHz. These are especially harmful for the winding insulation of rotating machines, because under this condition the inception field strength may be enhanced substantially and thus the PD activity, too, as discussed previously. This is underlined by the measuring example shown in Fig. 7.82, which refers to discharges between two circular-shaped magnet wires, each of 1.2 mm in diameter and coated by an insulating film of approx. 30 μm in thickness.

Applying repetitive voltage surges of 3.4 kV in magnitude (green trace), which was measured by means of a low-capacitive field probe (Lemke 2016), the maximum pulse charge scattered around 10 nC (pink trace). This comparatively high magnitude is due to the fact that most of the PD events ignited in the crest region of

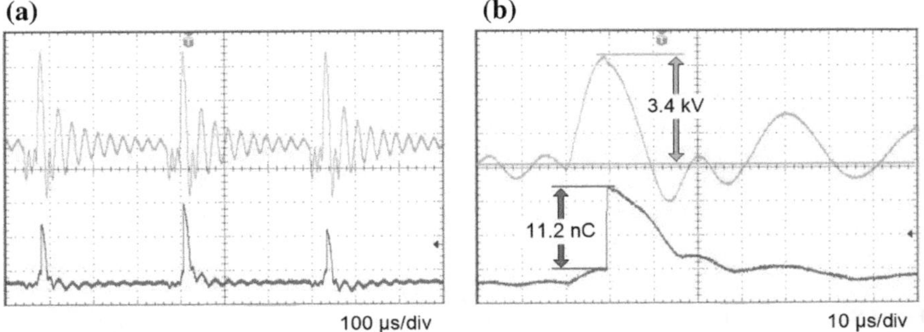

Fig. 7.82 Charge pulses (pink trace) of discharges between two circular shaped magnetic wires under repetitive commutating surges (green trace), recorded at time scale of 100 μs/div (**a**) and 10 μs/div (**b**), (explanation in the text)

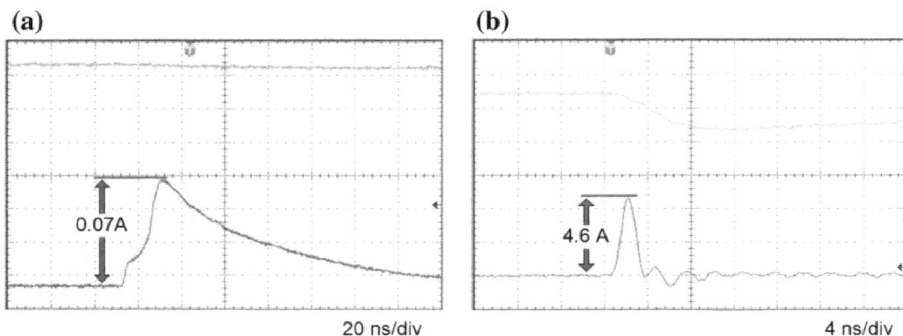

Fig. 7.83 Current pulses (pink trace) of discharges igniting in a 60 μm air gap between two circular shaped magnetic wires under commutating surges (green trace), recorded at time scale of 20 ns/div (**a**) and 4 ns/div (**b**), explanation in the text

the applied voltage surges, where the actual inception voltage exceeded considerably the static inception voltage. Further experimental studies revealed that the shape of the captured current pulses as well as the peak values were randomly distributed over an extremely wide range, often between some tens and some thousands of mA, see Fig. 7.83.

In this context it is noteworthy that usually so-called twisted pairs are employed to qualify the turn/turn insulation of random wound rotating machines under commutating surges. For this purpose a pair of film-insulated magnet wires is twisted together over a length of about 120 mm (IEEE Std. 522:2004). Using such test samples for apparent charge measurements according to IEC 60270:2000 it has to be taken into account that only a low measuring sensitivity is achievable due to the comparatively high capacitive load current, as exemplarily shown in Fig. 7.84a. To enhance the signal-to-noise (S/N) ratio, one option would be an increase of the

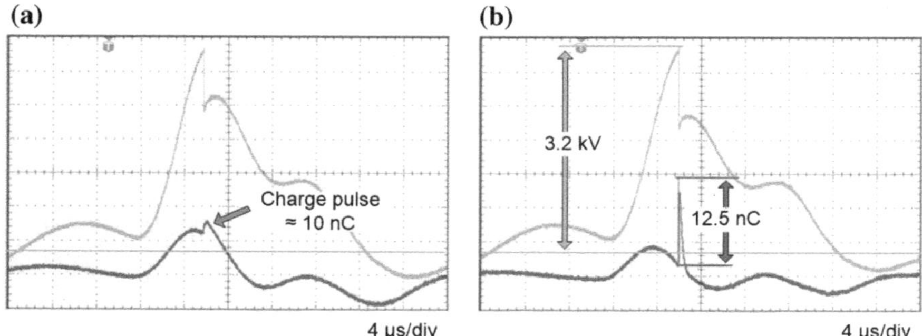

Fig. 7.84 Charge pulses (pink trace) superimposed on the capacitive load current caused by the high frequency commutating surges (green trace). Test conditions: 60 μm air gap between twisted pairs of magnetic wires; the current was integrated by means of a wide-band PD measuring system where the lower cut-off frequency was adjusted to 20 kHz (**a**) and to 1.8 MHz (**b**), explanation in the text

lower cut-off frequency, which should be chosen significantly above the frequency spectrum representative for the commutating surges. This is underlined by the measuring example shown in Fig. 7.84b, where the lower cut-off frequency of the measuring system was adjusted to $f_1 \approx 1.8$ MHz. Of course, this is in conflict with IEC 60270:2000, which recommends measuring frequencies below 1 MHz to measure the PD quantity apparent charge. Thus, as an alternative non-conventional methods could also be used, such as the PD detection in the UHF range or to evaluate the associated acoustic and optic signals. However, under this condition the PD activity can only be evaluated qualitatively but not quantitatively, as has already been pointed out in Sects. 4.7 and 4.8. Generally it can be stated that PD testing of inductive components stressed by commutating surges requires new approaches. For more information on this issue see also the Technical Brochure TB 703 published by Cigre WG D1.43 in 2017.

Chapter 8
Tests with Combined and Composite Voltages

Abstract In power systems, the over-voltage stresses of insulations are often combinations of the operational voltage with over voltages. This can be neglected as long as the over-voltage value includes the contribution of the operational voltage. It cannot be neglected when the insulation between phases or of switching devices is considered. In that case, the resulting voltage is the combination of two voltage stresses on three-terminal test objects. In other cases, the stressing voltage is composed of two different voltage components, e.g. in certain HVDC insulations, as a composite voltage of AC and DC components. This chapter is related to the definition, generation and measurement of combined and composite test voltages on the basis of IEC 60060-1:2010 and IEEE Std. 4. Also some examples for tests with combined and composite voltages are given. It should be mentioned that these voltages are sometimes called "hybrid" or "superimposed" voltages, also the summarizing term "mixed" voltages is in use. This book follows the terminology of the Standard IEC 60060-1 (2010).

8.1 Combined Test Voltage

The definitions of combined and composite test voltages are related to the position of the test object to the test voltage sources. When the test object is arranged between the two test systems, a *combined voltage* stresses the test object via two different HV terminals and to ground (three-electrode test arrangement). When the two test systems are directly connected, a *composite voltage* is generated stressing the test object from one HV terminal to ground (two-electrode test arrangement). In both cases, each test system must be protected against the voltage generated by the other system by means of an element that lets pass its own voltage and blocks the voltage of the other one (coupling/protecting element).

© Springer Nature Switzerland AG 2019
W. Hauschild and E. Lemke, *High-Voltage Test and Measuring Techniques*,
https://doi.org/10.1007/978-3-319-97460-6_8

8.1.1   Generation of Combined Test Voltages

The combined voltage appears between the two HV terminals of a *three-terminal test object* with the third terminal grounded (Fig. 8.1a). Typical three-terminal test objects are disconnectors and circuit breakers. Also the insulation between phases in three-phase systems, e.g. metal-enclosed busbars of GIS and three conductor cables, form a three-terminal test object. For a HV test of a GIS, two phases are connected each to one HV source, the third phase and the enclosure are grounded.

The test object is stressed by the difference of the two single voltages (Fig. 8.1a: $V_c = V_{AC} - V_{SI}$). Figure 8.1b shows as an example the combined test voltage of an AC and a SI voltage. As long as the test object withstands the combined voltage stress V_c, it separates the two HV sources from each other. But when it breaks

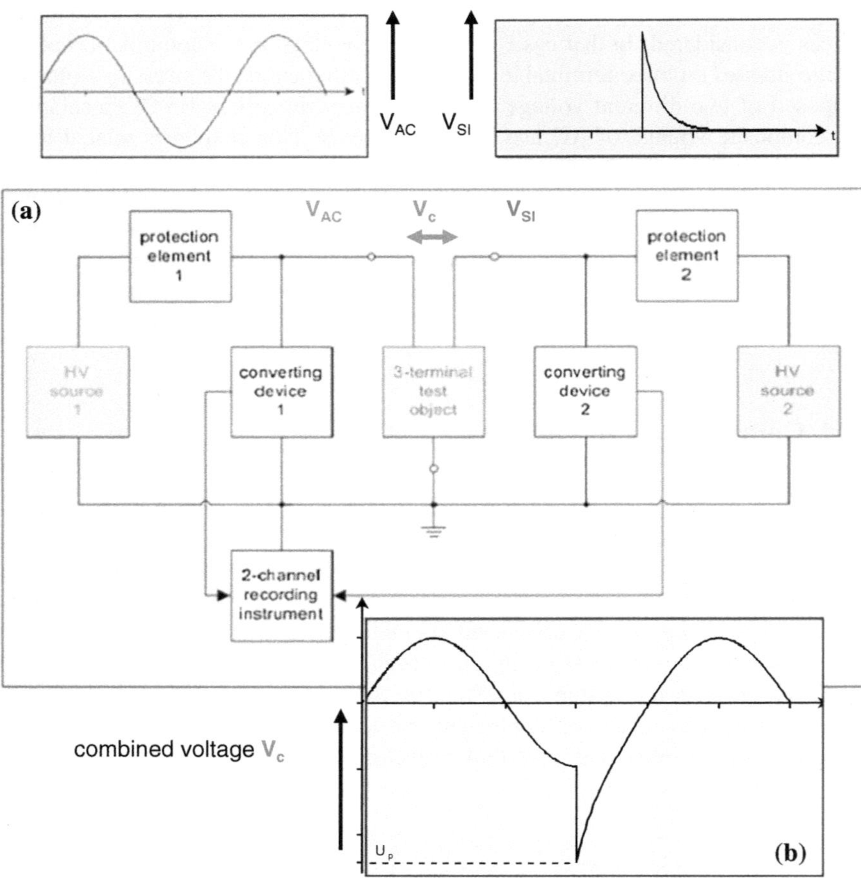

Fig. 8.1 Generation and measurement of combined voltages. **a** Schematic circuit diagram. **b** Combined test voltage as a difference of two voltages

down, each test system is also stressed by the voltage of the opposite HV source. Then, it must be protected by a suited protection element that blocks the other voltage, at least down to an acceptable voltage stress. But this element must also couple—and not block—the voltage of the protected HV source. This means that elements must be applied that have different impedances for different voltages. For instance, an inductor has no impedance for DC voltage, but an high impedance when stressed with LI voltages. Therefore, it may be used for protecting a DC voltage generator. It must be considered that the coupling/blocking elements influence the two components of the combined voltage. Therefore, the measurement of the two voltages has to be made after the coupling/protecting elements in parallel to the ground insulation of the test object (Fig. 8.1a). Compared with the impedance of the test object, the impedances of the coupling/protecting elements should be low.

Table 8.1 summarizes the characteristics of *coupling/protecting elements*. The preferred application of an element is shown in the first row, a second application with different parameters of the element is given in brackets. The arrows in brackets indicate a low ($\downarrow$) or high ($\uparrow$) value of the parameter (L, R, C). For instance, a capacitor of large capacitance (low impedance for power frequency) couples AC voltage of power frequency but one of low capacitance (high impedance) may block it. Triggered switches can be switched to positions "closed = coupling" or "open = protecting" and can consequently widely be applied. It has to be considered that a—possibly triggered—spark gap requires a certain voltage for ignition (see Fig. 7.6b). When it is reached, the resulting voltage jumps to that value (Fig. 8.8). After an impulse stress, it does not easily extinguish at an AC or DC voltage which is short circuited. Non-triggered gaps have the disadvantage of the dispersion of their breakdown voltage and that it is difficult to reach standard LI and SI voltage shapes. The alternative coupling element to the sphere gap is a capacitor, in the superposition it maintains the two voltages as they are (e.g. Fig. 8.5). Capacitors are expensive because their capacitances should be remarkably higher than that of the test objects (Felk et al. 2017).

Table 8.1 Coupling/protecting elements for combined/composite test voltage circuits

Test voltage elements	DC voltage	AC voltage	SI voltage	LI voltage
Inductors (L)	Coupling	Coupling (L$\downarrow$) (Protecting, L$\uparrow$)	Protecting (L$\uparrow$) (Coupling, L$\downarrow$)	Protecting
Resistors (R)	Coupling (R$\downarrow$) (Protecting, R$\uparrow$)	Coupling (R$\downarrow$) (Protecting, R$\uparrow$)	Protecting (R$\uparrow$) (Coupling, R$\downarrow$)	Coupling (only low R$\downarrow$)
Capacitors	Protecting	Coupling (C$\uparrow$) (Protecting, C$\downarrow$)	Coupling (C$\uparrow$) (Protecting, C$\downarrow$)	Coupling
Switches as triggered gaps, semiconductors	Coupling or protecting	Coupling or protecting	Coupling or protecting	Coupling or protecting

Example When a DC/LI combined voltage shall be applied, the right element between DC voltage test system and test object is an inductor, because it couples the DC voltage and protects against LI voltage. Between the LI test system and the test object, a capacitor is the right element, it couples the LI voltage and protects the impulse generator against a DC voltage stress. Also the application of a (-possibly triggered-) sphere gap can be taken into consideration.

As mentioned before, the coupling/protecting elements influence the voltage generation of both voltage sources, and also the two sources show interactions. As a result, the combined voltage has not the shape as expected. The test of a disconnector shall be considered: The AC/SI combined test voltage shall be generated by a test transformer and an impulse generator (Fig. 8.2a). The AC voltage is dropped down if the AC source is not stiff enough (Fig. 8.2b: 20%). If no powerful AC test system is available, a *supporting capacitor* $C_a \gg C_t$, larger than the test object capacitance C_t in parallel to AC source and test object, reduces the *voltage drop* remarkably (Fig. 8.2c: <5%) (Cui et al. 2009). When a combined (or composite) voltage test is planned, it is strongly recommended to analyse the test circuit by a suitable equivalent circuit.

8.1.2   Requirements to Combined Test Voltages

The test voltage value of the combined voltage is the maximum potential difference between the two HV terminals of the test object. Its tolerance, this means the difference between the specified value and the recorded value shall be within ±5% of the specified value. This includes that also a voltage drop does not exceed 5%. For each voltage component, the requirements mentioned above in the relevant Chaps. 3, 6 and 7 have to be applied. Furthermore, the *time delay*, this is the time difference between the two maxima of the voltage components, must be considered

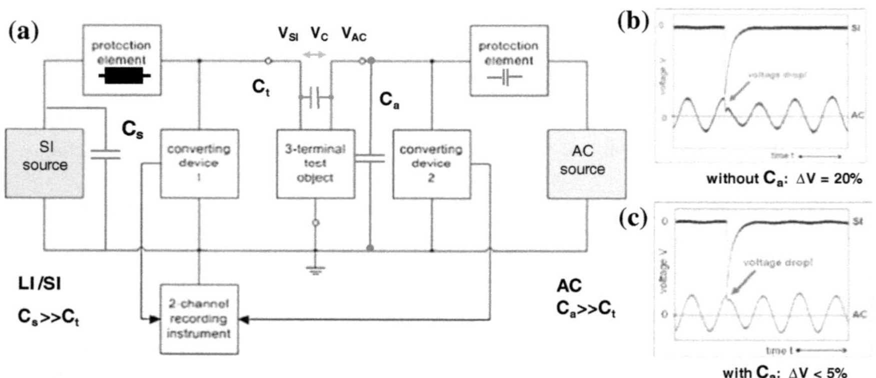

Fig. 8.2 Interaction between the two voltage sources. **a** Schematic circuit diagram. **b** Drop of the AC voltage without supporting capacitor. **c** Drop of the AC voltage with supporting capacitor C_a

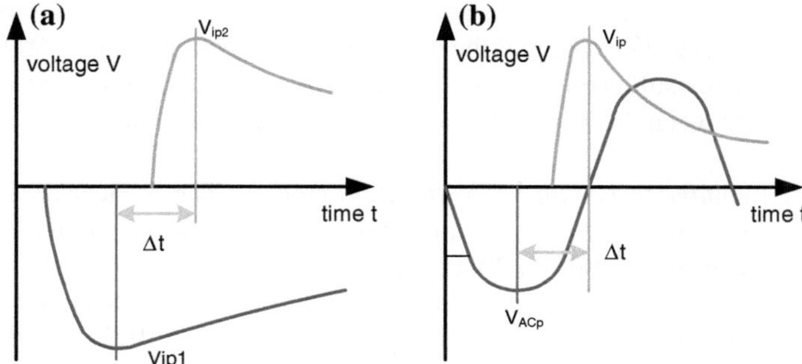

Fig. 8.3 Time delay of combined and composite voltages. **a** For two impulse voltages. **b** For an impulse voltage and an AC voltage

(Fig. 8.3). The tolerance of the time delay is 0.05 T_p, with T_p as the longer front parameter of the two voltages involved (where T_p is the LI front time or the SI time to peak or a quarter of an AC period).

8.1.3   Measurement of Combined Test Voltages

The two HV test systems require a HV measuring system each for the adjustment of their output voltages that contribute to the combined or composite voltage. These measuring systems must be able to record also the interactions between the two HV test systems.

The stressing combined voltage acts between the HV terminals of the test object (Fig. 8.1). A usual measurement of this voltage is difficult because there is no earth potential involved. IEC 60060-1:2010 allows therefore the calculation of the combined test voltage from the measurement of its two voltage components: Each of the two voltage dividers shall be arranged as near as possible to its relevant HV terminal of the test object. The two voltages are recorded, and the combined voltage is calculated as its difference. The uncertainty estimation for the measurement of the calculated combined voltage must consider the influences of the test object and of voltage drops over the connection/blocking elements the calculated one (Fig. 8.1b), should be displayed using an identical time scale.

8.1.4   Examples for Combined Voltage Tests

Disconnector testing: The testing of EHVAC disconnectors and phase-to-phase air insulation with a combined voltage of AC and SI components is the classical example of a combined voltage test. The interaction between the test voltage sources may cause a voltage drop as considered above in Sect. 8.1.1 (Fig. 8.2). Garbagnati et al. (1991) found that the atmospheric corrections of IEC 60060-1 (see Sect. 2.1.2) deliver sufficient results when they refer to that component of the test voltage value between the HV terminals that causes the maximum of the combined voltage.

Combined voltage tests with DC voltage component: The broader application of HVDC transmission systems will require test voltages that can be understood as combined voltages (Gockenbach 2010). There have been early investigations about the combination of DC voltage and oscillating SI voltage (Fig. 8.4) in preparation of test systems for the Russian HVDC transmission (Lämmel 1973). Meanwhile LI/DC voltage investigations are extended to compressed gas insulations of N_2 and SF_6 (Wada et al. 2011). They show that a DC voltage component below 50% of the test voltage value has little influence on the breakdown voltage.

8.2 Composite Voltages

8.2.1 Generation and Requirements

The connection of two different test voltages to one terminal generates a composite test voltage because of their superposition at that point (Fig. 8.5a). The connection is realized with suitable coupling/protecting elements (Table 8.1). In opposite to

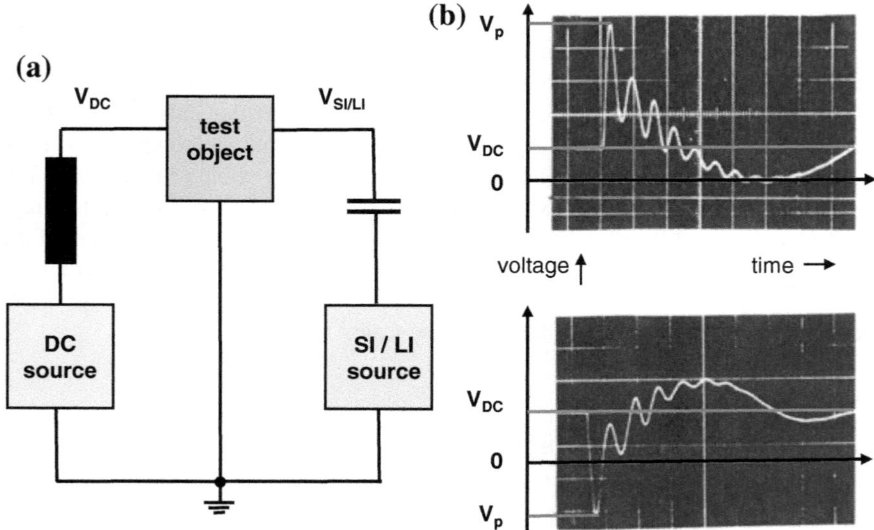

Fig. 8.4 DC/SI combined voltage. **a** Circuit with coupling/connecting elements. **b** OLI voltage superimposed on DC voltage

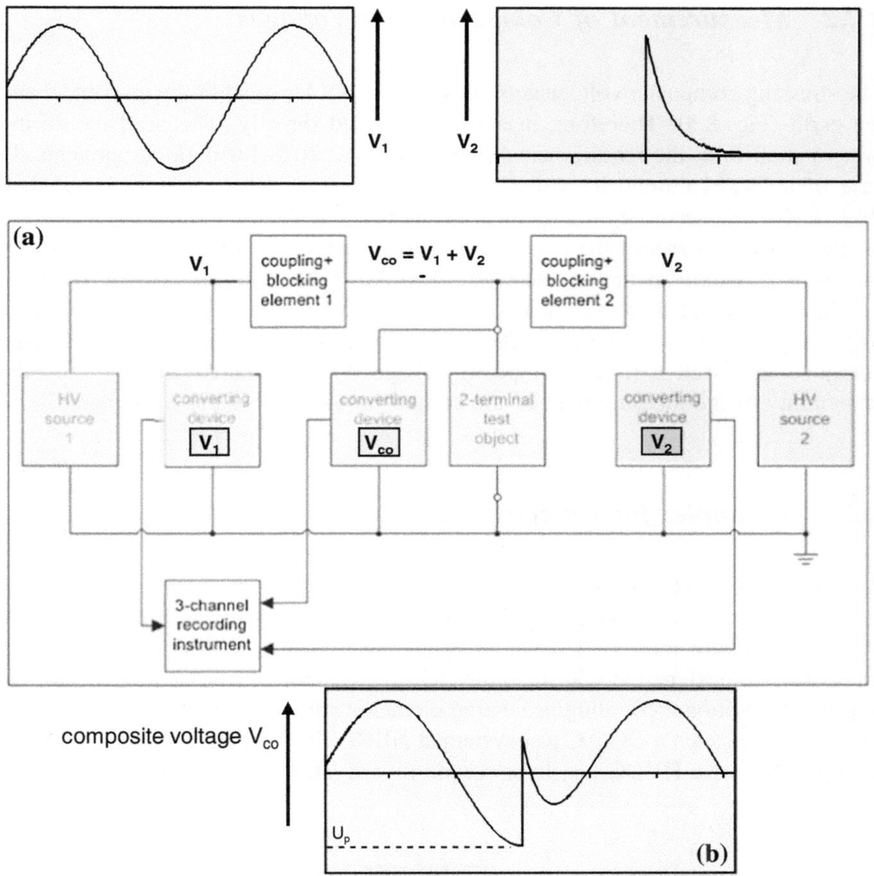

Fig. 8.5 Generation and measurement of composite voltages. **a** Schematic circuit diagram. **b** Composite voltage as the sum of the two voltages

combined voltages, the composite voltage is the sum of the two components (Fig. 8.5b: $V_{co} = V_1 + V_2$). If the two voltage sources are connected together, the interaction between the two HV sources including their coupling/protecting elements may play a role and should be analysed.

The test voltage value of the composite voltage is the maximum absolute value at the test object and shall meet the specified value within ±5%. Also any voltage drop shall not exceed 5%. The time delay is defined as for combined voltages (Fig. 8.3) and should be again within ±0.05 T_p (T_p as defined in Sect. 8.1.1). For the single-voltage components, the requirements in the relevant chapters of this textbook shall be applied.

8.2.2   Measurement of Composite Test Voltages

The stressing composite voltage acts between the HV terminal of the test object and the earth (Fig. 8.5). Therefore, it can be measured directly. The used measuring system shall fulfil the requirements of IEC 60060-2:2010 for both components. In case of a DC/LI composite voltage, e.g. an universal divider has to be applied (Fig. 2.10). The separate measurement of the two voltage components is necessary for the precise control of the two test voltage generators and for the verification of the correct relation of the two test voltages. All three voltages shall be recorded synchronously and displayed with an identical time scale (Fig. 8.5).The voltage measuring systems shall be calibrated for the measurement for the related voltage components as well as for the composite voltages to be measured. Based on that the uncertainty of the measurement of the composite voltages shall be estimated.

8.2.3   Examples for Composite Voltage Tests

A composite DC/AC test voltage can be generated when the smoothing capacitor of the DC generator is grounded via the HV winding of a test transformer (Fig. 8.6). The DC generator must be fed via an insulating transformer (Fig. 6.8). If the test transformer is designed to withstand the DC stress in case of a breakdown of the test object, no additional coupling/protecting elements are required. The photograph in Fig. 8.6 shows such a DC/AC test system at NIIPT, St. Petersburg, Russia, used for the operation of a HVDC test line, corona investigations and other basic research

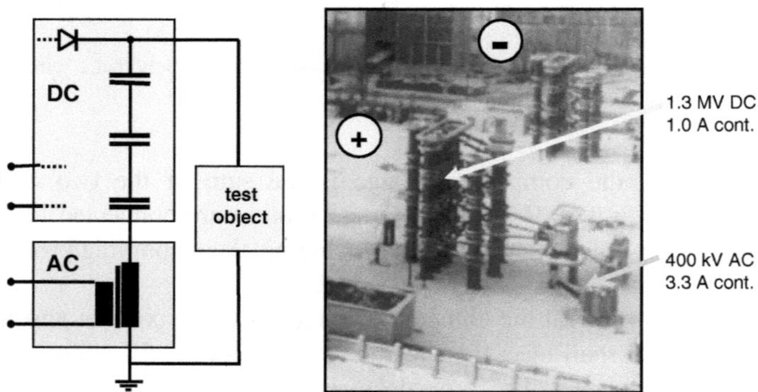

Fig. 8.6 Test system for DC/AC composite voltage up to ±1300 kV DC and 400 kV AC (The principle is shown for one polarity only)

work. Both voltage components can be adjusted separately. For the application of composite DC/LI respectively DC/SI test voltages, the basic circuit (Fig. 8.5) with coupling/protecting elements according to Table 8.1 shall be used.

With respect to the characteristics of HVDC insulation (see Chap. 6, especially Sects. 6.2.2.2, 6.2.3, 6.3), composite voltages are mandatory for the testing of HVDC insulations. This shall be considered for *HVDC gas-insulated systems* (Cigre JWG D1/B3.57 (2017)) with the sensitive field strength distribution in the surrounding of the spacers. Figure 8.7 (Hering et al. 2017) shows the situation of the electrostatic field at AC voltage and of the streaming field at DC voltage just before a transient LI or SI stress occurs. For an HVAC test object—with lowest field strength at the outer electrode—the transient stress will change the intensity of the field distribution only. In case of the HVDC insulation—with the space charge-dominated high steady-state field strength at the outer part of the spacer— the transition of the surface-charge dominated streaming field to a capacitively-dominated quasi-electrostatic field must be considered. The latter is related to field strength enlargements, especially if the polarity of the transient stress is opposite to the steady-state conditions before (Fig. 8.8b, d).

NOTE Fig. 8.8 is a proposal of the CIGRE JWG D1/B3.57 for the present draft of a CIGRE Brochure. It uses Fig. 8.8 for definitions of LI test voltages and defines separately a DC voltage pre-stress. It underlines the high LI stress when the two voltage components are of opposite polarity. But it seems to be better to consider the composite voltage because the two stresses act commonly.

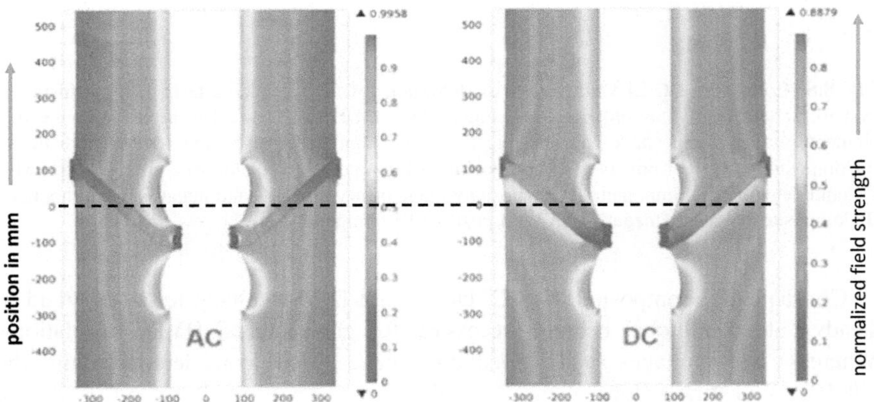

Fig. 8.7 Field conditions in an HVAC and an HVDC gas-insulated system before a transient stress (Hering et al. 2017). Consider the higher field strength on the outer side of the spacer at DC stress

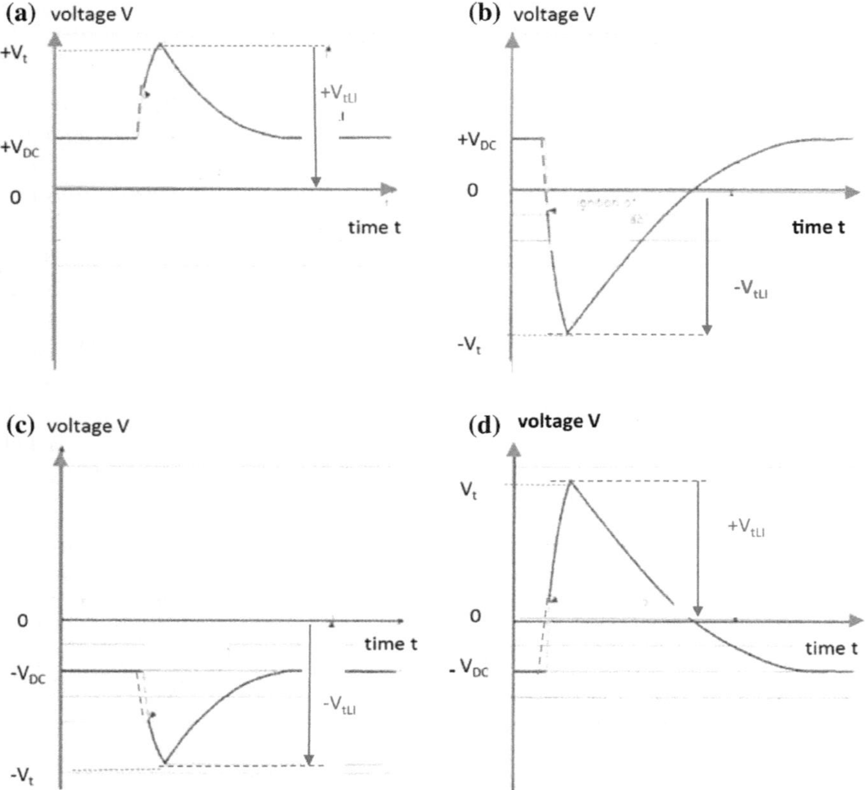

Fig. 8.8 Composite DC/LI voltage tests as definition for the LI voltage tests of a gas-insulated system (schematically) as proposed in CIGRE JWG D1/B3.57 2017. The blocking/connecting element is a triggered spark gap, causing a voltage jump when triggered (dotted blue lines). **a** Composite voltage of positive DC and positive LI component. **b** Composite voltage of positive DC and negative LI component. **c** Composite voltage of negative DC and negative LI component. **d** Composite voltage of negative DC and positive LI component

Consequently composite DC/AC, DC/LI and DC/SI voltage tests (DC field in steady-state conditions) become necessary for gas-insulated HVDC insulations. Whereas test voltages and procedures have been considered within the CIGRE JWG, first values of the test voltages and test durations for a *prototype installation test* (see Sect. 6.2.3.1) are proposed by Neumann et al. (2017). It should be mentioned that this prototype insulation test is also a composite HVDC and load

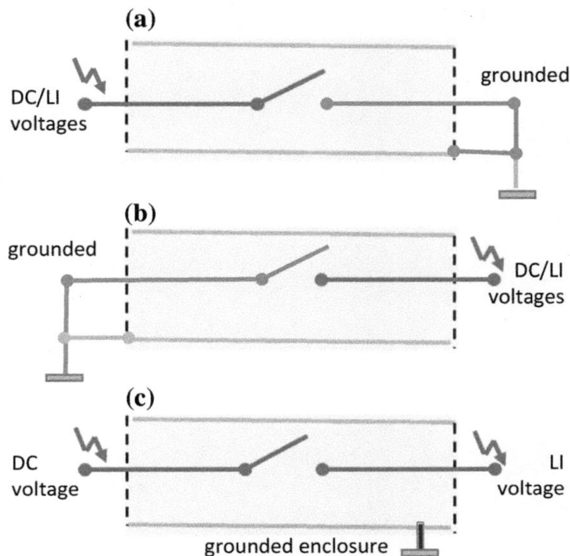

Fig. 8.9 Testing of a the switching device of a SF$_6$-insulated system (schematically). **a** DC/LI or SI composite voltage test, right side grounded. **b** DC/LI or SI composite voltage test, left side grounded. **c** DC/LI or SI combined voltage test

current test, using a suited current source for operating on HV potential. The testing of an open switching device of a gas-insulated HVDC system includes both, composite voltage test (Fig. 8.9a, b) and for the gap itself a combined voltage (Fig. 8.9c) test.

Chapter 9
High-Voltage Test Laboratories

Abstract Efficient HV testing, research work or students training requires well-designed HV laboratories. This chapter is related to the planning of HV laboratories. The basis is a clear analysis of the requirements to the laboratory and corresponding selection of the HV test systems. This includes a general principle of the control and measuring systems, the internal data evaluation and communication structures. The planning of test buildings or test rooms depends strongly on the objective of the laboratory and of the available funds. The general principles for the grounding and shielding, for power supply, transportation and auxiliary equipment are explained. An important part of the planning is a safety system which guarantees both safety for the operators and reliable, quick testing. Some specialities to outdoor test laboratories and updating of existing test fields are submitted.

9.1 Requirements and Selection of HV Test Systems

9.1.1 Objective of a Test Field

The term *"HV test laboratory"* may range from a small single room with test equipment of few kilovolts rated voltage up to huge UHV laboratory complexes with several test fields of different test areas. An optimum planning of a HV laboratory well adapted to the objective of a company or institution is the basis for its later smooth operation. The users of HV test fields can be subdivided into the following groups:

- manufacturer and repair shops of equipment for power systems (*"equipment provider"*),
- companies of electric power generation, transmission and distribution (*"utilities"*),
- research institutes and HV test service provider (*"service provider"*),
- measurement and calibration service institutions, national labs (*"calibration provider"*),
- universities and technical schools, education and training ("institutions").

© Springer Nature Switzerland AG 2019

W. Hauschild and E. Lemke, *High-Voltage Test and Measuring Techniques*,

https://doi.org/10.1007/978-3-319-97460-6_9

The performed HV tests can be subdivided as follows:

- routine tests on new or repaired HV equipment (*"routine tests"*),
- type tests on newly developed equipment (*"type tests"*);
- tests for research and development (R&D) of new HV equipment (*"development tests"*),
- tests for development of HV measurements and calibrations (*"calibration"*),
- tests for practical training and demonstrations (*"educational tests"*).

The different kinds of HV tests are related to the different users of HV tests in Table 9.1. The darkness of a field shall indicate its importance for the user, dark blue means most important, light blue means useful, but not necessary, and white usually not necessary. It should also be mentioned that some users should be able to perform combinations of tests. Many universities with well-equipped HV laboratories perform HV research or even type testing. Last but not least, HV laboratories are very attractive and may support the image of an institution remarkably.

The combination of education and research fits well together, whereas *routine testing* and research work would disturb each other. A routine test is a part of the production and has to follow the technological flux in the company. Short and smooth transportation from the previous workshop to the routine test field and from there to the next station is as important as a simple and quick test process. For an equipment provider, it might be useful to have different routine test fields related to its significantly different products, but in minimum separated between routine tests and type/development tests.

Utilities have to investigate very different, service-aged equipment and only in special cases new equipment. Therefore, a laboratory for multi-purpose application might be optimum. Service provider must also be very flexible and specialized in type tests and research/development tests.

Table 9.1 Objectives of HV tests (explanations in the test)

	Routine tests	Type tests	Tests for R & D	Calibrations	Educational tests
Equipment provider	X	X	X	–	–
Utilities	X	X	X	–	–
Test service provider	X	X	X	X	–
Calibration provider	–	–	X	X	–
Institutions	–	X	X	X	X

9.1.2   Selection of Test Equipment

For the different kinds of tests, different test systems and different *special accessories* are necessary. Tables 9.2 and 9.3 give an overview on the required test systems and accessories.

As in Table 9.1, the dark blue colour indicates necessary equipment, the light blue useful equipment. The mentioned voltage values in Table 9.2 indicate the highest rated voltages of test equipment, selected according to the procedure described in the example of Sect. 1.2 (Figs. 1.5 and 1.6). They are only necessary for UHV power equipment. In most other cases, the rated voltages of test equipment

Table 9.2   Test systems for the different types of HV tests

	AC test systems	DC test systems	LI/LIC/SI test systems	HV systems for combined and composite tests
Routine tests	< 1200 kV	<1500 kV	< 2000 kV	–
Type tests	< 1500 kV	<2000 kV	< 4000 kV	Resulting from the columns of AC, DC & impulse voltages
Development tests	< 1500 kV	<2000 kV	< 4000 kV	As for type tests
Calibrations	< 400 kV	< 400 kV	< 800 kV	–
Educational tests	< 200 kV	< 300 kV	< 800 kV	–

Table 9.3   Special accessories for the different types of HV tests

	Shielding for PD/dielectric. measurement	Artificial rain equipment	Pollution chamber	Climatic chamber	Oil tank	Compress. gas tank	Corona cage
Routine tests	X	X	–	–	X	X	–
Type tests	X	X	X	X	X	X	X
Development tests	X	X	X	X	X	X	X
Calibrations	X	–	–	–	–	–	–
Educational tests	X	–	–	–	X	–	–

depend on the test objects of highest rated voltage (Tables 1.2 and 1.3) or the targets of the HV research.

For *routine tests* (Table 9.2), the HV test systems must be sufficient for the products to be tested now and within the coming 10 years. It is recommended to avoid any over-dimensioning. AC voltage of power frequency is by far the most important voltage for routine tests. It is more and more connected with PD measurement. Therefore, the test rooms should be shielded or a shielded test chamber should be applied (Table 9.3; see also Fig. 9.32). Also a metal-enclosed test system (Fig. 3.47) is useful for metal-enclosed power equipment (GIS). If the AC voltage test system is metal-enclosed, the whole circuit is well shielded and does not require a shielded test room. It can be arranged in the workshop itself. Impulse voltage testing is required for routine tests of only few equipment, e.g. for power transformers. Then, related LI/SI test systems belong to the scope of supply. Routine tests with DC voltages are very seldom at the moment, but will become more and more necessary with the broader HVDC application.

For *type and development tests* and for HV research work, all types of test voltages and most of the special accessories are required (Tables 9.1, 9.2 and 9.3). Their rated values must be carefully selected considering the possible development within the next 2 decades.

Voltage calibrations (Table 9.2) can be performed at reduced voltages of >20% of the rated voltage of the measuring systems to be calibrated. Therefore, the rated voltages for calibration are about 20% of the rated voltages for type testing (Table 9.2) and small test systems are sufficient (Fig. 9.1, courtesy of TÜBITAK, Gebze, Turkey). Calibration requires always low electromagnetic noise; therefore, test rooms for calibration should be shielded.

Educational tests (Table 9.2) can be performed at high voltages up to some hundred Kilovolts (Fig. 9.2, courtesy FH Mittweida, Germany). Separate small rooms or areas for each HV test are recommended to enable several training groups to work in parallel without disturbing each other (Prinz 1965; Mosch et al. 1974; Hauschild and Fahd 1978; Kind and Feser 1999; Schwarz et al. 1999).

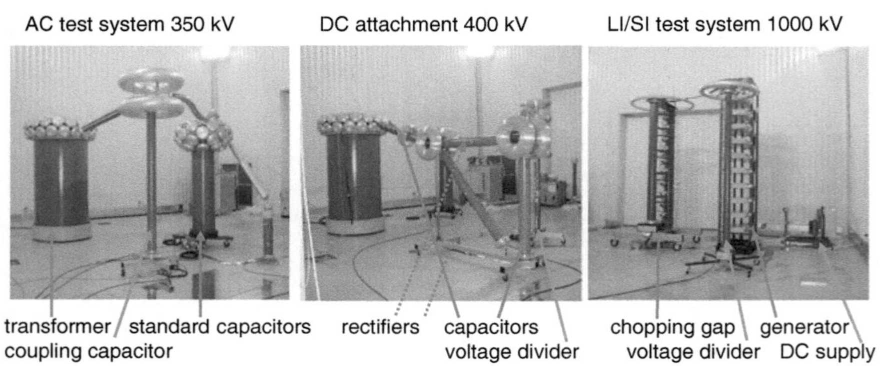

Fig. 9.1 Small HV test systems in a calibration laboratory

Fig. 9.2   Test areas for students training

Note For the selection of the rated data of test systems, special accessories and test areas, it is recommended to establish a list of power equipment to be tested with their rated electrical parameters, their test voltage values, test conditions, dimensions and weight. Also the standards which shall be applied should be summarized for the design of the laboratory as described in the following.

9.1.3   Clearances and Test Area

The size of a test room can be defined after the necessary HV test systems have been selected and the size of the largest test objects has been estimated. The necessary clearances between test objects and any grounded or energized structure have been discussed in Sect. 2.1.2 and are given in Fig. 2.1. This clearance is necessary to avoid any influence on the voltage distribution at the test object. In opposite, the distance in air between a HV test system and its surroundings must be selected in such a way that it withstands not less than about 120% of the rated voltage V_r of the test system.

In the test voltage range up to 600 kV (peak), the simple calculation of the air distance d can be performed based on the voltage demand 5 kV/cm of the positive streamer discharge in air (Fig. 9.3, dotted, blue curve):

Example A 400 kV AC test circuit shall be arranged. Which distance d is required to guarantee withstand between its components and the grounded metallic fences and walls? The rated voltage is a rms value; therefore, the stressing peak value is given by

$$V_{\text{peak}} = \sqrt{2} \cdot 400\,\text{kV} = 566\,\text{kV}$$

and the distance follows to

$$d > (1.2 \cdot 566\,\text{kV})/5\,\text{kV/cm} = 136\,\text{cm}.$$

It is decided that the distance to the metallic fence shall be 140 cm.

For higher voltages, the increase of the breakdown voltage (V_{50}) distance characteristic becomes remarkably lower due to the leader discharge (Fig. 9.3; Carrara and Zafanella 1968). According to the withstand voltage (V_{01}, green curve) of the positive rod-to-plane arrangement a clearance of about 16 m to other objects would be required for a 2MV SI or AC (peak) test voltage. Because of the high cost of the space in a HV laboratory, a much shorter distance must be reached by the application of suited *shielding and control electrodes*. Therefore, the impression of a UHV laboratory is determined by huge electrodes (Fig. 9.4).

There are many publications related to the design of HV electrodes, e.g. Moeller et al. (1972), Feser (1975), Mosch et al. (1979), Lemke et al. (1983) and Hauschild (1995), and electric fields can precisely be calculated now. The big problem remains to select the acceptable critical field strength. For PD-free electrodes, the concept of the streamer inception (Bürger 1976; Engelmann 1981; Dietrich 1982) can be used; otherwise, also the streamer–leader transition is applicable. For large electrodes, the streamer inception leads via a heavy streamer discharge which is immediately transferred into a leader discharge. The size of the electrode influences—in addition to the field strength distribution—also the probability of surface defects. This causes a distribution function of the inception field strength

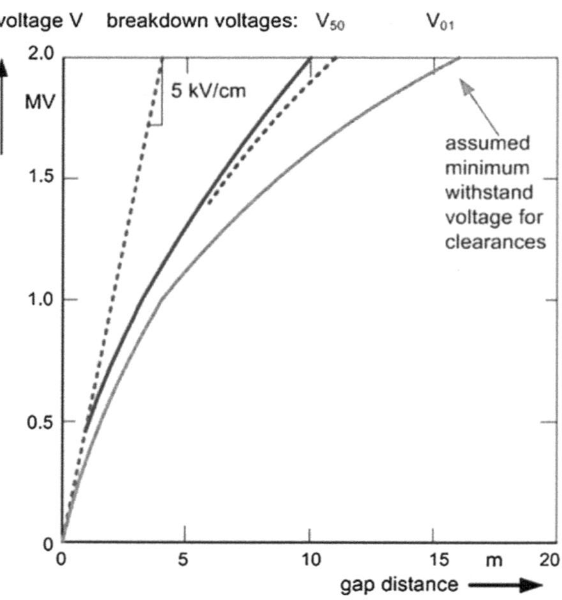

Fig. 9.3 Withstand voltage of a positive rod–plane gap in air

Fig. 9.4 Toroid electrodes in the UHV test laboratory of HSP Cologne (Germany)

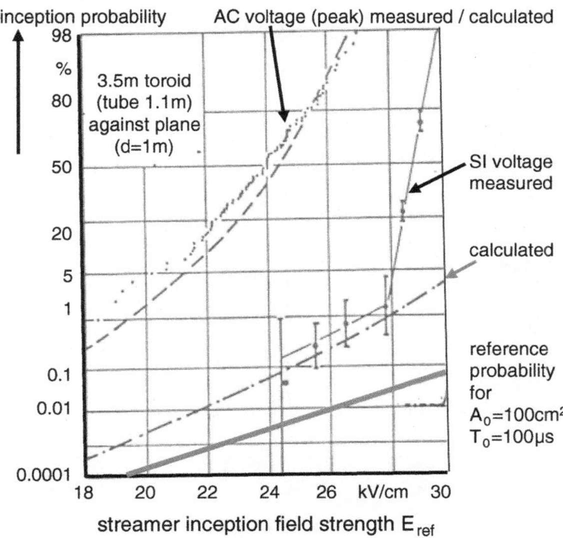

Fig. 9.5 Performance function of the streamer inception field strength in air (area element of 100 cm^2)

with a low increase for low probabilities and a steep one for high probabilities (Fig. 9.5). The dimensioning of electrodes is based on the low probabilities under consideration of the enlargement rule (See Subclause 2.4.7, Hauschild and Mosch 1992; Hauschild 1995). Usually, the dimensions of the electrodes are determined in an iterative process of field calculation and application of the streamer or leader inception criteria as described in the literature. Considering the enlargement of the electrode area and the prolongation of stressing time, the acceptable maximum field strength for a large electrode can be assumed to be in a range of 10–15 kV/cm for AC and DC peak voltage and of 15–20 kV/cm for SI voltage.

For test voltages below 2000 kV (peak), smooth electrodes like spheres and especially toroids are available. The *double toroid* (Fig. 9.4) is an ideal shielding element, because the necessary connections can be performed in the field shadow of the two rings.

For higher test voltage, composite electrodes are applied. Huge toroids can be realized by *cylinder segment electrodes* (Fig. 3.15). These are cylinder elements welded together to a toroid (Fig. 9.6). If the electrode is correctly designed, the higher field strength at the welded joint is related to a small area and the danger of a discharge would be not higher than at the larger area of the cylinders of lower field strength.

So-called *polycon electrodes* are sometimes huge spheres consisting of many metal plates fixed on a spherical scaffold (Fig. 9.7). The design of the polycon electrodes is well developed (Singer 1972; Hauschild et al. 1987; Schufft 1991).

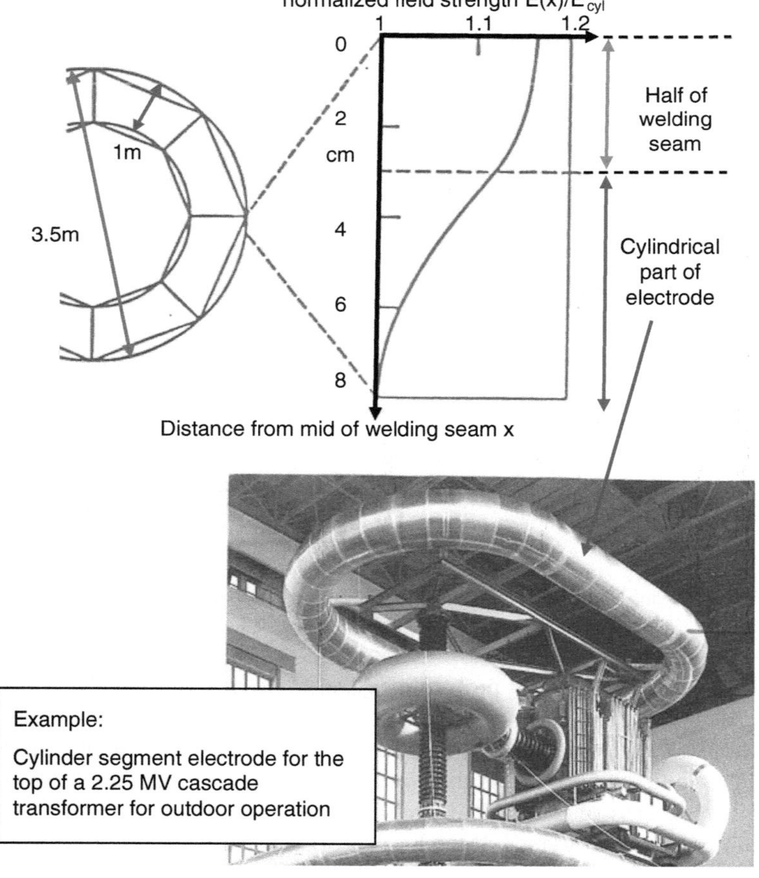

Fig. 9.6 Cylinder segment electrode for a 2.25-MV AC voltage cascade transformer

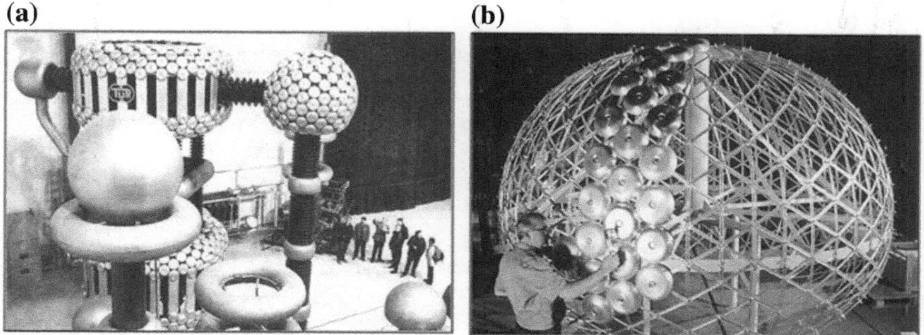

Fig. 9.7 Electrodes for UHV AC voltage components. **a** Different polycon electrodes and full-sheet sphere, respectively, toroid electrodes (EHV Laboratory of Dresden Technical University). **b** Framework of a polycon electrode (diameter 3 m)

9.1.4   Control, Measurement and Communication

The controls in a HV laboratory should not only be related each to a single HV test system (Fig. 2.8), but also to their interaction among each other and with auxiliary equipment. It is recommended that all control and measuring devices are equipped with industrial personal computers (IPC) which are connected to a common bus system. This can even be made in such a way that a certain HV circuit can be controlled from any of the IPCs in the control room (Fig. 9.8). Also external test data, e.g. the atmospheric conditions, data of wet or pollution tests, etc., and the correct function of the safety system shall be recorded. The test engineer should use the computer-aided evaluation of the test and the preparation of the test record.

The *IPC control and measuring system* enables automatic operation of the HV test system, which guarantees a better reproducibility of the test parameters, the direct recording and the evaluation of the measured data. Therefore, it may shorten the test duration and safe manpower. This is especially important for the testing of mass products (possibly in combination with other non-high-voltage tests), for lifetime testing and for large-scale tests for research and development. A fully automatic test procedure cannot be recommended for HV tests on valuable, single equipment as power transformers or GIS bays. In this case, a computer-aided procedure with decisions from the operator is optimum.

The computer system of the laboratory can be connected with the local area network (LAN) of the company or institution. Most important is the connection between the HV test field and test fields for other tests and measurements. This enables a common test record for the product. Furthermore, the status of the actual test can also be observed from any other place, e.g. a customer room during an acceptance test or a student's gallery in a university during HV demonstrations. A connection to the Internet enables remote service from the supplier (see Sect. 2.2).

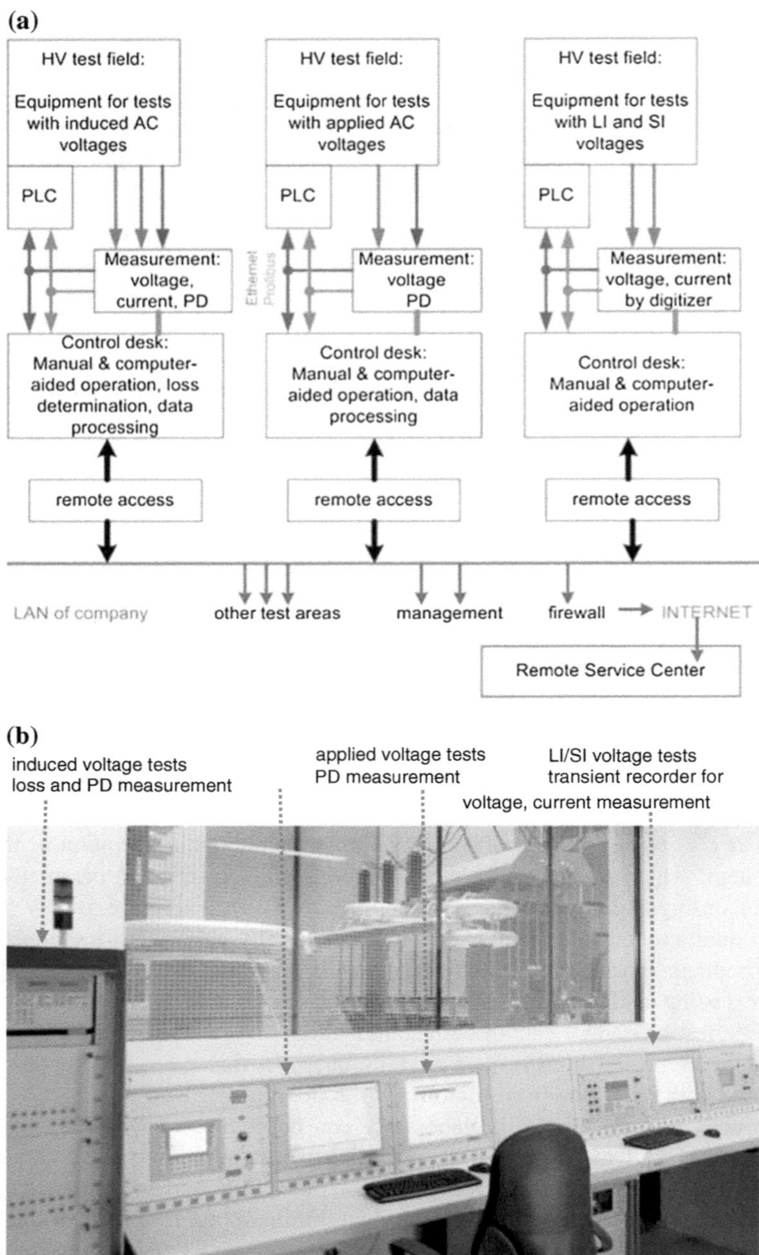

Fig. 9.8 Control room of a transformer test field (Courtesy Siemens AG, TBD Dresden). **a** Principle circuit diagram. **b** Control desk and rack

9.2 HV Test Building Design

There are many publications on the design of HV test laboratories, and their study is always useful for HV laboratory planning. It is impossible to mention here all of them, but the following section considers—in addition to the experience of the authors—especially the books of Prinz et al. (1965), Hylten-Cavallius (1988) and Schwarz et al. (1999) as well as publications by Läpple (1966), Mosch et al. (1974), Hauschild and Fahd (1978), Krump and Haumann (2011) and Hopke and Schmidt (2011). Each HV laboratory must be designed according to the special demand of the later user. This section will give suggestion for details to be considered during the planning of new or the refurbishment of existing HV laboratories.

9.2.1 Required Rooms and Principle Design

Minor details in a HV laboratory make the work of the test engineers efficient and easy. Good planning is therefore the necessary precondition for later smooth operation of the laboratory. Therefore, the application determines the basic design of the HV laboratory.

For *routine testing*, the HV test field may be a single test room or only a test area at the end of the production area. Then, the selected HV test systems and the dimensions of the test objects determine the size of the room. As an example, Fig. 9.9 shows such a quite compact, well-shielded test field for routine and type tests on power transformers up to 245 kV. It includes the facilities for induced and applied AC voltage tests and for LI/LIC/SI voltage tests in a room of only $L \times W \times H = 18.3$ m $\times$ 13.3 m $\times$ 12.3 m (Hopke and Schmidt 2011). Such a test field must be well adapted not only to the requirements of the HV tests, but also to the demand of the production and the flow of products. This includes the kind of transportation of the test objects (e.g. air cushion, rails, crane), the necessary space and clearances for the HV tests as well as the necessary conditions of measurement (e.g. of PD) in an industrial environment. The test room must be completed by a shielded control room and a power supply room.

A *universal HV laboratory* for research, development and training requires many test and auxiliary rooms. The size of the largest laboratory room is determined by the highest required test voltages. It is useful to have a second universal test room for equipment of clearly lower rated voltage. For training, smaller test rooms are recommended with one or two test areas each. Furthermore, special laboratories for cable testing, pollution testing, calibrations as well as for oil and solid insulation are often required. The selection of rooms shall be explained based on the following example:

Example (Fig. 9.10) A universal EHV laboratory shall be erected as the National HV Laboratory of a country. It has been decided to erect the laboratory near the campus of an

Fig. 9.9 Test field for routine and type tests on power transformers. Courtesy Siemens AG, TBD Dresden

university and to use it also for student's training. The largest equipment to be tested is for a rated voltage of 550 kV. Therefore, the highest test voltages are 1550 kV LI, 1705 kV LIC, 1175 kV SI (Table 1.2) and 680 kV AC (IEC 60076-3:2012). According to the principles described in Sect. 1.3, the following rated values of the largest HV test systems are selected: 3000 kV impulse voltage (LI; LIC: SI) with a highest SI output voltage of 1800 and 1000 kV (AC). Additionally, it is decided to have a DC voltage test system of 1000 kV (extendable up to 2000 kV) to meet future requirements on DC testing. After consideration of the necessary clearances and the necessary space for the test objects, an EHV laboratory of a length of 40 m, a width of 30 m and a height of 25 m was selected (Fig. 9.10). The test systems are arranged in three corners of the EHV hall. The control room (yellow) is in the fourth corner.

The HV laboratory shall be used for testing power equipment of rated voltage up to 145 kV. It is therefore equipped with an 800-kV impulse test system and a 400-kV AC test system. The related control room (Fig. 9.10) is in a corner. The three MV labs are planned for students training. Each of them shall be equipped with two HV test bays for voltages in the range up to 100 kV.

The laboratory is completed by a number of special test rooms. There is a special HV *laboratory for cable testing* with a big door to outside for the cable drums. As the other two laboratories, this laboratory including its control room is well shielded. The cable test laboratory is completed by an open-air field for pre-qualification tests of cables. The related control room with a window to outside is in the mezzanine. The area in front of the EHV

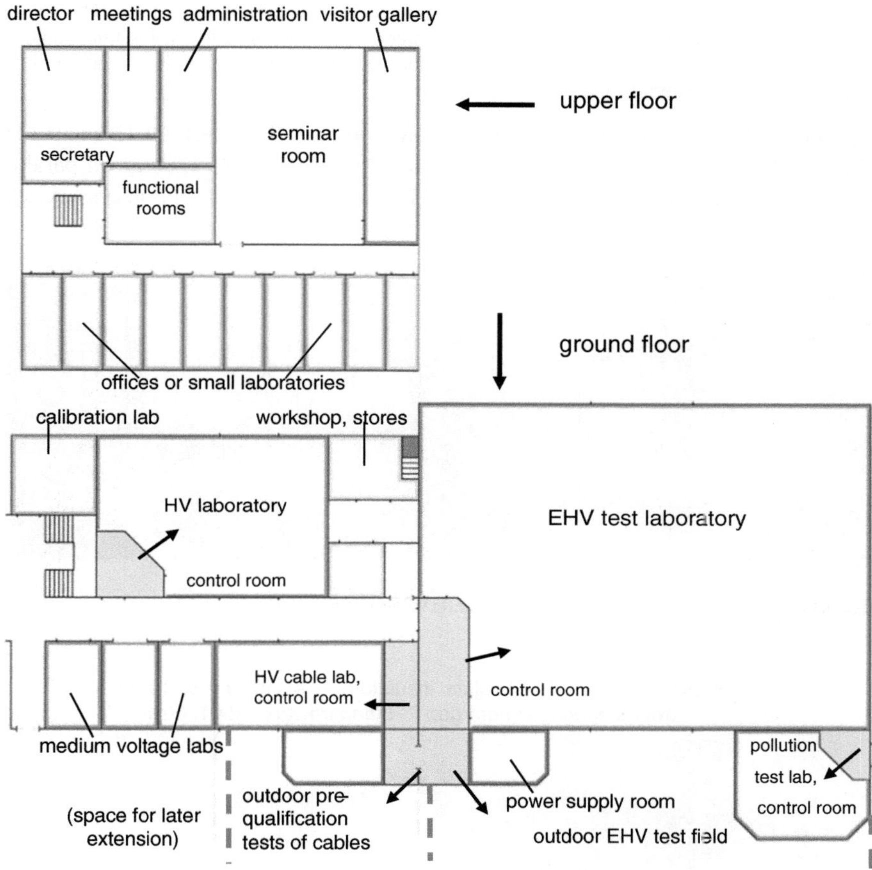

Fig. 9.10 Planning of the arrangement of rooms of an universal HV laboratory

hall is kept empty for a later outdoor EHV or UHV test field. Also a related control room in the mezzanine is planned for the future. A *pollution test laboratory* can be connected via a bushing to the AC test system of the big hall (Alternatively, the room is big enough for its own powerful test transformer. Also an air-conditioned *calibration laboratory* is part of the planning (left).

There are many auxiliary rooms required, such as the *power supply room*(s), stores, workshops, seminar and meeting rooms, offices for the director and his staff. Additional functional rooms as IT server room, kitchens, washrooms, etc., should be planned. The arrangement of these rooms shall be explained with the continuation of the above example:

Usually, a HV laboratory cannot have too many store rooms. Possibly, the one between the EHV and the HV laboratory is not sufficient. In opposite to that, there is a mechanical workshop in the ground floor and an electronic workshop in the mezzanine. The area under the calibration laboratory can be used as a meeting room. In the upper floor (Fig. 9.10), a seminar room enables the teaching of up to 60 students. The teaching room is connected with a visitors' gallery which enables the observation of experiments in the EHV hall (similar to Fig. 9.11). Additionally, the upper floor is used for all offices, for functional rooms and for small laboratories, for example for mechanical or chemical investigation of solid insulating materials and insulating oil.

Rectangular test rooms can be considered as optimum (Figs. 9.11, 9.12 and 9.13). They can be well subdivided into two separate test areas by *safety fences* for parallel testing. The relation between width W and length L depends on the necessary test

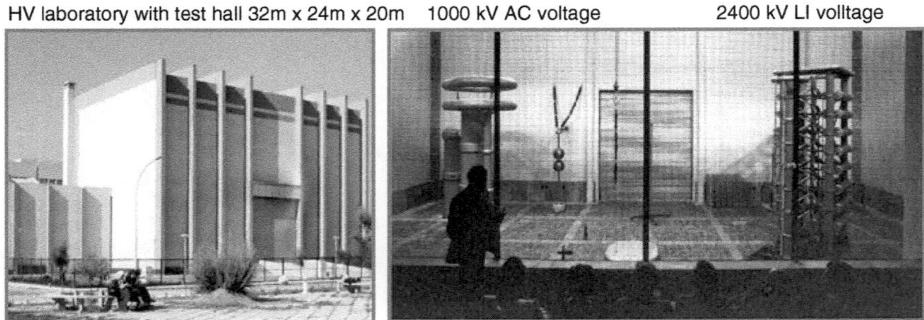

Fig. 9.11 Building and students gallery at the HV laboratory of Damascus University

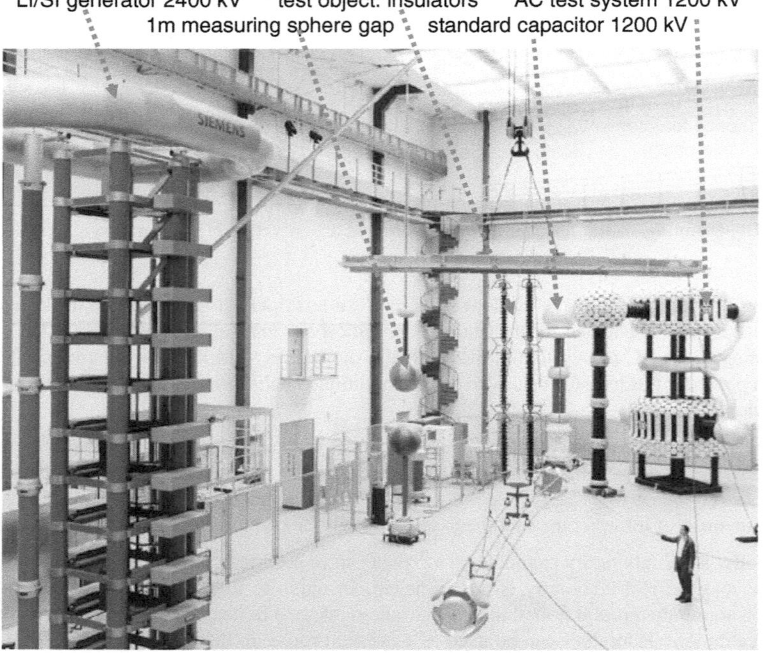

Fig. 9.12 EHV laboratory of Dresden Technical University (building erected in 1930)

Fig. 9.13 EHV laboratory of Cottbus Technical University (established in 1998)

voltages. If only AC and impulse voltages are required, the relation $W/L \approx 0.5 \ldots$ 0.6 seems to be optimum. In case of the three voltage sources (AC; DC and impulse), the relation $W/L \approx 0.7 \ldots 0.8$ is better adapted. Large test objects are arranged in between the test systems. Today, the HV generators can be placed on *air cushions* (see Sect. 9.2.5.3) and moved to the optimum place for testing or to a corner when not used. Smaller components of the HV test circuit can be equipped with wheels and also be placed at optimum positions for testing or stored on an empty area.

Usually, the largest *clearances* to walls and ceiling are required at the top of the test systems. Therefore, a limitation of the clearance which is sometimes applied by a parabolic cross section (Fig. 9.14a) cannot be recommended, but a rectangular one (Fig. 9.14b). The lower demand of clearance near its floor can be used for additional small built-in test areas of lower test voltages or for stores within a rectangular structure.

HV laboratories are impressive technical rooms. Therefore, an advanced planning should not only consider the technical aspects, but also a certain aesthetic planning. The three-dimensional design offers best opportunities for the aesthetic planning (Fig. 9.15, which is related to Fig. 9.4). The balanced planning of the colours between HV test systems, wall, ceiling and floor is highly recommended. The control room is often the place where also visitors appear who conclude from their impression to the quality of the HV testing. Therefore, the control desk (or table or rack) should be clearly arranged, of uniform design (even for test systems

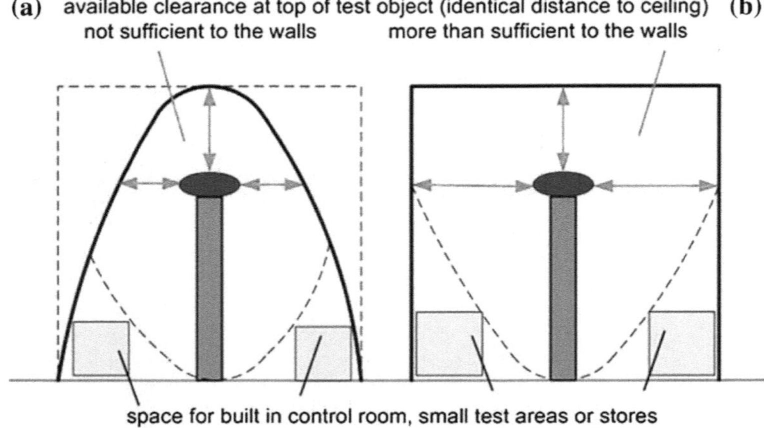

Fig. 9.14   Cross sections of HV laboratories

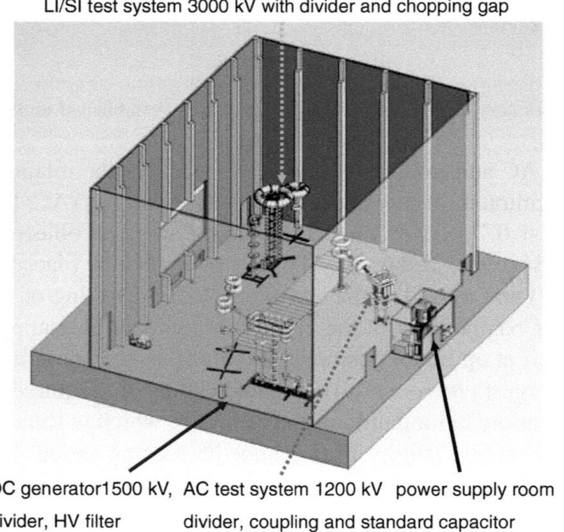

Fig. 9.15   Three-dimensional planning of a HV laboratory

of different suppliers) and well placed in the room. The height of a desk should correspond to the lower frame of the window (Fig. 9.8). There should be additional space in the desk for later extensions by different instruments, to avoid the poor impression of additional stand-alone device on a desk with built-in instruments.

9.2.2   Grounding and Shielding

Building grounding, test field grounding, electromagnetic shielding and earth return of a test circuit are often mixed up, not only theoretically, even worse in real test field operation. First, the different terms shall be defined by their descriptions:

Building grounding: Each HV laboratory building must be grounded, mainly for its lightning protection. The building grounding consists of the building's steel reinforcements, the foundation earth electrodes and possibly additional earthing rods. In a complex of buildings, the different building groundings are connected. The building grounding takes no care for external noise reduction.

Test field grounding: The upper layers of the ground are influenced by the earth return of power-electronic-driven machines and equipment in a factory. In brief, they are not free from noise signals. Therefore, a conductive contact of the test field grounding with the upper layers of soil and with the building grounding should be avoided. The test field grounding shall be realized by earthing rods which dip some metres into the ground water and which are electrically insulated along their first metres. The effective ground resistance of the parallel connection of all earthing rods should not exceed $2\,\Omega$.

Electromagnetic shielding: An electromagnetic field caused by noise signals may disturb measurements in a HV circuit (Chap. 4). According to the famous observation of Faraday, no electric field exists inside a closed metallic container. Therefore, HV laboratories are shielded against the penetration of electromagnetic fields by a closed metallic structure which is separated from both the soil and the building grounding and only at one point connected to the test field grounding.

Earth return: Each HV test circuit must be connected to the test field grounding at one point near to the voltage divider or the test voltage generator. It should have its own earth return of lowest possible inductance (see Sect. 7.1.2.3). Only in special cases, it can be recommended to use the test field grounding as the earth return.

Building grounding, test field grounding, electromagnetic shielding and the earth return(s) of the single test systems must be connected **only at one point** to guarantee a common stationary ground potential and to avoid any grounding loop which may act as an antenna for noise signals. For power cables from the regulators to the test room, a cable trench along the walls with metallic frames and covers is necessary. Frames and covers must be connected to the shielding of the floor. The connection of the floor shielding to that of the walls has to be made via the cable trench.

When a HV laboratory is erected, the mechanical *design of the floor* is related to the required maximum load by test generators and test objects. Also the kind of transportation, e.g. by air cushions, must be taken into consideration. The steel reinforcement of the concrete forms one part of the shielding of the floor. It must be isolated from the soil by a suited plastic foil, the single steel rods must be welded together, and a wide mesh of metal band, also welded to the reinforcement steel, completes the shielding of the floor (Fig. 9.16a). The basic concrete is usually finished up by a layer of final concrete which might contain a special fibre-glass reinforcement. The final concrete requires a very fine surface or an epoxy coating for air cushion transportation (see Sect. 9.2.5.3).

The surface of the floor is interrupted by *earthing boxes* and *connection boxes* (Fig. 9.16b). The metallic earthing boxes are connected to the floor shielding. The internal part is isolated from that and connected to the earthing rods and/or their connections by isolated copper bars. Furthermore, inside the steel reinforcement of the floor, metallic tubes between the HV test systems and the control room are provided for control and measuring cables. They are accessible via the connection boxes. Even measuring instruments can be arranged in a connection box, for example the PD measuring impedance and the PD measuring instrument in a box near to the coupling capacitor (Fig. 9.17).

The *shielding of the walls* shall enable multi-functions: In addition to the electromagnetic shielding, it shall be a heat isolation and a sound absorber. A perfect shielding of >100 dB up to 100 MHz—as usual for EMC shielding of computer centres and EMC test areas—is not necessary for a HV test room. The PD measurement according to IEC 60270 is usually performed at frequencies up to 1 MHz, and therefore, a damping ≤ 100 dB up to 5 MHz is sufficient for most laboratories. This can be reached with standard steel panels of two layers of galvanized steel (Fig. 9.18). The lower layer (Fig. 9.18a) carries the heat isolation (black, e.g. rock wool) which is at the same time the sound absorber. The upper layer is made of perforated steel panels and fixed to the lower panels (Fig. 9.18b, c). The panels of each layer among one another as well as of the two layers with each other should overlap and carefully screwed together for a reliable electric contact. Instead of screwing, welding of points every 50–100 cm is even more reliable.

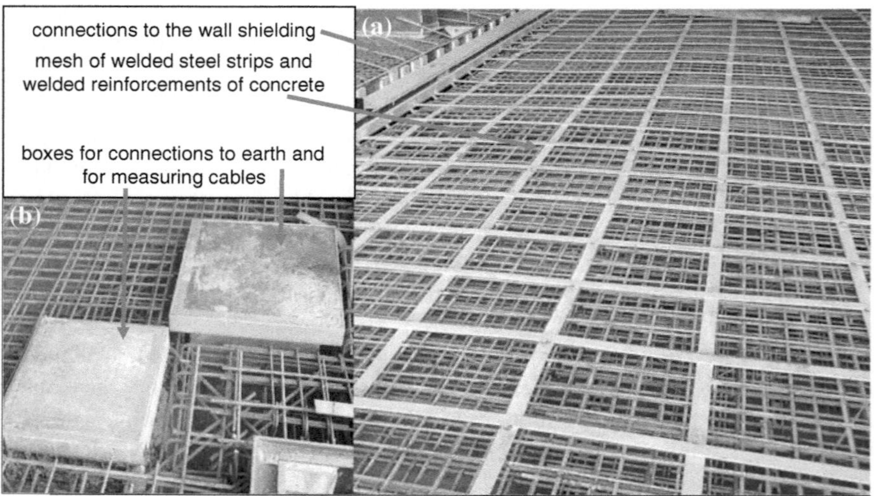

Fig. 9.16 Shielding of the floor of a HV laboratory. **a** Preparation of the shielding of the floor. **b** Earthing and connection boxes

Fig. 9.17 Connection box with a measuring instrument. Courtesy of HSP Cologne

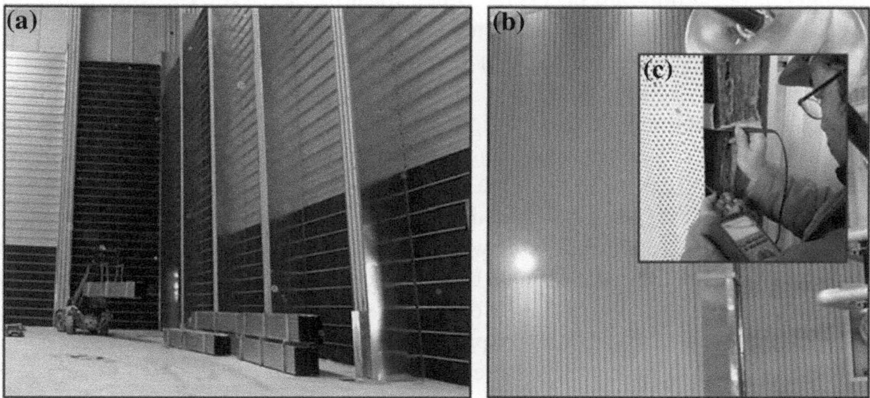

Fig. 9.18 Shielding of the walls (Courtesy of HSP Cologne). **a** Lower layer carrying the heat isolation (*black*). **b** Upper layer of perforated steel panels. **c** Check of electrical contact

The *shielding of the ceiling* (Fig. 9.19) may follow the same principle as that of the walls. But the ceiling usually contains panels of the heating system, the lighting system and the air conditioning (ventilation and aeration). The heating panels are also of steel and can be used as a part of the shielding. The shielding panels may be

Fig. 9.19 Shielding of the ceiling including heating panels, lighting and air conditioning. Courtesy of Siemens AG, TBD Dresden

screwed or welded to the heating panels. For the lamps and the air conditioning, openings in the ceiling are necessary which must be covered by a wire netting of steel wires with welded crossing points (width of meshes <30 mm). Such a wire netting is also recommended for windows, for example of the control room or a visitor's gallery.

It can be recommended to design a shielding also for the *control room* (Fig. 9.8b). In that case, the shielding can be made of expanded steel under the plaster of walls and ceiling. The single sheets of expanded steel must overlap and be welded together. For windows and doors, the sheets should be welded to their metallic frames. The openings of the windows should be covered with glass and the mentioned wire netting which has to be welded to its frame. It is useful when also the power supply room is shielded (for details see 9.2.3).

The most sensitive and expensive parts of a shielding are the *shielded doors*. For very sensitive PD measurements, for example of extruded cables down to very few picocoulombs, special doors with metallic feeder contacts which are pressed to the frame of the door are available. In many cases, such doors are not required. Then, sliding doors of steel or roller shutter doors with metallic lamellae—both moving in overlapping frames—are economic alternatives.

It should not be forgotten that all supplies (water, compressed air, oil, etc.) entering into the test field and carrying the building ground must be isolated and inside connected to the shielding of the test room. Otherwise, they may carry noise signals into the shielded test area. Electric supplies must be filtered.

9.2.3   *Power Supply and High-Frequency Filtering*

The power for a complete HV laboratory is usually supplied from a medium-voltage network (Fig. 9.20). One or several three-phase distribution transformers in a nearby substation should be used for that purpose. Their ratings depend on the equipment to be supplied. One has to take into consideration the maximum required test power (active and reactive components), the supply power for control and measuring systems, the power for all auxiliary equipment (see Sect. 9.2.4) and an extra charge for later extensions.

Power cables connect this external power supply with the *power supply room*. This room should be arranged close to the HV test hall (e.g. Figure 9.10) and should also be shielded in connection with the shielding of the related test room (Figs. 9.20 and 9.21). The connection of the external power cables to the equipment for distribution and regulation in the power supply room is realized via high-frequency (HF) power filters attached to the shielding of this room (Fig. 9.22). The power control room contains the equipment explained in Sect. 2.2 (Fig. 2.5) for the "power supply":

- switching cubicles with main and operation switches, measuring equipment for primary voltages and currents, protection devices and control components (Fig. 2.8) like programmable logic controllers (PLC);

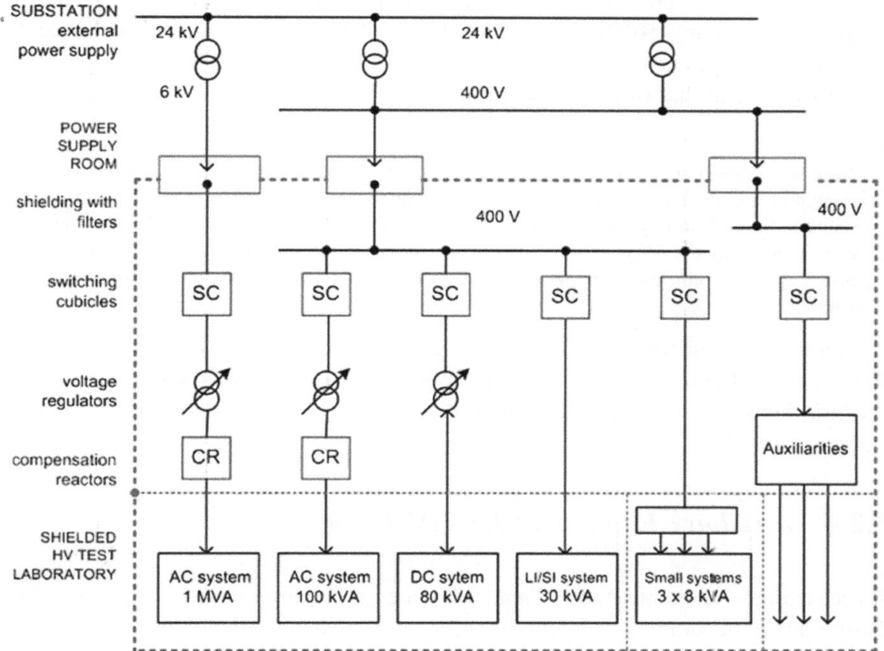

Fig. 9.20 Schematic power supply of a HV test laboratory

Fig. 9.21 Shielded power supply room. Courtesy HSP Cologne

- regulators, one for each HV test system and
- compensation reactors and/or capacitor banks, to reduce the demand of external supply power.

  The size of this power supply room depends on the mentioned single components and their arrangement. The *HF-filter* consists mainly of a L-C network (Fig. 9.22a). Conducted noise signals of frequencies remarkably higher than the power frequency 50/60 Hz are blocked by the inductances and conducted to ground via the capacitances. The frequency characteristic of the HF filters shall be adapted to that of the shielding.

Note If the power supply room is not shielded, the HF filters must be attached between the regulators and the test generator at the shielding of the test room. Then, a separate filter is required for each test system and each of the auxiliary equipment. For the AC test system, even a larger filter is necessary because the compensation reactors are in the—unshielded—power supply room. The shielding of the power supply room simplifies the filtering and the wiring.

Communication lines—as long as they are not realized by fibre optic links—must also be filtered. As HF filters for power connection, also such for communication lines are available on the market.

9.2.4   Auxiliary Equipment for HV Testing

If accessories—as mentioned in Table 9.3—are used, consequences for the building design must be drawn. It might be necessary to have supply tubes for water, compressed air, etc. into the shielded area [test room(s), control room(s), power supply room(s)] from a machinery room or the unshielded outer areas; these

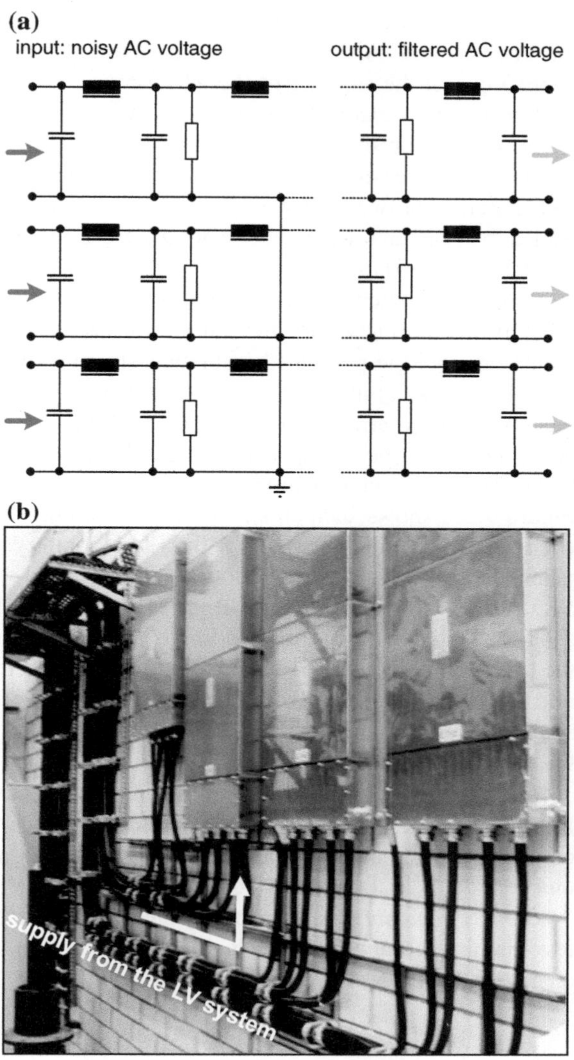

Fig. 9.22 HF filter for the power supply. **a** Principle circuit of low- and medium-voltage filters. **b** Arrangement of HF filters at the wall of an inside shielded power supply room. Courtesy HSP Cologne

connections must be isolated. This means that all metallic tubes must be interrupted by suited insulating tubes where they enter the shielding.

The size of the *artificial rain equipment* (see Sect. 2.1.3) must be adapted to the size of the largest HV equipment to be tested. This determines also the demand of water of the required conductivity, the water processing tank and the water connection from there to the rain equipment. Furthermore, the floor in the area of wet tests must be equipped with water drainage. This means it should have a certain slope which may come in conflict with the air cushion transportation system. At a too high slope, the air cushions with their load may run away by themselves.

Therefore, no air cushion transportation in that area should be allowed or the slope must be very low. A slope below 1 cm/m might be necessary, but it must follow the instructions of the supplier of the air cushions.

The pollution chamber must be equipped with a spray system, connections to water and compressed air, as well as drainage in the floor. The bushing for the pollution chamber must withstand the worst pollution conditions inside the chamber and therefore be well selected. Because the conductivity of the pollutant is controlled by salt and water, all materials used in the pollution chamber must be resistant to corrosion.

Using of a trench in a HV laboratory: *Tanks for insulating liquids*, usually for mineral oils, are necessary for testing transformer bushings and for R&D experiments on liquid-impregnated insulation structures. They are connected to oil (or liquid) processing units with related reservoirs outside the shielded area. It can be recommended to arrange the test tanks under the floor in a trench. Also other HV components can be arranged in a trench, especially when they include grounded tanks. This may even include test transformers or standard capacitors (Fig. 9.23). Figure 9.23a shows such an arrangement of a multifunctional capacitor which may even act as a central electrode for the HV laboratory. The trench must be carefully planned together with the floor shielding and the grounding system (see Sect. 9.2.2). All walls and the floor of the trench should be shielded. The tank should be isolated from the shielding of the floor. During a test, the tank is part of the test circuit and must be included into the ground return of the test circuit. If it is not used, it should be directly connected to the earthing system, but never to the shielding system. The trench must be equipped with well-fitted covers which close the trench not only mechanically, but also close the shielding.

Tanks for compressed gases are used for testing bushings from air to SF_6 and for R&D experiments with gas-insulated structures. They are connected to gas processing units for conditioning the used gases. The arrangement of gas-filled test tanks is identical to those of liquids (Fig. 9.23).

Climatic chambers should be adapted to the principles described here for HV test circuits, grounding and shielding. The design of a commercially available chamber determines the necessary measures.

9.2.5 Auxiliary Equipment and Transportation Facilities

9.2.5.1 Lighting

Usually, HV test laboratories are designed without windows for natural light. Natural light might be useful in very special cases, e.g. when the room is also used for production or larger assembling work. Then, the windows must be supplied with a metallic frame connected to the shielding and covered by the above mentioned wire nettings with welded crossing points, also welded to the frames. Darkening of the windows should be possible.

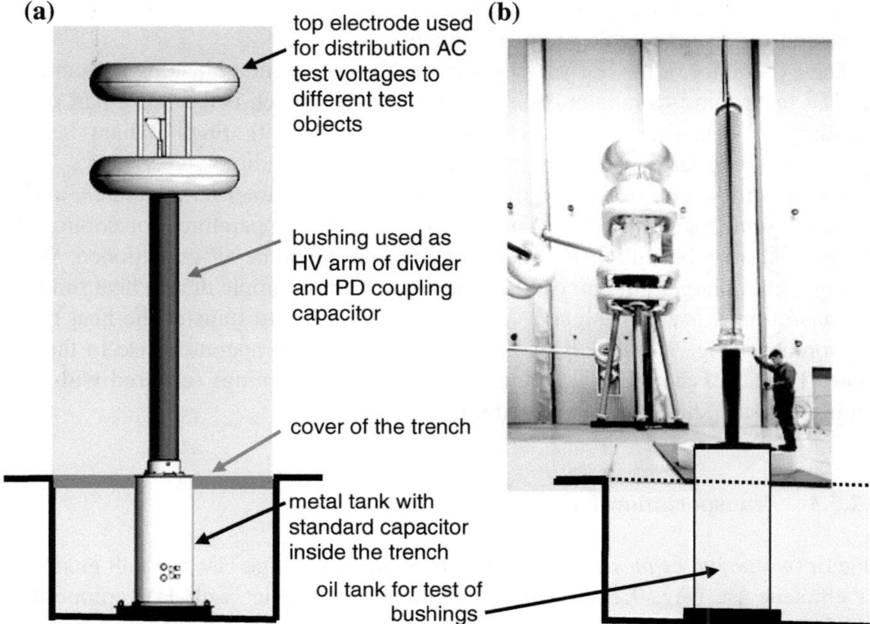

Fig. 9.23 Use of a trench in a HV laboratory. **a** Arrangement of a multifunctional capacitor. **b** Tank for testing bushings. Courtesy HSP Cologne

Artificial *lighting*—the standard case—should be realized with incandescent lamps in all shielded rooms—and not with fluorescent lamps which may cause noise signals during sensitive PD measurement. For many observations and photographs, it would be helpful if the light can be dimmed. Because this is realized with a power electronic adjustment, another source of noise signals, the lamps should be outside the shielding behind wire nettings (Fig. 9.19). Preferably, the lamps can be arranged at the ceiling. A separate emergency lighting is necessary. The local requirements and safety rules have to be considered.

9.2.5.2   Heating, Ventilation and Air Conditioning

HV test equipment should be specified for a temperature range not smaller than between 10 and 35 °C; for the control and measuring systems, a minimum range between 15 and 30 °C is required. The installation of a comfortable *heating system* in a large HV test hall seems to be only necessary when the minimum temperature of 10 °C cannot be guaranteed for a longer period than few days per year. In opposite to that, it is recommended to have an air-conditioned control room with a temperature in the range of 20 °C. The power supply room requires no heating as long as the temperature remains above 5 °C. Small HV test rooms where HV test equipment and the control and measuring system are in one common room should be heated for an acceptable room temperature.

The experience of the last years has shown that an optimum heating system for a big HV test room is a radiant ceiling heating system which is used as a part of the shielding as described above (Sect. 9.2.2 and Fig. 9.19). Such radiant heating systems of quite large panels are usual for industrial buildings today.

For countries with tropic conditions or very high summer temperatures, a ventilation system for the test hall which uses low night temperatures for cooling the HV test hall may be sufficient. Smaller test rooms should be air-conditioned. When during a test, a larger amount of heat is generated, for example during heat run tests of transformers, a well-designed ventilation system has to transfer the heat to the environment (Fig. 9.19). This avoids an unacceptable temperature rise in the test room. The necessary ventilators are arranged behind openings (covered with wire nettings) of the shielding of the ceiling.

9.2.5.3 Transportation Facilities

One or two *portal cranes* covering the whole area of a large HV test hall guarantee its efficient use (Fig. 9.24a). This may come into conflict with HV components which could be fixed at the ceiling, for example voltage dividers, central electrodes or rectifiers. In some laboratories, this had been done; consequently a single-rail crane had to be applied (Fig. 9.24b; Prinz et al. 1965, courtesy of TU Munich). The comparison of the two solutions shows clearly that in minimum for industrial HV test laboratories, the portal crane is by far superior to the one-rail crane. With two or three portal cranes in one HV test hall, a compromise between the two solutions is reached. Figure 9.25 (Krump and Haumann 2011) shows a test room with three cranes; the crane in the middle carries the large, PD-free connection electrode (1200 kV rms.), between the central connection point and the test object, whereas the cranes of both sides are available for transportation and fixing of test objects. In addition to the portal crane, lifts with a single rope from a hole in the ceiling can be taken into consideration. They are for fixing a test object only during a test. Such lifts and hand-operated cranes are also useful in small test rooms.

The maximum load of a crane depends on the maximum load of a test object, in some cases also of the weight of the heaviest HV test component, whatever is heavier. For the small hand-operated cranes and lifts, a load up to 1000 kg should be sufficient. It should be mentioned that available transportation by air cushions can reduce the maximum necessary load of a crane.

Air cushions are a very helpful tool for the ground transportation in a HV laboratory. On the one hand, they can be used for the generators and HV components (see Sect. 9.2.1), on the other hand for the transportation of heavy test objects as, e.g. power transformers or GIS. The air cushions replace the transportation on rails as it has been applied for generators or test objects in the past quite often. The rails interrupt the smooth floor and may even influence the

Fig. 9.24 Cranes in HV laboratories. **a** Portal crane (Courtesy HSP Cologne). **b** Single-rail crane

optimum erection of test circuits. The ground transportation of smaller HV components can be made on wheels. Also transportation by the crane is often applied.

Well-selected air cushions, usually arranged under the base frame of the generator, can carry up to the highest loads required for HV test systems. Number and size of the single-air-cushion modules depend on the load and the load distribution. But they require a well-levelled, smooth and stable floor. The surface can be a fine concrete, in most cases covered with a special epoxy resin. Air cushions can damage the floor if it shows gaps and cracks or is not smooth enough. Usually, the air pressure is 0.2 MPa, in special cases up to 0.4 MPa. It is recommended to order HV test equipment with frames and with the necessary stability suitable for air cushion transportation (Fig. 9.26). When the movement of air cushions suddenly stops due to an obstacle, the mechanic stress to the insulating supports, for example of an impulse voltage generator, is considerable.

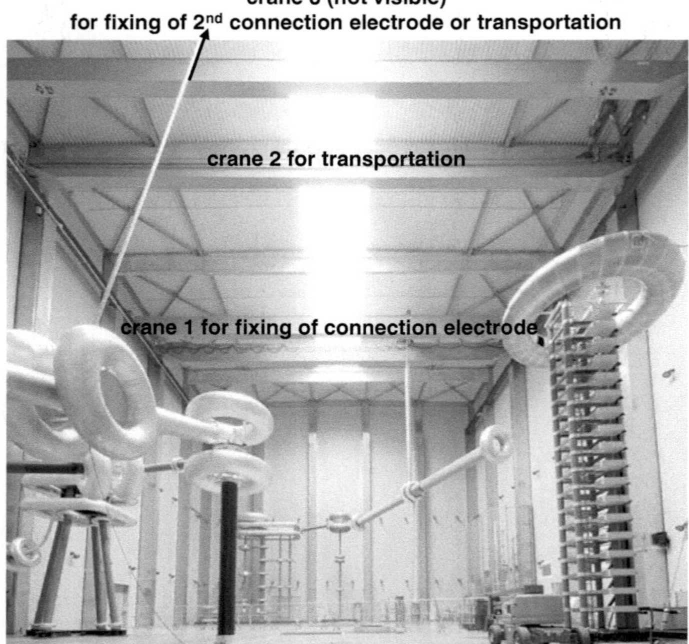

Fig. 9.25 UHV laboratory with three cranes. *Courtesy* of HSP Cologne

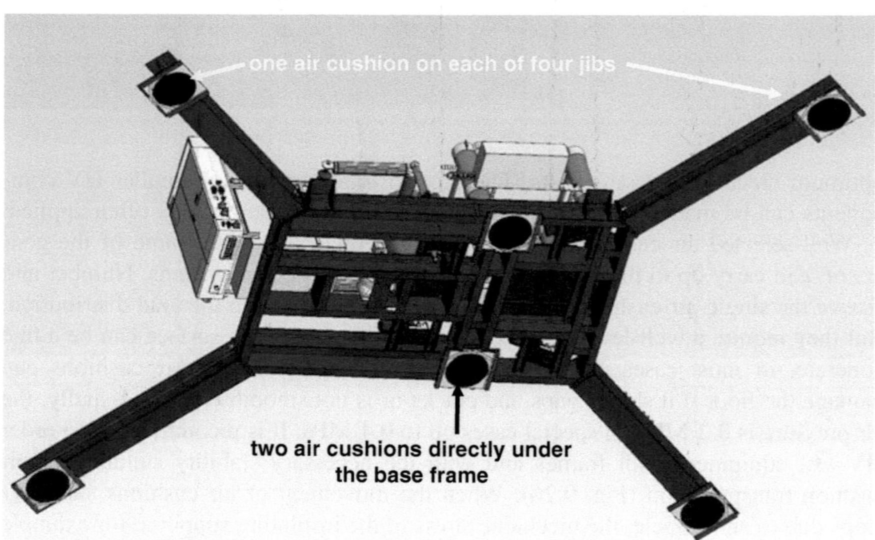

Fig. 9.26 Frame of a test generator with air cushions

9.2.5.4   Technical Media

The *electric power supply* is described in Sect. 9.2.3. In the single test and control rooms, a sufficient number of LV sockets should be arranged. This includes the sockets in the control desk or rack for instruments. Everywhere, sockets are necessary for electric tools, spot lights or additional instruments. Again, all electric power lines which enter shielded rooms must be filtered.

Data communication and *telephone* lines should be as far as possible realized by fibre optic links. They do not require filtering. All wire lines for communication must be filtered.

Compressed air is necessary for the air cushions (see Sect. 9.2.5.3), but also certain tools as well as pollution tests require compressed air. The necessary pressure depends on the local demand. The metallic tubes for compressed air must be isolated when they enter the shielded rooms.

A *water supply* is necessary for artificial rain and pollution equipment (see Sect. 7.2.4), for other special tests and for cleaning. Therefore, few water taps might be arranged in a big and only one in smaller test rooms.

Other technical media as oil and other liquids, SF_6 and other gases should be supplied via installed facilities if there is a permanent demand (see Sect. 9.2.4).

9.2.5.5   Fire and Environmental Protection

The *fire and environmental protection* in a HV laboratory has to follow both, the national laws and the technical knowledge on HV testing. Special consideration requires insulating oil and other inflammable liquids. This includes also oil-filled HV test components and equipment of power supply, as e.g. the mayority of test transformers, reactors, regulators, etc. One should consider that a leak of a related tank should not pollute the nature, the soil or the water. But a leak in the tank of a modern HV test component is very seldom and if it occurs it would remain very small. Consequently a volume which may overtake a certain part of the oil after a small leak should be taken into consideration. In some cases even cable ducts are used for that purpose. The application of the hard rules for a collector for all the oil of the related power transformer *cannot be recommended*. A power transformer in a power system—which may explode and dramatically burn due to heavy internal failures caused by over voltages. In opposite to that, test transformers and other test equipment are usually stressed remarkably below their rated voltage. The HV tests are connected with sensitive PD measurements, which show not only the defects of the test object, but also of all components of the test circuit. This means, a fault inside oil-filled equipment is very early indicated, the related component has to be taken out of operation. There is practically no danger of a powerful fault in an oil-filled test transformer or reactor, with dramatic tearing up the tank followed by fire. Nevertheless, a realistic concept for fire and environmental protection must be established during the planning of a HV laboratory and approved on the basis of the technical knowledge and the relevant local rules.

For tests of insulations with SF_6 gas, one has to consider that SF_6 is a green-house gas and a totally proof gas handling system must be planned. The GIS-enclosure shall fulfill the technical requirements of high-pressure containers.

9.2.6   Safety Measures

High electric fields, generated by high voltages between energized electrodes and all earthed objects in the HV test rooms, are very dangerous: When the electric field exceeds the dielectric strength of the surrounding air, electrical discharges appear with currents up to kiloamperes. But the rate of accidents in HV laboratories is low, because the operators are conscious of the high risk when the "safety concept" is not carefully considered. The *safety concept* includes all technical matters related to the test laboratory as a whole and related to the single HV test systems as well as the instructions for the personnel. The latter includes the general behaviour in HV testing and the instructions for the operation of test equipment.

There is no special IEC Document on safety in HV test laboratories; the IEC Publication 62061:2005 is partly applicable to the control- and power-feeding equipment. IEEE Guide 510 recommends a practice for safety in high-voltage and high-power testing. There is also the European Standard EN 50191 (2000) on the erection and operation of electrical test systems. The hints in this subsection cannot release the users from applying these internationally accepted documents, standards and special national rules. Also the instructions and hints of the suppliers of HV test systems shall be considered.

9.2.6.1   Safety in HV Test Fields and Areas

A *HV test field* is a room with one or several fenced-in *HV test areas* and related control areas, completed by a power supply area. A *HV test laboratory* may contain several HV test fields (Fig. 9.10). The basis for the safety of a whole test field is its correct grounding and shielding as described in Sect. 9.2.2. The recommended *current return* of a test circuit is only connected at one point to the grounding system. In special cases, the grounding system may be used as the current return. The shielding shall never be used as a current return.

In a HV test area, the safe *clearances* (see Sect. 9.1.3) of HV components to the walls, the fences or the ceiling as well as to earthed or energized objects in neigh-bouring fenced-in test areas shall be selected according to the maximum test voltages which can be generated there. If several fenced-in test areas are in one test field, metallic *safety fences* with a minimum height of 2 m and a maximum mesh size of 4 cm shall be applied. The single elements of the fence shall be electrically connected to each other and connected to the earthing system only at one point. A complete loop of fences shall be avoided. If the HV components in a test area are rearranged, interactions to neighbouring fenced-in areas must be considered (Fig. 9.27).

Fig. 9.27 Safety fences with warning signs and lamps, safety loop and emergency-off switch

The HV test area must be marked from outside by a *warning sign* and by *warning lamps* indicating the condition within the area: The "red" light indicates the HV test circuit is degrounded and possibly energized. The "green" light means the HV test circuit is grounded, the area is safe for entering (see Sect. 7.2.6.3). Warning signs must be arranged at all doors into the test area. In addition to the signs and lamps, the test area must be equipped with a safety loop and emergency-off switches.

A *safety loop* is a ring main integrated into the walls, fences and doors of a test area. The loop must be closed, before the test system is degrounded and energized. Contacts at the door shall ensure that the loop is closed. As soon as the loop is opened anywhere, the emergency-off switch has to operate and the test system shall be grounded (see Sect. 9.2.6.2). *Emergency-off switches* shall be large, red push-buttons mounted preferably on a yellow background at the related control desk or rack, near the doors of the test areas and in observation areas. The actuation of an emergency-off switch results in the operation of the power (main) switch and operation switch as well as of the earthing equipment of the relevant HV test system. It has to be checked whether also all HV sources of the whole test field shall be switched off. It should not deenergize the lighting system (in minimum an emergency lighting must remain).

Control and observation areas are outside the test area and its safety loop. Both shall be equipped with warning signs and lamps and emergency-off switches. For the power supply room, the relevant national regulations for the erection of electrical substations shall be applied.

9.2.6.2 Safety of HV Test Systems

There is a close connection between the safety system of a HV test area and that of the related HV test system. Both work together, and if the first fails, the second cannot operate. Therefore, a safety check of the related safety equipment should precede each test.

A HV test generator is energized in two steps: first, the power (main) switch supplying the energy from the grid up to the power supply unit, and then, the operation switch connects the generator (see Sect. 2.2). The two switches are part of the safety concept of a HV test system (Fig. 3.1).

The earthing of the HV test circuit shall guarantee a safe stationary operation, whereas the earth return has to minimize transient phenomena, e.g. after a breakdown of the test object (see Sect. 9.2.2). If not in use, the HV components must be discharged and permanently grounded. For that, they are equipped with *discharging switches* and *earthing switches*. The switches (Fig. 9.28a) can be realized by earthed rods with an electric or hydraulic drive. They open the grounding when the main switch is switched on. They close the earthing, e.g. by gravity or by a spring, when the power switch opens and the voltage is off after a test or an emergency off.

Earthing ropes (Fig. 9.28b) consist half of a flexible metallic cord and half of an insulating rope. In case of grounding the HV component, the metal rope connects the HV electrode and all intermediate electrodes with the earthing system. In case of HV potential at the component, the metallic rope is replaced by an insulting rope. Both ropes are moving by a motor drive. One should look for reliable earthing ropes; otherwise, they might become the weak point of the equipment.

Earthing bars for manual operation (Fig. 6.14) are often applied for smaller HV test systems. They are from insulating material with a handle at one side and a hook

(a)   **(b)**

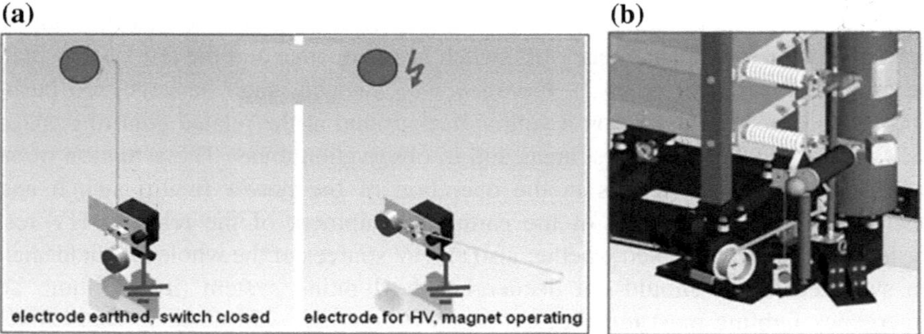

Fig. 9.28 Earthing equipment. **a** Earthing switch (earthing by gravity). **b** Earthing rope with motor drive

at the opposite site. The hook is connected to the earthing system by a flexible cord. It is recommended to arrange earthing and discharging bars at the door to the test area not to forget their application after a HV test.

Capacitors are charged during operation, keep the charge—at least partially—after switching off the test system and must be discharged, shorted and earthed, also when they are not in service. The discharging is performed by resistors in series with earthing bars, earthing switches and earthing ropes.

The *control* of a HV test system realizes not only the HV test procedures, but activates also the *safety functions*. This includes the interactions between position of the power switch and the operation switch with the status of the safety loop, the signal lamps and the earthing equipment. Furthermore, the control system shall react in case of breakdown of the test object, opening of the safety loop or emergency-off application. All these functions must be periodically checked according to the relevant instructions of the supplier of the test systems. Table 9.4 shows a timetable for such instructions as a brief example.

9.2.6.3   Operation of HV Test Systems

A HV laboratory should have a *"safety concept"* which describes the mentioned safety measures for test fields, test areas and test systems. Furthermore, it should include the following *safety instructions* for operating HV test systems and also the list of necessary inspections of safety-relevant equipment (Table 9.4). On the basis of the safety concept, the personnel should be instructed about the danger at HV tests according to national rules, but at least annually.

Before a HV test starts, the test engineer should take the following actions:

1. Final check of earth connections and clearances of the test set-up!
2. Check that no other personnel is in the test area, take off all manually operated earthing devices!
3. Close the safety loop!

Table 9.4 Survey of inspections of the safety equipment

Inspected equipment	Inspected component	Reference	Interval of inspections
Earthing system of the test area	Grounding resistance; earth connection boxes	9.2.2 Measurement of the ground resistance	<5 years
Test systems	Earthing bars, switches and ropes	9.2.6.1 Safe connections	<1 year Observations before each test
Safety equipment including controls	Safety loop, signal lamps, emergency-off switch	9.2.6.1	Check before each test

4. Switch on the control power for the control and measuring system (green lamps "on")!
5. Switch on the power (main) switch (red lamps "on")!
6. Warn by horn and loudspeaker "Attention, high voltage is switched "on" (automatic earthing "off")!
7. Switch on the operation switch and start the test. When a test is terminated the following should be done:

   1. Reduce the voltage to a level <50% or to zero!
   2. Switch off the operation switch (automatic discharging and earthing moves to "on", also red lamp still "on".)!
   3. Switch off the power (main) switch (red lamp "off", green lamp "on")!
   4. Open the safety loop and perform the manual discharging and earthing, if necessary!
   5. Switch off the control power, the test system is out of operation (green lamp "off")!

The described steps of operation shall also be part of the instructions for the personnel working in the HV laboratory. The instructions shall include all international and national rules on safety in electrical testing, the information of the supplier of the test systems and the own experience. The instructions may include demonstrations in the laboratory. They should be performed in minimum once a year and shall be recorded.

9.3   Outdoor HV Test Fields

When an *outdoor test field* completes a traditional "indoor" laboratory (compare Fig. 9.10), the test voltages might be transferred to outside via bushings or big doors. Even complete HV test systems can move on air cushions to the open-air test area. Outdoor test fields are usually equipped for research and development of EHV and UHV overhead transmission lines including the related open-air substations. In most cases, they are connected to test lines of some hundred metres length to test the design of towers, the arrangement of the conductors or to measure corona losses.

The HV test systems are placed on a concrete area which gives also space for tests on substation equipment. This area may be applicable for air cushions (see Sect. 9.2.5.3) and should be well connected to roads for the transportation of test objects. Transportation and assembling work must be done by mobile cranes. The HV test systems placed outdoor have a special design for the local climatic situation. Sometimes, the local conditions are the reason to have an outdoor test field, e.g. on high altitude to perform tests under low air density. Pollution of the test systems is a challenge which needs both good design and permanent maintenance.

For LI and SI voltage testing, impulse generators of more or less indoor design are arranged in an insulating tower for weather protection (Fig. 9.29). The air inside such a tower is conditioned to avoid dew on inner surfaces due to too low outer temperatures. Also too high inner temperatures shall be avoided. The rated voltage of outdoor impulse generators exceeds usually 3MV.

For AC voltage testing, metal-tank transformers (see Sect. 3.1.1) are well suited for both single-transformer application and transformer cascades (Fig. 9.30). They have proven high reliability for that application, because the design uses the experience with power transformers. The rated voltages of AC test systems are usually above 1MV.

For DC voltage testing on test lines, quite powerful generators (of rated voltages up to 2MV and currents of several 100 mA) are required (Fig. 9.31). It is very difficult to design the outdoor DC insulation of such HV test components.

The safety measures in outdoor laboratories may follow the principles described in 9.2.6, but have to consider the broad range of atmospheric influences as temperature, humidity, rain, snow, ice, sandstorm, etc. This requires larger clearances. The earthing system of an outdoor laboratory can be made as for indoor laboratories (see 9.2.2), but the steel reinforcement of the concrete should be welded and used—in addition to earthing rods—as an area earthing, no isolation from the soil should be made. It has to be connected to the earthing system of the indoor laboratory only at one point.

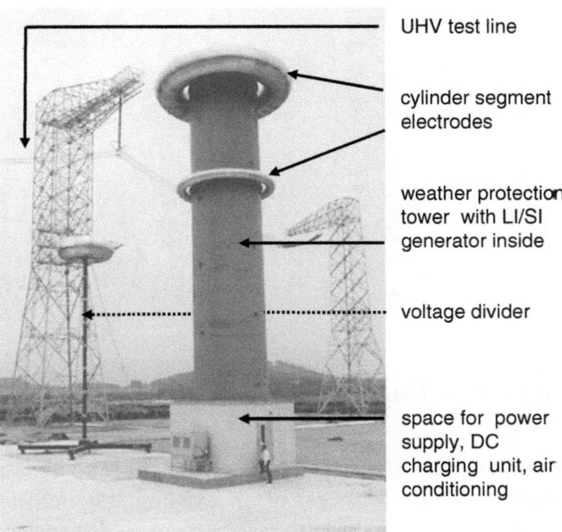

Fig. 9.29 Outdoor impulse voltage generator 4000 kV/400 kJ. Courtesy KEPRI Korea

 Outdoor
transformer cascade 1800 kV/
1.25 A. Courtesy of
Siemens AG, Berlin

Fig. 9.30 Outdoor transformer cascade 1800 kV/1.25 A. Courtesy of Siemens AG, Berlin

Fig. 9.31 Outdoor DC generator 1300 kV/1 A

9.4 Updating of Existing HV Test Fields

The erection of a new HV laboratory is a big investment. The updating (sometimes also called upgrading or refurbishment) of an existing test field is often an economically and technically acceptable alternative to a new test laboratory. In the simplest case, it is related to the replacement of complete HV test systems, the addition of new or the exchange of old components. The updating of existing HV test rooms is often more difficult. Viewpoints for updating existing HV test fields are given in the following. Generally, the updating of existing HV test fields shall follow the principles described in this chapter for new test fields as far as possible (see Sects. 9.1, 9.2, 9.3).

9.4.1 Updating of HV Test Systems

The lifetimes of HV and power components are usually higher than those of the components of control and measuring systems. Sometimes, also regulating transformers or thyristor controllers have a shorter lifetime than generators for high AC, DC or impulse voltages. Therefore, updating is often related to the control and measuring units. It cannot be recommended to replace only some of these components and to operate with a mixture of new and old control and measuring instruments. The control and measuring system should enable easy and reliable control, safe data recording and evaluation, communication within local data networks and remote service as shown in Fig. 2.8 and described in Sect. 9.1.4. The necessary interfaces between digital control and measurement systems and the components of the power circuit should be individually established depending on the design and age of the equipment to be refurbished.

If a new HV test system—possibly of higher rated voltage—shall be arranged in an existing laboratory, necessary clearances must be taken into consideration (see Sect. 9.1.3). In case of limited clearance between HV components (generator, divider, connections, etc.) and neighbouring grounded or energized objects, the application of larger control electrodes can be taken into consideration. This requires the calculation of the conditions of the electric field between the HV components and their surroundings. Additionally, the necessary clearance for test objects must be guaranteed (Fig. 2.1). To save space in the test area, tank-type test transformers and reactors can be arranged outside with their HV bushing into the test room (Fig. 9.21). For the safety requirements, see Sect. 9.2.6.2.

9.4.2 Improvement of HV Test Rooms

The safety system has to follow the principles explained in Sect. 9.2.6.1 completely. Any reduction in the safety requirements is not acceptable. When auxiliary equipment (Sect. 9.2.5) is modified or improved, the reliable operation of the test systems including the necessary sensitivity of PD and dielectric measurement must not be influenced: Clearances should sufficiently be maintained, and shielding effects should not be reduced.

Most expenditure is necessary to improve the grounding and shielding of a HV test field. This is often necessary because withstand tests are more and more completed by sensitive PD and/or dielectric measurement. Often, an unshielded test room is not longer sufficient for these monitored withstand tests. Then, consequences for filtering the supply power, for shielding and improved grounding become unavoidable.

Fig. 9.32 Shielded cabin in a routine test field

Usually, old grounding rods are corroded and must be replaced to reach an effective ground resistance of the order of 1Ω. The grounding should be independent from the shielding (Sect. 9.2.2), but in older test fields, the grounding is realized by a combination of earthing rods and an area grounder covering the whole test field area. This means grounding and shielding are combined in the floor. It should be investigated whether it is necessary to separate grounding and shielding in the floor (Sect. 9.2.2) or not. The shielding of the floor is not as important as that of the walls and the ceiling, which shall be performed as described above (Figs. 9.18 and 9.18). If the floor grounding is considered to be necessary, one had to put an insulation foil over the area grounder, followed by a suited metal mesh or metal panels forming the floor shielding and a protection layer, usually of concrete with an upper layer of epoxy resin for air cushion transportation.

In some cases, it should be considered whether the shielding of the whole test field is necessary or the application of a shielded cabin (Fig. 9.32) is sufficient. Inside such a self-carrying cabin the lowest PD noise levels can be reached. If the space of such a commercially available cabin is sufficient, no expensive shielding of the whole test area is required. A similar effect can be reached with a metal-enclosed test system (Fig. 3.41). A perfect shielding must be completed with a perfect filtering of all voltages penetrating into the shielded area (Sect. 9.2.3).

Chapter 10
High-Voltage Testing on Site

Abstract High-voltage (HV) tests on-site are performed for two different reasons: First, to assure the insulation integrity after new equipment or systems have been assembled on site, and second, to assess the condition of aged equipment, e.g. after maintenance and repair. The first category covers commissioning tests where classical HV withstand voltage tests are nowadays combined more and more with non-destructive dielectric tests, such as $C/\tan\delta$ and PD measurements to demonstrate the necessary quality and reliability. Such tests complete the quality assurance tests carried out in the factory and should follow their philosophy based on insulation coordination. The aim of the second group of tests, which are commonly referred to as diagnostic tests, is to assess the insulation condition as well as the remaining life time of service-aged equipment. For diagnostic purposes, usually a set of tests and measurements is performed. HV tests after repair are in between these both categories, because the repaired part is new, but the other insulation of the equipment or system is service-aged. After a consideration of the general requirements to HV test systems used on site, the applied test voltages according to IEC 60060-3:2006 are introduced. Moreover, the chapter provides examples for both quality acceptance withstand tests and diagnostic tests on site, which are usually applied for compressed-gas-insulated equipment (GIS), cable systems, power transformers and rotating HV machines.

10.1 General Requirements to HV Test Systems Used on Site

10.1.1 Quality Acceptance Tests

High-voltage (HV) testing is not only related to tests in a factory. The whole life cycle of insulation is accompanied by HV tests (Fig. 10.1). The quality of equipment and systems which are finally assembled on site is carefully verified by quality acceptance tests of their components in the factory (*routine tests*). These HV withstand tests are completed by quality acceptance tests on site, which verify the sufficient quality of both, transportation and on-site assembling, for reliable

© Springer Nature Switzerland AG 2019

W. Hauschild and E. Lemke, *High-Voltage Test and Measuring Techniques*,

https://doi.org/10.1007/978-3-319-97460-6_10

operation (*commissioning tests*). All quality acceptance tests shall follow the same principles that the test voltages shall represent stresses in service and shall be well reproducible. After successful tests (Fig. 10.1, yellow area), the equipment can be commissioned and handed over from the supplier to the user.

The user of the equipment may arrange the *in-service monitoring* of the insulation (Cigre TF D1.02.08 (2005)), e.g. by online, non-conventional PD measurement using built-in sensors. The *monitoring* applies to the trend of suited parameters (possibly the combination of different PD measurands) for the condition assessment. If an indicator value (combination of parameters) reaches a magnitude for warning, an off-line *diagnostic test* might be useful for the identification and location of the defect. The diagnostic test should repeat some principles of the quality test, but this is not absolutely necessary. It should include a set of measurements at different measuring positions and should be able to deliver information about the endangering of the insulation by a detected defect. The off-line diagnostic test with an external voltage source enables measurements at different voltage levels including the determination of inception and extinction voltages. Therefore, it delivers more information than the in-service (online) monitoring. Nevertheless, both monitoring and diagnostic tests complete each other and play a common role for *condition assessment* (Fig. 10.1, blue area).

It is very useful to compile all tests and checks e.g. on a cable system and its components in a "*life cycle record*" (Fig. 10.1). This record supplies information on trends of diagnostic indicator values. Quality acceptance tests and diagnostic tests have developed during the last 20 years quite independent on each other. It seems to be necessary to overcome this difference and to consider all HV tests with one general view based on the physical phenomena and the requirements of the practice.

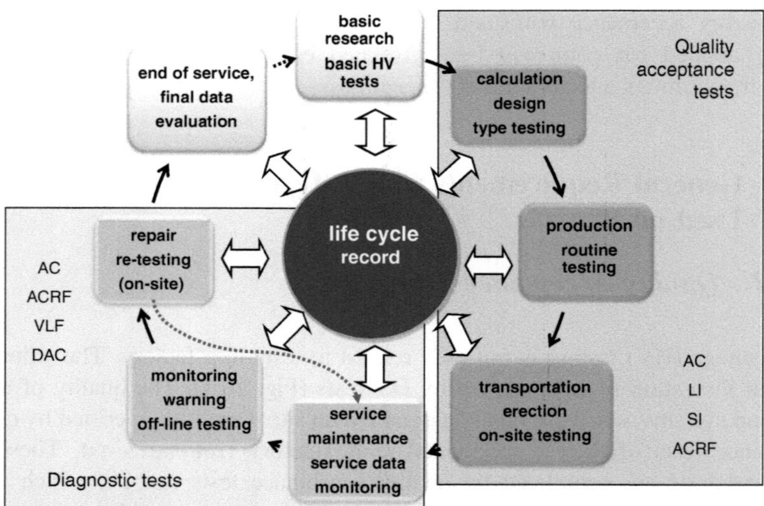

Fig. 10.1 HV tests and measurements in the life cycle of equipment

The *insulation co-ordination* (see Sect. 1.2) of the insulation of the components of a power system shall guarantee its reliable operation as well as the protection of the operating personnel and important equipment. This is realized by protection levels of protecting devices (arresters) and test voltage levels for the equipment. The diversity of defects in insulations prevents to establish general equivalence between different test voltages. ***Therefore, the test voltages applied shall represent typical stresses in service*** (Hauschild 2013). This principle is the basis for the verification of insulation coordination by test voltages. Consequently, it is also mandatory for quality testing on new equipment including quality acceptance test in the field (Fig. 10.1, yellow area). The typical stresses in service, and the related test voltage according to horizontal standards, are explained in Sects. 1.1 and 1.2. Test voltages for field tests have wider tolerances than those for laboratory tests.

The following general requirements to test voltages for quality acceptance tests are well considered for many years:

(a) Test voltages shall represent stresses in service.
(b) Test voltages shall be reproducible within defined tolerance limits for their parameters. The tolerances reflect both the feasibility of the test voltage generation and the dispersion of real stresses in service.
(c) All quality acceptance tests shall be comparable to each other, because the different tests are related to the common system of quality control.
(d) Quality acceptance tests require a clear pass/fail criterion and a comparable test procedure. The results of "direct" withstand tests do not require explanations, and an "indirect" test based on a measurement requires a limit which has to be agreed in a standard or a contract.

Withstand tests on non-self-restoring insulation require a detailed consideration:

The test voltage value and the test duration should be selected in such a way that a healthy insulation (Fig. 10.2a, green line) is not influenced, whereas one with a serious defect breaks down (orange line). It seems to be wrong to slander a withstand test as "destructive". If there is a defect which is dangerous for operation, then the test object should break down. For example, a failure of a cable system with an internal defect is even welcome during a withstand test and much better than one during service. A withstand test of service-aged cable systems may take potential lifetime. To indicate that a withstand test has not harmed the insulation, it should be combined with the measurement of a suited measurand, e.g. PD's or $\tan\delta$. The principle of such a *PD-monitored withstand test* procedure is shown in Fig. 10.2b (Cigre TF D1.33.05, 2012):

The "upwards procedure" of a PD-monitored withstand test should start on a voltage step which shall be PD-free (operational voltage or slightly above) and goes up with another two steps to the withstand voltage level. Also, at that level, a PD measurement is recommended. Then, the "downwards procedure" repeats the same voltage steps in opposite sequence. In addition to the withstand test and PD pass/fail criteria, the comparison of the PD characteristics at identical voltage steps before and after the withstand level completes the information on the condition of the test

(a)

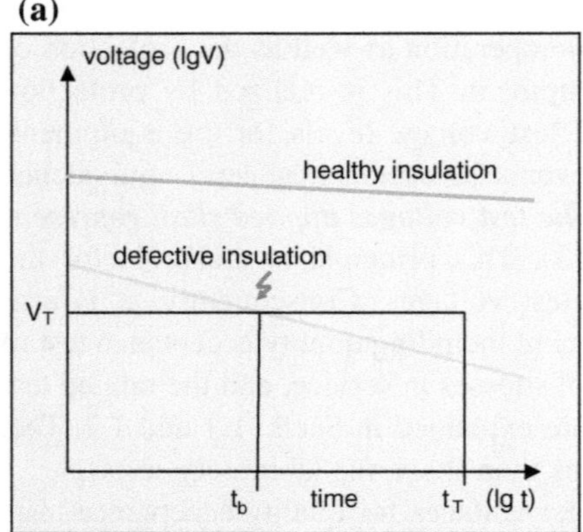

(b)

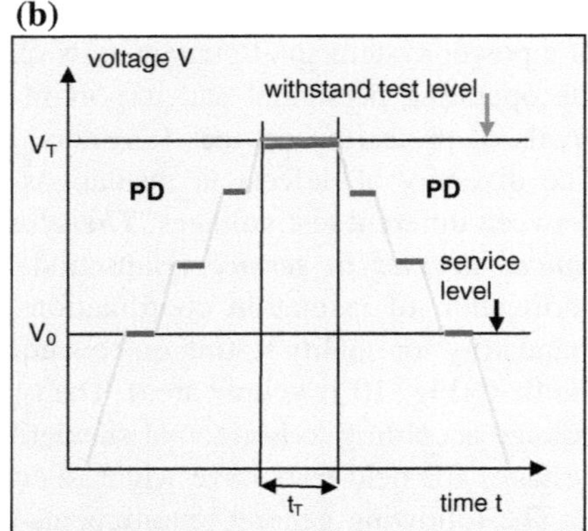

Fig. 10.2 Background and procedure of a PD-monitored withstand test. **a** Lifetime characteristic (schematically). **b** Example of the procedure for a monitored withstand test

object. It is an indication that the withstand test has not damaged the equipment or system under test, when the PD characteristic of the "downwards procedure" is very similar to that in the "upwards procedure".

A quality acceptance test is always connected with important decisions. For example, a successful type test is the technical condition for starting the production of a HV component. A new equipment or system must pass the quality acceptance test on site before it can be handed over to the user. This means that also the quality acceptance test justifies the fulfilment of a contract.

10.1.2 Diagnostic Tests

Diagnostic tests are performed for condition assessment with the aim to estimate the remaining lifetime of the tested equipment or system. The result of such a test is the classification of the test object according to their performance, for example to the classes

"safe and reliable",

"keep under observation",

"insufficient, repair or replace".

This cannot be done by a single withstand test, and it usually requires a set of tests and measurements and the consideration of the trend of the related data according to the *life cycle record*. These may be PD measurements of different characteristics at different voltage levels, e.g. inception and extinction voltages, apparent charge or repetition rate (IEEE Guide P400™-2012). PD measurements deliver information about weak points in the insulation. When methods of dielectric response are applied, one gets an overall—or integral—condition assessment of the cable system. Such methods are measurements of the dissipation factor, of polarization/depolarization current, of leakage current or of recovery voltage (IEEE Guide P400™-2012). Also, monitored withstand tests as described above (Fig. 10.2b) may deliver valuable information. The service-aged insulation should be stressed with test voltages higher than the operational voltage, but lower than the test voltages before commissioning.

In opposite to the mentioned quality acceptance tests, there are no general rules or standards when diagnostic tests and measurements must be performed. The related decisions are a matter of the users of the equipment or systems. Decisions for diagnostic tests may be made

- according to fixed time schedules,
- after a certain increase in defects,
- after certain overload of the power system,
- after a warning from the monitoring.

If a system has been repaired or sections or components are exchanged, the quality of the repair or extension work should be tested. The problem for the test is the service-aged part of the system which cannot simply be stressed as a new one. In such cases, also principles of diagnostic tests might be applied to avoid an overstress of the service-aged part. Only if the considered equipment is relatively new, a quality acceptance test as described above can be recommended.

10.1.3   Overall Design of Mobile HV Test Systems

The overall design of a mobile HV test system has to consider the set of criteria listed in the following:

The selection of appropriate test voltages has the highest priority. The selected test voltage influences the mechanical design and the power demand remarkably. For quality acceptance tests, they should be related to the test voltages required by insulation coordination. For diagnostic tests, they should be selected in conjunction with measurements which are efficient for condition assessment.

Weight and compactness of the HV test system are important for the optimum *weight-to-test power relation* of the system, for the acceptable size for transportation and for arrangement at the site of the test.

Both transportability and *assembly* influence the handling of the test system. It shall be equipped with accessories for transportation, robust against mechanical shocks and well protected against environmental conditions which may endanger the function of the test system (rain, snow, ice, very low and very high temperatures, sand storms, etc.). There should be few assembling work on site until the test system is ready for tests. Smaller test systems—ready for the connection of the test object via a shielded cable—can be arranged inside a van, trailer or container completely (Fig. 10.3a; see Sect. 10.2.1.1). The HV components (exciter

(a)   **(b)**

(c)   **(d)**

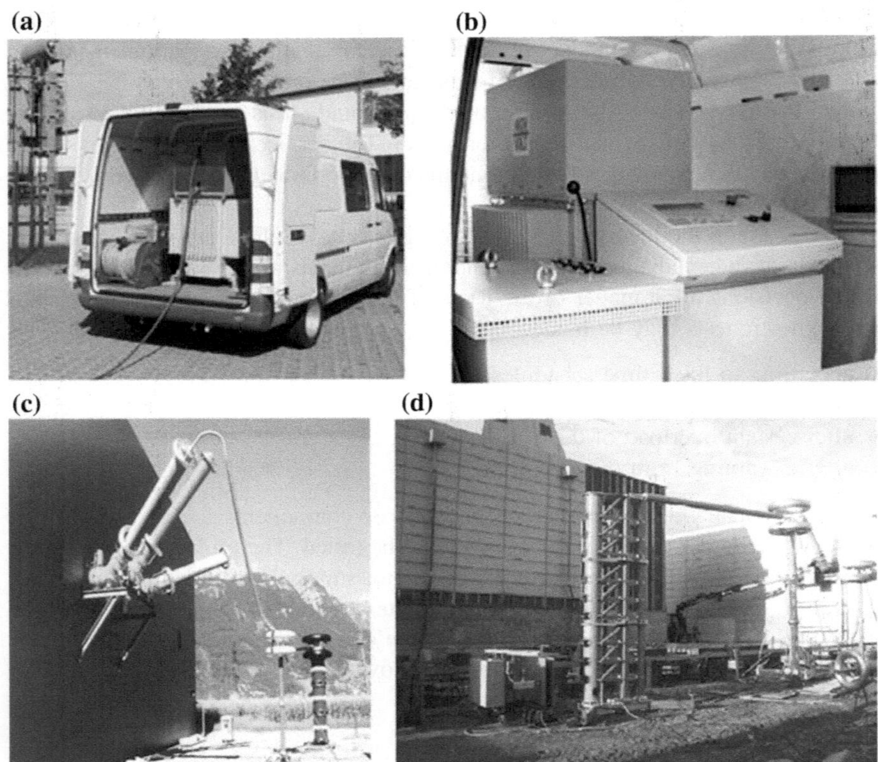

Fig. 10.3 Photographs showing the principle design of HV on-site test systems. **a** Outside and inside view of a van with an ACRF test system for distribution cables. **b** Assembled ACRF test system with three modular reactors. **c** Assembled impulse test system for 900 kV cumulative charging voltage and 1600 kV OLI output voltage. Courtesy of Siemens AG, Berlin

transformer, reactor, voltage divider) of the shown ACRF test system are metal-enclosed to fulfil safety requirements (Fig. 10.3b). Heavy and large systems —as, e.g. for testing HV and UHV cable systems—are arranged on a trailer (Fig. 10.4). The total weight of HV test system, trailer and truck must not exceed the permitted value for roads (e.g. in Europe 42 t) to avoid special transportation. A second possibility is the use of components which can easily be assembled on site. Figure 10.3c shows an ACRF test system with three modular reactors and a separate voltage divider. Figure 10.3d shows a test circuit for OLI voltages (see Sect. 10.2.1.3), consisting of a DC charging unit, three generator modules for 300 kV each, the HV inductance connecting the generator to the voltage divider/ (basic load capacitor) and the test object.

The power demand for testing shall be an optimum relation between the reactive test power and the active power demand from a voltage source. Considering the capacitive load of the test objects, this can be established by the application of

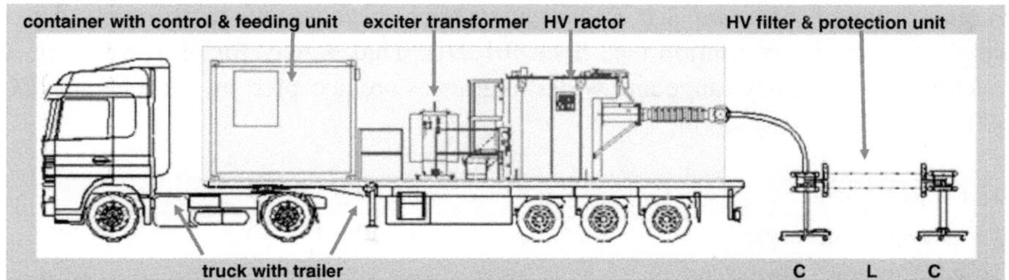

Fig. 10.4 Mobile ACRF test system for HV and EHV cable systems

oscillating circuits and the principle of resonance. When the power supply on site is a *Dieselgenerator set*, it must be considered that it is not a "stiff" source. Its rated power should exceed the necessary active power remarkably.

The control and measuring system of the HV test system should be easy to handle. A computer system is recommended. In case of unexpected events, e.g. breakdown of the test object, failure of the test system and interruption of feeding power, this would save all data of the test and all measuring results.

Between the mentioned technical requirements, a compromise must be found for the selection of a suited HV test system. It cannot be expected that one system is optimum for all rated voltages and all kinds of off-line tests, especially when the cost is taken into consideration, too.

10.2   Test Voltages Applied on Site

Whereas the test voltages for quality acceptance tests shall be related to the quality tests in factory, test voltages for diagnostic tests do not require such a stringent selection. The priority is the interpretation of the set of results based on knowledge rules for the applied voltage. Therefore, additional special voltages can be applied. The following chapter considers test voltages for field testing as described in the relevant IEC Standard 60060-3:2006 or in the IEEE guide P400^{TM}-2012.

10.2.1   Voltages for Withstand Tests

10.2.1.1   Alternating Voltage of the Power Frequency Range (HVAC)

AC voltage of power frequency (see Chap. 3) is considered as the most important test voltage, especially if the withstand test is PD-monitored. Therefore, it serves as reference for many test voltages applied in the field. Power frequency test voltage for laboratory testing is defined with a frequency range from 45 to 65 Hz (IEC 60060-1:2010). For field testing, AC voltages are often generated by mobile, *frequency-tuned resonant (ACRF) test* systems because of their much better weight-to-test power relation (see Sect. 3.1.2.4). That is why for on-site testing, a much wider frequency range and wider tolerances are accepted by the relevant IEC 60060-3:2006:

Test voltage value	Peak/$\sqrt{2}$
Test voltage frequency	10–500 Hz
Tolerance of test voltage value	$\pm3\%$ up to 1 min $\pm5\%$ for >1 min
Relation of peak to rms value	Within $\sqrt{2} \pm 15\%$
Uncertainty ($k = 2$) of	
–Peak voltage measurement	5%
–frequency measurement	10%

The single apparatus committees apply reduced frequency ranges; for example, the range 20–300 Hz (Fig. 10.5) is recommended for field tests on cable systems (IEC 62067:2011; IEC 60840:2011). Numerous practical field tests show that the test frequency remains in the quite narrow range between 30 and 100 Hz very often. It shall be demonstrated that the application of AC test voltages of the power frequency range is very important.

As an example, Fig. 10.6 (Schiller 1996; Gockenbach and Hauschild 2000) shows withstand voltages of models of *extruded cable insulation* depending on the frequency. The red line represents the withstand voltage of a technical perfect insulation, whereas the blue line characterizes heavy artificial defects. Real defects in extruded cable insulation cause withstand voltages between the two lines. Each type of defect influences the breakdown process in a different way. This prevents

Fig. 10.5 AC voltages for on-site tests of extruded cable systems

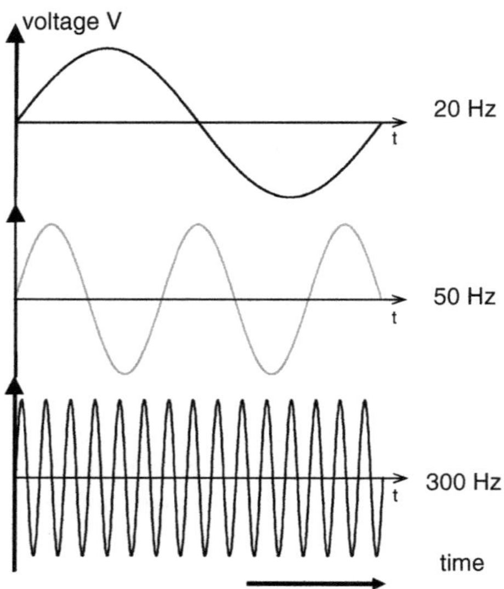

the application of a general equivalence factor between two different test frequencies. The deviations between the breakdown voltages at 20–300 Hz and at 50/60 Hz remain in an acceptable small range <10% (Fig. 10.6, hatched area: "AC"). Also, PD patterns are very little influenced by the frequency in the usual frequency range (Fig. 10.7; Schreiter et al. 2003). This is the reason for the wide and successful application of AC voltages of variable frequency for quality acceptance tests of HV and EHV cables in the field for more than 20 years.

As a second example, the on-site testing of *SF₆-insulated switchgear (GIS)* shall be considered. Their most dangerous defects are *free-moving particles*. Figure 10.8 (Mosch et al. 1979) shows the motion of such particles in a horizontal coaxial model at different voltages: At DC voltage, the particle lifts off and reaches immediately the inner electrode. Then, a nearly regular motion between the electrodes is observed, which causes the breakdown when the position of particle is near to the inner electrode at a sufficiently high voltage. The LI voltage is too fast for causing any

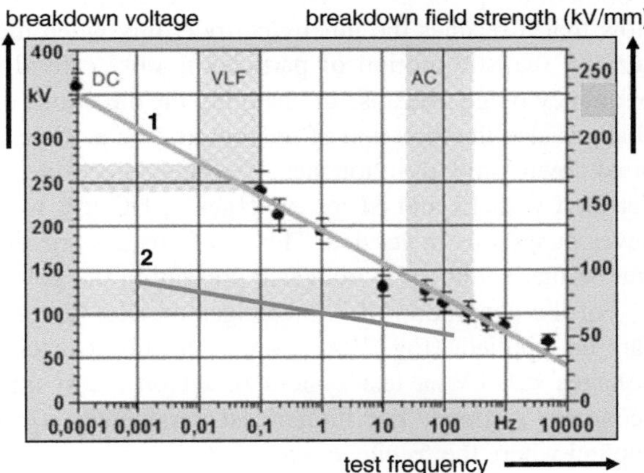

Fig. 10.6 Withstand voltage of XLPE cable models and test frequency (*1*) without artificial defects and (*2*) with artificial defects

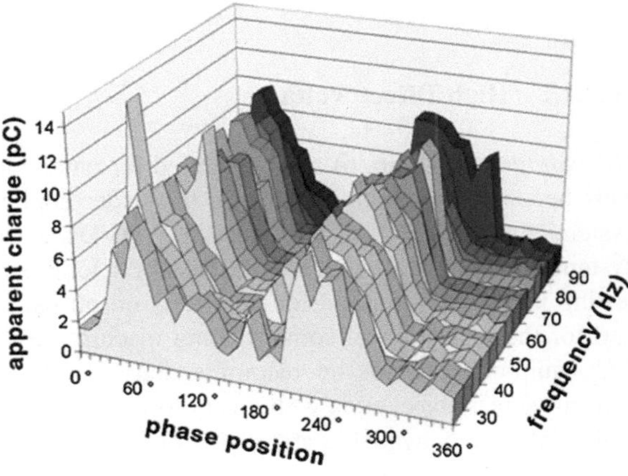

Fig. 10.7 PD pattern of XLPE models at different test frequencies

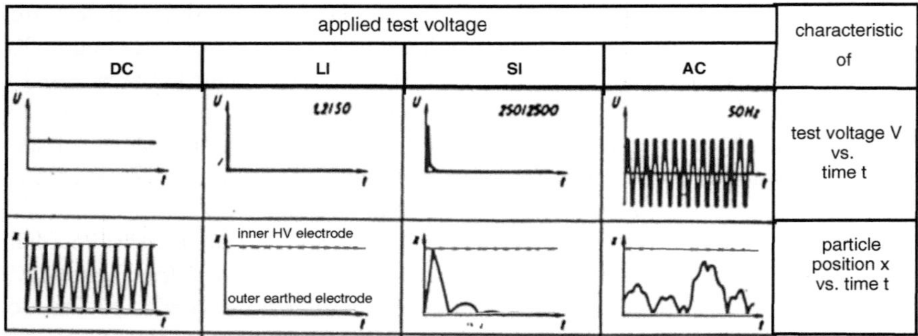

Fig. 10.8 Position–time characteristics of spherical particles between coaxial electrodes at different voltage waveforms

motion of the particle. Even at SI voltage, the mechanical motion of the particle is so slow that it reaches the inner electrode only when the voltage has been reduced to zero. A realistic motion of particles is only caused by AC voltage of the power frequency range. Because of changing the directions of the electric field, the particle changes also the direction of its motion. The particle is hovering and may cause the breakdown only if it touches the inner electrode. Whereas particles cannot be detected with LI and SI test voltages, a DC test would detect particles which are never dangerous in service. This confirms the principle that the only efficient tests can be made with AC voltages representing the power frequency voltage.

For the generation of AC voltages on site, the principles explained in Sect. 3.1 can be applied (Fig. 10.9). When testing capacitive objects <0.5 μF by test voltages <20 kV, a test system based on a test transformer should be the most economic solution. For higher test parameters, a resonant test system has to be applied where the *frequency-tuned circuits* have a much better weight-to-test power ratio than the inductance-tuned systems (Table 3.2; Fig. 10.9). The design of the transformers and reactors depends strongly on the test objects and will be discussed in Sect. 10.3.

10.2.1.2   High Direct Voltage

DC voltage (see Chap. 6) can be applied for on-site testing of HVDC components. This becomes more important with the wider application of HVDC transmission systems based on HVDC overhead lines, HVDC extruded cables and gas-insulated systems. For historic reasons and based on long experience (Fig. 10.10), HVDC testing is also still applied for field tests on oil-paper-insulated AC cable systems and for the insulation of some rotating machines as well. The DC testing of other AC equipment cannot be recommended. A DC generator can charge a large capacitive test object even with a very low current in the range of few milli-Amps if sufficient time is available and if this duration is not critical for the test object.

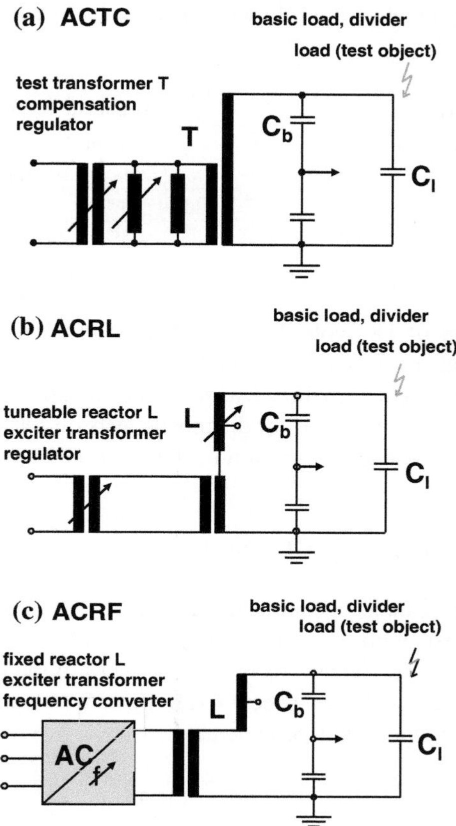

Fig. 10.9 AC test voltage generation and approximate weight-to-test power ratio. **a** Based on transformers (about 15 kg/kVA). **b** Based on inductance-tuned resonant circuits (about 5 kg/kVA). **c** Based on frequency-tuned resonant circuits (about 1.5 kg/kVA)

Fig. 10.10 Historical HVDC test system rated 250 kV for oil-paper-insulated cable systems made by Koch and Sterzel (1936)

IEC 60060-3:2006 requires the following parameters of DC voltages for on-site tests:

Test voltage value	Arithmetic mean value
Tolerance of test voltage value	$\pm 3\%$ up to 1 min
	$\pm 5\%$ for >1 min
Ripple factor	$\leq 3\%$
Uncertainty of voltage measurement	$\leq 5\%$
Uncertainty of ripple measurement	$\leq 10\%$

For the generation of DC voltages, usually modular HVDC test systems are applied. They can be air-insulated or—for higher current—oil-insulated, see Figs. 6.1 c, 6.9 and 6.10). For testing internal or clean external insulation, a HVDC test system of very low current is sufficient, but for diagnostic tests on service-aged or polluted insulation, test systems of higher current must be taken into consideration.

10.2.1.3 Impulse Voltage (LI, OLI, SI, OSI)

Impulse voltages (see Chap. 7) cannot replace tests with continuous voltages on site. They complete tests with AC or DC voltages. IEC 60060-3:2006 allows on site not only aperiodic impulse voltages of wider tolerances of the time parameters, but also oscillating impulse voltages (Sect. 7.1.3, Figs. 7.16 and 7.17). Parameters within the following ranges shall be generated:

	LI/OLI	SI/OSI
Test voltage value	Peak	Peak
Tolerance of test voltage value	$\pm 5\%$	$\pm 5\%$
Front time/time to peak	0.8–20 μs	20–400 μs
Time to half-value	40–100 μs	1000–4000 μs
Frequency	15–500 kHz	1–15 kHz
Uncertainty (voltage measurement)	$\pm 5\%$	$\pm 5\%$
Uncertainty (time/frequency measurement)	$\pm 10\%$	$\pm 10\%$

Oscillating lightning impulse (OLI) and oscillating switching impulse (OSI) voltages are applied for their much higher utilization factor (see Sect. 7.1.3). For the oscillating impulses (OLI, OSI), the time parameters are determined from the enveloping curve. For their generation see Sect. 7.1.3 and for mobile use Fig. 10.3d.

LI and OLI test voltages are useful for the detection of fixed defects in GIS. For GIS, a related OLI acceptance test is recommended on site if the PD measurement at AC voltage is not sensitive enough. Till now, power transformers are tested with aperiodic LI voltages after repair on site, oscillating impulse voltages are not yet applied for these tests.

10.2.2   Voltages for Special Tests and Measurements

IEC 60060-3:2006 distinguishes between the test voltages in relation to IEC 60060, and this means voltages for withstand tests for quality acceptance (Sect. 10.2.1) and those for special (= diagnostic) tests. The latter are *very-low frequency* (*VLF*) voltages and damped AC (DAC) voltages considered in the following. Additional voltages for special tests can be specified by relevant IEC Technical Committees.

10.2.2.1   Very Low Frequency Voltage

To avoid the high power demand when cables are tested with 50/60 Hz voltages, a test voltage of a very low frequency (VLF), preferably of 0.1 Hz, had been introduced in the early 1980s (Nelin et al. 1983; Boone et al. 1987). This reduces the power demand to 0.2% compared with 50 Hz (Fig. 10.11a). The frequency range of VLF test voltages is fixed to 0.01–1 Hz by IEC 60060-3:2006, respectively, by the IEEE Guide 400.2. The standards accept very different wave shapes between sinusoidal and rectangular for on-site testing (Fig. 10.11b) if the following additional requirements of a test are fulfilled:

Test voltage value	Peak (for some application rms value)
Tolerance of test voltage value	±5%
Frequency range	0.01–1 Hz
Uncertainty of voltage measurement	$\leq 5\%$
Uncertainty of frequency measurement	$\leq 10\%$

VLF test voltages are far from stresses in service, but VLF test systems (Fig. 10.12) are well introduced for medium-voltage cables. They are compact and of low power demand.

The comparison between 0.1 and 50 Hz (Fig. 10.11a) delivers an impression what it means for the discharge process when the power frequency in service is 500 times higher. The much lower repetition rate of stresses with polarity reversals causes a different discharge mechanism, which leads to higher breakdown voltages. Figure 10.6 shows that the breakdown voltage of a healthy insulation at 0.1 Hz is about twice of that at 50 Hz (red line)! Consequently, the applied VLF test voltages must be 50% higher than those at power frequency.

Because of their compactness and low power demand, VLF voltages are well introduced for diagnostic tests and even for quality acceptance tests on medium-voltage cables (Fig. 10.12a). According to the European Standard EN HD 620 S (1996), the test voltage value is the rms value, with the consequence that the peak value of a sinusoidal VLF test voltage shall be by a factor $\sqrt{2}$ higher than that of a rectangular VLF voltage. Their application to quality acceptance tests of extruded HV and EHV AC cables cannot be recommended. VLF voltage is not a stress, which appears under operation of a cable. In combination with the higher

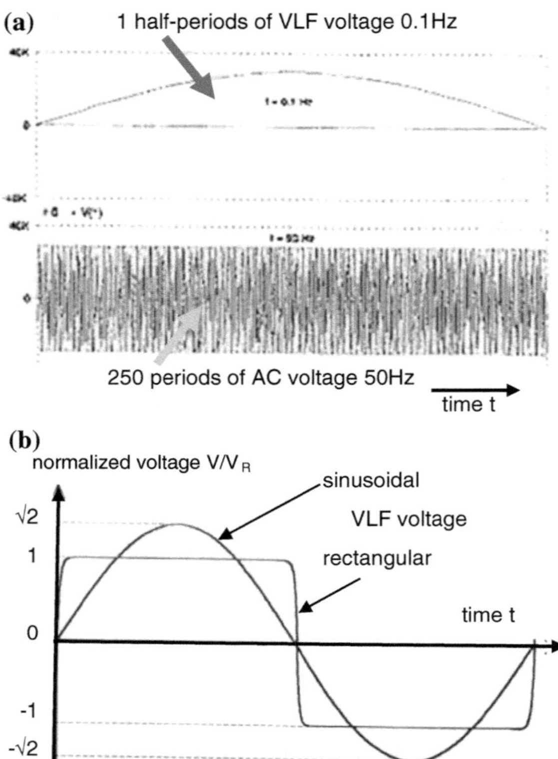

Fig. 10.11 VLF test voltages and power frequency AC voltage. **a** Comparison 0.1 and 50 Hz. **b** Sinusoidal and rectangular VLF voltages of identical rms value

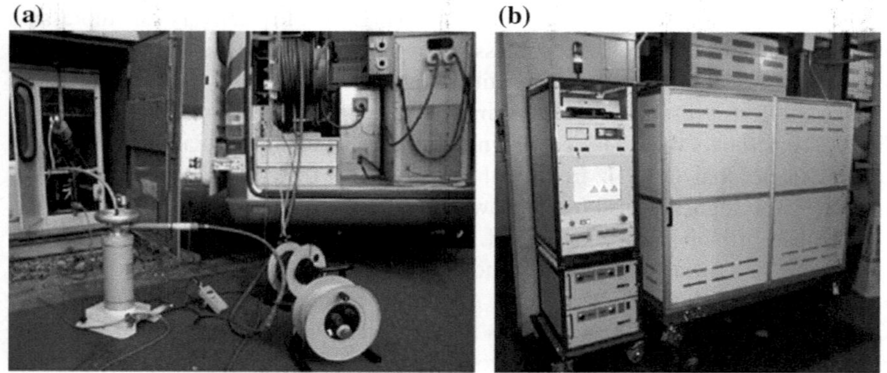

Fig. 10.12 VLF test system for 60 kV. **a** Mobile test system for land cables up to few µF. **b** Test system for submarine cables up to 25 µF. Courtesy of BAUR AG, Austria

stress in HV and EHV cables, the test result may not correspond to test results with power frequency voltages.

10.2.2.2   Damped Alternating Voltage

As already carried out in Sect. 7.6.2 PD diagnosis tests at impulse voltages have been introduced in the 1980s to ensure the integrity of the insulation of extruded power cables. For this purpose the cable insulation was initially subjected to SI and OSI voltages (Lemke et al. 1987; Plath 1994). However, the main obstacle of such tests is that the utilization factor of the test facility decreases at increasing cable length due to the fact that the charge required for energizing the cable capacitance must be stored previously in the impulse capacitor of the SI or OSI generator (see Sect. 7.1.3 and Fig. 7.19). To overcome this crucial problem, the use of the cable capacitance itself as the impulse capacitor has been proposed (Lemke and Schmiegel 1995; Gulski et al. 2000). This results in PD diagnosis tests using so-called damped alternating current (DAC) test voltages (Sect. 7.1.3) Under this condition the utilization factor approaches 100%. However, to create the damped alternating voltage used for the PD detection it has to be taken into account that the cable insulation is pre-stressed by a slowly rising DC voltage ramp where the time-to-crest varies between some seconds and more than one minute. In this context it should be mentioned that the time parameters of the pre-stressing DC voltage ramp (Fig. 10.13) are not specified, i.e. the relevant standard IEC 60060-3:2006 considers only the damped oscillating voltage, where the following requirements are defined:

Test voltage value	Peak value
Tolerance of test voltage value	$\pm 5\%$
Frequency	20–1000 Hz
Damping of subsequent peaks	$\leq 40\%$
Uncertainty (voltage measurement)	$\pm 5\%$
Uncertainty (time/frequency measurement)	$\pm 10\%$

As mentioned previously, the cable insulation is initially stressed by an unipolar DC voltage ramp and just thereafter by a fast polarity reversal, where the peak value of the voltage sweep of opposite polarity is only marginally attenuated, if compared to the peak value appearing just prior the DAC test generator is triggered. The frequency of the DAC voltage depends on both, the test object capacitance and the inductance of HV inductor used for discharging this capacitance [see Eq. (7.6)]. Provided, the oscillating frequency ranges between 1000 and 20 Hz, the polarity reversal would vary between 0.5 and 25 ms, respectively.

The attenuation of the oscillating voltage is not only governed by the dielectric losses dissipating in the cable capacitance but also by the ohmic resistances of the test facility components, such as the HV reactor and the solid state HV switch. Despite of the above mentioned attenuation, the DAC voltage causes for a low

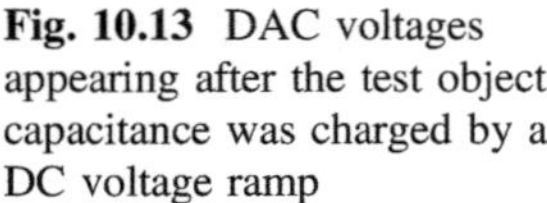

Fig. 10.13 DAC voltages appearing after the test object capacitance was charged by a DC voltage ramp

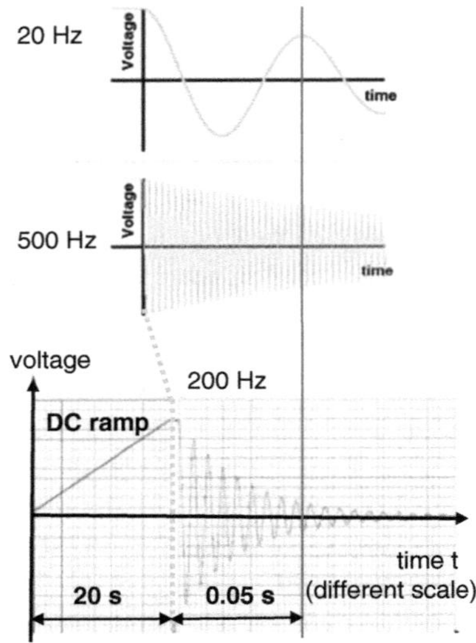

number of cycles a voltage stress in the cable insulation similar to that occurring under power frequency test voltage. This is the reason why the DAC voltage specified in IEC 60060-3:2006 is well suited for PD measurements. As mentioned previously, the peak value of the first voltage sweep is almost identical with that of the DC voltage ramp. The duration of the pre-stressing DC ramp, which may vary under practical condition between some seconds and more than one minute, depends on both, the rated current of the DC voltage supply and the capacitance of the test object, and thus on the length of a cable system under test (Fig. 10.14).

The generation of DAC voltage, even if used for PD testing of large test object capacitances, requires only few HV components (Fig. 10.15): a DC generator, a HV reactor (inductance), and a basic capacitive load, which can be used as a voltage divider and a coupling unit to capture the PD signals from the power cable under test. The main benefit of DAC test facilities is the compact design and low-weight, which enables an easy transportation and fast assembling on-site. Practical experiences revealed, however, that damped alternating voltages are well suited for on-site PD diagnosis tests but not for withstand voltage tests, carried out as commissioning tests after installation (Sect. 10.1.1) due to the following obstacles:

- The DAC voltage is associated with a DC charging ramp which is not representative for any stress under service conditions.
- The duration of the ramp depends on the test object capacitance and is thus not reproducible for different cable systems.
- The frequency of the DAC voltage depends on the test object capacitance and is thus also not reproducible for different cable systems.
- The damping rate of the DAC voltage depends on the losses in the test circuit and is hence also not reproducible.

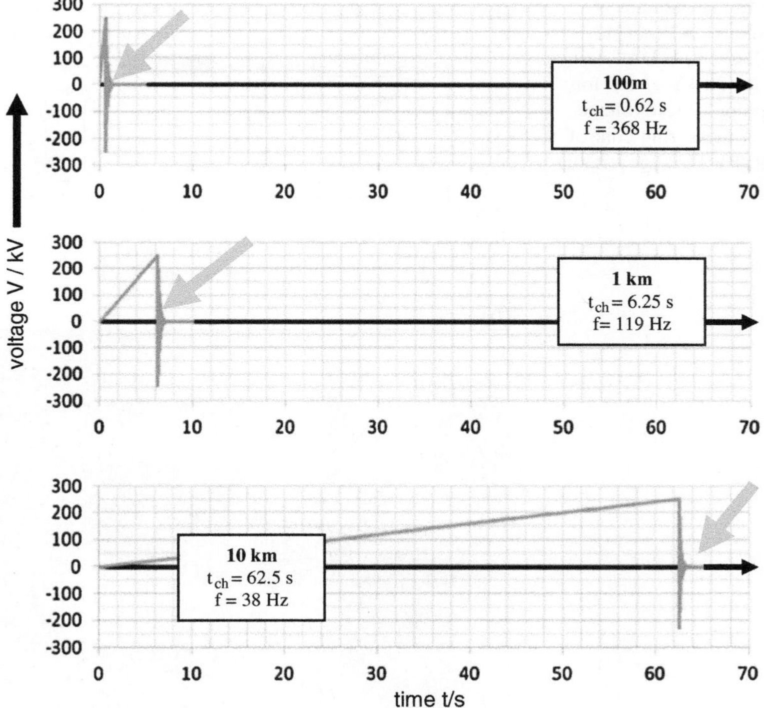

Fig. 10.14 DC voltage ramp and DAC voltage displayed for different cable lengths of cable systems

- The sequence of DAC pulses with DC ramps (Fig. 7.19c)—proposed for replacing AC withstand tests—may cause cumulative space charge effects and can thus not represent AC voltage of power frequency.

Nevertheless, IEC 60060-3:2006 considers the DAC voltages as a reasonable voltage for diagnostic on-site tests. In combination with PD measurement, it is an interesting tool for condition assessment, especially for medium-voltage cables (Gulski et al. 2007; Cigre TF D.1.33.05). A PD-monitored withstand test might be applicable for quality acceptance tests of some medium-voltage equipment. For HV and EHV equipment and systems of higher operational stress, the pre-stress of the DC ramp followed by the fast polarity reversal of the DAC voltage causes test results which are hardly comparable to those at AC voltage of power frequency. As DAC represents in principle an impulse voltage, DAC testing should never be applied to GIS (compare Fig. 10.8).

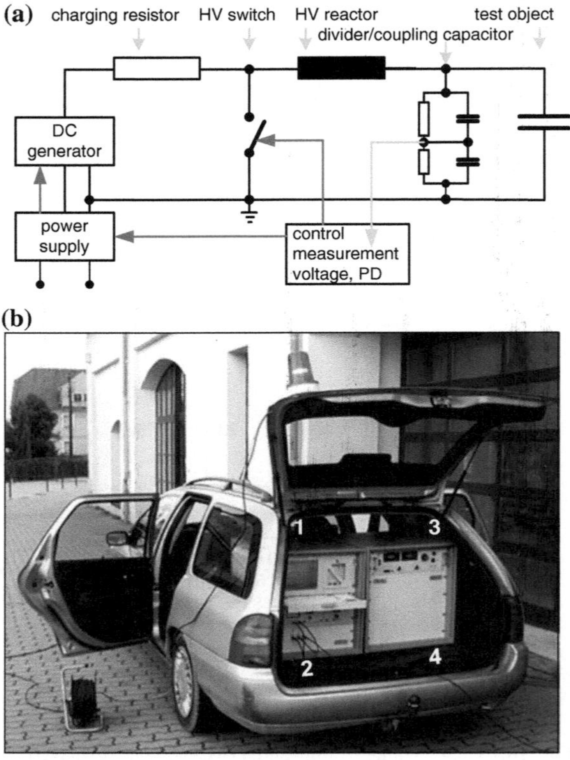

Fig. 10.15 Generation of DAC voltage. **a** Principle circuit diagram. **b** DAC test system for 30 kV (*1* PC for PD mapping, *2* test voltage control, *3* measuring control and *4* 30-kV DAC generator)

10.3 PD Measurement and Diagnostics on Site

The phenomena of partial discharges and the procedures applied for their measurement are identical under both laboratory and on-site conditions, as described in Chap. 4 and will hence not repeated in the following. The random characteristics of partial discharges under direct voltage stress can be found particularly in Sect. 6.5, which would make efficient on-site PD tests under HVDC more difficult, as well as less sensitive and time-consuming. That's why instead of DC voltages the use of AC voltage is often recommended for HVDC equipment, e.g. for commissioning tests of gas-insulated HVDC systems (CIGRE JWG D1/B57 (2017)). In the following some specific aspects of PD diagnostic tests under on-site conditions will briefly be considered.

As known, the insulation of HV apparatus suffers from ageing not only due to the permanent applied service voltage, but also caused by temporary and transient over-voltages as well as thermal and mechanical stresses. As a consequence, weak spots in the insulation might be initiated, which become the source of partial discharges. To avoid an unexpected outage associated with serious consequences not only for the supplier and customer but also for the environment, the asset management based on diagnostic PD measurement is becoming increasingly of importance.

To recognize a progressive insulation deterioration as early as possible, the traditional *time-based maintenance* is nowadays more and more replaced by the *scheduled condition-based maintenance*. The *insulation condition assessment* is especially focused on strategic important HV equipment. The diagnostic tools commonly used to judge the insulation integrity can be classified as global (integral) and selective (local) methods (Fig. 10.16). As the global insulation degradation causes finally also weak spots and thus PD events, which have to be considered as a precursor for an ultimate breakdown, *PD-monitored withstand voltage tests* have become an indispensable tool for the maintenance and replacement policy of power equipment.

Performing PD measurements under on-site conditions, the PD transients radiated from the test object are commonly received via the electromagnetic field using capacitive or inductive *PD couplers* (also referred to as *PD sensors* or even PD antennas). The captured signal is acquired by nowadays available advanced digital PD monitoring systems to provide the asset manager with reliable information on the insulation condition, required to decide for further actions (Fig. 10.17). In this context it must be underlined that besides technical aspects also economic consequences have to be taken into consideration. So a holistic approach for continuous PD monitoring is desired only for strategic important and expensive EHV/UHV apparatus. Medium-voltage equipment, however, must not be PD-monitored permanently but only inspected from time to time.

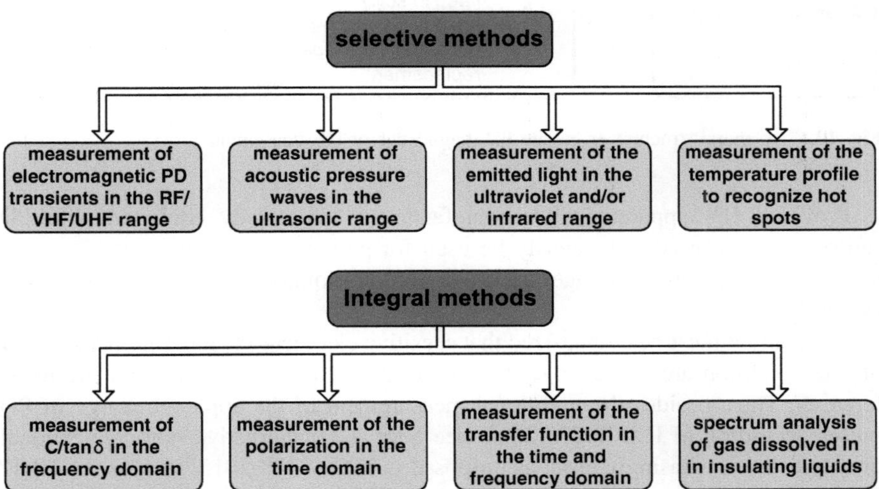

Fig. 10.16 Survey on diagnostic measurements used for insulation condition assessment of HV apparatus under on-site condition

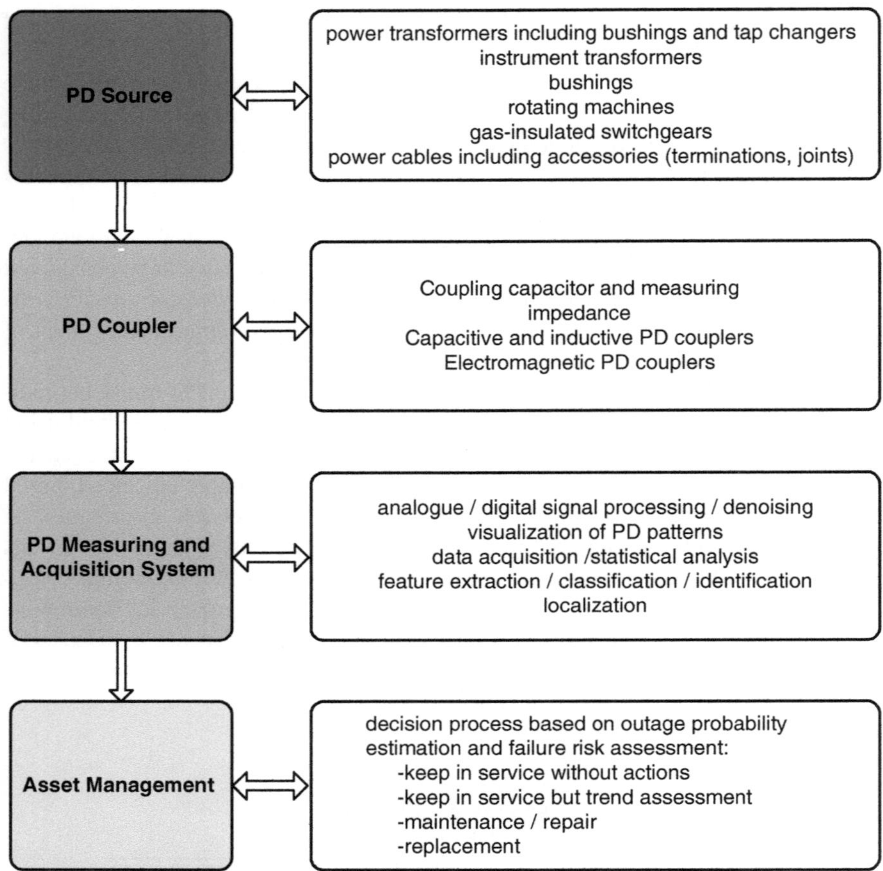

Fig. 10.17 General structure of on-site PD diagnostics of HV equipment

Provided HV apparatus after manufacturing are already equipped with PD sensors, these can advantageously be used for periodical or permanent PD monitoring under operation voltage and even for PD-monitored acceptance tests after assembled on-site.

Generally it must be emphasized that sensitive electrical PD measurements under on-site condition are a challenge due to the always-present electromagnetic interferences. Thus, besides the traditional measurement of the apparent charge of PD pulses according to IEC 60270:2000 (see Sect. 4.3), alternative non-conventional principles are increasingly used, as proposed in IEC 62478:2015, such as the *UHV/ VHF PD measurement* and the *acoustic PD measurement* (see Sects. 4.7 and 4.8). The measuring examples considered in the following are addressed to the *decoupling* of electromagnetic PD transients as well as to the acquisition of the captured PD transients (Fig. 10.18).

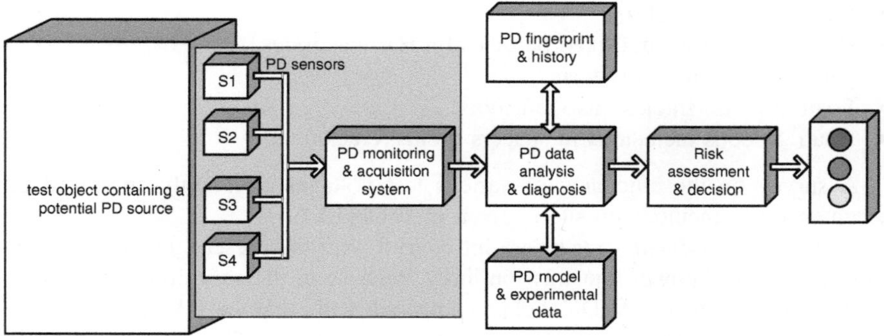

Fig. 10.18 Major components of a PD monitoring system for periodical and continuous measurements

10.4 Examples for On-Site Test

Prior a HV equipment is put into service, numerous electrical, mechanical, thermal and functional tests must be performed under on-site condition, as specified in the relevant IEC and IEEE standards and described in numerous technical papers of various Cigre working groups (IEEE Draft P1861TH/D1 2011, Cigre WG 33 TF 04 2000; Cigre WG D1.33 TF 05 2012). Besides the traditional withstand tests under different kinds of test voltages, commonly referred to as *quality acceptance tests* or even *commissioning tests*, also non-destructive *diagnostic tests* are often performed, such as C/tanδ measurements as well as PD measurements, where some specific aspects of the latter method will be highlighted in the following.

10.4.1 Testing of Gas-Insulated Systems (GIS, GIL)

10.4.1.1 Some Basics

Gas-insulated switchgears (GIS), introduced in the 1960s, consist of easily transportable units, which are assembled on-site to a complete substation. The return of experience (Cigre JW 33/23.12 1998) shows a low dielectric failure rate in the order of 1 per 100 bay-years operation. The reason for that is the early introduction of quality acceptance tests under on-site condition, where withstand voltage tests are combined with preventive PD measurements. Nevertheless, about 35% of the dielectric failures appearing in service are caused by an insufficient assembling work. From a physical point of view (Mosch and Hauschild 1979; Mosch et al. 2 1979; Cigre TF D1.33.05 (2012)), most of the recognized defects can be classified into

- free-moving particles,
- sharp protrusions and fixed particles on HV electrodes,
- particles sticking on spacers,
- floating parts (shields, also left tools),
- small gaseous inclusions in spacers (voids, cracks).

A survey on the efficiency of various test voltages commonly employed for acceptance test methods on site is given in Table 10.1.

HVDC gas-insulated systems are introduced with the present upcoming installations of HVDC power transmission links. Because of the very different behavior of gas insulations in HVDC fields compared with that in HVAC fields (see Sect. 6.2.3.1), the test philosophy and related standards are still under development. There is a trend to confirm the correct design of an HVDC GIS by very detailed type tests and prototype installation tests (Neumann et al. 2017), but to apply only AC test voltages for both quality acceptance tests, routine tests in factory and commissioning tests on site (CIGRE JWG D1/B3.57 2017). Behind this provisional decision are the very time-consuming HVDC tests and the circumstances that HVDC insulations must withstand transient and AC stresses (see e.g. Sect. 8.2.3), too, as well as the long and very good experience with testing of HVAC GIS. Therefore in the following HVAC on-site tests are preferably considered.

10.4.1.2   Acceptance Tests on HVAC GIS

The Standard IEC 62271-203:2010 (see also IEEE Standard C 37.122:2ß010) relevant for HVAC GIS recommends either a PD-monitored AC withstand test (procedure B) or—if the sensitivity of the PD measurement is not sufficient because of a comparatively high bachground noise level (q_N >5 pC)—the combination of an AC withstand test and a LI/OLI withstand test (procedure C) which applies for GIS of 245 kV and above. For voltages <245 kV, an AC withstand test is considered as sufficient (procedure A). This standard defines the test voltages (slightly different from IEC 60060-3:2006) and the test procedures as follows:

AC voltage	Frequency 10–300 Hz; withstand for 1 min (followed by a PD measurement at 1.2 U_r with requirements $q \leq 10pC$)
LI/OLI voltage	Front time $T_1 = 0.8–8$ µs, for OLI voltage ≤ 15 µs, withstand to three impulses of each polarity;
SI/OSI voltage	Time to peak $T_p = 0.15–10$ ms, only applicable if no AC source is available

On-site tests of HVAC gas-insulated systems with DC voltage are simply wrong and might even cause defects due to very different PD phenomena triggered for instance by particles and space charges deposited at spacers. The recommended test voltage values are given in Table 10.2.

Whereas medium-voltage AC GIS can be tested by transformers, all GIS for higher-rated voltages are tested by resonant test systems. With respect to the lower

Table 10.1 Efficiency of test methods for on-site quality acceptance tests of HVAC GIS

Test procedure Kind of defect	AC withstand	PD-monitored AC withstand	LI/OLI withstand	SI/OSI withstand
Free-moving particle	X	X	o	o
Sharp protrusions and fixed particle	o	X	X	x
Particle on spacer	o	X	X	x
Floating parts (left tools)	x	X	x	o
Defects in spacers	x	X	x	x

X very efficient, *x* less efficient, *o* not efficient

Table 10.2 Preferred test voltage values for HVAC GIS on-site tests (IEC 62271-203)

Rated voltage of GIS kV (rms)	AC withstand voltage kV (peak/$\sqrt{2}$)	LI/OLI Withstand voltage kV (peak)	SI/OSI Withstand voltage kV (peak)
72.5	120	260	–
123	200	440	–
170	270	600	–
245	380	840	–
362	425	940	760
420	515	1140	840
550	560	1240	940
800	760	1680	1140
1200	960*	2040*	1440*

The star values (*) are calculated by 0.8 times the rated insulation levels

weight, resonant test systems of variable frequency (ACRF test systems) are preferred. A frequency higher than about twice the power frequency enables the inductive voltage transformers that may remain at the GIS during the on-site test.

There are two principle designs of ACRF test systems:

Modular ACRF systems consist of cylinder-type reactors (Fig. 10.19, each 230 kV, 3A, 200H) which can be switched in series for higher voltages and in parallel for higher current. Then consequently, also the inductance and the natural frequencies are changed. Figure 10.20 shows the resulting load–frequency characteristics of the three reactors (Hauschild et al. 1997). The parallel (*p*) or series (*s*) connections are not only useful for the adaptation of voltage and current, but also for that of the test frequency.

The air-insulated test system requires a bushing of the GIS for the connection of the ACRF test system. The capacitor provides the voltage divider, the basic load and the coupling capacitor for PD measurement (IEC 60270:2000). If the GIS is not equipped with a bushing, a test bushing or a cable adapter must be added. The connection between the reactors and the capacitor is realized by a blocking impedance. But the PD measuring circuit is exposed to the environment. Therefore, electromagnetic noise signals may penetrate and reduce the sensitivity of the PD measurement.

Fig. 10.19 GIS test with a modular 680-kV ACRF test system. Courtesy of Siemens AG, Berlin

A *metal-enclosed ACRF test system* of lightweight SF_6-insulated reactors avoids the penetration of radiated noise: It can be flanged directly to the GIS under test (Fig. 10.21), causing a shielded HV circuit. When the electromagnetic UHF PD measurement is applied, no further components must be connected to the GIS. For PD measurement, according to IEC 60270:2000, the system includes a coupling capacitor (Fig. 10.21). The operation characteristic of such a system is shown in Fig. 3.25 and explained in Sect. 3.1.2.3. Two SF_6-insulated reactors can also be connected in series or in parallel (Fig. 10.22) when suited adapters of the enclosure are available (Pietsch et al. 2005).

Example A 245-kV GIS with a capacitance of $C_{GIS} = 2.2$ nF shall be tested after assembling with a metal-enclosed ACRF test system $V_r = 460$ kV, $I_r = 1.5$ A and $L = 720$ H. For the test circuit, a quality factor $Q = 50$ can be assumed. According to Table 10.2, the test voltage is 380 kV. Determine the test frequency, the necessary test current, the reactive test power and the necessary feeding power! Is the test system suited for the test? Can a coupling capacitor $C_k = 1.2$ nF be added for PD measurement? Can the inductive voltage transformers remain on the GIS during the test?

$$\text{Test frequency} \quad f_t = \frac{1}{2\pi\sqrt{L \cdot C_{GIS}}} = 126.5 \text{ Hz}$$

$$\text{Test current} \quad I_t = 2\pi f_t \cdot C_{GIS} \cdot U_t = 0.66 \text{ A}$$

$$\text{Reactive test power} \quad S_t = I_t \cdot U_t = 250 \text{ kVA}$$

$$\text{Active feeding power} \quad P_t = S_t / Q = 5 \text{ kW}$$

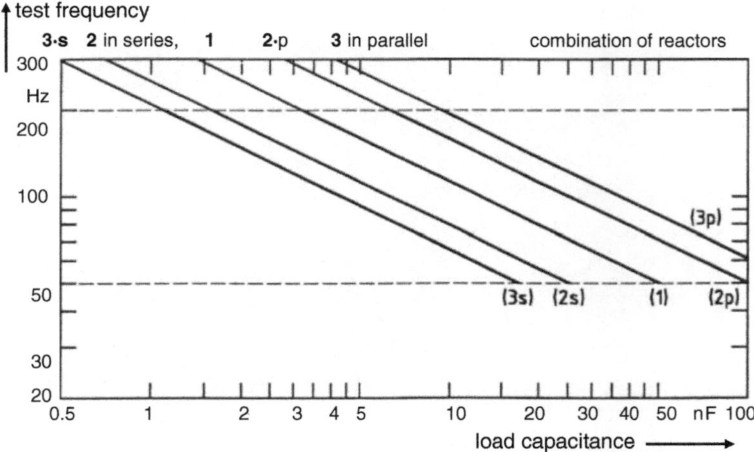

Fig. 10.20 Frequency–load characteristics of the combinations of three modular reactors

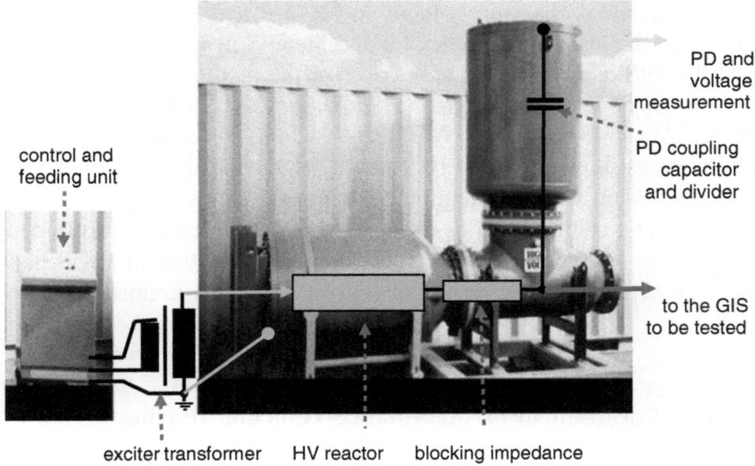

Fig. 10.21 Metal-enclosed ACRF test system of SF_6-insulated components

The additional coupling capacitor increases the capacitance to $C = C_t + C_k = 3.4$ nF, which reduces the frequency to $f_t = 101.8$ Hz and increases the test current to $I_t = 0.83$ A. In both cases, the test current is lower than the rated current of the ACRF test system and the test frequency is in the accepted range. Because the test frequency is more than the doubled power frequency, the inductive voltage transformers may remain at the GIS during the test.

Acceptance tests with *impulse voltage* are usually performed with oscillating voltages (see Sect. 7.1.3 and Figs. 7.16, 7.17 and 7.18). As shown in Table 10.2, the testing with OLI voltages is of practical importance, when no sensitive PD measurement can be applied. The OLI application enables generators with

Fig. 10.22 ACRF testing of a large GIS using two reactors in parallel

lower-rated cumulative charging voltage, lower size and lower weight. The connection between the test voltage generator and the GIS under test requires always a bushing at the GIS as also mentioned above for the modular ACRF test system.

For acceptance testing of *gas-insulated lines* (*GIL*) (IEC 61640:1998; Cigre JWG 23/21/33-2003; Cigre JWG B3/1.09-2008) of considerable capacitance, AC test voltages as shown in Table 10.2 should be applied. The higher capacitances of GIL in comparison with GIS require ACRF test systems of higher power. This can be generated by using the series connection of ACRF test systems as described in Sect. 10.4.2 for testing HV/EHV cable systems. In addition to a withstand test, the acceptance depends on the results of sensitive PD measurement (Okubo et al. 1998).

10.4.1.3   PD Measurement for Acceptance Tests and Diagnostics

PD diagnosis tests of GIS and GIL are important not only for acceptance tests but also to assess the insulation condition from time to time (periodical PD monitoring), or even continuously under service voltage (permanent *PD monitoring*). The general requirement is a "PD-free" insulation, which has to be proven at high measuring sensitivity. To discover severe defects in GIS or GIL including cavity discharges in spacers, a measuring sensitivity in the pC range is desired. Under on-site condition this is achievable only by the use of non-conventional methods, such as the detection electromagnetic PD transients in the UHF range (IEC 62478:2015) or the measurement of the acoustic emission as presented already in the Sects. 4.6 and 4.7.

To capture the very fast PD transients from the test object, different kinds of *UHF sensors* have been developed, such as mobile window sensors and fixed disc/cone sensors (Fig. 4.63), as well as field grading sensors (Boggs et al. 1981). As also carried out in Sect. 4.7, from a physical point of view, the non-conventional UHF method cannot be calibrated in terms of pC. Therefore, the "pass/fail" criterion is based on a sensitivity check, where the UHF signal magnitude in terms of mV is compared with the apparent charge magnitude in terms of pC gained by using the IEC 60270 method under laboratory conditions. In this context it seems worth to notice that the phase-resolved PD patterns are often very similar for both the UHF and the IEC method. Therefore, the classical PD pattern recognition is also applicable to identify and classify typical PD defects recognized by the UHF method, such as free-moving particles, protrusions fixed on the electrodes, floating metallic parts as well as cavities due to voids and cracks in spacers (Mosch and Hauschild 1979; Kranz 2000).

After a potential PD defect has been encountered, the site of origin must be known to assess the risk for an unexpected breakdown. A common procedure for this is the time-of-flight measurement. As the PD signal is travelling in both directions of the GIS compartment, which occurs at a velocity close to the light velocity (300 mm/ns), the oscilloscope applied should have a rise time below the ns range to get an appropriate temporal and spatial resolution.

Using the test arrangement illustrated in Fig. 10.23, where the PD signal is caused by a fixed metallic particle at site P, the time difference between both pulses arriving the sensors S_1 and S_2 is given by

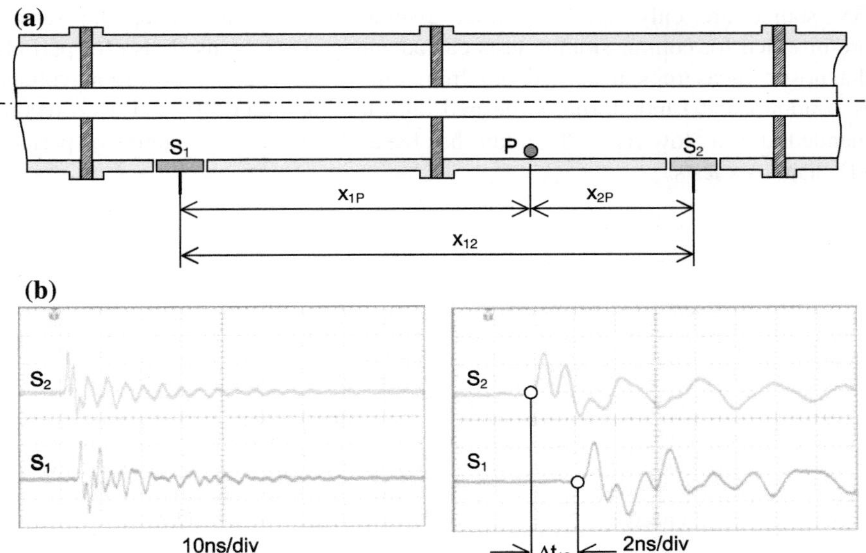

Fig. 10.23 Principle of PD fault localization based on the time-of-flight measurement. **a** Significant geometric parameters. **b** Oscilloscopic records gained from a GIS dummy

$$\Delta t_{12} = \frac{x_{1P} - x_{2P}}{v_g}.$$

From this follows for the distance of the PD site P from the right sensor S_2

$$x_{2p} = \frac{1}{2}\left(x_{12} - v_g \cdot \Delta t_{12}\right).$$

The measuring example shown in Fig. 10.23 refers to a distance $x_{12} = 100$ cm between the both sensors. Inserting the values $\Delta t_{12} \approx 2$ ns and $v_g \approx 30$ cm/ns one gets a distance $x_{P2} = 20$ cm between the right sensor S_2 and the PD source.

Propagating along the GIS/GIL, the PD signal is strongly attenuated by the spacers separating the different GIS or GIL compartments. This effect can also be used to localize the PD site, sometimes referred to as *PD sectionalizing* method.

Another promising tool to localize the PD site is the *acoustic PD detection*, particularly when the measuring system is triggered by the electromagnetic signal radiated from the PD source, see Sect. 4.8. A practical example is shown in Fig. 10.24, which refers to a free-moving metallic particle under AC test voltage having a frequency of 68 Hz. Typical for such kinds of PD defects is that the repetition rate of the detectable pulses is comparatively low, often below that of the test frequency, where the pulses are not correlated with the test frequency (Fig. 10.24). The use of an array of *acoustic transducers* could also be helpful for the localization of PD defects because the emitted acoustic signal is also subjected to strong attenuation when travelling along the GIS/GIL compartment.

Practical experiences revealed that on-site PD tests of GIS/GIL with a separate HV source are only necessary after assembling or repair. Thus the sensors pre-installed for commissioning tests can advantageously be used also for periodic diagnostic measurement carried out from time to time under operation voltage. A continuous PD monitoring is desired only for important GIS and even recommended after a "low-risk" PD failure has been identified in the course of periodic PD diagnosis tests.

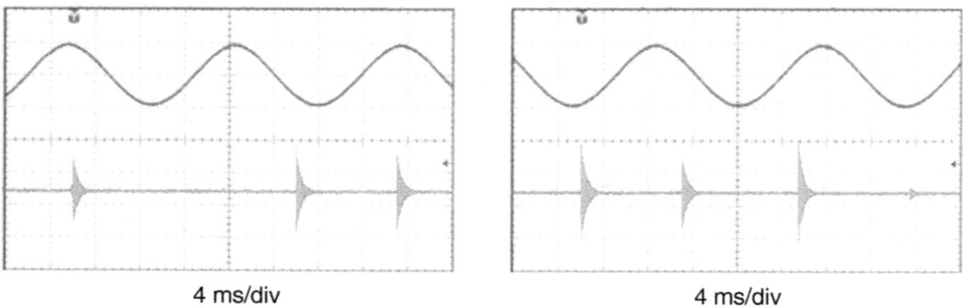

Fig. 10.24 Acoustic PD signal caused by a free-moving particle in a GIS subjected to AC test voltage of variable test frequency (68 Hz)

10.4.1.4 On-Site Testing of HVDC Gas-Insulated Systems

For *HVDC GIS*, the experience of using HVDC tests is very limited and consequently not recommended, but allowed. Routine tests in factory and commissioning tests on site of *HVDC gas-insulated systems* shall be efficient and not time consuming. Both tests are under discussion, but wether a CIGRE Brochure nor an IEC Standard is available. The CIGRE JWG D1/B3.57 (2017) recommends at the present stage the application of AC voltages according to Table 10.2 (Apply identical conductor-to-ground voltages!). These HVAC tests shall be PD monitored.

There might be the following reasons for this recommendation: PD monitored HVAC tests are very sensitive against most typical defects (see Sect. 10.4.1.1), there are decades of experience with HVAC tests combined with PD measurements on gas-insulated systems and last but not least a very detailed type test is performed for HVDC gas-insulated systems (see Sect. 6.2.3.1). But the behavior of particles at DC and AC voltage is very different. Therefore the future experience with routine and on-site commissioning tests will show if the HVAC tests are sufficient or they should be completed by HVDC withstand tests.

10.4.2 Testing of Cable Systems

The application of power cable systems for transmission and distribution of electric power has a long tradition. So the first cables insulated with liquid-impregnated-paper (LIP) were introduced more than 100 years ago. Since the 1960's these were more and more replaced by extruded, cross-linked polyethylene insulated (XLPE) cables. Meanwhile, both types of power cables are available up to a rated voltage of 550 kV. For economic and environmental reasons, extruded cables become more and more dominant not only for AC but also for DC transmission networks (Fig. 1.2). The following considerations focuses mainly on XLPE cable systems installed in AC power networks.

On-site testing of power cables has also a long history. Table 10.3 provides an overview on the kinds of test voltages recommended for after-laying tests.

10.4.2.1 History of DC Voltage Testing of AC LIP Cable Systems

As known, the conductivity of *oil-paper insulation* is much higher than that of high-polymeric insulation. Therefore, the radial voltage distribution in coaxial, oil-paper insulated cables is qualitatively comparable under both AC and DC stress, which holds true even in the vicinity of dielectric imperfections. Furthermore, LIP insulation is quite resistant against partial discharges. Additionally, decades ago, only mobile *DC voltage test systems* were available (see Fig. 10.10). Therefore, if AC cable systems had to be tested, exclusively DC test voltages were applied. The

Table 10.3 Characteristics of different test voltages for on-site testing of extruded cable systems

Test voltage IEC 60060-3; IEEE 400 characteristic	DC	AC 50/60 Hz	ACRF 20...300 Hz	VLF 0.01...1 Hz	DAC DC ramp; 20... 500 Hz; damping <40%
Representation of stresses in service	Identical for DC cables, very different for AC cables	Identical for AC cables	Close to AC 50/60 Hz stress	Very different from the AC 50/60 Hz stress	Impulse voltage, much different from AC 50/60 Hz stress
Description of voltage	Unipolar and continuous voltage (reference for all DC cables)	Alternating voltage (reference for all AC cables)	Alternating voltage; in the range of the reference frequency	Very slowly alternating; far from reference frequency	Direct voltage ramp followed by damped oscillation
Discharge and breakdown process	Slow and determined by space charges	Typical for AC of power frequency	Very similar to AC of power frequency	Different, higher test voltage values necessary	Different, charge accumulation during ramp, then polarity reversal
Reproducibility at different cable systems	Perfect	Perfect	Acceptable, frequency mainly within 30... 100 Hz	Good	Poor, changes in • ramp duration, • frequency, • damping
Generation in the field	Easy	For HV/EHV: impossible, MV: few seconds	Good for MV cables, acceptable for HV/EHV cables	Good for MV cables, questionable for HV cables	Good for MV cables
Recommended for acceptance testing	All DC cables; (AC cables with LIP insulation traditionally)	Only for AC medium-voltage cables	All AC cables from MV to EHV	Partly applied for MV cables, not for HV cables	Partly applied for MV cables, not for HV cables
Recommended for diagnostic testing	All DC cables	MV AC cables	All AC cables from MV to EHV	MV cables	MV AC cables

comparatively high DC withstand voltage test levels up to 4 V_0 and above are related to the limited sensitivity of defects at DC voltage. For details, see IEEE Std. 400.1™-2007.

With respect to condition assessment, LIP-insulated AC cable systems are nowadays mainly tested with AC, VLF or even DAC voltages, as will be considered in the following sections. To assess the integral ageing of the LIP insulation, dissipation factor and/or dielectric response measurements are commonly performed, while PD measurements are carried out to identify local defects, which occur mainly in the accessories, such as joints and terminations.

10.4.2.2   Testing of Medium-Voltage AC Cable Systems with Extruded Insulation

The relevant IEC Standard 60502:1997 mentions that *acceptance test* of cable systems up to $U_R = 30$ kV must only be performed if "required" in the contract and considers the following three three alternatives:

1. a DC voltage test at 4 V_0 for 15 min (which cannot be recommended because of different voltage distributions at DC compared with AC as well as additional space charge effects in the cable insulation);
2. an AC voltage withstand test at 1.7 V_0 for 5 min;
3. the application of the rated voltage for 24 h on the no-loaded cable. Very often no test or method 3 is applied, because in contrast to contrast to alternative 2 it does not require a separate HV test system. The IEEE Guide 400™-2012 considers PD-monitored AC/ACRF and VLF withstand tests as "useful", but the application of any acceptance test is left to the contract between user and supplier. According to practical experience, quality acceptance testing does not play an important role for *medium-voltage cable systems*. This is related to their relatively high reliability due to the design of cable and accessory insulation of low dielectric stress compared with HV/EHV cables. Unlike that, the first generation of cross-linked polyethylene (XLPE) MV cables was not protected against the penetration of water causing the phenomenon of *"water trees"* and this means the enhanced development of partial discharges in the electric field under the influence of water. This endangered in the past many thousands of kilometers of cable systems worldwide and stimulated the development of further diagnostic tools.

Diagnostic tests remain very important with respect to the long-time utilization of cable systems and the mentioned water tree problems. When the research work had shown that DC test voltage creates dangerous space charges (Montanari 2011; Choo et al. 2011) and is hence not an acceptable tool for the identification of water trees, *VLF testing* has been developed as an alternative due to the low power demand and lightweight equipment (Boone et al. 1987), see Sect. 10.2.2.1 and Fig. 10.12. The withstand voltage test has later been completed by dissipation factor measurement and also by PD measurement. The measurements of Schiller

(1996) (Fig. 10.6) are sometimes taken as an indication for the high sensitivity of VLF withstand voltages against defects. This might be correct for the investigated large defects, but not for the full scale of possible defects from small to large, see Sect. 10.2.1.1. The large number of diagnostic VLF tests and related measurements, performed in the recent 25 years, delivered experience for the condition assessment. The latest experiences of VLF tests on cable systems up to 69 kV are summarized in the IEEE Guide 400.2-2012.

The great difference between the VLF test frequency and the power frequency led to the introduction of damped alternating voltages (see Sect. 10.2.2.2) and its combination with PD measurement (Fig. 10.15). The damping of the oscillations is even used for the estimation of the dissipation factor (Houtepen et al. 2011).

DAC withstand testing with a sequence of DC voltage ramps followed by attenuated AC oscillations (Fig. 10.25 see also 10.14)—as taken into consideration by a draft of IEEE 400.4 (2013)—cannot be recommended as mentioned in Sect. 10.2.2.2. DAC voltage is an introduced tool for diagnostic testing of medium-voltage cable systems. There are good reasons for applying ACRF voltages also for diagnostic tests (Weck 2003).

10.4.2.3 Testing of HV and EHV AC Cable Systems with Extruded Insulation

There are good reasons for a more careful selection of test voltage values and procedures for extruded HV and EHV cables: Their design of is based on comparatively high field strengthes of 15 to more than 20 kV/mm and differs thus substantially from that for MV cables which is in the order of 5 kV/mm. The breakdown process is governed by space charges depending on the voltage value

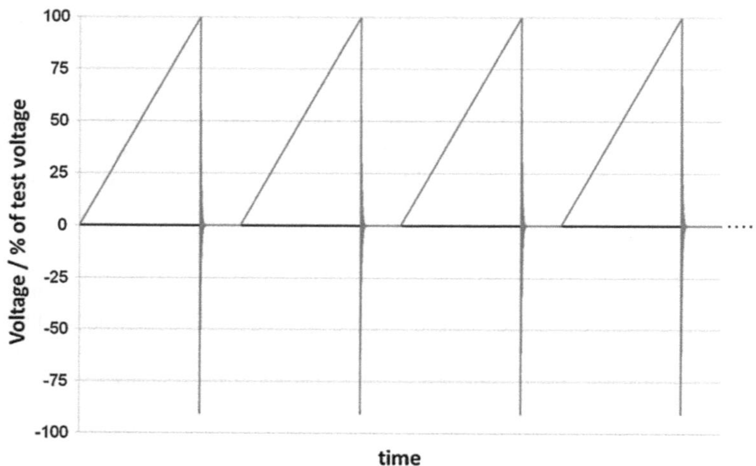

Fig. 10.25 Sequence of DAC voltages including the DAC ramps

and the test frequency (Cavallini and Montanari 2006; Nyamupangedengu and Jendrell 2012; Pietsch 2012). Consequently, for testing, the much higher-stressed insulations of extruded HV and EHV cables require a more stringent selection of the test voltage values and their frequency (Table 10.3: ACRF 20 to 300 Hz)).

For *quality acceptance test* on site, the standards IEC 60840:2011 for HV cables (rated voltage 50–150 kV) and IEC 62067:2011 for EHV cables (rated voltage 150–500 kV) recommend AC voltages in the frequency range 20–300 Hz, according to Table 10.4.

All test voltages shall be applied for one hour. As a certain alternative if a separate mobile test system is not available, the standards allow a 24-h check with the system voltage V_0 (!). This should be related to the practice in some special cases. By a large majority, an ACRF test is performed for important cable systems. For EHV cable systems ($V_m \geq 245$ kV), a range of test voltage values $V_{tmin} \leq V_t \leq 1.7\,V_0$ is proposed. A certain trend to higher test voltages within this range can be observed. The test voltage value to be applied is always a matter of agreement between supplier and user of the cable system. There is a strong opinion for the application of the last column of Table 10.4 in the future.

The HVAC acceptance test of an extruded cable system can only be performed with an ACRF test system (Hauschild et al. 2002, 2005), and alternatives cannot supply the necessary test power on site or their test voltage shape does not fulfill the requirements of an acceptance test (Table 10.3). If very long or even super-long cable systems shall be tested, several ACRF test systems can be combined as shown in the following example.

Example A 400-kV cable system of a length of 22 km and a capacitance of $C_t = 4.9$ µF shall be tested in a Middle East country. The test voltage of 260 kV and 20 to 300 Hz to be applied for one hour has been agreed between supplier and user. The environmental temperature at the place of the test can be remarkably above 30 °C. The super-long cable system requires the combination of some single ACRF test systems.

Table 10.4 AC withstand voltages for acceptance tests on extruded cable systems (IEC 60840:2011 and IEC 62067:2011)

Highest voltage for equipment V_m/kV	Range of nominal voltages V_n/kV	Reference voltage line to ground V_0/kV	Test voltage value according to IEC Standard V_{tmin}/kV	Test voltage value according to $V_t = 1.7\,V_0$/kV
52	45–47	26	52	–
72.5	60–69	36	72	–
123	110–115	64	128	–
145	132–138	76	132	–
170	150–161	87	150	–
245	220–230	127	180	216
300	275–287	160	210	272
362	330–345	190	250	323
420	380–400	220	260	374
550	500	290	320	493

There are several test systems available in that region: three systems for 260 kV with in total four reactors of $L = 16.2H/83A$ and one system for 160 kV with two reactors of 23H/55A each. A suitable combination of the reactors must be found, to meet the requirements such as the test voltage levels, the test frequency and the test power. Because of their limited voltage, the two 160 kV reactors must be connected in series. Then, the test current should be—also for thermal reasons - as low as possible, which means that the test frequency should be only slightly above 20 Hz. The series connection of the 160 kV reactors suggests that also the other reactors should be arranged accordingly (Fig. 10.26a). Then, the total inductance of the test system and the frequency are

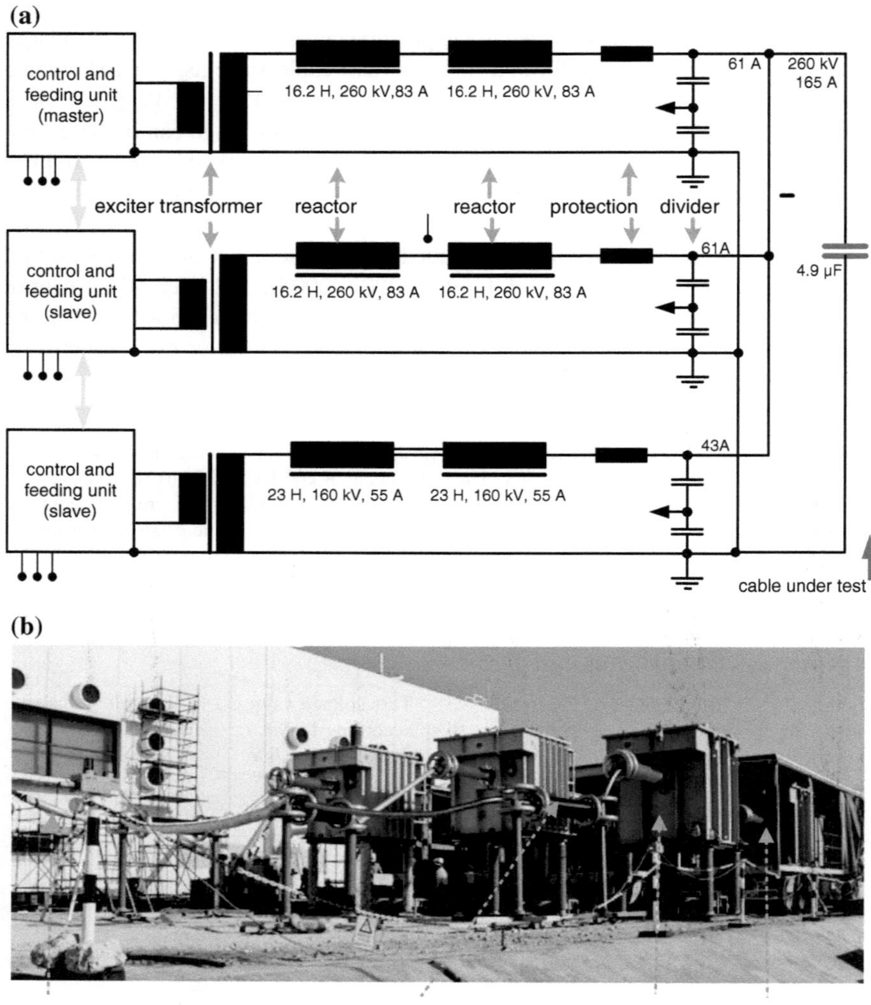

Fig. 10.26 Quality acceptance test of a 400-kV/22-km cable system of 4.9 µF. **a** Simplified circuit diagram. **b** Test arrangement. Courtesy of CEPCO, Saudi Arabia

$$L_t = \frac{(16.2 + 16.2)\mathrm{H} \cdot (23 + 23)\mathrm{H}}{2 \cdot (16.2 + 46)\mathrm{H}} = 12\mathrm{H},$$

$$f_t = 1/(2\pi\sqrt{L_t \cdot C_t}) = 21 \text{ Hz}.$$

With the known frequency and the test voltage, the test current and the test power can be calculated for

$$I_t = 2\pi \cdot f_t \cdot C_t \cdot U_t = 168 \text{ A},$$

$$S_t = I_t \cdot U_t = 43.7 \text{ MVA}.$$

If the test would be performed at 60 Hz, an equivalent test power S_{60} = (60 Hz/21 Hz) · 43.7 MVA = 125 MVA would be required. Even if the current can be supplied by two frequency converters, all frequency converters are used to keep the load for each of them low. The three reactor branches are grounded via their frequency converters and connected in parallel (Fig. 10.26a). All converters are connected in parallel to the power supply. The control of one converter operates as the master and the other controls as slaves. A quality factor of Q = 80 is usually assumed to calculate the feeding power $P_t = S_t/Q$ = 550 kW, which is necessary for the compensation of the losses in the test arrangement. The real test (Fig. 10.26b) has confirmed the pre-calculated data except for the assumed quality factor, because the real one was Q = 141, which means the required feeding power approached only P_t = 310 kW. The real test has been a PD-monitored withstand test. The monitoring was realized by a non-conventional PD measurement with sensors in the cable joints and terminations as described in Sect. 10.4.2.5.

When *diagnostic tests* on HV/EHV cable systems are performed, usually the procedure of the PD-monitored acceptance test cycle is repeated with a reduced withstand voltage level, but identical PD reference voltage levels before and after the withstand level shall be applied (compare Fig. 10.2b). Especially the insulation of the joints and terminations may age. Therefore, the trend of the PD level measured by means of sensors is an indication of the condition of joints and terminations.

10.4.2.4   Testing of HVDC Cable Systems with Extruded Insulation

Different to the capacitive field distribution at AC voltage, a DC field is resistively controlled. Steady state conditions are reached only after quite long times, often under the influence of space charges. This has to be taken into account by choosing test voltages.

The extremely high resistivity of the extruded material, mainly XLPE, enables the generation of very stable space charges, which determine the local field strength (AC: electrostatic field; DC: streaming field) in general and especially in the environment of insulation defects. The resistivity depends strongly from the temperature causing different field strength conditions on cold cables and loaded cables (A capacitive voltage distribution is much less influenced by the temperature.) A dangerous field enhancement occurs in case of transient processes (charging, overvoltages, especially polarity reversals). A *HVDC cable system* has to be designed for both, stationary DC stress and the mentioned transient stresses. For a better control of the space charges, the material for HVDC cables has a different characteristic based on additives compared with that of HVAC Cables. This has been explained by the field strength distribution of extruded HVDC and HVAC cables in Sect. 6.2.3.1 with Fig. 6.15.

Whereas pre-qualification and type tests have to verify that cables and systems withstand all types of stresses, quality acceptance tests in the factory (routine tests) and on site (tests after installation) have to consider the acceptable effort for the realization of tests. The routine test program includes a HVDC test with $V_t = 1.85$ V_0 for 60 min and recommends an HVAC test combined with PD measurement (when allowed according to the cable design; V_0 is the conductor-sheet voltage). After installation a test with $V_t = 1.45 \cdot V_0$ is required for 15 min, but an additional HVAC test with PD measurement at the accessories (Sect. 10.4.2.6) would supply additional confidence.

10.4.2.5 Testing of AC and DC Submarine Cable Systems with Extruded Insulation

Submarine cable systems are characterized by their huge length up to more than many 10 km. *HVAC cable systems* are used for the connection of islands, wind-parks or oil platforms, their characteristic length attains some 10 km, corresponding to a test object capacitance up to 15 µF. The required very long cables are manufactured in factories close to the coast and—after routine test of production lengths (and possibly joining several of them to a super-long cable)—directly reeled to the laying ship (Figs. 3.45 and 10.27). The Cigre Working Group B1.27 has investigated conditions, voltages and procedures for testing *HVAC submarine cables* (Cigre WG B1.27 2012). Even for the AC routine tests in the factory, AC test voltages of variable frequency between 10 and 500 Hz are accepted with respect to the high capacitances of the test objects (Karlstrand et al. 2005). The extended frequency range down to 10 Hz enables the reduction in the test power compared with 20 Hz to its half, but the breakdown mechanism remains similar as for power frequencytest voltage.

After installation, an *AC quality acceptance test* identical to that of land cables, but with the wider frequency range, shall be performed. The test voltages (duration 1 h) are selected according to Table 10.4. If the cable system is too long for testing with the specified values, a reduced test voltage with a longer duration can be agreed between supplier and user. If a suitable test voltage source is not available, a check with the nominal voltage for 24 h can also be agreed. Special voltages such as VLF and DAC cannot be recommended for tests after installation (Cigre WG B1.27 2012).

HVDC cable systems can transmit a much higher power than AC cables. Therefore, they become very important for future power transmission systems including submarine applications. HVDC cable systems of several hundreds of kilometers are under planning (Fig. 1.3). The latest developments of extruded DC

Fig. 10.27   After-laying test at 174 kV of a submarine cable (150 kV, 62 km, 13.8 µF). **a** Laying ship. Courtesy of ABB Karlskrona **b** Cable end prepared for the single-phase test, **c** Test voltage generation using five test systems (260 kV, 83A each) in parallel

insulations offers very economic perspectives. The testing of super-long DC cable systems, corresponding to a test object capacitance up to the order of 100µF is a challenge for the test techniques (Pietsch et al. 2010). Because HVAC voltage testing seems to be impossible, a test with DC voltage, perhaps completed with a *polarity reversal* (changeover time in the order of 1 min), could be applied. However, a PD measurement of the required sensitivity cannot be reached because of the attenuation and dispersion of PD signals when travelling along the long cable (see below Sect. 10.4.2.6 and Cigre TF D1.33.05 2012). It is proposed by the Cigre Working Group 21.01 (2003) to perform in addition to the HVDC test an HVAC test on the production lengths and to apply PD monitoring of joints and terminations in the system.

The HVAC test on production lengths (e.g. 30 km!) should be performed by using an ACRF test system operating in the extended frequency range 10–500 Hz. As the sheet resistance of an HVDC cable is significantly higher than the conductor resistance and this contributes to the losses in the test circuit remarkably. Therefore, in case of testing a HVDC cable, the quality factor of the test circuit is lower than that for HVAC cable of identical length. For the latter, only the ACRF test system itself determines the losses (Hauschild et al. 2005).

As the parameters and values specified for routine tests in factory are not yet clear, recommendations for the quality test after installation of HVDC cable systems are even more difficult. Even tests with VLF and DAC voltages are discussed. But the applicability and feasibility of VLF and DAC tests in the EHV range is still very questionable (Pietsch et al. 2010; Cigre WG B1.27 2012). Thus further research work is needed for both quality testing and diagnostic testing.

10.4.2.6 PD Testing of Cable Systems

On-site PD measurements of power cables became of interest in the late 1980s, when paper-insulated, lead-covered cables (PILC) were increasingly replaced by cross-linked polyethylene-insulated (XLPE) power cables. The reason for that is the fact that the high-polymeric insulation is very sensitive to partial discharges. Practical experiences revealed that PD magnitudes as low as few pico-Coulombs might already cause an immediate breakdown. This refers in particular to extruded *HV/EHV cables* due to the comparatively high operational field strength, which is being higher than 10 kV/mm. Considering *medium-voltage cables*, however, these are less sensitive to PD magnitudes in the pC range due to the significantly lower operational field strength. Consequently, the philosophy for on-site diagnosis PD tests of extruded power cables is different for both MV and HV/EHV cables, as will be considered more in detail in the following.

Medium-voltage cable systems: The length of MV cable systems exceed very seldom 5 km, i.e. the maximum capacitive load of the test facility is commonly limited below 1 μF. The circuit applied for periodical PD tests under on-site condition is well comparable with that applied under laboratory condition (IEC 60270:2000; IEC 60885-3:1988). That means after the operation voltage has been switched off and the three phases of the cable line have been disconnected from the power network, each phase is individually PD-tested. For this purpose, a coupling capacitor in series with a measuring impedance is connected to one cable end to capture the PD signal and transmit it to the PD measuring system, in order to record the phase-resolved PD patterns, see Fig. 4.17a. After calibrating the complete measuring circuit in terms of pC, the cable is energized up to the desired test voltage. In case of potential PD failures, a localization of the site of origin and a PD mapping described in Sect. 4.4 is required for further decisions.

On-site PD diagnosis tests of MV cables have been introduced in the 1980s, when XLPE distribution cables—installed about a decade before—showed the phenomenon of *"water trees"* (Auclair et al. 1988). However, at that time, no ACRF test systems were available and AC test systems based on conventional test transformers (ACT) or tuneable reactors (ACRL) were too heavy and too expensive for on-site application. Therefore, alternative test voltages were introduced, such as very low frequency (VLF) and damped AC (DAC) voltages specified in IEC 60060-3:2006. However, VLF voltage, originally applied for diagnostic withstand tests and loss factor measurements, is not so well suited for PD measurement because of the long duration of a single period (see Sect. 10.2.2.1; Fig. 10.11. The main benefit of the *DAC voltage* (see Sect. 10.2.2.2; Fig. 10.13 as well as Sect. 7.6) is that significant PD quantities, such as the inception/extinction voltage and the pulse charge magnitude, are in a certain relation to that values measurable at power frequency. Further benefits are the low power demand and the easy mobility due to the low weight of the DAC test facility (Fig. 10.15). Therefore, the use of DAC voltages for PD diagnosis tests of MV cables is nowadays widely accepted.

Applying this type of test voltage, it seems sufficient to acquire only those PD pulses occurring during the first positive and/or negative DAC voltage sweep resulting from a negative and/or positive DC voltage ramp. A practical measuring example to provide the *PD map* is shown in Fig. 10.28, which refers to an extruded MV power cable of 1200 m in length (20 kV rated voltage). As can be seen, the PD events appeared within the time interval elapsing between the peak-to-peak values, i.e. during the first negative voltage sweep, which attains about 3 ms. As indicated in Fig. 10.28b, only those PD pulses occurring during the first negative DAC voltage sweep have been extracted to create the PD map depicted in Fig. 10.28c, This shows three typical clusters indicating the position of three defective cable joints, where the PD activity scattered between some 100 pC and more than 1000 pC. Therefore, it was decided to replace the identified joints. Thereafter, the PD diagnosis test has been repeated and no any harmful PD signals were detected.

To specify the PD test parameters, it is a common practice to express the pre-stressing DC voltage and thus the peak value of the DAC voltage in terms of the conductor-to-ground voltage of the cable network (IEEE Guide 400.4 2015), which is denoted in Table 10.5 as V_0. Practical experiences revealed that the scattering of the inception voltage is comparatively low if compared with continuous power frequency AC voltage. This is likely due to the DC pre-stress causing a minimizing of the statistical time lag required for the availability of an initiatory electron. Hence only few DAC voltage applications at each test level are required from a statistical point of view. A survey on the test parameters initially recommended for on-site PD diagnosis tests of MV cables is given in Table 10.5.

HV/EHV cable systems: As mentioned already previously, PD testing of extruded HV/EHV cables requires a very high *detection sensitivity* due to the fact that PD events in the pC range might already cause an immediate breakdown of high-polymeric dielectrics. This is forced by the very high operational field strength, which is substantially greater than the design field strength of MV cables. As a consequence, on-site PD tests of HV/EHV power cables after laying or repair shall be performed under continuous AC voltage, which is usually generated by ACRF test systems.

Using the classical PD measuring circuit according to IEC 60270:2000, however, the PD detection sensitivity desired for HV/EHV cables is commonly not achievable. This is not only due to the always-present ambient electromagnetic noises but also due to the strong attenuation and dispersion of the PD pulses, if travelling from the PD source to the cable ends. As a typical measuring example consider Fig. 10.29a, which refers to a 2000 pC calibrating pulse injected in the remote end of a 450-m-long cable, where both the direct and reflected pulses are displayed. Based on such experimental studies the PD detection sensitivity versus the cable length can be estimated, as exemplarily shown in Fig. 10.29b (Lemke 2003).

Assuming, for instance, that for a cable of 1 km in length a minimum PD level of 10 pC would be detectable under noisy on-site condition, the PD threshold level achievable for cable systems of 5 and 10 km in length would increase up to approx. 50 and 150 pC, respectively. Of course, this is not acceptable for extruded HV/EHV

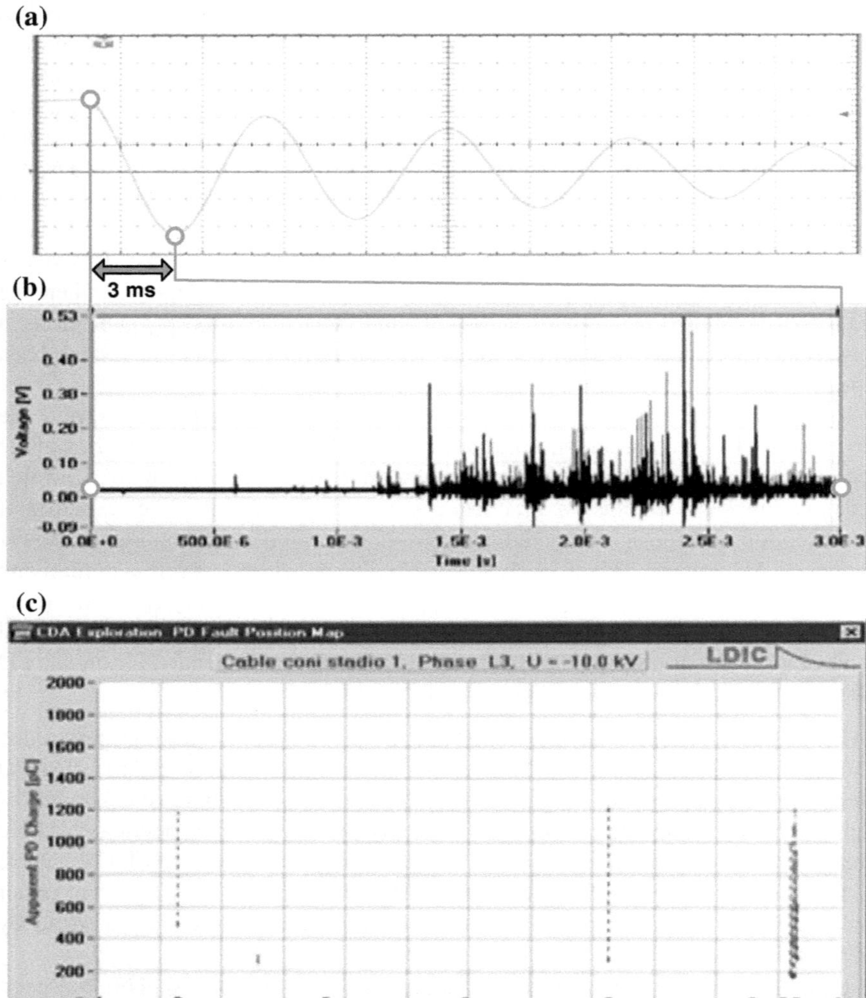

Fig. 10.28 PD mapping of a 30-kV XLPE cable (length, 1,200 m) at DAC test voltage. a DAC test voltage, 24 kV peak-to-peak value. b PD pulse train during the first negative voltage sweep. c PD fault position map showing the position of the three identified joints

cables, as discussed above. Thus, on-site PD diagnosis tests are usually only performed to prove the integrity of the cable accessories, where the PD signal is captured in the VHF/UHF range using inductive, capacitive or even electromagnetic PD couplers, as has already been treated in Sect. 4.7. The test philosophy is based on the fact that HV/EHV cables are carefully PD-tested in the factory after manufacturing, so that critical PD failures have to be expected only due to an

Table 10.5 Parameters recommended for on-site PD diagnosis tests of MV cables at DAC test voltages (Lemke et al. 2003)

Cable insulation	Test voltage level (V_0)	Voltage applications	Accepted PD level (pC)
XLPE, aged	0.7	5	<50
	1.0	5	<50
	1.5	5	<50
	2.0	5	<50
XLPE, new	3.0	20	<20
Oil-paper, aged	1.0	10	<2000
	1.5	10	<2000
	2.0	10	<2000
Oil-paper, new	3.0	20	<500

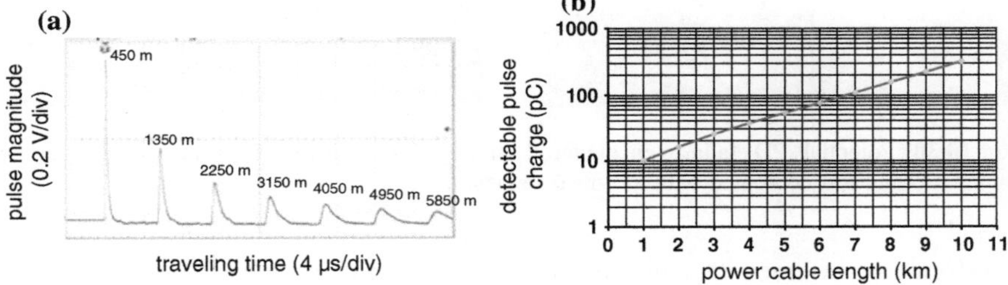

Fig. 10.29 Attenuation of PD pulses travelling along an extruded HV cable. **a** Attenuated PD pulses. **b** Detection sensitivity versus the cable length

inproper assembling of the cable accessories, such as terminations and joints (Fig. 10.30). The concept applied for this purpose is in principle comparable with that employed for PD monitoring of power transformers, as will be described below. That means the PD couplers initially used for commissioning tests remain installed in order to use these for a periodic or even a continuous PD monitoring.

10.4.3 Testing of Power Transformers

Different to GIS and cable systems, which are finally assembled and tested on site, power transformers are completed and carefully tested in factory and commonly not under on-site conditions. Up to a certain transformer size, only those parts exceeding the transportation profile, such as the bushings, conservator and other similar components exceeding the transportation profile are dismounted. As on-site assembly was quite limited in the past, no need for on-site acceptance tests of power

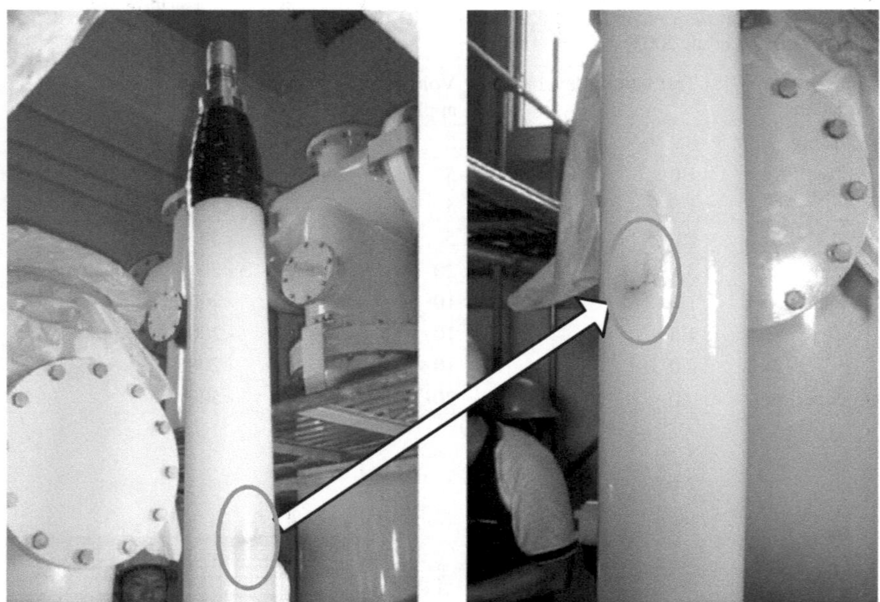

Fig. 10.30 Potential PD failure encountered in a 380-kV power cable termination during commissioning test combined with PD measurement

transformers was seen (Kachler et al. 1998). Meanwhile, the situation has been changed, and large units are assembled on site, especially in areas of insufficient transportation possibilities (Yamagata and Okabe 2009; Ohki 2010). Because of increasing transportation costs, the available technology for on-site assembly is also used for repair and refurbishment. Transformer manufacturers provide also the related service including HV testing (Siemens 2007; ABB 2006). On-site testing for quality acceptance and diagnostics of power transformers is well introduced.

10.4.3.1   Quality Acceptance Testing

IEC 60076-3:2012 states for tests of transformers having been in service: "Any transformer that is to be regarded as complying with this standard in the same way as a new transformer (for example following a warranty repair or complete rewind and refurbishment intended to restore the transformer to 'as new' condition) shall be subject to all routine tests required by this standard at 100% of the required test voltages (Table 10.6) after the repair or refurbishment is complete". This statement, which does not distinguish between repair in a factory and *on-site repair*, would also be applicable to power transformers assembled on site. In case of a repair, repaired parts should be tested at a level between 80 and 100% of the original test voltage (Table 10.6). New parts should be tested at 100%. The induced voltage acceptance test combined with PD measurement (AC-IVPD test, see Sect. 3.2.5; Table 3.6 and Fig. 3.51) shall be performed at the same voltage level as required for the routine test. As a consequence of these requirements of IEC 60076-3, HV on-site test systems have to generate test voltages with parameters as specified for routine tests in factory.

A detailed consideration shows that ACIT test systems of variable frequency based on *static frequency converters* are applicable (Table 10.6, last line) (Hauschild et al. 2006; Martin and Leibfried 2006; Werle 2007; Thiede et al. 2010). As shown before, these test systems have remarkable advantages not only for factory tests, but also for mobile application (e.g. see Tables 3.2 and 3.3 and Sect. 10.1.3). They are also in the scope of IEC 60060-3:2006 and enable optimum conditions for HV tests on site (Fig. 10.31). Besides the applied and induced voltage withstand tests, the systems shall also be applicable for no-load and load-loss (short-circuit impedance) measurements. The HVAC test system must be selected according to the requirements of the test and the characteristic of the test object, especially its necessary test power demand:

Applied voltage withstand test	The transformer under test is a capacitive load (<50 nF), and the test (f >40 Hz) can be performed by an ACRF test circuit with rated current of some Amps
Induced voltage withstand test	In most cases, the transformer is a linear, mixed resistive–capacitive load, and the test can be performed with a frequency converter-based ACIT test system (f >100 Hz) and requires relatively low test power
No-load loss measurement	The transformer is a nonlinear, mixed resistive–capacitive load in most cases, and the test can be performed with a frequency converter based ACIT test system (f = 50/60 Hz) of relatively low power but sufficient stability against harmonics
Load-loss measurement	The transformer is a linear resistive–inductive load, and the test can be performed with a powerful frequency converter based ACIT test system (f = 50/60 Hz) generating a fine, sinusoidal wave shape

The load–loss test (determination of short-circuit impedance) determines the power selection of the test system. (A heat run test has a similar test power demand.) The today available frequency converters can supply an active power in the order of 4000 kW or even more and a reactive power in the order of 8000 kVA (Fig. 10.32). For higher output parameters, several systems can be connected in parallel. If a higher reactive power is required, a capacitor bank must be added. Three-phase test systems can also operate in a single-phase mode.

The *applied voltage test* is a single-phase test between the short-circuited HV side and the grounded short-circuited low-voltage side. The excitation voltage of the resonant circuit can be generated by the frequency converter for the tests with induced voltages. But a small separate converter completes an independent ACRF test system, which might be even more practicable for separate application.

The components of a stationary frequency converter system are well arranged in a 40-feet container which fits to the dimensions for standard road transportation (Fig. 10.33). It includes the frequency converter, the step-up transformer, the control room, the space for HVHF filters and measuring equipment for voltage and current.

Table 10.6 Selected routine test voltages of power transformers (IEC 60076-3:2012)

Test voltage Highest voltage for equipment V_m/kV	Applied voltage AC withstand test V_a/kV for 1 min (Fig. 10.31b)	Induced voltage AC withstand and PD test V_{in} and V_{PD}/kV (Fig. 10.31a)	LI and LIC voltage withstand test V_{LI} and V_{LIC}/kV (3 impulses each)	SI voltage withstand test V_{SI}/kV (3 impulses each)
123	230	128 and 112	550 and 605	460
145	360	151 and 132	650 and 715	540
245	460	255 and 224	1050 and 1150	850
420	630	437 and 382	1425 and 1570	1175
550	680	572 and 501	1675 and 1845	1390
800	–	832 and 728	2100 and 2310	1675
1200	–	1248 and 1092	2250 and 2475	1800
Recommended HV test system	ACRF test system of f >40 Hz (see 3.1.2.3 and 3.2.5)	ACIT test system with frequency converter f >100 Hz (see 3.1.3 and 3.2.5)	Impulse test systems for aperiodic LI and SI voltages according to IEC 60060-1	

U_m is a phase-to-phase value; all test voltages are phase-to-ground values

(a)

coupling capacitors power transformer under test   container with frequency converter, compensation, step-up transformer

(b)

Fig. 10.31 On-site transformer tests. **a** Mobile system with frequency converter for induced voltage tests. **b** Mobile ACRF test system for applied voltage test

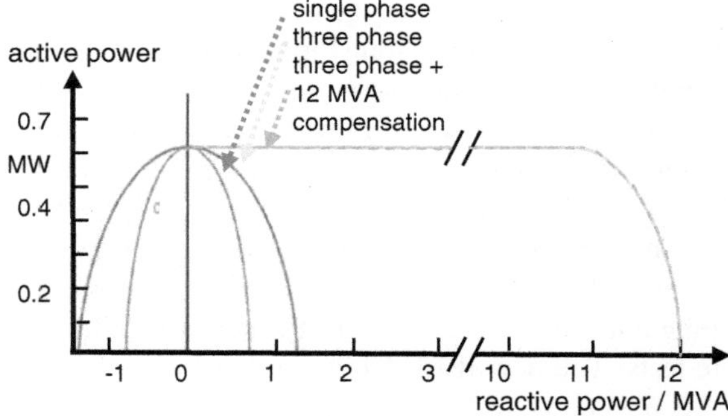

Fig. 10.32 Active and reactive test power of a 600 kW frequency converter without and with a capacitor bank (schematically)

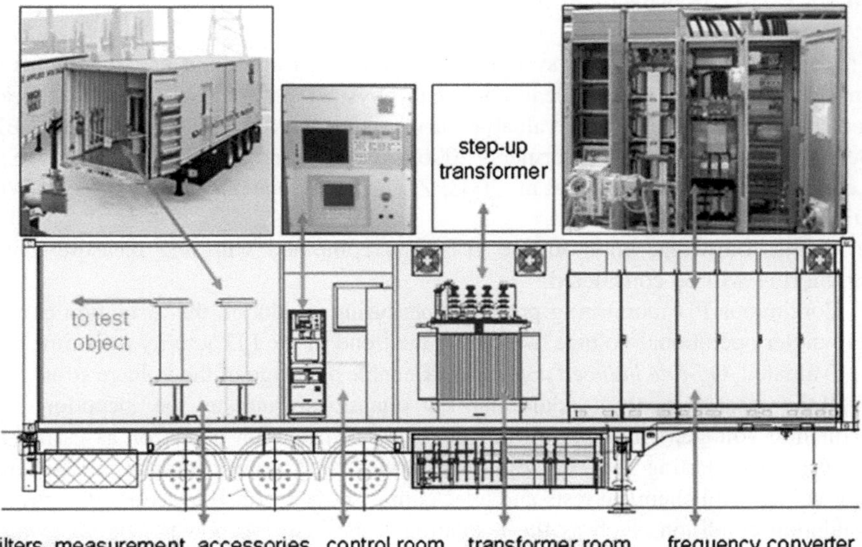

Fig. 10.33 Design of a mobile transformer test system

Additionally, to the AC voltage tests, IEC 60076-3:2012 requires aperiodic impulse voltages for LI and SI on-site application. The oscillating impulse (OLI; OSI) voltages which are generated with a much higher utilization factor are not accepted for transformer testing on site. This means that an impulse generator for on-site testing of UHV power transformers would be a quite huge and a not-easy-to-handle unit (see Sects. 7.1.2.4 and 7.2.1.3). It seems to be questionable whether LI test voltages above 2 MV can be generated on site. It can be assumed that reduced test voltage levels or OLI voltages will be applied in future.

In all the applications of the very stringent requirements of IEC 60076-3:2013 for on-site acceptance tests of power transformers, we could find their limits of practicability. Therefore, the real necessary tests should be considered, and the required parameters of the test system defined and agreed between supplier and user. The most important insulation test seems to be the PD-monitored, induced voltage withstand test which is also important for diagnostics. When an utility plans to operate a system for induced voltage tests, it seems to be not optimum to order the frequency converter for testing their largest power transformer. Rather the later extension by combination of two or more test systems should be taken into consideration for rare tests. The new IEEE Guide on diagnostic field testing on power transformers (IEEE 62-PC 57.152-2012) recommends also test systems based on frequency converters, which might also be helpful when acceptance and diagnostic tests have to be agreed.

10.4.3.2 PD Measurement in HV Tests and PD Monitoring

The very complex insulation system of a power transformer and its components is an object of numerous electrical, chemical, physical and thermal tests and checks for condition assessment. A valuable survey on this issue is found in the "ABB Service Handbook for Transformers" 2006 and even in numerous publications (e.g. Leibfried et al. 1998; Singh et al. 2008; Zhang and Gockenbach 2008; Cigre WG D1.29 (2017).

In the following, only off-line HV tests combined with PD measurements/monitoring will be considered.

Continuous PD monitoring provides measuring results on the insulation condition under operational voltage. Moreover the trend of the PD activity over time can be evaluated. *Off-line induced voltage tests* enable a change of the voltage stress and thus the measurement of additional PD quantities, such as the inception and extinction voltage, which are informative for the insulation condition assessment.

Just prior starting the HV test of a service-aged power transformer, several low-voltage and chemical tests and checks must be performed to assess the current insulation condition, such as the insulation resistance, the power factor, the turn ratio, the moisture content as well as dissolved *gas-in-oil analysis*. The HV test is only recommended when the measured parameters are within their acceptable tolerance band. Its procedure should be similar to that used for PD-monitored acceptance tests (Fig. 3.51). If useful for diagnostic purpose, this procedure can be modified accordingly. For insulation condition assessment, power transformers are commonly inspected periodically and even continuously PD monitored from time to time, provided a severe PD activity was encountered, e.g. indicated by spectral gas-in-oil analysis.

For periodic PD monitoring, the transformer is commonly disconnected from the power network and excited by a separate power supply (today usually by a frequency converter, as already presented in Sect. 3.1.3). To capture the PD transients, the *bushing tap PD coupling mode* in compliance with IEC 60270:2000 can be considered as the most convenient method (Fig. 4.19 and Fig. 10.34). Practical experiences revealed that the ambient noise level is usually not higher than some hundreds of pico-Coulombs, which is usually tolerated for PD tests of liquid-immersed power transformers.

The main benefit of the traditional IEC method is that the PD measuring circuit can be calibrated in terms of pC, which ensures well-reproducible test results and thus a comparison with fingerprints established by quality acceptance tests in the factory and during commissioning tests on site. Another benefit of the bushing tap coupling mode is the capability to perform multi-terminal, phase synchronous measurements in order to create *PD star diagrams*. As already treated in Sect. 4.6, such graphs are a promising tool for the identification of typical insulation failures as well as for the recognition of disturbing noises, which can thus be cancelled.

The philosophy of periodical on-site PD measurements is based on the assumption that the insulation of liquid-immersed power transformers ages only marginally over a comparatively long time. Nevertheless, the development of weak spots cannot always be excluded within the time span between two periodic inspections. Thus, it cannot be excluded that an abnormal PD activity remains undiscovered and could thus initiate a potential insulation failure triggering an unexpected outage. Therefore, strategic important EHV/UHV transformers are nowadays continuously PD-monitored. As this occurs under service condition, the measurement is often strongly interfered by ambient noises which may exceed tens of nano-coulombs. A promising tool to eliminate such disturbances is the application of the UHF/VHF technology, see Sect. 4.7. For this purpose, different kinds of UHF PD couplers have been developed in the past, such as the *drain-valve sensor* and the *plate-hatch sensor*, as shown in Fig. 10.35 (Markalous 2006; Coenen et al. 2007).

Due to the strong attenuation of UHV signals if propagating from the PD source to the sensor, the use of several UHF PD couplers is highly recommended even for PD monitoring of single-phase transformers. The installation of a multiple sensor array offers also the possibility for a so-called *dual-port performance check* (Coenen et al. 2007). For this purpose, an UHF signal of known parameters is injected in one of the installed sensors, and the amplitude–frequency spectrum obtained from all other sensors is recorded. Additionally, the measuring sensitivity should be verified, preferably to be performed under factory condition. This can be carried out based on comparative PD studies by applying both the traditional IEC 60270 method and the UHF method. For this purpose the signal radiated from an artificial PD source implemented previously in the transformer tank is measured simultaneously by both methods. Recording the pulse charge in terms of pC and the UHF signal in terms of mV, the threshold PD level can roughly be estimated, and in some cases, it can be investigated whether there exists a sufficient correlation between the measurands expressed in terms of pC, respectively, in terms of mV even if the magnitudes of both signals are scattering over an extremely wide range.

Performing continuos PD monitoring, it has to be taken into account that the complete *PD data stream* cannot be stored for a very long time period due to the limited memory capacity of the PD monitoring system. Thus it is a common

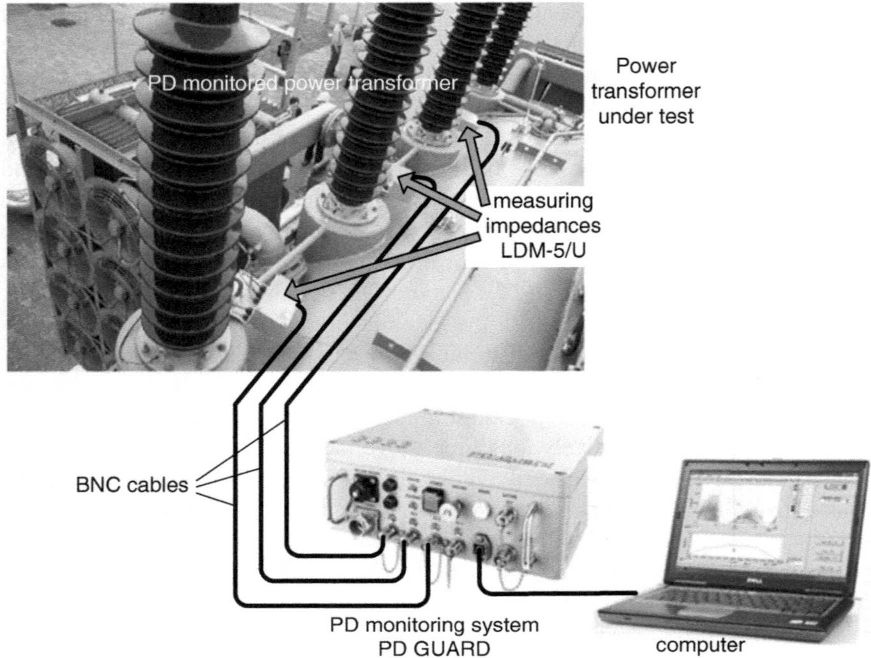

Fig. 10.34 Set-up for periodical on-site PD monitoring of power transformers using the bushing tap coupling mode in compliance with IEC 60270:2000. Courtesy of Doble Lemke

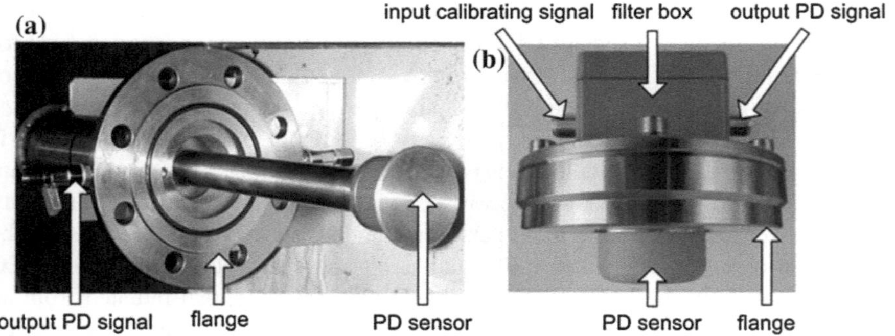

Fig. 10.35 Drain-valve sensor (**a**) and plate-hatch sensor (**b**) developed for continuous PD monitoring of power transformers in the UHF range. Courtesy of Doble Lemke

practice to install a simple "PD system" in the vicinity of each power transformer to be monitored, as illustrated in Fig. 10.36. Moreover, the raw PD data stream is commonly not stored permanently but rather from time to time due to the fact that the transformer insulation ages comparatively slowly, as mentioned above. Only in

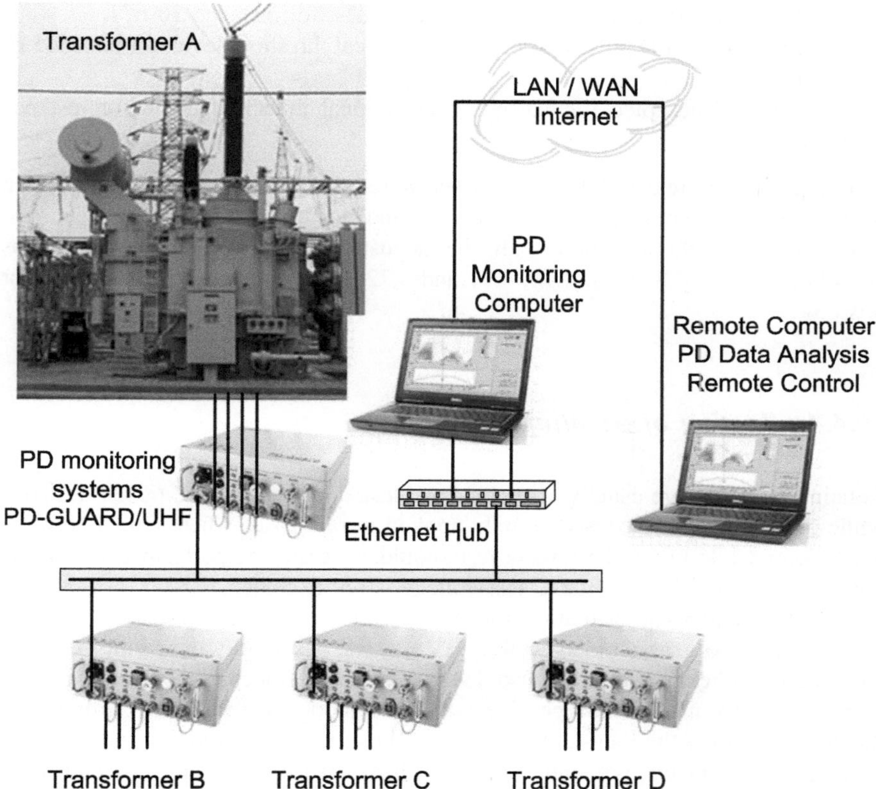

Fig. 10.36 Layout of a continuous PD monitoring system. Courtesy of Doble Lemke

case of a suddenly high PD activity this will be recorded permanently—to recognize any trends immediately—whereas the phase-resolved and time-dependent PD pulse trains are stored and acquired only for repetitive preselected time intervals, for instance, during 5-min periods repeating on an hourly basis.

If desired, the characteristic PD patterns can be observed in the control room of the substation, as usual also in case of factory tests. Recognizing drastically rising trends, however, the data can be transmitted via Internet to permit an investigation by experts for further decisions. Today's PD monitoring systems are also equipped with features to generate a warning or even an alarm (see Fig. 10.18). Based on practical experience, an appropriate set of parameters should be chosen to avoid an alarm release due to sporadic PD events or stochastically appearing interferences. Appropriate parameters for triggering an alarm relay are, for instance

- threshold level for critical PD pulse magnitudes, for instance 10 mV,
- mean number of PD pulses exceeding the critical threshold level, for instance 10 per minute,
- time interval accepted for exceeding the critical threshold level, for instance 10 min.

If an alarm was released, further actions are required, for instance, performing an off-line induced voltage test to identify and localize potential PD sources. Combining the PD detection with the acoustic emission (AE) technique, as described in Sect. 4.8 (Figs. 4.70, 4.71 and 4.72) is also a well promising tool for this task.

10.4.4 *Testing of Rotating Machines*

Rotating machines are usually assembled and tested in factory (IEC 60034-1:2010), while on-site acceptance tests are only applied in rare cases when agreed between manufacturer and user. The on-site test should generally be performed under AC voltage, however, the use of DC test voltage is not excluded.

The degradation of insulating materials used for rotating machines, such as hydro- and turbo-generators as well as HV motors, is not only the result of a high local electrical field strength but also due to thermal and mechanical stresses caused in particular by numerous repetitive starts and stops as well as by strong load fluctuations during the long-term operation. Thus, besides *loss factor* and *recovery voltage measurements*, preventive *PD tests* have also become an important diagnosis tool for assessing the insulation condition of rotating machines (IEC 60034-27:2006; Cigre TF D1.33.05 2012). The test philosophy and the layout commonly used for periodic and continuous monitoring of rotating machines is very similar to that already presented above for on-site testing of power transformers. As an example, Fig. 10.37 shows harmful PD defects in the overhang section of a rotating machine, which have been detected in the course of periodical PD monitoring under operation voltage.

Using the non-conventional PD detection in the UHF/VHF range, it has to be taken into consideration that the PD signal is strongly attenuated if travelling from the PD source to the PD couplers. This is not only caused by the high winding inductance but also due to the close electromagnetic coupling between the individual stator bars providing the three phases.

Performing on-site tests of rotating machines, different HVAC sources are applicable. When the maximum test voltage applied remains below 20 kV and the required test power below 150 kVA, a conventional test transformer can be used. If higher test voltage and test power are desired, the use of resonant test circuits is recommended. As the dissipation factor depends on the frequency by definition, inductance-tuned ACRL test systems (Fig. 10.38) are well suited for both *dissipation factor measurement* and *PD detection*.

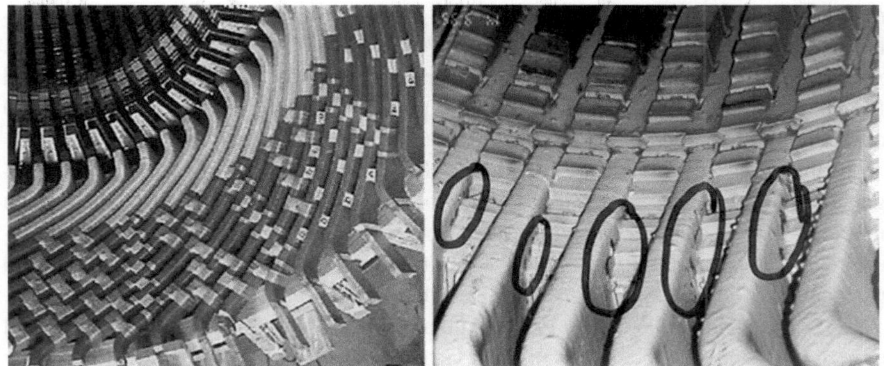

Fig. 10.37 Overhang section of a rotating machine after manufacturing (*left*) and failures identified by periodical PD monitoring (*right*). Courtesy of Doble Lemke GmbH

Fig. 10.38 Mobile ACRL test system (30–50 kV, 350 kVA) used for a generator test. Courtesy of TU Graz, Austria

Besides the dissipation factor itself also the change of this quantity at increasing test voltage ($\Delta\tan\delta/\Delta v$) provides a valuable tool to assess the insulation condition. Measurements at different test frequencies indicated that this parameter is nearly frequency independent (Fig. 10.39a). Therefore, *ACRF test systems* (Fig. 10.3a) can also be used for off-line, on-site testing, particularly if combined with $\tan\delta$ and PD measurements which are considered to be the most informative ones for the insulation condition assessment (IEC 60034-27:2006; IEEE 1434-2000), (Fig. 10.39b). As can be deduced from the measuring example shown in Fig. 10.39c, the measured PD charge versus test voltage shows the typical "ionization bend" and different PD charge magnitudes under increasing respective decreasing test voltage.

In this context it should be mentioned that *PD monitoring* of turbo- and hydrogenerators became an indispensable tool for the insulation condition assessment in the 1980s when besides the PD-resistant epoxy–mica insulation vacuum-pressure-impregnated (VPI) epoxy insulation was introduced, as this ages rapidly if exposed permanently to PD's in the nC range (Henriksen et al. 1986; Kemp 1987; Fruth et al. 1989; Grünewald and Weidner 1994). At that time, the PD signal has been decoupled either via so-called slot couplers or even via capacitive couplers of comparatively low capacitance, commonly below 100 pF. As a frequency spectrum below about 10 MHz cannot be captured by the use of such kinds of PD couplers, the

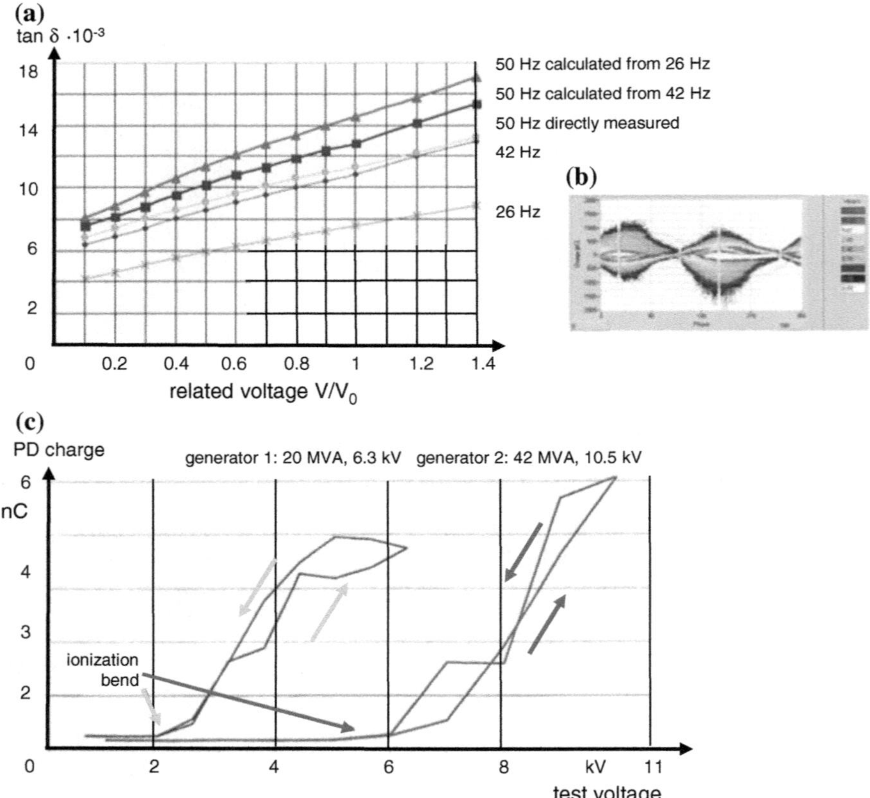

Fig. 10.39 Measured characteristics of two generators (Cigre TF D1.33.05 2012). **a** Dissipation factor depending on test voltage. **b** PD pattern (recorded at 42 Hz, comparable to 50 Hz). **c** PD charge of two different generators depending on the test voltage

high-frequency transients radiated from end-winding discharges as well as from discharges in gaseous cavities due to delamination, can often not be detected. This is due to their strong attenuation if traveling from the PD source to the PD couplers, which decreases the frequency content of the PD signal considerably. Apparently, this was the reason why numerous VPI-insulated rotating machines failed in the past, even if permanently PD monitored in service using slot couplers or low-capacitive couplers, both receiving a frequency spectrum only above 10 MHz

The capability and limits of different kinds of *PD couplers* can simply be proven by injecting calibrating signals in the neutral points of rotating machines and measuring the signal response at the terminals providing the three phases $L1$, $L2$ and $L3$, which are connected to the power network. This is underlined by the measuring example shown in Fig. 10.40, which refers to a 300 MVA motor/ generator operating in a pumped storage power plant. Here repetitive calibrating pulses as high as 10 nC were injected into the neutral point. Using a 100 pF PD coupler the received signal magnitude was below the background noise level which was about 1 mV. Increasing the coupling capacitance from originally 100 pF up to 2 nF, however, the signal magnitude attained about 20 mV (Fig. 10.40), which was well detectable by means of conventional PD monitoring systems, if designed according to IEC 60270:2000. In this context it is worth to notice that the PD level of rotating machines under operation condition can commonly not be kept below some nC, which means that a PD detection sensitivity in the nC range is fully acceptable.

Performing periodical PD diagnostics on-site, the rotating machine under test is usually excited by a mobile test voltage source (Fig. 10.38), where the individual phases are tested in turn, while the other both phases are connected to the grounded stator. Another option is illustrated in Fig. 10.41, which refers to a PD test under on-line condition. That means the three phases are simultaneously energized by the three-phase operational voltage and therefore subjected to the actual service stress.

Under this condition, however, the measurement is commonly extremely disturbed by high commutation pulses as exemplarily shown in Fig. 10.42a. This refers to a 300 MVA motor/generator, rated 21 kV, where the noisy commutating pulses originated in the six-pulse converters of the auxiliary power supply

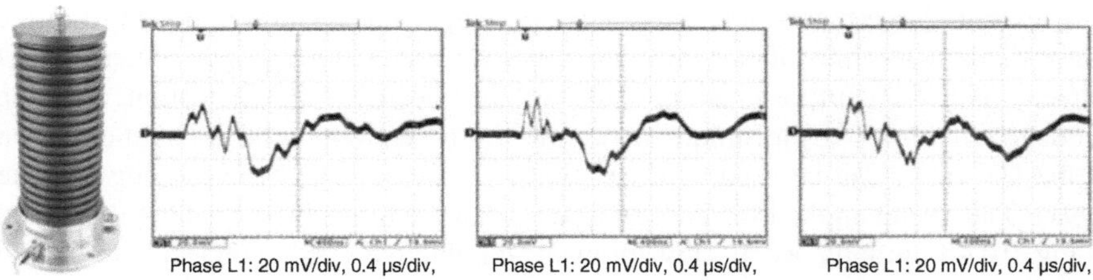

Fig. 10.40 PD coupler equipped with a 2 nF/24 kV capacitor (left) and measured pulse response (right) gained by injecting of 10 nC calibrating pulses in the neutral point of a 300 MVA motor/ generator. Courtesy of Doble Lemke

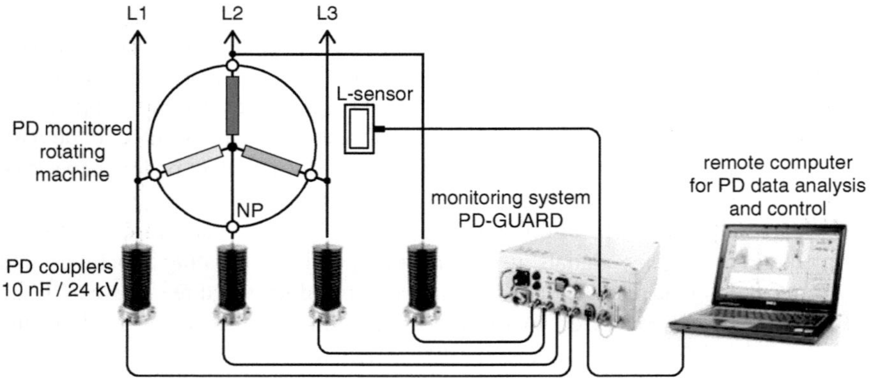

Fig. 10.41 Set-up for diagnostic online PD tests of rotating machines under service voltage

providing the static excitation of the rotor. As can be deduced from Fig. 10.42a, the disturbing pulses approach magnitudes up to about 10 nC. However, due to their fixed phase position these can clearly be distinguished from real PD events, which are scattering around 5 nC. Thus, a noise suppression, for instance by gating, is not necessary while observing the PD patterns on the computer screen. However, evaluating the PD trend a noise canceling seems to be a reasonable solution. Thus, for continuous PD monitoring, an automatic *noise gating* is highly recommended, as described in Sect. 4.5. For this purpose the disturbing commutation pulses itself can be used, which could be captured by an *inductive sensor* of frame antenna type if installed adjacent to the slip rings of the motor/generator. The feasibility of this option has successfully been proven in the course of preventive PD measurements carried out in numerous pumped storage and hydroelectric power plants. A typical measuring example for this is shown Fig. 10.42. As can be seen, the phase shift between the consecutive PD clusters scatters neither around 90 degrees nor around 180 degrees, which has to be expected for commonly occurring phase-to-ground discharges, but rather around 120 respective 240 degrees, which is typical for discharges in the insulation between two phases. This can readily be deduced from the potential diagram of a three-phase network and was also confirmed by a localization of the PD site and a following visual inspection.

Repairing large hydro- and turbogenerators, the rotor is often removed from the stator. This offers the opportunity for a convenient localization of potential PD defects. For this purpose, additionally to the electrical PD detection also the acoustic emission (AE) technique as well as the measurement of the light emission in the UV range can be used. Another option is the so-called PD probing (Lemke 1991). The set-up applied for the latter method and a typical test result is illustrated in Fig. 10.43. Initially, a capacitive or even an inductive sensor connected to the PD probe was moved along the stator bar under investigation to localize the critical

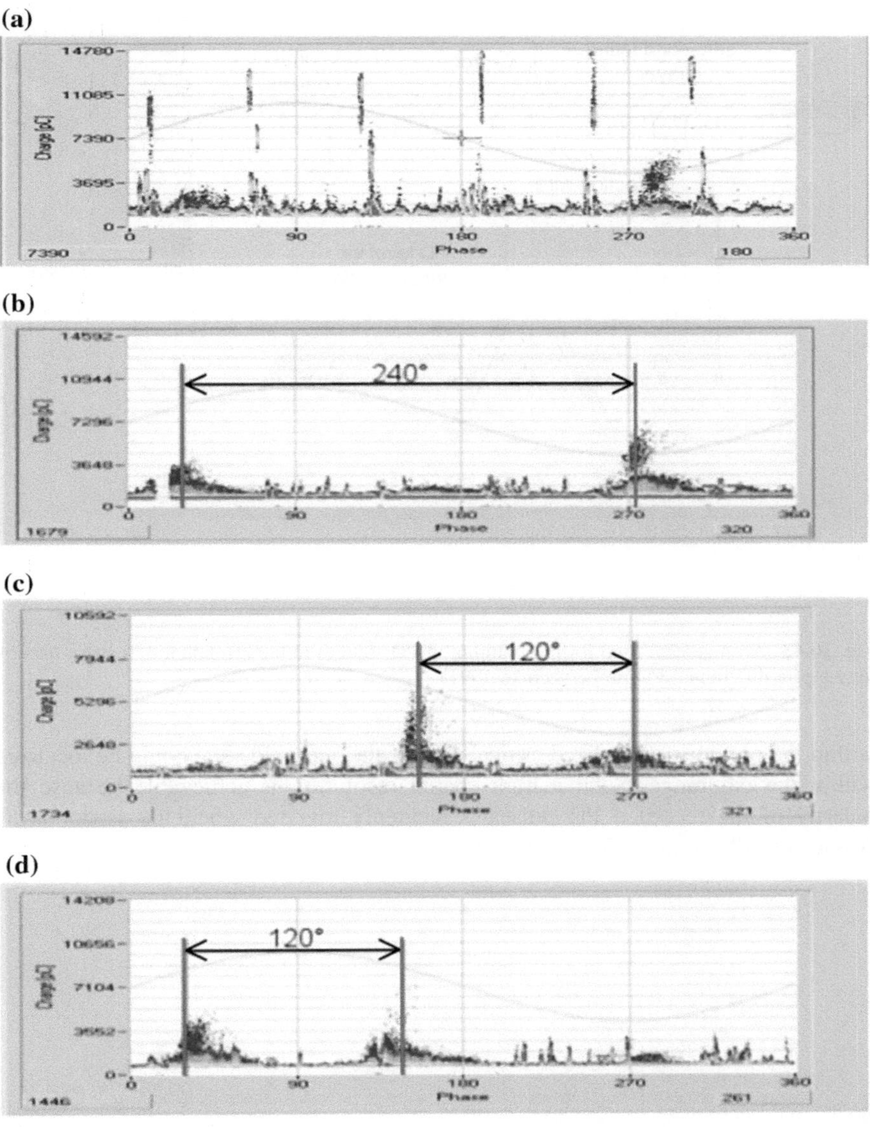

Fig. 10.42 PD signatures recorded for of a 300 MVA motor/generator under service condition. **a** Phase *L*1 showing PD clusters and superimposed noises originating in a 6-pulse AC/DC converter of the auxiliary power supply. **b** Phase *L*1, noise gating activated. **c** Phase *L*2, noise gating activated. **d** Phase *L*3, noise gating activated

region showing the highest PD activity. Thereafter, two capacitive sensors were adapted to both the inverting and non-inverting input of the potential-free operating differential PD probe. If the sensor array was slowly moved along the stator bar

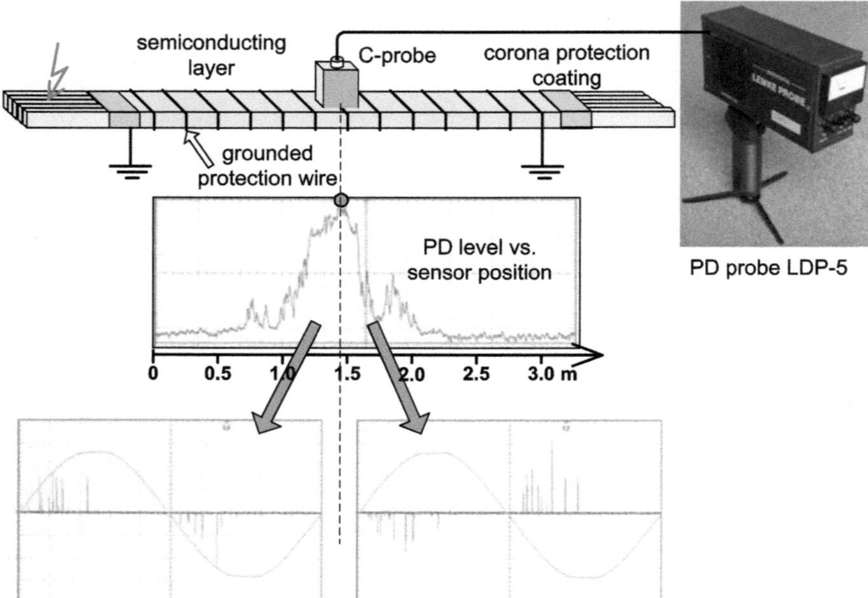

Fig. 10.43 Arrangement used for pin-pointing of PD sources in the insulation of a stator bars by means of a differential PD probe (Lemke 1991)

within the previously localized critical region, the actual PD site could be localized within the cm range. Such a high spatial resolution is achievable because the polarity of the recorded PD pulses is suddenly inverted when the sensor array crosses the PD source (Fig. 10.43).

Biography of W. Hauschild

Wolfgang Hauschild

Wolfgang Hauschild received the diploma degree in 1965, the Ph.D. in 1970 and the *habilitation* (university lecturing qualification) in 1976 from the Technical University (TU) of Dresden, Germany. In 2007, he became *doctor honoris causa* of the Technical University of Graz, Austria. From 1966 to 1979, he was a researcher of TU Dresden managing a research group on SF_6 insulation. In 1976/1977, Dr. Hauschild was a guest professor at Damascus University, Syria, and responsible for the erection of a large HV laboratory there. In 1980, he moved to industry and has been in leading positions of HV test equipment production in Dresden, from 1990 to 2007 as the Technical Director of HIGHVOLT Prüftechnik Dresden GmbH. After retirement, he is still active as a consultant. Dr. Hauschild was German speaker to IEC TC 42 (HV test technique) from 1995 to 2009 and was a member of IEEE, VDE and CIGRE. He published three books and numerous papers on HV engineering, especially HV testing.

Dr. Wolfgang Hauschild
DRESDEN
Germany
E-Mail: whauschild@t-online.de

© Springer Nature Switzerland AG 2019
W. Hauschild and E. Lemke, *High-Voltage Test and Measuring Techniques*,
https://doi.org/10.1007/978-3-319-97460-6

Biography of E. Lemke

Eberhard Lemke

Eberhard Lemke graduated from the Technical University (TU) Dresden, in 1962, where he was involved in research and education in the field of high-voltage engineering for more than three decades. He received the Ph.D. degree and Dr.sc. techn. degree in 1967 and 1975, respectively. In 2010, he was awarded an honorary doctorate (Dr.h.c) from the Technical University Graz, Austria. From 1978 until 1981 he joined a power cable factory in Germany. During this time, he developed the so-called Lemke Probe using the non-conventional field coupling mode for PD diagnostics of HV apparatus in service. In 1987, he was appointed Professor at TU Dresden and founded the company Lemke Diagnostics GmbH in 1990, which manufactures among others instruments for PD diagnostics of HV equipment. Eberhard Lemke is author and coauthor of several textbooks. He published numerous technical papers, holds various patents and is active in several national and international organizations, such as VDE, CIGRE, IEC and IEEE.

Prof. Dr. Eberhard Lemke
DRESDEN
Germany
E-Mail: elemke.37@gmail.com

© Springer Nature Switzerland AG 2019
W. Hauschild and E. Lemke, *High-Voltage Test and Measuring Techniques*,
https://doi.org/10.1007/978-3-319-97460-6

References

ABB. (2006). *Service handbook for transformers*. ABB Management Services LTD/Transformers.

Abbasi, A., et al. (2014). Pollution performance of HVDC SiR insulators at extra heavy pollution conditions. *IEEE Transaction on Dielectrics and Electrical Insulation, 21*(2), 721–728.

Achillides, Z., Georghiou, G. T., & Kyriakides, E. (2008). Partial discharges and associated transients: The induced charge concept versus capacitive modeling. *IEEE Transaction on Dielectrics and Electrical Insulation, 15*(6), 1507–1516.

Achillides, Z., Danikas, M. G., & Kyriakides, E. (2017). Partial discharge modeling and induced charge concept: Comments and Criticism of Pedersen's model and associated measured transients. *IEEE Transactions on Dielectrics and Electrical Insulation, 24*(2), 1118–1122.

Achillides, Z., Kyriakides, E., & Georghiou, G. E. (2013). Partial discharge modeling: An improved capacitive model and associated transients along medium voltage distribution cables. *IEEE Transaction on Dielectrics and Electrical Insulation, 20*(3), 770–781.

Albiez, M., & Leijon, M. (1991). *PD measurements in GIS with electric field sensor and acoustic sensor*. 7th ISH, Dresden, paper No. 75.08.

Allibone, T. E., & Dring, D. (1972). Influence of humidity on the breakdown of sphere and rod gaps under impulse voltages of short and long wave fronts. In *Proceedings IEE* (Vol. 119, pp. 1417–1422).

Allibone, T. Z., Achillides, Z., Kyriakides, E., & Georghiou, G. T. (2013). Partial discharge modeling: An improved capacitive model and associated transients along medium voltage distribution cables. *IEEE Transaction on Dielectrics and Electrical Insulation, 20*(3), 770–781.

Anderson, J. G. (1956). Ultrasonic detection and location of electric discharges in insulating structures. *AIEE Transactions, 75*, 1193–1198.

Anderson, J. M. (1971). Wide frequency range current transformers. *Review of Scientific Instruments, 42*, 915–926.

Arman, A. N., & Starr, A. T. (1936). The measurement of discharges in dielectrics. *Journal of the Institution of Electrical Engineers, 79*(67–81), 88–94.

Arora, R., & Mosch, W. (2011). *High voltage and electrical insulation engineering*. Hoboken: Wiley.

Asner, A. (1969). Progress in the field of measuring very high fast transient surge-voltages. *BBC-Mitteilungen, 47*, 239–267.

Asner, A. M. (1974). *High-voltage measuring techniques*. New York: Springer. (in German).

Auclair, H., Boone, W., & Papadopulos, M. S. (1988). *Development of a new after laying test method for high voltage power cables*. CIGRE Session Paris, France, paper 21–06.

© Springer Nature Switzerland AG 2019

W. Hauschild and E. Lemke, *High-Voltage Test and Measuring Techniques*,

https://doi.org/10.1007/978-3-319-97460-6

Azizian Fard, M., et al. (2017). Partial discharge behavior under operational and anomalous conditions in HVDC systems. *IEEE Transactions on Dielectrics and Electrical, 24*(3), 1494–1502.

Bach, R. (1993). *Investigation to the on-site testing of medium voltage cables using different voltage shapes*. Ph.D. Thesis TU Berlin (in German).

Bachmann, H., et al. (1991). *Hardware and software for computer-aided impulse voltage tests*. 7th ISH, Dresden, Paper 5.14.

Baer, C., Bärsch, R., Hergert, A., & Kindersberger, J. (2016). Evaluation of the retention and recovery of hydrophobicity of insulating materials in HV outdoor applications und AC and DC stresses with the dynamic drop test. *IEEE Transactions on Dielectrics and Electrical, 23*(1), 294–303.

Bailey, C. A. (1966). A study of internal discharges in cable insulation. *IEEE Transactions on Electrical Insulation, 31*(2), 360–366.

Balzer, G., et al. (2004). *Evaluation of failure data of HV circuit-breakers for condition-based maintenance*. CIGRE Session Paris Report A3-305.

Barsch, R., Jahn, H., & Lambrecht, J. (1999). Test methods for polymeric insulation materials for outdoor HV insulation. *IEEE Transaction on Dielectrics and Electrical Insulation, 9*, 668–675.

Bartnikas, R. (2002). Partial discharges-their mechanism, detection and measurement. *IEEE Transaction on Dielectrics and Electrical Insulation, 9*(5), 763–808.

Bartnikas, R., & Levi, H. R. (1969). A simple pulse-height analyzer for partial discharge rate measurements. *IEEE Transactions on Instrumentation and Measurement, IM-18*, 341–345.

Beigert, M., Henke, D., & Kranz, H-G. (1991). *Isothermal relaxation current measurement, a destruction free tracing of pre-damage at synthetic compounds*. 7th ISH, Dresden Paper 72.05.

Beinert, J., Kadry, E. A., & Schuppe, W. (1977). The role of PD measurement for the detection of defects in extruded medium-voltage cables. *Elektrizitatswirtschaft, 76*(26), 925–928. (in German).

Bellaschi, P. L. (1933). The measurement of high surge voltages. *Transactions of the American Institute of Electrical Engineers, 52*(2), 544–552.

Bellaschi, P. L. (1934). Heavy surge currents generation and measurement. *Transactions of the American Institute of Electrical Engineers, 53*(1), 86–94.

Bellaschi, P. L., & Teague, W. L. (1935). Sphere-gap characteristics of very short impulses. *The Electric Journal, 32*(3).

Bilinski, E., et al. (2017).*Reactor and test arrangement for realization of HV tests*. Application for a German Patent. Registration-No.2017P00011 DE (in German).

Bellm, H., Kuechler, A., Herold, J., & Schwab, A. J. (1985). Rogowski coils and sensors for the magnetic field for the measurement of transient currents in the nanoseconds range. *Archiv fur Elektrotechnik, 68*(part I), 63–74, (part II), 69–74.

Bengtsson, T., Kols, H., & Jonsson, B. (1997). Transformer PD diagnosis using acoustic emission technique. In *10th ISH Conference Proceeding* (Vol. 4, pp. 115–119). Montreal, Canada.

Bergman, A., et al. (2001). Demonstration of traceability in high-voltage tests by means of a record of performance. *Electra, 199*, 35–43.

Berlijn, S. (2000). Influence of lightning impulses to insulating systems. *Dissertation*, Graz Technical University.

Berlijn, S., et al. (2007). *Manual evaluation of lightning impulses according to the new IEC 60060-1*. 15th ISH, Ljubljana.

Bernard, G. (1989). Application of Weibull distribution to the study of power cable insulation. *Electra, 127*, 75–83.

Bernasconi, F., Zaengl, W., & Vonwiller, K. (1979). *A new HV-series resonant circuit for dielectric tests*. 3rd ISH Milan, Report 43.02.

Beyer, J. (2002). *Space charge and partial discharge phenomena in HVDC devices*. Ph.D. Thesis, TU Delft, The Netherlands.

Beyer, M. (1978). Possibilities and limits of PD measurement and localization—basics and measuring systems. *ETZ-A, 99*(2), 96–99, (3), 128–132 (in German).

Beyer, M., & Borsi, H. (1977). PD measurement on HV cables—reasons for failures and possibilities for improvement. *Elektrizitatswirtschaft, 76*(26), 931–936. (in German).

Beyer, M., Boeck, W., Moeller, K., & Zaengl, W. (1986). *High-voltage technology: Theoretical and practical basics.* Berlin: Springer. (in German: Hochspannungstechnik).

Beyer, M., Borsi, H., & Hartje, M. (1987). Some aspects about possibilities and limitations of acoustic PD measurements in insulating fluids. In *5th ISH* (pp. 1–4). Braunschweig, Germany.

Binder, L. (1914). About switching processes and electrical travelling waves. *ETZ, 35,* 177–203. (in German).

Black, I. A. (1975). *A pulse discrimination system for discharge detection in electrically noisy environments.* 2nd ISH Zurich, paper 3.2-02.

Blake, W. (1870). On a method of producing, by electric spark, figures similar to those of Lichtenberg. *American Journal of Sciences and Arts, II-49,* 289.

Boggs, S. A., Ford, G. L., & Madge, R. C. (1981). Coupling devices for the detection of partial discharges in gas-insulated switchgear. *IEEE Transactions on Power Apparatus and Systems, 10,* 3969–3973.

Boggs, S. A., & Stone, G. C. (1982a). Fundamental limitations in the measurement of corona and partial discharge. *IEEE Transactions on Electrical Insulation, 17*(2), 143–145.

Boggs, S. A., Pecena, D. D., Rizzett, S., & Stone, G. C. (1987). Limits to partial discharge detection-effect of sample and defect geometry. In L. G. Cristophorou (Ed.), *Gaseous dielectrics.* Oxford: Pergamon Press.

Boggs, S. A., & Stone, G. C. (1982b). Fundamental limitations in the measurement of corona and partial discharges. *IEEE Transactions on Electrical Insulation, 17*(2), 143–150.

Bolza, A., et al. (2002). *Prequalification test experience on EHV XLPE cable system.* CIGRE Session Paris, Report 21-104.

Boning, P. (1938). Remarkable relations of anomalous currents, loss factor, apparent capacitance and the return voltage of insulating materials. *Zeitschrift fur technische Physik, 109,* 241–247. (in German).

Boone, W., Damstra, G. C., Jansen, W. J., & de Ligt, G. (1987). *VLF HV generators for testing cables after laying.* 5th ISH Braunschweig, Paper 62-04.

Burger, W. (1976). Beitrag zum Entladungsverhalten groBflachiger, schwach gekrummter Elektroden in Luft bei groBen Schlagweiten [Contribution to the discharge mechanism of large, slowly bended electrodes in air of lang gap distances]. *Dissertation,* Dresden Technical University.

Burawoy. (1936). The delay of sparks at very short impulse voltages. *Archiv fur Elektrotechnik, 16,* 186–219 (in German).

Burstyn, W. (1928). Losses in layer insulation. *ETZ, 49,* 1289–1291. (in German).

Carrara, G., & Dellera, L. (1972). Accuracy of an extended up-and-down method in statistical testing of insulation. *Electra, 23,* 159–175.

Carrara, G., & Hauschild, W. (1990). Statistical evaluation of dielectric test results. *Electra, 133,* 109–131.

Carrara, G., & Zafanella, L. (1968). *UHV Laboratories: Switching impulse clearance tests.* IEEE Power Summer Meeting, project no. 68 CP 692-PWR.

Cavallini, A., Contin, A., Montanari, G. C., Psini, G., & Puletti, F. (2002). Digital detection and fuzzy classification of partial discharge signals. *IEEE Transaction on Dielectrics and Electrical Insulation, 5*(3), 335–348.

Cavellini, A., Contin, A., Montanari, G. C., & Puletti, F. (2003). Advanced PD interference in on —field measurements. Part I: Noise rejection. *IEEE Transaction on Dielectrics and Electrical Insulation, 10*(2), 23–30.

Cavellini, A., Contin, A., Montanari, G. C., & Puletti, F. (2003). Advanced PD interference in on-field measurements. Part II: Identification of defects in solid insulation systems. *IEEE Transaction on Dielectrics and Electrical Insulation, 10*(3), 528–538.

Cavallini, A., & Montanari, G. C. (2006). Effect of supply voltage frequency on testing of insulation systems. *IEEE Transaction on Dielectrics and Electrical Insulation, 13,* 111–121.

CENELEC Study Group. (2010). *Technical guidelines for HVDC grids*. Minutes of meeting, November 11, 2010.

Charlton, E. E., et al. (1939). *Journal of Applied Physics, 10*, 374 (cited after Kuffel, Zaengl, Kuffel 2006).

Chen, S., & Czaszejko, T. (2011). Partial discharge test circuit as a spark gap transmitter. *IEEE Electrical Insulation Magazine, 27*(31), 36–43.

Choo, W., Chen, G., & Swingler, S. G. (2011). Electric field in polymeric cable due to space charge accumulation under DC and temperature gradient. *IEEE Transaction on Dielectrics and Electrical Insulation, 18*(2), 596–606.

Christen, T. (2014). Electrical insulation for modern HVDC systems—a challenge for research and development.*ETG—Mitglieder information,* July 2014, pp. 20–23 (in German).

Chubb, L. W., & Fortescue, C. (1913). Calibration of the sphere gap voltmeter. *Transmission AIEE, 32,* 739–748.

Cigre JWG 33/23.12. (1998). Insulation co-ordination of GIS: Return of experience, on-site tests and diagnostic techniques. *Electra, 176,* 67–97.

Cigre JWG 23/21/33. (2003). *Gas-insulated transmission lines (GIL).* Technical Brochure No. 218.

Cigre JWG B3/B1. (2008). *Application of long, high-capacity gas-insulated lines in structures.* Technical Brochure No. 351.

Cigre TF 33.03.04. (2000). Proposed requirements for HV withstand tests on-site. *Electra, 195,* 13–21.

Cigre TF 33.04.01. (2000). *Polluted insulators: A review of current knowledge.* CIGRE Technical Brochure No. 158.

Cigre TF D1.02.08. (2005). *Instrumentation and measurement for in-service monitoring of high-voltage insulation.* Technical Brochure No. 286.

Cigre WG D1.29. (2017). *Partial discharges in transformers.* Technical Brochure No. 676.

Cigre TF D1.33.05. (2012). *HV on-site testing with PD measurement.* Technical Brochure No. 502.

Cigre JWG D.1/B.3.57. (2017). *Recommendations for dielectric testing of gas-insulated HVDC systems up to 550 kV.* Draft Technical Brochure.

Cigre WG 21.03. (1969). Recognition of discharges. *Electra, 11,* 61–98.

Cigre WG 33.03. (1998). Measurement of very fast front transients. *Electra, 181,* 71–91.

Cigre WG D1.33. (2010). *Guidelines for unconventional partial discharge measurements.* Technical Brochure No. 444.

Cigre WG B1.23. (2012). *Recommendations for testing of long AC submarine cables with extruded insulation for system voltage above 30 to 500 kV.* Technical Brochure 490.

C.I.S.P.R. (1977). Specification for radio interference measuring apparatus and measurement methods. In IEC (Ed.), *International. Special committee on radio interference.* Document no. 16.

Cockcroft, J. D., & Walton, E. T. S. (1932). Experiments with high velocity ions. *Proceedings Royal Society, London, Series A, 136,* 619–630.

Coenen, S., Tenbohlen, S., Markalous, S., & Strehl, T. (2007). *Performance check and sensitivity verification for UHF PD measurements on power transformers* (pp. 157–264). CIGRE SC A1 & D1 Joint Colloquium, Gyeongju, Korea.

Cousineau, D. (2009). Fitting the three-parameter Weibull distribution: Review and evaluation of existing and new methods. *IEEE Transaction on Dielectrics and Electrical Insulation, 10*(1), 281–288.

Creed, F., Kamamura, T., & Newi, G. (1967). Step response of measuring systems for high impulse voltages. *IEEE Transactions on Power Apparatus and Systems, 86*(11), 1408–1420.

Crichton, G. C., Karlsson, P. W., & Pedersen, A. (1988). A theoretical derivation of the transients related to partial discharges in ellipsoidal voids. In *Conference Record IEEE, International Symposium on Electrical Insulation (ISEI)* (p. 238). IEEE Publication 88CH2594-0-DEI.

Crichton, G. C., Karlsson, P. W., & Pedersen, A. (1989). Partial discharges in ellipsoidal and spherical voids. *IEEE Transaction on Dielectrics and Electrical Insulation, 24,* 335–342.

Csepes, G., Hamos, I., Schmidt, J., & Bognar, A. (1994). *A DC expert system (RVM) for checking the refurbishment efficiency of high voltage oil-paper insulating system using polarization spectrum analysis in range of long-time constants.* CIGRE Session Paris, Paper 12-206.

Cui, D., et al. (2009). *Discussion on insulation levels and dielectric test technology requirements of AC UHV transmission and transformation equipment in China.* 16th ISH Johannesburg, paper A-23.

Dakin, T. W., & Malinaric, P. J. (1960). A capacitive bridge method for measuring integrated corona-charge transfer and power loss per cycle. *Power Apparatus and Systems, Part III. Transactions of the American Institute of Electrical Engineers, 79*(3), 648–653.

Davis, R., Bowdler, G. W., & Standring, W. G. (1930). The measurement of high voltages with special reference to the measurement of peak voltages. *Journal IEE, London, 68,* 1222.

Dawson, G. A., & Winn, W. P. (1965). A model of streamer propagation. *Zeitschrift fur Physik, 183,* 159–171.

Dennhardt, A. (1935). Reason and measurement of the high-frequency noise caused by insulators. *Elektrizitatswirtschaft, 34,* 15. (in German).

Densley, R. J. (1979). Partial discharges under direct-voltage conditions. *Engineering dielectrics,* 1. In R. Bartnikas & E. J. McMahon (Eds.), *Corona measurement and interpretation* (Vol. STP669). Philadelphia: ASTM.

Devins, J. C. (1984). The physics of partial discharges in solid dielectrics. *IEEE Transaction on Electrical Insulation, 19,* 475–495.

Diaz, R. R., & Segovia, A. A. (2016). A physical approach of the test voltage function for evaluation of the impulse parameters in lightning impulse voltages with superimposed oscillations and overshoot. *IEEE Transactions on Dielectrics and Electrical Insulation, 23*(5), 2738–2746.

Dietrich, M. (1982). Dimensioning of large electrodes for HV test equipment of the UHV range. *Dissertation,* Dresden Technical University, 1982 (in German).

Dong, B., Jiang, X., Hu, J., Shu, L., & Sun, C. (2012). Effects of artificial polluting methods on AC flashover of composite insulators. *IEEE Transactions on Dielectrics and Electrical Insulation, 19*(2), 714–722.

Dorison, E., & Aucourt, C. (1984). *After laying tests of HV and EHV cables.* Jicable Versailles. Paper BV-5.

Eager, G. S., & Bader, G. (1967). Discharge detection in extruded polyethylene insulated power cables. *IEEE Transactions on Power Apparatus and Systems, 86*(1), 10–34.

Eager, G. S., Bader, G., & Silver, D. A. (1969). Corona detection experience in commercial production of power cables with extruded insulation. *IEEE Transactions on Power Apparatus and Systems, 86*(4), 342–346.

Edwards, F. S., & Smee, J. F. (1938). The calibration of the sphere spark gap for voltage measurement up to one million volts (effective) at 50 cycles. *Journal of the Institution of Electrical Engineers, 82*(1938), 655–669.

Eleftherion, P. M. (1995). Partial Discharge. Part XXI: Acoustic emission-based PD source location in transformers. *IEEE Electrical Insulation Magazine, 11*(6), 22–26.

Elserougi, A., et al. (2015). A HV pulse-generator based on DC-to-DC converters and capacitor-diode voltage multipliers for water treatment application. *IEEE Transactions on Dielectrics and Electrical Insulation, 22*(67), 3290–3298.

Elmore, W. C. (1948). The transient response of damped linear networks with particular regard to wideband amplifiers. *Journal of Applied Physics, 19*(1), 55–63.

Elsner, R. (1939). Measurement of steep HV impulses by voltage dividers. *Archiv fur Elektrotechnik, 33*(1), 23–40. (in German).

Elstner, G., et al. (1983). *Powerful DC and mixed voltage testing equipment up to 2.25 MV for outdoor installation.* 4th ISH Athens (1983) paper 51.04.

Emanuel, H., Kalkner, W., Plath, K. D., & Plath, R. (2002). Synchronous three-phase PD measurement on power transformers on site and in the laboratory. In *ETG Conference on Diagnostik elektrischer Betriebsmittel* (in German).

Engelmann, E. (1981). *Contribution to the discharges on large electrodes with defects in air.* Thesis Dresden Technical University 1981 (in German).

EN, C. S. (2000) *0.50191 Erection and operation of electrical test equipment* (German version DIN EN 50191-VDE0104).

EN HD 620 S. (1996/A3: 2007). *Power cables—Part 620: Distribution cables with extruded insulation for voltages rated 3.6/6 kV to 20.8/36 kV.*

Fan, J., & Li, P. (2008). *Effect of sandstorm on external insulation.* Presentation at the TC 42 Meeting, Sao Paulo.

Farneti, F., Ombello, F., Bertani, E., & Mosca, W. (1990). *Generation of oscillating waves for after-laying test of HV extruded cable links.* CIGRE Session Paris, France, paper 21-10.

Farzaneh, M. (2014). Insulator flashover under icing conditions. *IEEE Transactions on Dielectrics and Electrical Insulation, 21*(4), 1997–2009.

Farzaneh, M., & Chisholm, W. A. (2014). 50 Years in icing performance of outdoor insulators. *IEEE Electrical Insulation Magazine, 30*(1), 14–24.

Felk, M., et al. (2017). Protection and measuring elements in the test setups of the superimposed test voltage. In *20th ISH Buenos Aires/Argentina.*

Feser, K. (1973). Extension of the trigger range of multi-stage impulse generators for the generation of SI voltages. *ETZ, 94(3),* 171–174 (in German).

Feser, K. (1974). Problems of the generation of SI voltages in the test field. *Bulletin SEV, 65(6),* 496–506 (in German).

Feser, K. (1975). Dimensioning of electrodes in the UHV range illustrated with the example of toroid electrodes for voltage dividers. *HAEFELY Publication E-130 !975 and ETZ-A, 96,* 206–210 (in German).

Feser, K. (1981). HV tests of metal-enclosed, gas-insulated substations. *Bulletin SEV, 72(1),* 19–26 (in German).

Feser, K. (1997). *Thoughts to the test and measuring techniques at steep impulse voltages.* HIGHVOLT Kolloquium, Dresden, Paper 1.4 (pp. 37–40) (in German).

Feser, K., & Pfaff, W. R. (1984). A potential free spherical sensor for the measurement of transient electric fields. *IEEE Transactions on Power Apparatus and Systems, PAS-103,* 2904–2911.

Feser, K., & Hughes, R. C. (1988). Measurement of direct voltage by rod-rod gap. *Electra, 117,* 23–34.

Finkelmann, J. (1936). *Electrical breakdown of different gases under high pressure.* Ph.D. Thesis Technische Hochschule Hannover (in German).

Frank, H., Hauschild, W., et al. (1983). *HVDC testing generator for short-time polarity reversal on load.* 4th ISH Athens, paper 51.05.

Frank, H., Schrader, W., & Spiegelberg, J. (1991). *3 MV AC voltage testing equipment with switching voltage extension—Technical concept, first operation, results.* 7th ISH Dresden, paper 52.04.

Fromm, U. (1995a). Interpretation of partial discharges at DC voltage. *IEEE Transactions on Dielectrics and Electrical Insulation, 2*(5), 761–770.

Fromm, U. (1995). *Partial discharge and breakdown testing at high DC Voltage.* Ph.D. Thesis TU Delft, The Netherlands.

Frommhold, L. (1956). The potential of the charge of electron avalanches within parallel plates and side effects. *Zeitschrift for Physik, 145*(3), 324–340. (in German).

Fruth, B., Liptak, L., Ullrich, L., Dunz, T., & Niemeyer, L. (1989). Ageing of rotating machine insulation—Mechanisms, measurement technique. In *Proceedings of the 3rd International Conference on Conduction and Breakdown in Solid Dielectrics* (pp. 597–601).

Fruth, B., & Gross, D. (1994). Phase resolving partial discharge pattern acquisition and spectrum analysis. In *Proceedings of the ICPDAM* (pp. 578–581). Brisbane NSW, Australia, 94CH3311-8.

Fuhr, J., Haessig, M., Boss, P., Tschudi, D., & King, R. A. (1993). Detection and location of internal defects in the insulation of power transformers. *IEEE Transactions on Electrical Insulation, 28*(6), 1057–1067.

Fujimoto, N., Boggs, S. A., & Madge, R. C. (1981). *Electrical transients in gas-insulated switchgear.* Transactions on the March 1981 Meeting of the Canadian Electric Association.

Fujimoto, N., Boggs, S. A., & Madge, R. C. (1981b). Coupling devices for the detection of partial discharges in gas-insulated switchgear. *IEEE Transactions on Power Apparatus and Systems, 100*(8), 3369–3973.

Gabor, D. (1926). Oscillographic records of travelling waves. *Archiv fur Elektrotechnik, 16,* 296–298.

Gänger, B. (1953). *Electrical breakdown of gases (book in German).* Berlin, Göttingen, Heidelberg: Springer.

Garbagnati, E., et al. (1991). The influence of atmospheric conditions on the dielectric strengths of phase-to-phase insulation when subjected to switching impulse. CESI Publication.

Garnacho, F., et al. (2014). K-factor test voltage function for oscillating lightning impulses in non-homogenous air gaps. *IEEE Transactions on Power Delivery, 29*(5), 2254–2260.

Garnacho, F. (2010, February). *K-factor results for air dielectric medium. Presentation to CIGRE Working Group D1.36.*

Garnacho, F., et al. (1997). Evaluation procedure for lightning impulse parameters in case of waveforms with oscillations and/or an overshoot. *IEEE Transactions on Power Delivery, 12* (2), 640–649.

Garnacho, F., et al. (2002). Evaluation of lightning impulse voltages based on experimental results —proposal for the revision of IEC 60060-1 and IEC 61083-2. *Electra, 204,* 31–37.

Gemant, A., & v. Philippoff, W. (1932). Spark gap with pre-capacitor. *Zeitschrift fur Technische Physik, 13*(9), 425–430 (in German).

Ghorbani, H., et al. (2017). Electrical characterization of extruded DC Cable insulation—The challenge of scaling. *IEEE Transactions on Dielectrics and Electrical Insulation, 24*(24), 1465–1470.

Gockenbach, E. (2010). Voltage shapes in HVDC systems—static and dynamic loads. ETG Fachtagung: Isoliersysteme bei Gleich- und Mischfeldbeanspruchung, Cologne paper 1.2 (in German).

Gockenbach, E., & Hauschild, W. (2000). The selection of the frequency range for HV on-site testing of extruded cable systems. *IEEE Insulation Magazine, 16*(6), 11–16.

Gockenbach, E., et al. (2007). *Challenges on the measuring and testing techniques for UHV AC and DC equipment.* IEC/CIGRE UHV Symposium Beijing, Paper 4-2.

Goosens, R. F., & Provoost, P. G. (1946). The registration of high impulse voltages by a cathode oscilloscope. *Bulletin SEV, 37,* 175–184. (in German).

Graybill, H. Q., Cronin, J. C., & Field, E. J. (1974). Testing of gas insulated substations and transmission systems. *IEEE Transactions of Power Apparatus and Systems, PAS-93(1),* 404–413.

Grey Morgan, C. (1965). Fundamentals of electric discharges in Gases, Vol. II: Physical electronics. In A. H. Beck (Ed.), *Handbook of vacuum physics.* Oxford: Pergamon Press.

Greinacher, H. (1920). Generation of direct voltage of multiple amount of an AC voltage. *Bulletin SEV, 66* (in German).

Gross, D. (2011). Locating partial discharge using acoustic sensors. In *HIGHVOLT KOLLO-QUIUM '11 Conference Proceedings* (pp. 99–106).

Gronefeld, P. (1983). A very low frequency 200 kV generator as precondition for testing insulating materials with 0.1 Hz alternating voltage. In *4th International Symposium on High Voltage Engineering (ISH),* Athens, Greece, paper 21.02.

Grunewald, F. (1921). Characteristics of open-air insulators under high-frequency voltages. *ETZ, 42,* 1377. (in German).

Grunewald, P., & Weidner, J. (1994). *Possibilities and experience with off- and on-line diagnosis of turbine generator stator winding insulations.* CIGRE-Session, Paris, France, paper 11-206.

Gubanski, S. M., Boss, P., & Csepes, G., et al. (2002). *Dielectric response methods for diagnostics of power transformers. Electra, 202,* 25-3 (Report of CIGRE TF 15.01.09).

Gulski, E. (1991). *Computer-aided recognition of partial discharges using statistical tools*. PhD Thesis, Delft University Press.

Gulski, E., Smit, J. J., Seitz, P. N., & Tuner, M. (1999). *On-site diagnostics of power cables using oscillating wave test system*. 11th ISH London, UK, paper 5.112.

Gulski, E., Smit, J. J., van Breen, H., de Vries, F., Seitz P. P., & Petzold, F. (2000). Advanced PD diagnostics of medium-voltage power cables using oscillating wave test system. *IEEE Electrical Insulation Magazine, 16*(2).

Gulski, E., et al. (2007). Dedicated on-site condition monitoring of HV power cables up to 150 kV. In *8th International Power Engineering Conference*, Singapore.

Gutman, I., & Derfalk, A. (2010). Pollution tests for polymeric insulators made of hydrophobicity material. *IEEE Transaction on Dielectrics and Electrical Insulation, 17*, 384–393.

Gutman, I., et al. (2014). Development of time- and cost-effective pollution test methods for different station insulation options. *IEEE Transactions on Dielectrics and Electrical Insulation, 21*(6), 2525–2530.

Hagenguth, J. H. (1937). *Short time spark-over of gaps. Transactions of the American Institute of Electrical Engineers, 56*, 67–76.

Hagenguth, J. H., et al. (1952). Sixty cycle and impulse sparkover of large gap spacings. *Transactions AIEE Part III, 71*, 455–460.

Hague, B. (1959). *Alternating-current bridge methods* (5th ed.). London: Pitman & Sons.

Hallstrom, J. (2002). *A calculable impulse voltage calibrator*. Acta polytechnica scandinavia, Electrical Engineering, Series No. 109.

Hallstrom, J., Li, Y. & Lucas, W. (2003). *High accuracy comparison measurement of impulse parameters at low voltage levels*. 13th ISH Delft, paper 432.

Harrold, R. T. (1975). Ultrasonic spectrum signatures of under-oil corona sources. *IEEE Transactions on Electrical Insulation, EI-10(4)*, 109–112.

Harrold, R. T. (1976). The relationship between ultrasonic and electrical measurement of under oil corona sources. *IEEE Transactions on Electrical Insulation, EI-11(1)*, 8–11.

Harrold, R. T. (1996). Acoustic theory applied to the physics of electrical breakdown in dielectrics. *IEEE Transactions on Electrical Insulation, EI-21(5)*, 781–792.

Hauschild, W. (1970). *About the breakdown of insulating oil in a non-uniform field at SI voltages*. Ph.D. Thesis, Dresden Technical University (in German).

Hauschild, W. (1995). *Engineering the electrodes of HV test systems on the basis of the physics of discharges in air*. 9th ISH Graz, invited paper.

Hauschild, W. (2007). A common view on high-voltage testing and insulation diagnostics. In *Proceedings of HIGHVOLT Kolloquium '07,0.7-13* (also: CIGRE D1.33 Colloquium Gyeongju, 2007).

Hauschild, W. (2013). Critical review of voltages applied for quality acceptance and diagnostic field tests on HV and EHV cable systems. *IEEE Electrical Insulation Magazine, 29*(2), 16–25.

Hauschild, W.,& Fahd, I. (1980). Installation of a HV laboratory at the faculty of mechanical and electrical engineering of Damascus University. *Elektrie, 32*(3), 124–127(in German). *Monthly Technical Review, 24*(1), 4–11 (in English).

Hauschild, W., Kuttner, H., & Thummler, K. (1981). Systems for the wideband PD measurement in HV insulations. *Elektrie, 35*(7), 353–357. (in German).

Hauschild, W., & Mosch, W. (1992). *Statistical techniques for high-voltage engineering*, (Statistik fur Elektrotechniker, Berlin: Verlag Technik Berlin, 1984). IEE Power Series 13. London: Peter Peregrinus Ltd.

Hauschild, W., & Steiner, T. (2009). *The design of HVLI tests fort he improvement of the k-factor function*. CIGRE D1.33 Meeting Budapest.

Hauschild, W., Rausendorf, S., & Schufft, W. (1987). *Calculation of field strength and streamer inception voltage for multi-segment electrodes of UHV test equipment*. 5th ISH Braun-schweig, paper 33.10.

Hauschild, W., Wolf, J., & Spiegelberg, J. (1987). *Calculation of the pollution test characteristic of a powerful DC voltage generator*. 5th ISH Braunschweig, paper 62.03.

Hauschild, W., Spiegelberg, J., & Lemke, E. (1997). *Frequency tuned resonant test systems for HV on site testing of SF$_6$ insulated apparatus*. 10th ISH Montreal, (Vol. 4, pp. 457-460).

Hauschild, W., Schufft, W., & Spiegelberg, J. (1997). *Alternating voltage on-site testing and diagnostics of XLPE cables: The parameter selection of frequency-tuned resonant test systems*. 10th ISH Montreal, (Vol. 4, pp. 75-78).

Hauschild, W., Schierig, S., & Coors, P. (2005). *Resonant test systems for HV testing of super-long cables and gas-insulated transmission lines*. 14th ISH Beijing paper J-02.

Hauschild, W., Thiede, A., Leibfried, T., & Martin, F. (2006). Static frequency converters for HV tests on power transformers. In *High Voltage Symposium Stuttgart* (in German).

Hauschild, W., et al. (1982). The influence of stochastic processes on the breakdown of slightly non-uniform fields in SF$_6$. *Z. elektr. Informations- und Energietechnik, 12*(4&5), 289–318, 385–403 (in German).

Hauschild, W., et al. (1991). *Breakdown voltage characteristic of long rod-to-plane air gaps at bipolar oscillating switching voltages*. 7th ISH Dresden paper 42.25.

Hauschild, W., et al. (1993). *Computer-aided performance tests and checks for LI voltage measuring systems*. 8th ISH Yokohama, paper 53.01.

Hauschild, W., et al. (2002). *The technique of AC on-site testing of HV cables by frequency- tuned resonant test systems*. CIGRE Session, Report 33-304.

He, L., & Gorur, R. S. (2016). Source strength impact analysis on polymer insulator flashover under contaminated conditions and a comparison with porcelain. *IEEE Transactions on Dielectrics and Electrical Insulation, 23*(4), 2189–2195.

Henriksen, M., Stone, G. C., & Kurtz, M. (1986). Propagation of partial discharge and noise pulses in turbine generators. *IEEE Transactions on Energy Conversation. EC-1*(3).

Herb, R. G., Parkinson, D. B., & Kerst, D. W. (1937). The development and performance of an electrostatic generator operating under high air pressure. *Physical Review, 51*(75).

Hering, Maria, et al. (2017). Field transition in gas-insulated HVDC systems. *IEEE Transactions on Dielectrics and Electrical Insulation, 24*(3), 1608–1616.

Hering Maria, Riechert, U. & Tenbohlen S. (2017). Gas-insulated systems for the HVDC transmission. *ETG—Mitglieder information, August 2017*, pp. 18–21.

Hinow, M. (2011). *Optimized test field for power transformer testing*. HIGHVOLT Kolloquium (Dresden) (pp. 123–126).

Hinow, M., Hauschild, W., & Gockenbach, E. (2010). Lightning impulse and overshoot evaluation proposed in drafts of IEC 60060-1 and future UHV testing. *IEEE Transaction on Dielectrics and Electrical Insulation, 17*(5), 1628–1634.

Hinow, M., & Steiner, T. (2009). *Influence of the new k-factor method of IEC 60060-1 on the evaluation of LI parameters in relation to UHV testing*. 16th ISH Cape Town, paper G-14.

Hoof, M., & Patsch, R. (1994). *Analyzing partial discharge pulse sequences: A new approach to investigate degradation phenomena*. In *IEEE Symposium on Electrical Insulation (ISEI)*, (pp. 327–331). Pittsburgh, USA.

van Hoove, C., & Lippert, A. (1973). Measurements related to the electric strength of polyethylene and HV cables. *Elektrizitatswirtschaft, 71*, 630–635. (in German).

van Hoove, C. (1993). Partial discharges in cable accessories. In D. Konig & Y. N. Rao (Eds.), *Partial discharges in electrical power apparatus* (pp. 173–181). Berlin: VDE-Verlag.

Hopke, F., & Schmidt, M. (2011). Factory test field for power transformers based on a static frequency converter. *HIGHVOLT Kolloquium Dresden* (pp. 127–132).

House, H., Waterton, F. W., & Chew, J. (1979). *1000 kV standard voltmeter*. 3rd ISH Milan, Italy, paper 43.05.

Houtepen, R., et al. (2011). Estimation of dielectric loss using damped AC voltage. *IEEE Insulation Magazine, 27*(3), 14–19.

Howels, E., & Norton, E. T. (1978). Detection of partial discharges in transformers using acoustic emission techniques. *IEEE Transactions on Power apparatus and Systems, PAS-97(5)*, 1538–1546.

Hughes, R. C., et al. (1994). Traceability of measurement in HV tests. *Electra, 155*, 91–101.

Hylten-Cavallius, N. (1957). Impulse tests and measuring errors. *ASEA-Journal, 5,* 75–84.

Hylten-Cavallius, N. (1988). *High-voltage laboratory planning.* Switzerland: Haefely.

ICEA T-24-380. (2006). *Standard for partial discharge test procedure.*

IEC 60034-1. (2010). *Rotating electrical machines—Part 1: Rating and performance.*

IEC 60034-27. (2006). *Rotating machines—Part 27: Off-line partial discharge measurements on the stator winding insulation of rotating electrical machines.*

IEC 60038. (2009) *IEC standard voltages.*

IEC 60052. (1960). *Recommendations for voltage measurement by means of sphere gaps.*

IEC 60052 Ed. 3. (2002). *Voltage measurement by means of standard air gaps.*

IEC 60060-1. (1989). *High-voltage test techniques, Part 1: General definitions and test requirements.*

IEC 60060-1. (2010). *High-voltage test techniques, Part 1: General definitions and test requirements.*

IEC 60060-2. (2010). *High-voltage test techniques, Part 2: Measuring systems.*

IEC 60060-3. (2006). *High-voltage test techniques, Part 3: Definitions and requirements for on-site testing.*

IEC 60071-1. (2006). *Insulation co-ordination—Part 1: Definitions, principles and rules.*

IEC 60071-2. (1996). *Insulation co-ordination—Part 2: Application guide.*

IEC 60071-2. (2010). *Amendment to Part 2.*

IEC 60071-5. (2002). *Insulation co-ordination—Part 5: Procedures for high-voltage direct current (HVDC) converter stations.*

IEC 60076-3. (2000). *Power transformers - Part 3: Insulation levels, dielectric tests and external clearances, Annex A: Application guide for partial discharge measurements during AC voltage withstand test on transformers according to 12.2, 12.3, and 12.4.*

IEC 60076-3. (2013). *Power transformers—Part 3: Insulation levels, dielectric tests and external clearances in air.*

IEC 60099-4. (2009). *Surge arresters—Part 4: Metal-oxide surge arrestors without gaps for AC systems.*

IEC 60143-1. (2004). *Series capacitors for power systems—Part 1: General.*

IEC 60250. (1969). *Recommended methods for the determination of the permittivity and dielectric dissipation factor of electrical insulating materials.*

IEC 60270. (2000). *HV test techniques—Partial discharge measurement.*

IEC 60270. (2015). *Amendment A1 (CSV, Consolidated Version).*

IEC 60502. (1997). *Power cables with extruded insulation for rated voltages from 1 to 30 kV.*

IEC 60505. (2011). *Evaluation and qualification of electrical insulation systems.*

IEC 60507. (1991). *Artificial pollution tests on high-voltage insulators to be used in AC systems.*

IEC 60840. (2011). *Power cables with extruded insulation for rated voltages from 30 to 150 kV.*

IEC 60885-3. (2003). *Electrical test methods for electric cables—Part 3: Test methods for partial discharge measurements on lengths of extruded power cable.*

IEC 61083-1. (2001). *Instruments and software used for measurement in high-voltage impulse tests—Part 1: Requirements for instruments.*

IEC 61083-2. (2013). *Instruments and software used for measurement in high-voltage impulse tests—Part 2: Requirements for software for tests with impulse voltages and currents.*

IEC 61083-3. (2012). *Draft—Instruments and software used for measurements in high-voltage and high-current tests—Part 3: Requirements for instruments for tests with alternating and direct voltages and currents.*

IEC 61180. (2014). *HV test techniques for low-voltage equipment-test and procedure requirements, test equipment.*

IEC 61245. (2015). *Artificial pollution tests on high-voltage ceramic and glass insulators to be used on D.C. systems.*

IEC 61245. (2013). *Artificial pollution tests on HV insulators to be used on DC systems.* Draft 36/329/CD.

IEC 61259. (1994). *Gas-insulated metal-enclosed switchgear for rated voltages 72.5 kV and above—Requirements for switching of bus charging currents by disconnectors.*

IEC 61640. (1998). *Rigid high-voltage gas-insulated transmission lines for rated voltage of 75 kV and above.*

IEC 61934. (2006). *Electrical measurement of partial discharges during short rise time repetitive voltage impulses.*

IEC 62061. (2005). *Safety of machinery—functional safety-related electrical, electronic and programmable electronic control systems.*

IEC 62067. (2006). *Power cables with extruded insulation and their accessories for rated voltages above 150 kV up to 500 kV—test methods and requirements.*

IEC 62271-203. (2010). *HV switchgear and control gear,-Part 203: Gas-insulated, metal-enclosed switchgear for voltages above 52 kV.*

IEC 62475. (2010). *High-current test techniques: Definitions and requirements for test currents and measuring systems.*

IEC 62478. (2015). *High voltage test techniques—Measurement of partial discharges by electromagnetic and acoustic methods.*

IEC 62478. (2013). *High voltage test techniques—Measurement of partial discharges by electromagnetic and acoustic methods (CDV).*

IEC TC 115: HVDC. (2010). *Transmission for DC voltages above 100 kV.* Strategic Business Plan, IEC 115/35/INF.

IEEE Std.4. (1995). *IEEE standard techniques for high-voltage testing.*

IEEE P5^{TM}/D006. (2013). *Draft trial-use standard for high-voltage testing techniques. 200 9* (new edition harmonized with IEC 60060-1:2010 is expected for 2014).

IEEE 62-PC57.152. (2012). *Guide for diagnostic field testing of fluid filled power transformers, regulators and reactors.*

IEEE Std.400™. (2012). *IEEE Guide for field testing and evaluation of the insulation of shielded power cable systems rated 5 kV and above.*

IEEE Std.400.1. (2007). *Guide for field testing of laminated dielectric, shielded power cable systems rated 5 kV and above with high direct current voltage.*

IEEE P400.2 ^{TM}. (2012). *Draft Guide VLF for field testing of shielded power cable systems using very low frequency (VLF).*

IEEE Standard 400.4. (Draft 2012). *Guide for field testing of shielded power cable systems rated 5 kV and above with damped alternating current (DAC) voltage.*

IEEE Std. C37.122. (2010). *Standard for HV gas-insulated substations rated 52 kV and above.*

IEEE Std C57.113. (2010). *IEEE Recommended practice for partial discharge measurement in liquid-filled power transformers and shunt reactors.*

IEEE Std. C57.127. (2007). *IEEE Guide for the detection and location of acoustic emissions from partial discharges in oil-immersed power transformers and reactors.*

IEEE 510. (1983). *Recommended practice for safety in high-voltage and high-power testing.*

IEEE Std.1313.1. (1996). *Standard for insulation coordination—Definitions, principles and requirements.*

IEEE 1434. (2000). *Trial-use guide to the measurement of partial discharges in rotating machinery.*

IEEE P1861^{TM}/D1. (2012). *Draft Guide for on-site acceptance tests of electric equipment and commissioning of 1000 kV AC and above system.*

ISO/IEC Guide 98-3. (2008). *Uncertainty of measurement—Part 3: Guide to the expression of uncertainty in measurement* (=GUM:1995).

Illias, H., Chen, G., & Lewin, P. L. (2011a). Modeling of partial discharge activity in spherical cavities within a dielectric material. *IEEE Electrical Insulation Magazine, 27*(1), 38–45.

Illias, H., Chen, G., & Lewin, P. L. (2011b). Partial discharge behavior within a spherical cavity in a solid dielectric material as a function of frequency and amplitude of the applied voltage. *IEEE Transaction on Dielectrics and Electrical Insulation, 18*(2), 432–443.

Jiang, X., Shu, L., et al. (2008). Positive switching impulse performance and voltage correction of rod-plane air gaps based on tests at high-altitude site. *IEEE Transaction on Power Delivery, 24* (1).

Jiang, X., et al. (2008). Switching impulse flashover performance of different types of insulators at high altitude sites of above 2800 m. *IEEE Transaction on Dielectrics and Electrical Insulation, 15*(5), 1340–1345.

Jiang, X., et al. (2009). Study on AC pollution flashover performance of composite insulators at high altitude sites of 2800–4500 m. *IEEE Transaction on Dielectrics and Electrical Insulation, 16*(1), 123–132.

Jiang, X., et al. (2010). Equivalence of influence of pollution simulating methods on DC flashover stress of ice-covered insulators. *IEEE Transactions on Power Delivery, 25*(4), 2113–2120.

Jiang, X., et al. (2011). DC flashover performance and effect of sheds configuration on polluted and ice-covered insulators at low pressure. *IEEE Transaction on Dielectrics and Electrical Insulation, 18*, 97–105.

Jouaire, J., & Sabot, A., et al. (1978). *HV measurements—present state and future development.* Revue General de l'Electricite, Special Number.

Judd, M. D., Farish, O., & Hampton, B. F. (1996). The excitation of UHF signals by partial discharges in GIS. *IEEE Transaction on Dielectrics and Electrical Insulation, 3*, 213–228.

Judd, M. D., Cleary, G. P., & Bennoch, G. J. (2002). Applying UHF partial discharge detection to power transformers. *IEE Power Engineering Review, 57–59.*

Kachler, A. J. (1975). *Contribution to the problem of impulse voltage measurement by means of sphere gaps.* 2nd ISH Zurich (pp. 217–221).

Kachler, A. J., Kroon, C., & Machado, T. (1998). *Pro and contras of on-site testing on power transformers and reactors.* Cigre Session Paris paper 12-201 (see also: The discussion to that paper by W. Hauschild: A necessary clarification to the activities of Cigre WG 33.03 related to HV on-site tests. Cigre 1998).

Kapcov, N. A. (1955). *Electrical phenomena in gases and vacuum.* Berlin: Deutscher Verlag der Wissenschaften. *(in German).*

Karlstrand, J., Henning, G., Schierig, S., & Coors, P. (2005). *Factory testing of long submarine cables using frequency-tuned resonant systems.* CIRED Turin.

Kaufhold, M., Kalkner, W., Obralic, R., & Plath, R. (2006). *Synchronous 3-phase partial discharge detection on rotating machines.* CIGRE Session Paris, paper D1-105.

Kaul, G., Plath, R., & Kalkner, W. (1993). Development of a computerized loss factor measurement system, including 0.1 Hz and 50/60 Hz. In *8th International Symposium on High Voltage Engineering*, Yokohama, Japan, paper 56.04.

Kawamura, T., Nagai, K., Seta, T., & Naito, K. (1984). *DC pollution performance of insulators.* CIGRE Session Paris, Report 33-10.

Keller, A. (1959). Constancy of capacitance of compressed gas capacitors. *ETZ-A, 80,* 757–761. (in German).

Kemp, I. J. (1987). Calibration difficulties associated with PD detectors in rotating machines. In *Proceedings of IEEE Electrical Insulation Conference,* Chicago, USA.

Kind, D. (1957). The formative area at impulse voltage stress of electrode arrangements in air. *Dissertation,* Technical University Munich, 1957 (in German).

Kind, D. (1961). Basics of measuring equipment for corona—insulation tests. *ETZ-A, 84,* 781–787. (in German).

Kind, D. (1974). Recommendations for the performance of HV tests on GIS and GIL. *ETZ-A, 95 (11),* 588–589 (in German).

Kind, D., & Salge, J. (1965). On the generation of switching impulse voltages using HV test transformers. *ETZ-A, 86(11),* 588–589 (in German).

Kind, D., & Shihab, S. (1969). Partial discharges in solid insulating material subjected to high direct voltages. *ETZ-A, 90,* 476–468 (in German).

Kind, D., & Feser, K. (1999). *High-voltage test technique* (2nd English Edn). Vieweg and SBA Publishers (First German Edn. 1972).

Kind, D., et al. (2016). Voltage- time characteristics of air gaps and insulation coordination—Survey of 100 Years Research. *International Conference on lightning protection. Esteril, Portugal.*

Kindersberger, J. (1997). *Why plastic compound insulators require test procedures different from ceramic insulators?* 2nd HIGHVOLT Kolloquium, Dresden 1997, Paper 1.5 (pp. 41–51) (in German).

King, R. W. P. (1983). The conical antenna as a sensor or probe. *IEEE Transactions on Electromagnetic Compatibility, 25,* 8–13.

Kleinwachter, H. (1970). The influence-E-meter used as an electrostatic amplifier of extremely high amplification and its application as a sensitive measuring instrument. *Archiv technisches Messen, 413,* R62–R64. (in German).

Kluge, A., et al. (2015). IGBT-based switching modules for Laser applications. *IEEE Transactions on Dielectrics and Electrical Insulation, 22*(4), 1954–1962.

Kohler, W. (1988). Voltage sources for pollution testing. *Dissertation,* University of Stuttgart, 1988 (in German).

Kohler, W., & Feser, K. (1987). *Test sources for DC pollution tests.* 5th ISH Braunschweig, paper 62.08.

Konig, D., & Rao, Y. N. (1991). *Partial discharges in electrical power apparatus.* Berlin: VDE-Verlag.

Koske, B. (1938). Tests of insulations of HV overhead lines under operation. *Elektrizitatswirtschaft, 36*(11), 291. (in German).

Kranz, H. G. (2000). Fundamentals in computer aided PD processing, PD pattern recognition and automated diagnosis in GIS. *IEEE Transactions on Dielectrics and Electrical Insulation, 7*(1), 12–20.

Kranz, H. G., & Krump, R. (1988). Computer aided partial discharge evaluation about the surface material of the PD source in gas-insulated substations. In *IEEE International Symposium on Electrical Insulation (ISEI),* Boston, USA.

Krefter, K. H. (1991). *Tests for condition assessment of medium cable systems.* Frankfurt a.M: VWEW-Verlag. (in German).

Kreuger, F. H. (1964), *Discharge detection in high voltage equipment.* London: Temple Press, London; New York: American Elsevier.

Kreuger, F. H. (1989), *Partial discharge detection in high voltage equipment.* London: Butterworth & Co.

Krug, W. (1929). *Investigation of the behaviour of impulse circuits using records by means of a cathode-ray oscilloscope.* ETZ 19, Ch. 4.2 (in German).

Krueger, M. (1989). *Field test of the insulation of cable systems of rated voltages 10 to 30 kV using VLF voltage 0.1 Hz.* PhD Thesis TU Graz (in German).

Krump, R., & Haumann, T. (2011). *Possibilities and limits of a modern HV test laboratory* (pp. 115–122). Dresden: HIGHVOLT Kolloquium.

Kreuger, F. H. (1989b). *Partial discharge detection in high-voltage equipment.* London: Butterworths.

Kubler, B., & Hauschild, W. (2004). New ways of HV testing of electric power apparatus including power transformers. In *Proceedings of Transform.*

Küchler, A. (2017). *HV Engineering—Basics—Technology—Application* (4th Edn in German). Berlin: Springer (1st Edn.1997).

Kuechler, A. (2009). *Hochspannungstechnik, Grundlagen-Technologie-Anwendungen* (3rd Edn in German). Berlin: Springer (1st Edn. 1997).

Kuechler, A., Dunz, T., Hinderer, A., & Schwab, A. (1987). *Transient field-distribution measurements with "electrical long" sensors.* 5th ISH Braunschweig Paper 32.08.

Kuan, J. T., & Chen. M. K. (2006). Parameter evaluation for lightning impulse with oscillation and overshoot using the eigensystem realization algorithm. *IEEE Transactions on Dielectrics and Electrical Insulation, 13*(6), 1303–1316.

Lemke, E. (2016). Using a field probe to study the mechanism of partial discharges in very small air gaps under DC voltage. *IEEE Electrical Insulation, 32*(4), 43–51.

Kuffel, E. (1956). The effect of irradiation on the breakdown of sphere gaps in air under direct and alternating voltages. In *Proceedings IEE* (Vol. 108, pp. 133–139).

Kuffel, E. (1961). The influence of nearby earthed objects and of the polarity of the voltage on the direct breakdown of horizontal sphere gaps. *Proceedings of the IEE-Part A: Power Engineering, 108,* 302–307.

Kuffel, E., Zaengl, W., & Kuffel, J. (2006). *High-voltage engineering: Fundamentals* (2nd Edn). Elsevier/Newness (First edition 1984 by Pergamon Press).

Kuhlmann, K., & Mecklenburg, W. (1935). Ohmic measuring resistor. *Bulletin SEV, 26,* 737. (in German).

Kupfmuller, K. (1990). Introduction into theoretical electrical technique. Ed. 13. Berlin, Heidelberg, New York: Springer (in German).

Kurrat, M. (1992). Energy considerations for partial discharges in voids. *ETEP, 2*(1), 39–44.

Kuschel, M., Plath, R., & Kalkner, W. (1995). *Dissipation factor measurement at 0.1 Hz as a diagnostic tool for service-aged XLPE-insulated medium voltage cables.* 9th ISH Graz, Paper 5156.

Kutschinski, G. S. (1968). Determination of the life time of oil-impregnated paper of capacitors based on the PD quantities. *ELEKTRIE, 22,* 183–186. (in German).

Lammel, J. (1973). Design of HV test systems for the superimposition of DC voltages with impulse voltages. *Dissertation,* Technical University of Dresden, 1973 (in German).

Lapple, H. (1966). The HV test field of the Schaltwerk of Siemens-Schuckert Werke. *Siemens-Zeitschrift, 40,* 428–435. (in German).

Lalot, J. (1983). Statistical processing of dielectric testing methods. *EDF Bulletin de la Direction des Etudes et Recherches-Series B,* (1/2), 5–30.

Lazarides, L. A. (2010). Negative impulse flashover along cylindrical insulating surfaces bridging a short rod-plane gap under variable humidity. *IEEE Transactions Dielectrics and Electrical Insulation, 17,* 1585–1591.

Lazarides, L. A., & Mikropoulos, P. N. (2011). Positive impulse flashover along smooth cylindrical surfaces under variable humidity. *IEEE Transactions Dielectrics and Electrical Insulation, 18,* 745–754.

Leibfried, T., et al. (1998). *On-line monitoring of power transformers—trends, new developments and first experiences.* CIGRE Paris Report 12-211.

Lemke, E. (1967). *Breakdown mechanism and breakdown vs. gap-distance characteristics of non-uniform air gaps at SI voltages.* Ph.D. Thesis, Technische Universitat Dresden (in German).

Lemke, E. (1967). The breakdown in in-homogenous fields in air at switching voltages. *Periodica Polytechnica, Electrical Engineering (Budapest),* 11(3), 229–239.

Lemke, E. (1968a). A principle for the measurement of impulse charges. *Periodica Polytechnica, Electrical Engineering, Budapest, 12*(1), 31–37. (in German).

Lemke, E. (1968b). Development of streamer discharges in ai rat positive SI voltages. *ELEKTRIE, 22*(4), 166–168. (in German).

Lemke, E. (1966). Electrical breakdown in air at switching voltages. *ELEKTRIE, 20*(5), 195–198. (in German).

Lemke, E. (1969). A new principle fort he wide-band PD measurement. *ELEKTRIE, 23*(11), 468–469. (in German).

Lemke, E. (1974). System for the PD measurement. *ELEKTRIE, 26,* 165–167. (in German).

Lemke, E. (1975). *Contribution to electrical measurement of partial discharges highlighting a wide-band procedure for evaluating the accumulated charge.* Habilitation Thesis, TU Dresden (in German).

Lemke, E. (1979). *A new method for PD measurements on polyethylene insulated power cables.* 3rd ISH Milano, paper 43.13.

Lemke, E. (1981a). A new procedure for PD measurement on long HV cables. *ELEKTRIE, 35*(7), 358–360. (in German).

Lemke, E. (1981b). Problems of the localization of PD defects in extruded HV cables. *ELEKTRIE, 35*(7), 360–362. (in German).

Lemke, E. (1987). *A new procedure for partial discharge measurements on the basis of an electromagnetic sensor.* 5th ISH Braunschweig, paper 41.02.

Lemke, E. (1989). *PD probe measuring technique for on-site diagnosis tests of HV equipment.* 6th ISH, New Orleans, paper 15.08.

Lemke, E. (1991). Progress in PD probe measuring technique. In *7th International Symposium on High Voltage Engineering (ISH) Dresden,* paper 72.01.

Lemke, E. (2004). Possibilities and limits of localization of PD failures in extruded power cables under on-site condition. In *Cologne: VDE/ETG Fachtagung "Diagnostik elektri- scher Betriebsmittel". pp. 209–213* (in German).

Lemke, E. (2012). A critical review of partial discharge models. *IEEE Electrical Insulation Magazine, 28*(6), 11–16.

Lemke, E. (2013). Analysis of the PD charge transfer in extruded power cables. *IEEE Electrical Insulation Magazine, 30*(1), 24–28.

Lemke, E., et al. (1983). *Dimensioning of electrodes for ultra high voltage.* 4th ISH Athens paper 44.01.

Lemke, E., Roding, R., & Weissenberg, W. (1987). On-site testing of extruded cables by PD measurements at SI voltages. In *CIGRE Symposium Vienna paper 1020-02.*

Lemke, L., & Schmiegel, P. (1991). *Progress in PD probe measuring technique.* 7th ISH Dresden, paper 72.02.

Lemke, E., & Schmiegel, P. (1995). Experience in PD diagnosis tests on site based on the PD probe technique. In *3rd Workshop & Conference on HV Technology.* IISc Bangalore/India, (pp. 199–203).

Lemke, E., & Schmiegel, P. (1995). *Complex Discharge Analyzing (CDA)—an alternative procedure for diagnosis tests on HV power apparatus of extremely high capacitance.* 8th ISH Graz, paper 56.17.

Lemke, E., RuBwurm, D., Schellenberger, L., & Zieschang, R. (1996). *Computer-aided system for PD diagnostics.* 7. Tagung "Technische Diagnostik", Merseburg, Germany (in German).

Lemke, E., Schmiegel, P., Elze, H., & RuBwurm, D. (1997). *Procedure for the evaluation of dielectric properties based on complex discharge analyzing (CDA)* (pp. 385–388). Montreal, Canada: ISEI.

Lemke, E., & Strehl, T. (1999). Advanced measuring system for the analysis of dielectric parameters including PD events. In *Electrical Insulation Conference (EIC/EMCW),* Cincin-nati, USA.

Lemke, E., & Strehl, T. (1999). Advanced measuring system for the analysis of dielectric parameters including PD events. In *5th International Conference on Insulated Power Cables (Jicable),* Versailles, France, paper A9.4.

Lemke, E., Strehl, T., & RuBwurm, D. (1999). *New developments in the field of PD detection and location in power cables under on-site condition.* 11th ISH London, UK.

Lemke, E., Strehl, T., & Boltze, M. (2001). *Advanced diagnostic tool for PD fault location in power cables using the CDA technology.* 12th ISH Bangalore, India, paper 6-46, (pp. 983–986).

Lemke, E., Gockenbach, E., & Kalkner, W. (2002). Measuring devices for the diagnostics of electrical equipment. *ETG-Fachbericht Diagnostik elektrischer Betriebsmittel, 87,* 25–32. (in German).

Lemke, E., Elze, H., & Weissenberg, W. (2003). *Experience in PD diagnosis tests of HV cable terminations in service using an ultra-wide-band PD probing in the real-time mode.* 13th ISH, Delft, the Netherlands, paper 11.20, (p. 339).

Lemke, E., Strehl, T., Singer, M., Schneider, M., & Hinkle, J. L. (2003). Practical experience in on-site PD assessment of XLPE and PILC distribution power cables using damped AC exciting voltages. In *Nordic Insulation Symposium (NORD-IS) Tampere*, Finland, pp. 47–54.

Lemke, E., Gulski, E., Hauschild, W., Malewski, R., Mohaupt, P., Muhr, M., et al. (2006). Practical aspects of the detection and location of partial discharges in power cables. [CIGRE Technical Brochure no.297]. *Electra, 226*, 63–70.

Lemke, E., Berlijn, S., Gulski, E., Muhr, M., Pultrum, E., Strehl, T., et al. (2008). *Guide for partial discharge measurements in compliance with IEC 60270*. [CIGRE Technical Brochure no.366].

Lemke, E., Strehl, T., & Markalous, S. (2008b). Ultra-wide-band PD diagnostics of power cable terminations in service. *IEEE Transactions on Dielectrics and Electrical Insulation, 15*(6), 1570–1575.

Les Renardieres Group (1974). Research on impulse measuring systems—Facing UHV measuring problems. *Electra, 35*.

Les Renardieres Group. (1977). Positive discharges in long air gaps at Les Renardieres. *Electra, 53*, 31–153.

Lefevre, A., Legros, W., & Salvador, W. (1989). *Dielectric test with oscillating discharge on synthetic insulation cables* (pp. 270–273). France: CIRED Paris.

Lewin, P. L., et al. (2008). Zero-phase filtering for LI evaluation: A k-factor filter for the revision of IEC 60060-1 and -2. *IEEE Transactions on Power Delivery, 23*(1), 3–12.

Li, P., et al. (2016). Influence of forest fire particles on the breakdown characteristic of air gaps. *IEEE Transactions on Dielectrics and Electrical Insulation, 23*(4), 1974–1984.

Lewis, I. A. D., & Wells, F. H. (1959). *Millimicrosecond pulse techniques*. Oxford: Pergamon Press.

Lichtenberg, G. C. (1777). *De nova method naturam ac motum fluidi electrici investigandi. Novi Commentarii Societatis Regiae Scientiarum Gottingae Tom 8*, p. 168 (Part 1) (For German text see Ostwald's Klassiker der exakten Wissenschaften, No. 246, Akademische Verlagsgesellschaft, Leipzig 1956).

Lichtenberg, G. C. (1778). *Commentationes SocietatisRegiae Scientiarum Gottingae Tom 1*, p. 65 (Part 2) (For German text see Ostwald's Klassiker der exakten Wissenschaften, No. 246, Akademische Verlagsgesellschaft, Leipzig 1956).

Lloyd, W. L., & Starr, E. C. (1928). Investigation of the AC corona using a cathode-ray oscilloscope. *ETZ, 49*, 1279. (in German).

Loeb, L. B. (1939a). *Fundamental processes of electrical discharges in gases* (p. 429). New York: Wiley.

Loeb, L. B. (1939b). *Basic processes of gaseous electronics*. London: Wiley.

Loeb, L. B., & Jaeger, G. (1906). Kinetic theory of *gases. Winkelmanns Handbuch der Physik, 3* (2). (Barth-Verlag Leipzig).

Long, W., & Nilsson, S. (2007, March/April) HVDC Transmission: Yesterday and today. *IEEE Power and Energy Magazine*, 22–31.

Lukaschewitsch, A., & Puff, E. (1976). PD measurement on long cables. *Zeitschrift fur praktische Energietechnik, 28*(2), 32–39. (in German).

Lundgard, L. E. (1992). Partial discharge. Part XII: Acoustic partial discharge detection-Fundamental considerations. *IEEE Electrical Insulation Magazine, 8*(4).

Lundgard, L. E., Hansen, W., & Dursun, K. (1989). *Location of power transformers using external acoustic sensors*. 6th ISH, New Orleans, USA.

Lundgard, L. E., Runde, M. P., & Skyberg, B. (1990). Acoustic diagnoses of gas insulated substations: A theoretical experimental basis. *IEEE Transactions on Power Delivery, 5*(4), 1751–1759.

Mahdjuri-Sabet. (1977). Transfer characteristic of textile resistor bands of low inductance in HV test circuits. *Archiv fur Elektrotechnik,59*, 69-73 (in German).

Malewski, R. (1968). New device for current measurement in exploding wire circuits. *Review of Scientific Instruments, 39*, 90–94.

Malewski, R. (1977). Micro-Ohm shunts for precise recording of short-circuit currents. *Transactions PAS-96*, 579–585.

Malewski, R., Corcoran, R. P., Feser, K., McComb, T. R., Nellis, C., & Nourse, G. (1982). Measurements of the transient electric and magnetic field components in HV laboratories. *IEEE Transactions on Power Apparatus and Systems, PAS-101*, 4452–4459.

Mann, N. R., et al. (1974). *Methods for statistical analysis and life data.* New York: Wiley.

Markalous, S. M. (2006). *Detection and location of partial discharges in power transformer using acoustic and electromagnetic signals.* Ph.D. Thesis Technische Universitat Stuttgart.

Markalous, S. M., Tenbohlen, S., & Feser, K. (2008). Detection and location of partial discharges in power transformers using electric and electromagnetic signals. *IEEE Transactions on Dielectrics and Electrical Insulation, 15*(6), 1576–1583.

Martin, F., & Leibfried, T. (2006). *An universal HV source based on a static frequency converter ISEI Toronto*, 420–423.

Maruyama, S., et al. (2004). Development of a 500 kV DC XLPE cable system. *Furukawa Review, No. 25, pp. 47–52.*

Marx, E. (1952). *HV test practicals/Hochspannungspraktikum (2nd Edn, in German).* Berlin: Springer.

Marx, E. (1926). Breakdown voltage of insulators depending on the voltage waveshape. *Hescho-Mitteilungen, Heft, 21*(22), 657. (in German).

Marzinotto, M., & Mazzanti. (2015). The statistical enlargement law for HVDC cable lines. Part 1: Theory and application to the enlargement length. Part 2: Application to the enlargement over cable radius. *IEEE Transactions on Dielectrics and Electrical Insulation, 22*(1), 192–2010.

Matsumoto, T., Ishii, M., & Kawamura, T. (1983). *The requirement of a DC source for tests on contaminated insulators.* 4th ISH Athens paper 62.03.

Maucksch, S., et al. (1996). Calibration of HV measuring systems for the mutual recognition of HV test results in Eastern and Western Europe. In *ERA Conference Milan.*

Maxwell, J. C. (1873). *A treatise on electricity and magnetism* (Vol. 1, 3rd Edn). Oxford: Clarendon Press, (Reprint by Dover, 1981, pp. 450–461).

Meek, J. M. (1940). A theory of spark discharges. *Physical Review, 57,* 722–728.

Meek, J. M., & Craggs, J. D. (1953). *Electrical breakdown of gases.* New York: Wiley (2nd Edn. 1978).

Meiling, W., & Stary, F. (1969). *Nanosecond pulse techniques.* Berlin: Akademie-Verlag.

Meinke, H., & Gundlach, F. W. (1968). *Handbook of high-frequency techniques.* New York: Springer. *(in German).*

Melville, D. R. G., Salvage, B., & Steinberg, N. R. (1965). Discharge detection and measurement under direct-voltage conditions: Significance of discharge magnitude. In *Proceedings IEEE* (Vol. 112, pp. 1815–1817).

Menke, P. (1996). *Optical current sensor of high accuracy using the Faraday effect.* Ph.D. Thesis, University Kiel.

Merkhalev, S. D., & Vladimirsky, L. L. (1985). *Requirement to and design of HV rectifiers for polluted insulation tests.* Meeting Cigre SC 33 (Budapest).

Mikropoulos, P. N., et al. (2008). Positive streamer propagation and breakdown in air: The influence of humidity. *IEEE Transaction on Dielectrics and Electrical Insulation, 15*(2), 416–425.

Millmann, J., & Taub, H. (1956). *Pulse and digital circuits.* New York: McGraw-Hill Book Co.

Moeller, J. (1975). *Metal-clad test transformer for SF_6-insulated switchgear.* 2nd ISH Zurich, Paper 21-09 (pp. 161–164).

Moeller, J., Steinbigler, H., & WeiB, P. (1972). *Field strength distribution on shielding electrodes for UHV test systems.* 1st ISH Munich, (pp. 36–41) (in German).

Mole, G. (1954). *The E.R.A. portable discharge detector.* CIGRE Session, Paris, France, No. 105, App. I.

Mole, G. (1970). Measurement of the magnitude of internal corona in cables. *IEEE Transactions on Power Apparatus and Systems, 89*(2), 204–212.

Montanari, G. C. (2006). Effect of supply voltage frequency on testing of insulation systems. *IEEE Transactions on Dielectrics and Electrical Insulation, 13*(1), 111–121 (see also the discussion to this publication by Hauschild W in *IEEE Transactions on Dielectrics and Electrical Insulation,* 1189–1191).

Montanari, G. C. (2011). Bringing an insulation to failure: The role of space charges. *IEEE Transactions on Dielectrics and Electrical Insulation, 18*(2), 339–364.

Montanari, G. C., & Cavallini, A. (2013). Partial discharge diagnostics: From apparatus monitoring to smart grid assessment. *IEEE Electrical Insulation Magazine, 29*(3), 8–17.

Morshuis, P. H. F. (1993). *Partial discharge mechanisms.* Ph.D. Thesis, Delft University.

Morshuis, P. H. F. (1987). Degradation of solid dielectrics due to internal partial discharges: Some thoughts on progress made and where to go now. *IEEE Transactions on Dielectrics and Electrical Insulation, 12,* 905–913.

Morshuis, P. H. F., & Smit, J. J. (2005). Partial detection at DC voltage: Their mechanism, detection and analysis. *IEEE Transactions on Dielectrics and Electrical Insulation, 12*(2), 328–340.

Mosch, W. (1969). The simulation of switching over-voltages in EHV systems by HV test equipment. *Wiss. Zeitschrift TU Dresden, 18*(2), 513–517. (in German).

Mosch, W., & Hauschild, W. (1979). *HV insulation with sulphur hexafluoride.* Berlin: Verlag Technik, Heidelberg: Huthig *(in German).*

Mosch, W., et al. (1974). The HV laboratory of the section of electrical engineering of Dresden technical university in education and research. *Wiss. Zeitschrift TU Dresden, 23(5),* 1125–1135 (in German).

Mosch, W., et al. (1979). *Dimensioning of screening electrodes for UHV test equipment based on a critical streamer intensity.* 3rd ISH Milan, paper 52.04.

Mosch, W., et al. (1979). *Phenomena in SF6 insulations with particles and their technical valuation.* 3rd ISH Milan paper 32.01.

Mosch, W., et al. (1988). Hochspannungsisoliertechnik. In E. Philippow (ed). *Taschenbuch Elektrotechnik* (Vol. 6, pp. 235–425). Berlin: VEB Verlag Technik Berlin (2nd Edn in German).

Mosch, W., et al. (1988). *Model of the creeping flashover of polluted insulators and direct voltage generators for pollution tests.* CIGRE Session Paris, Report 33-05.

Muller, K. (1934). About the measurement of characteristics of radio noise. *Veroffentlichungen auf dem Gebiete der Nachrichtentechnik, 2,* 159. (in German).

Muller, P. H. (1974). *Probability calculations and mathematical statistics—encyclopedia of stochastic.* Berlin: Akademie Verlag. (in German).

Muller, K. B. (1976). *On the performance of extruded PE-cables under high direct voltage long-term stress.* Ph.D. Thesis TH Darmstadt, Germany (in German).

Muller, U. (1927). Newer measurements by a Klydonograph. *Hescho-Mitteilun- gen, 37,* 1049. (in German).

Naderian, J. A., et al. (2011). Load-cycling test of HV cables and accessories. *IEEE Electrical Insulation Magazine, 27*(5), 14–28.

Nelin, G., Ryzko, H., & Kvarngreen, M. (1983). *Improved infra-frequency high-voltage test generator.* 4th ISH Athens, paper 52.07.

NEMA 107. (1987). *Methods of measurement of radio influence voltage (RIV) of high-voltage apparatus.* NEMA Publication No. 107.

Nemeth, E. (1966). *Non-destructive testing of insulations by discharging and recovery voltages.* 11th. Wiss. Konferenz Ilmenau (pp. 87–91) (in German).

Nemeth, E. (1972). Proposed fundamental characteristics describing dielectric processes in dielectrics. *Periodica Polytechnica, Budapest, 15*(4), 305–322.

Neumann, C., et al. (2017). Some thoughts regarding prototype installation tests of gas-insulated HVDC systems. *CIGRE Winnipeg Colloquium, Study Committees A3, B4 and D1, Contribution CIGRE D1-110.*

Nieschwitz, H. (1982). *Localization of partial discharges.* 50th VDE-Seminar (in German).

Nieschwitz, H., & Stein, W. (1976). PD measurement on HV power transformers—a tool of quality control. *ETZ-A, Heft 11* (in German).

Nyamupangedengu, C., & Jendrel, I. R. (2012). PD spectral response to variations in the supply frequency. *IEEE Transactions on Dielectrics and Electrical Insulation, 19*(2), 521–532.

Obenaus, F. (1958). Pollution flashover and creeping path. *Deutsche Elektrotechnik, 4,* 135–136. (in German: Fremdschichtuberschlag und Kriechweglange).

Ohki, Y. (2010). News from Japan: Advanced site assembly technologies for UHV transformers. *IEEE Electrical Insulation Magazine, 26*(2), 55–57.

Okabe, S., & Takami, J. (2011). Occurrence probability of lightning failure rates at substations in consideration of lightning stroke current waveforms. *IEEE Transaction on Dielectrics and Electrical Insulation, 18*(1), 221–231.

Okabe, S., & Takami, J. (2009). Evaluation of improved lightning stroke waveform using advanced statistical method. *IEEE Transactions on Power Delivery, 4,* 2197–2205.

Okabe, S., et al. (2013). Discussion on standard waveform in the LI voltage test. *IEEE Transactions on Dielectrics and Electrical Insulation, 20(1),* 147–156.

Okabe, S., et al. (2015). Uncertainty in k-factor measurement for lightning impulse voltage test. *IEEE Transactions on Dielectrics and Electrical Insulation, 22*(1), 266–277.

Okubo, H., et al. (1998). Insulation design and on-site testing method for a long distance, gas-insulated transmission line (GIL). *IEEE Electrical Insulation Magazine, 14*(6), 13–22.

Okubo, H. (1012). Enhancement of electrical insulation performance in power equipment based on dielectric material properties. *IEEE Transactions on Dielectrics and Electrical Insulation, 19* (3), 733–754.

Olearczyk, M., Hampton, R. N., et al. (2010, November/December). Notes from underground—cable fleet management. *IEEE Power and Energy Magazine,* 75–84.

Ortega, P., Waters, R. T., et al. (2007). Impulse breakdown voltages of air gaps: A new approach to atmospheric correction factore applicable to international standards. *IEEE Transaction on Dielectrics and Electrical Insulation, 14*(6), 1498–1507.

Palm, A. (1932). Schering measuring bridges. *Archiv fur Technisches Messen,* 921–923 (in German).

Park, E. H. (1947). Shunts and inductors for current measurements. *NBS Journal Research, 39,* 191–212.

Park, J. H., & Cones, H. N. (1956). Surge voltage breakdown of air in a non-uniform field. *Journal on Research of the National Bureau of Standardization, 56*(4), 201–224.

Paschen, F. (1898). About the necessary voltages fort the breakdown of air, hyrogen and carbon acid at different pressures. *Annalen der Physik, 273*(5), 69–96. (in German: Über die zum Funkenübergang in Luft, Wasserstoff und Kohlensäure bei verschiedenen Drücken erforderliche Potentialdifferenz).

Pattanadech, N., & Yutthagowith, P. (2015). Fast Curve fitting algorithm for parameter evaluation in lightning impulse test technique. *IEEE Transactions on Dielectrics and ElectricalInsulation, 22*(5), 2931–2936.

Pearson, J. S., Hampton, B. F., & Sellars, A. G. (1991). A continuous UHF monitor for gas—insulated substations. *IEEE Transactions on Electrical Insulation, 26*(3), 469–472.

Pedersen, A. (1986). Current pulses generated by discharges in voids in solid dielectrics. A field theoretical approach. In *IEEE International Symposium on Electrical Insulation (IESI),*IEEE Publication. 86CH2196-4-DEI, 112.

Pedersen, A. (1987). Partial discharges in voids in solid dielectrics, an alternative approach. In *Annual Report—Conference on Electrical Insulation and Dielectric Phenomena,* IEEE Publication 87CH2462-0, 58.

Pedersen, A., Crichton, G. C., & McAllister, I. W. (1991). The theory and measurement of partial discharge transients. *IEEE Transactions on Electrical Insulation, 26,* 487.

Pedersen, A., Crichton, G. C., & McAllister, I. W. (1995). Partial discharge detection: Theoretical and practical aspects. In *IEE Proceedings—Science, Measurement and Technology, 142,* 29.

Pedersen, A., Crichton, G. C., & McAllister, I. W. (1995b). The functional relation between partial discharges and induced charge. *IEEE Transactions on Dielectrics and Electrical Insulation, 2,* 535.

Peek, F. W. (1913). Law of corona and dielectric strength of air III. *Transaction of the AIEE II, 32,* 1767–1785.

Peek, F. W. (1915). The effect of transient overvoltages on dielectrics. *Transaction of the AIEE, 34,* 1915.

Peier, D., & Graetsch, V. (1979). *A 300 kV DC measuring device with high accuracy.* 3rd ISH Milan, Italy, paper 43.08.

Peschke, E. (1968). *Breakdown and flashover at high direct voltages in air.* Ph.D. Thesis Munich Technical University (in German).

Peschke, E. (1969). Influence of humidity on the breakdown and flashover behavior at high direct voltages in air. *ETZ-A, 90,* 7–13. (in German).

Petcharales, K. (1986). *Numerical calculation of breakdown voltages of standard air gaps (IEC 52) based on streamer breakdown criteria.* PhD thesis, ETH Zurich.

Pfeffer, A., & Tenbohlen, S. (2009). *Analysis of full and chopped lightning impulse voltages from transformer tests using the new k-factor approach.* 16th ISH Cape Town, paper A-10.

Pietsch, R., et al. (2003). *Optimized cable end termination system with water conditioning unit.* 13th ISH Delft, Paper.

Pietsch, R., et al. (2005). *SF₆-insulated, frequency-tuned resonant test system for spacer testing with AC voltage up to 1000 kV.* 16th ISH Beijing paper J-56.

Pietsch, R., Hinow, M., & Steiner, T. (2010). *Challenge to the HV test techniques for testing HVDC cables.* Kolloquium "Isoliersysteme" Cologne, ISBN 978-3-8007-3278-4 (in German).

Pietsch, R., Hauschild, W., et al. (2012). *High-voltage on-site testing with partial discharge measurement.* CIGRE Technical Brochure No. 502, Working Group D1.33.

Pietsch, R. (2012). *On-site testing of extruded AC and DC cables above 36 kV and up to 500 kV-Some thoughts about the physics behind it, standards and test techniques.* 2012 CIGRE Canada Conference, Montreal, Paper 196.

Pigini, A. (2010). Design of insulators under pollution. In *International Conference on development of a 1200 kV national test station* (pp. 265–280). New Delhi.

Pigini, A., et al. (1985, October). Influence of air density on the impulse strength of external insulation. *IEEE Transaction on PAS,* 104(10).

Pigini, A., et al. (2015). *Pollution tests on composite insulators: The Italian experience.* 19th ISH Pilsen, Paper 367.

Pilling, J. (1976). *A contribution tot he interpretation of life-time characteristics of solid insulations.* Habilitation Thesis Technische Universitat Dresden (in German).

Plath, R. (1994). Oscillating test voltages for on-site testing and PD measurement of extruded cables. *Dissertation,* TU Berlin, 1994 (in German).

Plath, K., Plath, R., Emanuel, H., & Kalkner, W. (2002). *Synchronous three-phase PD measurement on power on site and in the laboratory.* ETG-Fachtagung Diagnostik elektrischer Betriebsmittel, Berlin, paper 11 (pp. 69–72) (in German).

Plath, R. (2005). *Multi-channel PD measurements.* 14th ISH, Bejing, China.

Poleck, H. (1939). Measuring bridges for the measurement of capacitances and loss factors of grounded test objects. *Archiv fur Technisches Messen,* 921–951 (in German).

Pommerenke, D., Krage, I., Kalkner, W., Lemke, E., & Schmiegel, P. (1995). *On-site PD measurement on high voltage cable accessories using integrated sensors.* 9th ISH Graz/Austria.

Praehauser, T. (1973). PD measurement on HV apparatus using the bridge method. *Bulluetin SEV, 64,* 1183–1189. (in German).

Prinz, H., et al. (1965). *Fire, lightning and spark.* Munich: Verlag F. Bruckmann KG. (in German: Feuer Blitz und Funke).

Ramirez, M., et al. (1987). *Air density influence on the strength of external insulation under positive impulses: Experimental investigations up to an altitude of 3000 m a.s.l.* CIGRE WG 33.03. IWD.

Raske, W. (1937). Measuring dividers for high voltages—Part I: Resistive dividers. *Archiv fur Elektrotechnik, 31*(10), 653–666. (in German).

Raske, W. (1939). Measuring dividers for high voltages—Part II: Capacitive dividers. *Archiv fur Elektrotechnik, 1*(33) (in German).

Raske, W. (1939). Measuring dividers for high voltages—Part II: Mixed dividers. *Archiv fur technisches Messen, Z* 116-5 (in German).

Raether, H. (1939). The development of the electron avalanche to the spark channel. *Zeitschrift fur Physik, 112,* 464–489. (in German).

Raether, H. (1940). On the development of channel discharges. *Archiv fur Elektrotechnik, 34,* 49–51. (in German: Zur Entwicklung von Kanalentladungen).

Raether, H. (1941). On the formation of gas discharges. *Zeitschrift fur Physik, 117,* 375–524. (in German: Uber den Aufbau von Gasentladungen).

Raether, H. (1942). On the electrical breakdown of gases. *ETZ, 63,* 301–303. (in German: Uber den elektrischen Durchschlag in Gasen).

Raether, H. (1964). *Electron avalanches and breakdown in gases.* London: Butterworths.

Reichel, R. (1977). Influence of a parallel capacitance on the pollution flashover and its role for the determination of the pollution flashover voltage. *Dissertation*, Dresden Technical University (in German).

Reid, R. (1974). *High-voltage resonant testing.* IEEE PES Winter Meeting. Paper C74 038-6.

Renne, V. T., Stepanov, S. I., & Lavrov, D. S. (1963). Ionization processes in the dielectric of paper capacitors under direct voltage. *Elektricestvo,* 269–278 (in Russian).

Rethmeier, K., Kraetge, A., et al. (2008). Separation of superimposed PD faults and noise by synchronous multi-channel data acquisition. In *International Symposium on Electrical Insulation (ISEI)*, Toronto, Canada, (pp. 611–615).

Rethmeier, K., Obralic, A., Kraetge, A., Kruger, M., Kalkner, W., & Plath, R. (2009). *Improved noise suppression by real-time pulse-waveform analysis of PD pulses and pulse-shaped disturbances.* 16th ISH, Cape Town, South Africa.

Rethmeier, K., Kraetge, A., & Hummel, R. (2012). *About the influence oft the PD repetition rate on the apparent charge value in PD measurements according to IEC 60270* (in German). Kolloquium Diagnostik Elektrischer Betriebsmittel, Fulda (15-16.11.2012), VDE Verlag GmbH Berlin, Offenbach.

Reynolds, P. H. (1985). DC insulation analysis. A new and better method. *IEEE Transactions PAS, 104(7),* 1746–1749.

Rizk, F. A. (1981). Mathematical models for pollution flashover. *Electra, 78,* 71–103.

Rizk, F. A., & Bourdage, M. (1985). Influence of the AC source parameters on flashover characteristics of polluted insulators. *IEEE Transaction PAS, 104,* 948–958.

Rizk, F. A., & Nguyen, D. H. (1987). *Digital simulation of source-insulator interaction in HVDC pollution tests.* IEEE PES WM 168-8, New Orleans.

Robinson, R. A., & Silvia, M. T. (1978). *Digital signal processing and time series analysis.* San Francisco: Holden-Day.

Rodewald, A. (1969a). Transient processes in the Marx multiplier circuit after the ignition of the first switching gap. *Bulletin SEV, 60,* 37–44. (in German).

Rodewald, A. (1969b). Probability of ignition of the switching gaps in the Marx multiplier circuit. *Bulletin SEV, 60,* 857–863. (in German).

Rodewald, A. (1971). Marx multiplier circuit with supporting switching gaps for the extension of the trigger range. *ETZ-A, 92,* 56–57. (in German).

Rodewald, A. (1972). New principle of a triggered multiple chopping gap for all kinds of test voltages. *1st ISH Munich* (in German).

Rodewald, A. (2000). *Electromagnetic compatibility—Basics and practice*. Braunschweig: Friedr. Vieweg & Sohn. (in German).

Rodewald, A. (2017). The inductive component in common impedance coupling and ground bounce. *IEEE Electromagnetic Compatibility Magazine, 8,* 61–65.

Rodrigues Filho, J. G., et al. (2016). Very fast overvoltage waveshapes in a 500 kV gas- insulated switchgear setup. *IEEE Insulation Magazine, 32*(3), 17–23.

Rogers, E. C., & Skipper, D. J. (1960). Gaseous discharge phenomena in high-voltage DC cable dielectrics. *Proceedings IEE, Part A, 107,* 241–254.

Rogowski, W. (1913). About some applications of magnetic voltage measurement. *Archiv fur Elektrotechnik, 1,* 511–527. (in German).

Salvage, B. (1962). Electrical discharges in gaseous cavities in solid dielectrics under direct voltage conditions. In *Proceedings of International Conference on Gas Discharges and the Electricity Supply Industry* (pp. 439–446). Butterworth, London.

Salvage, B., & Sam, W. (1967). Detection and measurement of discharges in solid insulation under direct-voltage conditions. *Proceedings IEE, 114,* 1334–1336.

Satish, L., & Gururaj, B. I. (2001). Wavelet analysis for estimation of mean curve of impulse waveforms superimposed by noise oscillations and overshoot. *IEEE Transactions on Power Delivery, 16*(1), 116–121.

Sayah, A., Oussalah, N., & Boggs, S. A. (2016). Optimization of water terminations for testing solid dielectric cables. *IEEE Transactions on Dielectrics and Electrical Insulation, 23*(1), 61–69.

Schenkel, M., v. Issendorf, I., & Schering, H. (1919). *Bridge for loss measurement.* Tatigkeitsbericht der Physikalisch-Technischen Reichsanstalt, Berlin (in German).

Schering, H. (1919). *Bridge for loss measurement.* Braunschweig, Germany: Tatigkeitsbericht der Physikalisch-Technischen Reichsanstalt. (in German).

Schering, H. (1933). Determination of the high-voltage value during loss factor measurement by a bridge. *ETZ, 45,* 51. (in German).

Schering, H., & Vieweg, F. (1928). A measuring capacitor for highest voltages. *Zeitschrift fur Technische Physik, 9,* 442. (in German).

Schiller, G. (1996). The breakdown behavior of cross-linked polyethylene at different test voltages and pre-stresses. *Dissertation,* University of Hanover (in German).

Schmuck, F., Aitken, S., & Papailiou, K. O. (2010). Proposal for intensified inspection and acceptance tests of composite insulators as an addition to the Guidelines of IEC 61109 and IEC 61952. *IEEE Transactions on Dielectrics and Electrical Insulation, 17*(2), 394–401.

Schon, K. (2013). *High Impulse voltage measurement techniques.* Heidelberg, New York, Dordrecht, London: Springer. (English edition, in German 2010).

Schon, K. (1986). Concept of PD measurement at PD tests. *ETZ Archiv, 8*(9), 319–324. (in German).

Schon, K., & Schuppel, W. (2007). *Precision Rogowski coil used with numerical integration.* 13th ISH Ljubljana, paper T10-130.

Schrader, W. (1971). *Multi-stage high-voltage test system fort he generation of electrical impulse voltages.* DDR-Patent No. 86049, issued on 20.11.1971 (in German).

Schrader W et al. (1989) *The generation of switching impulse voltages up to 3.9 MV with a transformer cascade of 3 MV.* 6th ISH New Orleans paper 47.39.

Schrader, W. (2000). Parallel compensation of overshoot in lightning impulse voltage testing. *Private communication.*

Schreiter, F., Jilek, U., & Schufft, W. (2003). *Combined diagnostic and withstand test to upgrade the operational voltage of 10 kV cables.* 13th ISH Delft, paper P 08.07.

Schufft, W. (1991). *Considerations on the area effect at large electrodes for HV test equipment.* 7th ISH Dresden paper 54.03.

Schufft, W., et al. (2007). *Paperback of electrical power engineering.* Fachbuch- verlag Leipzig im Carl Hanser Verlag Munich (in German).

Schufft, W., & Gotanda, Y. (1997). *A new DC voltage test system with fast polarity reversal.* 10th ISH Montreal (Vol. 4, pp. 37–40).

Schufft, W., & Schrader, W. (1993). *A new Marx generator for the simulation of lightning impulse voltages and currents.* 8th ISH Yokohama, paper 79.03.

Schufft, W., et al. (1995). *Powerful frequency-tuned resonant test system for after laying tests of 110 kV XLPE cables 9th ISH Graz Volume.* Paper 4486.

Schufft, W., et al. (1999). *Frequency-tuned resonant test systems for on-site testing and diagnostics of extruded cables.* 11th ISH London, paper 5.335.P5.

Schuler, R. H., & Liptak, G. A. (1980*). A new method for HV testing of field windings on large rotating electrical machines.* CIGRE Session Paris Report 11-04.

Schulz, W. (1979) Free-hovering particles causing low breakdown voltages in air. *ETZ Archiv,* 123–126 (in German).

Schumann, W. O. (1923). *Electrical breakdown field strength of gases.* Berlin: Springer. (in German).

Schwab, A. J. (1971). Low-resistance shunts for impulse currents. *IEEE Transactions PAS-90,* 2251–2257.

Schwab, A. J. (1981). *High-voltage measuring technique* (2nd English Edn). Berlin: Springer (1st German Edn. 1969).

Schwarz, H., et al. (1999). Megavolts in Cottbus—planning and erection of a HV laboratory (book in English and German).

Schwaiger, A. (1923). *The theory of the electric strength.* Berlin: Springer. (in German).

Schwaiger, A. (1925). *About discharge processes on insulators.* Rosenthal- Mitteilungen, Heft 6, (pp. 1–23) *(in German).*

Seifert, J. M., Petrusch, W., & Janssen, H. (2007). A comparison of the pollution performance of long-road and disc type HVDC insulators. *IEEE Transactions Dielectrics and Electrical Insulation, 14*(1), 125–129.

Shannon, C. E. (1949). Communication in the presence of noise. *Proceedings IRE, 37,* 10–21.

Shi, H., et al. (2015). High-voltage pulse waveform modulator based on solid-state Marx generator. *IEEE Transactions on Dielectrics and Electrical Insulation, 22*(4), 1983–1990.

Shockley, W. (1938). Current to conductors induced by a moving point charge. *Journal of Applied Physics, 9,* 635.

Shu, Y. (2010, October). *Current HVDC development and standardization demand.* Presentation on the Plenary Meeting IEC TC 115 Seattle.

Seitz, P., & Osvath P. (1979). *Microcomputer controlled transformer ratio-arm bridges.* 3rd ISH, Milan, paper 43.11.

Siemens, A. G. (2007*). "Final electrical testing of transformers and reactors" and "Transformer life management".* Technical Brochures of Siemens AG, Power Transmission and Distribu-tion, Nuremberg.

Simmons, J. G., & Tam, M. C. (1973). Theory of isothermal currents and the direct determination of trap parameters in semiconductors and insulators. *Physical Review B,7*(8), 3706.

Simon, P. (2004). *Research of the characteristic parameters of the behaviour of dielectric media under non-standard impulses in high-voltage.* Doctoral Thesis, Polytechnical University of Madrid.

Singer, H. (1972). *The electric field of the polycon electrode.* 1st ISH Munich (pp. 59–66) (in German).

Singh, J., Sood, Y. R., Jarial, R. K., & Verma, P. (2008). Condition monitoring of transformers-bibliography survey. *IEEE Electrical Insulation Magazine, 24*(3), 11.

Sklenicka, V., et al. (CIGRE TF 33.04.09). (1999). Influence of ice and snow on the flashover performance of outdoor insulators. *Electra, 187,* 91–111.

Slama, M. E. A., et al. (2010). Analytical computation of discharge characteristic constants and critical parameters of flashover of polluted insulators. *IEEE Transactions Dielectrics and Electrical Insulation, 17,* 1764–1771.

Speck, J. (1987). *Statistical evaluation of test data of the aging of electrical apparatus*. 11. Scientific Conference of the Section Electrical Engineering, TU Dresden (in German).

Speck, J., et al. (2009). *Statistical estimation of the life time of solid insulations in consideration of defects*. 16th ISH Cape Town, Paper C-13.

Spiegelberg, J. (1966). *Contribution to the design and calibration of resistive LI voltage dividers*. PhD Thesis Technical University Dresden, Institute of HV Engineering (in German).

Spiegelberg, J. (1984). Powerful HVDC and composite voltage test systems for open air arrangement. *Elektrie, 38*(10), 368–371. (in German).

Spiegelberg, J. (2003). Highlights of HV test equipment manufactured in Dresden—Review of the last century. In *Proceedings HIGHVOLT Kolloquium*, pp.7–20 (in German).

Spiegelberg, J., et al. (1993). *A new series of resonant testing systems for cable testing*. 8th ISH Yokohama, paper 55.05.

Stanley, W. D. (1975). *Digital signal processing*. Reston.

Starke, H., & Schroder, R. (1928). Electrometer for the measurement of very high DC and AC voltages. *Archiv fur Elektrotechnik, 20*, 115–117. (in German).

Steenis, E. F., & van de Laar, A. M. (1989). Characterization test and classification procedure for water-tree aged medium voltage cables. *Electra, 125*, 88–101.

Steiner, T. (2011). *Standardization of digital recorders, IEC 61083-1, -2, -3 and -4*. HIGHVOLT Kolloquium Dresden, paper 1.5.

Storm, R. (1976). *Probability calculations-mathematical statistics-statistical quality control*. Fachbuchverlag Leipzig (in German).

Strauss, W. (1983). *Automatization of impulse voltage tests using microcomputers operating in real-time domain and transient recorders*. Ph.D. Thesis, Technical University Berlin (in German).

Strauss, W. (2003). Progress and calibration of digital recorders for HV impulse testing. *HIGHVOLT Kolloquium Dresden, paper, 3*, 3. (in German).

Strehl, T., Lemke, E., & Elze, H. (2001). *On-line PD measurement: Diagnostic tools on monitoring strategy for generators and power transformers*. 12th ISH, Bangalore, India, paper 6-72.

Strehl, T., & Engelmann, A. (2003). *Mobile Test system for insulation diagnostics of electrical equipment* (p. 18). Heft: ETZ. (in German).

Su, J., et al. (2016). An unified expression for enlargement law on electric breakdown strength of polymers under short pulses: Mechanism and review. *IEEE Transactions on Dielectrics and Electrical Insulation, 23*(4), 2319–2327.

Su, Z., et al. (2005). *The DC rain flashover of station insulators under contamination conditions*. 14th ISH Beijing paper D-59.

Sun, Z. Y., Liao, W. M., Su, Z. Y., & Zhang, X. J. (2009). Test study on the altitude correction factors of air gaps of ± 800 kV UHVDC projects. In *International Conference on UHV Power Transmission*, Beijing, Paper FP0557.

Swedish Power Circle. (2010). *Electricity for sustainable energy*. Presentation for the Swedish Academy of Engineering.

Szewczyk, M., et al. (2016). Determination of breakdown voltage characteristics of 1100 kV disconnector for modelling of VFTO in gas-insulated switchgear. *IEEE Transactions on Power Delivery, 31*(5), 2151–2158.

Taheri, S., Farzaneh, M., & Fofana, I. (2014). Improved dynamic model of DC arc discharge on ice-covered post insulator surfaces. *IEEE Transactions on Dielectrics and Electrical Insulation, 21*(2), 729–739.

Takami, J. (2007). Observation results of lightning currents on transmission towers. *IEEE Power Delivery, 22*, 547–556.

Tanaka, T., & Okamoto, T. (1978). A minicomputer-based partial discharge measurement system. In *IEEE International Symposium on Electrical Insulation (ISEI) Philadelphia, USA, Conference Records* 86–89.

Tanaka, T., & Okamoto, T. (1985). Micro-computer application to in-site diagnosis of XLPE cables in service. In *International Conference on Properties and Application of Dielectric Materials, Xian* (pp. 499–502).

Thiede, A., & Martin, F. (2007). *Power frequency inverters for HV tests*. HIGHVOLT Colloquium Dresden, paper 1.7 (pp. 57–61).

Thiede, A., Steiner, T., & Pietsch, R. (2010). *A new approach of testing power transformers by means of static frequency converters*. CIGRE Session Paris, Report D1.202

Thione, L. (1983). *Evaluation of switching impulse strength of external insulation* (p. 94). No: Electra.

Toepler, M. (1898). On sliding discharges along a clean glass surface. *Wied. Annalen der Physik and Chemie, 66,* 1061. (in German).

Townsend, J. S. (1915). *Electricity in gases*. Oxford: Oxford University Press.

Townsend, J. S. (1925). *Motion of electrons in gases*. Oxford: Clarendon Press.

Townsend, J. S. (1937). The equations of motion of electrons in gases. *Philosophical Magazine, VII*(23), 481.

Trichel, G. W. (1938). Mechanism of the negative point-to-plain corona near onset. *Physical Review, 54,* 1078–1084.

Tretter, S. A. (1976). *Introduction to discrete-time signal processing*. New York: Wiley.

Tsuboi, T., et al. (2010a). Weibull parameter of oil-immersed transformer to evaluate insulation reliability on temporary overvoltages. *IEEE Transactions Dielectrics and Electrical Insulation, 17*(6), 1863–1876.

Tsuboi, T., et al. (2010b). Experiment on multiple times voltage application to evaluate insulation reliability of oil immersed transformer. *IEEE Transaction Dielectrics and Electrical Insulation, 17*(5), 1657–1664.

Tsuboi, T., et al. (2010c). Transformer insulation reliability for moving oil with Weibull analysis. *IEEE Transactions Dielectrics and Electrical Insulation, 17*(3), 978–983.

Tsuboi, T., et al. (2011). Insulation breakdown characteristics of UHV class GIS for LI withstand voltage test waveform-k-factor value and front related characteristics. *IEEE Transactions on Dielectrics and Electrical Insulation, 18*(5), 1734–1742.

Tsuboi, T., Ueta, G., & Okabe, S. (2013). K-factor value and front-time related characteristics in negative polarity LI test for UHV-class air insulation. *IEEE Transactions on power delivery, 26* (2), 1148–1155.

Ueta, G., et al. (2010). Evaluation of overshoot rate of LI witstand voltage test waveform based on new base-curve fitting methods. 17(4), 1336–1344.

Ueta, G., Tsuboi, T., & Okabe, S. (2011a). Evaluation of overshoot rate of LI withstand voltage test waveform based on new base fitting methods-study on overshoot waveform in an actual test. *IEEE Transactions on Dielectrics and Electrical Insulation, 18*(3), 783–791.

Ueta, G., Tsuboi, T., & Okabe, S. (2011b). Evaluation of overshoot rate of LI withstand voltage test waveform based on new base fitting methods-study by assuming waveforms in an actual test. *IEEE Transactions on Dielectrics and Electrical Insulation, 18*(6), 1912–1921.

Ueta, G., Wada, J., & Okabe, S. (2011c). Evaluation of breakdown characteristics of CO_2 gas for non-standard LI waveforms- breakdown characteristics under single-frequency oscillating waveforms of 5.3 to 20.0 MHz. *IEEE Transaction on Dielectrics and Electrical Insulation,18* *(1)*, 238–245.

Ueta, G., Wada, J., & Okabe, S. (2011c). Evaluation of breakdown characteristics of CO_2 gas for non-standard impulse waveforms—method for converting non-standard LI waveforms into standard LI waveforms. *IEEE Transactions on Dielectrics and Electrical Insulation, 18*(5), 1724–1733.

Ueta, G., et al. (2012a). k-factor value and front related characteristics of UHV-class air insulation for positive polarity LI test. *IEEE Transactions on Dielectrics and Electrical Insulation, 19*(3), 877–885.

Ueta, G., et al. (2012b). Study on the k-factor function in the LI test for UHV power equipment. *IEEE Transactions on Dielectrics and Electrical Insulation, 19*(4), 1383–1391.

Van Brunt, R. J. (1992). Stochastic properties of partial discharge phenomena. *IEEE Transactions on Electrical Insulation, 26*(5), 902–948.

Vardeman, S. B. (1994). *Statistics for engineering problem solving.* Boston: IEEE Press-PWS Publishing Company.

Verma, M. P., & Petrusch, W. (1981). Results of pollution tests on insulators in the >1100 kV range and the necessity of testing in the future. *IEEE Transactions on Electrical Insulation, 3.*

Vilbig, F. (1953). *High-frequency measuring technique.* Munchen: Carl Hanser Verlag. (in German).

Wada, J., Ueta, G., & Okabe, S. (2011a). Evaluation of breakdown characteristics of N_2 gas for non-standard lightning impulse waveforms-breakdown characteristics under single-frequency oscillation waveforms and bias voltage. *IEEE Transactions on Dielectrics and Electrical Insulation, 18*(5), 1759–1766.

Wada, J., Ueta, G., & Okabe, S. (2011b). Evaluation of breakdown characteristics of CO_2 gas for non-standard lightning impulse waveforms und non-uniform electric field—breakdown characteristics for single-frequency oscillation waveforms. *IEEE Transactions on Dielectrics and Electrical Insulation, 18*(2), 640–648.

Wagner, K. W. (1912). On the measurement of dielectric losses using the alternating current bridge. *ETZ, 33,* 635–637. (in German).

Wagner, K. W. (1914). Explanation of dielectric relaxatuations by models of Maxwell. *Archiv fur Elektrotechnik, II*(9), 371–387. (in German).

Wagner, K. W. (1922). The physical nature of electrical breakdown in solid dielectrics. *Journal of American Institution of Electrical Engineering, 61,* 1034.

Wakimoto, T., Hallstrom, J., Cherukov, Y., Ishii, M., Lucas, W., Piiroinen, J., et al. (2007). High-accuracy comparison of lightning and switching impulse calibrators. *IEEE Transactions on Instrumentation and Measurement, 56,* 619–623.

Ward, B. H. (1997). Digital techniques for partial discharge measurements. *IEEE Transactions on Power Delivery, 7*(2), 469–479.

Ward, A. D., Exon, J., & La, T. (1993). Using Rogowski coils for transient current measurements. *IEE Engineering Science and Education Journal, 2,* 105–113.

Weck, K. H. (2003). *Condition assessment of LIP-insulated medium voltage cables—a contribution to the risk management of distribution cables.* HIGHVOLT Kolloquium '03, Paper 5.2 (pp. 159–166) (in German).

Weicker, W. (1927). Voltage measurements with the air gap. *Hescho-Mitteilungen, 31,* 899–902. (in German).

Weicker, W., & Hoercher, W. (1938). Fundamentals for the establishment of calibration data for sphere spark gaps. *ETZ, 59,* 1029–1064. (in German).

Shu, E. W., & Boggs, S. A. (2008). Dispersion and PD detection in shielded power cable. *IEEE Electrical Insulation Magazine, 21*(1), 25–29.

Werle, P., et al. (2006). *Repair and HV testing of power transformers on site.* ETG Fachtagung "Diagnostic elektrischer Betriebsmittel"/ETG Fachbericht 104, Kassel (in German).

Werle, P. (2007). *On-site tests of power transformers.* HIGHVOLT Kolloquium '07 paper 4.5, (pp. 143–149).

Whitehead, S. (1951). *Dielectric breakdown of solids.* Oxford: Clarendon Press.

Windmar, D., Gutman, I., & Jonsson, J. (2014). HVDC Pollution testing of insulation: Experience from service, laboratory and test station. *IEEE Transactions on Dielectrics and Electrical Insulation, 21*(6), 2496–2502.

Winter, A., et al. (2007). *A mobile transformer test system based on a static frequency converter.* HIGHVOLT Kolloquium '07, paper 4.4 (pp. 137–142).

Wolf, J., & Voigt, G. (1997). A new solution for the extension of the load range of impulse voltage generators. In *14th ISH Montreal* (Vol 4, pp. 363–366).

Wu, et al. (2009). *Uncertainties in the application of atmospheric and altitude corrections as recommended in IEC standards.* 16th ISH Cape Town, Paper A-15.

Witt, H. (1960). Response of low ohmic resistance shunts for impulse currents. *Elteknik, 3,* 45–47.

Yakov, S. (1991). *Statistical analysis of dielectric test results*. Electra Brochure No. 66.

Yamagata, Y., & Okabe, S. (2009). Utility's experience on design and testing for UHV equipment in Japan. In *2nd International Symposium on Standards for UHV Transmission*.

Yang, L., et al. (2012). Comparison of pollution flashover performance of porcelain long-rod, disk type and composite UHVDC insulators at high altitudes. *IEEE Transactions on Dielectrics and Electrical Insulation, 19*(1), 1053–1059.

Yin, F., & Farzaneh, M. (2016). Influence of AC electric field on conductor icing. *IEEE Transactions on Dielectrics and Electrical Insulation, 23*(4), 2134–2144.

Yu, Y.-Q., et al. (2007). Standardization of HVDC transmission field. In *IEC/CIGRE UHV Symposium Beijing*, Paper 5-3.

Yuan, Y., et al. (2015). Calculation of breakdown voltage of rod-plane gaps in the presence of water streams. *IEEE Transactions on Dielectrics and Electrical Insulation, 22*(3), 1577–1587.

Zaengl, W. (1965). A new divider for steep impulse voltages. *Bulletin, SEV, 57,* 1003–1017. (in German).

Zaengl, W., et al. (1982). *Experience of AC voltage tests with variable frequency using a lightweight on-site series-resonant device*. CIGRE Session (Paris) Report 23-07.

Zhang, G., Luo, C., & Pai, S.T. (1995). *Magneto-optical sensors for pulsed current measurements*. 9th ISH Graz, paper 7851.

Zhang, X., Gockenbach, E., et al. (2007). Estimation of the life time of electrical equipment in distribution networks. *IEEE Transactions on Power Delivery, 22*(1), 515–522.

Zhang, X., & Gockenbach, E. (2008). Asset management of transformers based on condition monitoring and standard diagnosis. *IEEE Electrical Insulation Magazine, 24*(4), 26–40.

Zhang, Z., et al. (2010a). Study of the influence on DC pollution flashover voltage on insulator strings and its flashover process. *IEEE Transactions Dielectrics and Electrical Insulation, 17* (6), 1787–1795.

Zhang, Z., et al. (2010b). Study on DC flashover performance of various types of long string insulators under low atmospheric pressure conditions. *IEEE Transactions on Power Delivery, 25*(4), 2132–2142.

Zhang, C., Wang, I., & Guan, Z. (2016). Investigation of DC discharge behavior of polluted porcelain post insulators in artificial rain. *IEEE Transactions on Dielectrics and Electrical Insulation, 23*(1), 331–338.

Zhao, I., et al. (2015). Correlation between volume effect and lifetime effect of solid dielectrics on nanosecond time scale. *IEEE Transactions on Dielectrics and Electrical Insulation, 22*(4), 1769–1776.

Index

© Springer Nature Switzerland AG 2019
W. Hauschild and E. Lemke, *High-Voltage Test and Measuring Techniques*,
https://doi.org/10.1007/978-3-319-97460-6

MIX
Papier aus verantwortungsvollen Quellen
Paper from responsible sources
FSC® C105338

Printed by Libri Plureos GmbH
in Hamburg, Germany